CADMIUM IN THE ENVIRONMENT

Part I: Ecological Cycling

CADMIUM IN THE ENVIRONMENT

Part I: Ecological Cycling

Edited by

JEROME O. NRIAGU

Canada Center for Inland Waters
Burlington, Ontario, Canada

A WILEY-INTERSCIENCE PUBLICATION

JOHN WILEY & SONS
New York • Chichester • Brisbane • Toronto

Library of Congress Cataloging in Publication Data:

Main entry under title:

Cadmium in the environment.

   (Environmental science and technology)
   "A Wiley-Interscience publication."
   Includes index.
   CONTENTS: pt. 1. Ecological cycling.
   1. Cadmium—Environmental aspects. 2. Cadmium-
Toxicology. I. Nriagu, Jerome O.

QH545.C3C33   574.5'2   79-25087
ISBN 0-471-06455-6 (v.1)

Printed in the United States of America

10 9 8 7 6 5 4 3 2 1

SERIES PREFACE
Environmental Science and Technology

The Environmental Science and Technology Series of Monographs, Textbooks, and Advances is devoted to the study of the quality of the environment and to the technology of its conservation. Environmental science therefore relates to the chemical, physical, and biological changes in the environment through contamination or modification, to the physical nature and biological behavior of air, water, soil, food, and waste as they are affected by man's agricultural, industrial, and social activities, and to the application of science and technology to the control and improvement of environmental quality.

The deterioration of environmental quality, which began when man first collected into villages and utilized fire, has existed as a serious problem under the ever-increasing impacts of exponentially increasing population and of industrializing society. Environmental contamination of air, water, soil, and food has become a threat to the continued existence of many plant and animal communities of the ecosystem and may ultimately threaten the very survival of the human race.

It seems clear that if we are to preserve for future generations some semblance of the biological order of the world of the past and hope to improve on the deteriorating standards of urban public health, environmental science and technology must quickly come to play a dominant role in designing our social and industrial structure for tomorrow. Scientifically rigorous criteria of environmental quality must be developed. Based in part on these criteria, realistic standards must be established and our technological progress must be tailored to meet them. It is obvious that civilization will continue to require increasing amounts of fuel, transportation, industrial chemicals, fertilizers, pesticides, and countless other products; and that it will continue to produce waste products of all descriptions. What is urgently needed is a total systems approach to modern civilization through which the pooled talents of scientists and engineers, in cooperation with social scientists and the medical profession, can be focused on the development of order and equilibrium in the presently disparate segments of the human environment. Most of the skills and tools that are needed are already in existence. We surely have a right to hope a technology that has created such manifold environmental problems is also capable of solving them. It is our hope that this Series in Environmental Sciences and Technology will not only serve to make this challenge more explicit to the established professionals, but that it also will help to stimulate the student toward the career opportunities in this vital area.

Robert L. Metcalf
Werner Stumm

PREFACE

The environmental toxicology of cadmium has attracted a lot of attention in recent years. While there are now a number of excellent monographs and reports on cadmium toxicity in human beings and animal species, few detailed attempts have been made to relate the toxicological aspects of the other biogeochemical features of the metal in the ecosystem. An objective of this volume has been to provide a comprehensive picture of current biological, chemical, geochemical, and clinical research pertaining to cadmium in the environment. The review chapters are written by acknowledged experts from a wide variety of scientific disciplines; indeed, the literature pertaining to the biogeochemistry of cadmium has become so vast that no single scientist can provide a detailed account of all the recent developments.

The volume has been divided into two parts. This Part describes the sources, distribution, mechanisms of transport, transformations, and flow of cadmium in the environment. Part II deals primarily with the biological and toxic effects of cadmium. Some overlapping of material between chapters is inevitable in dealing with ecosystems in which diverse processes are closely interlinked. No attempt has been made to reconcile differing views expressed in several chapters; such controversial issues seem to reflect the uncertain state of present knowledge in a rapidly developing field.

This transdisciplinary volume should be of interest to a wide spectrum of readers, notably environmental scientists and managers, ecologists, biologists, geochemists, nutritionalists, and toxicologists as well as industrial and occupational health officials. It provides the professional with a broad picture of recent research activities and the advanced student or new researcher with an authoritative selection of original literature on specific aspects of cadmium. The focus on contrasts between natural and contaminated ecosystems should make the volume a fundamental reading for anyone concerned about the quality of the environment in general and the environmental cadmium problem in particular.

It is a pleasure to acknowledge the support and cooperation of our distinguished group of contributors. Appreciation is also extended to Wiley-Interscience for invaluable editorial assistance.

JEROME O. NRIAGU

Burlington, Ontario, Canada
February 1980

CONTENTS

CONTENTS
PART II

CADMIUM IN THE ENVIRONMENT

Part I: Ecological Cycling

1

GLOBAL CADMIUM CYCLE

Jerome O. Nriagu

National Water Research Institute, Canada Centre for Inland Waters, Burlington, Ontario, Canada

AND AN OVERVIEW OF CADMIUM IN HUMAN BEINGS

Seiichi Yasumura

Downstate Medical Center, Brooklyn, New York, and research collaborator, Brookhaven National Laboratory, Upton, New York

David Vartsky

Kenneth J. Ellis

Stanton H. Cohn

Medical Research Center, Brookhaven National Laboratory, Upton, New York

HUMAN INFLUENCE ON THE GLOBAL CADMIUM CYCLE

HUMAN INFLUENCE ON THE GLOBAL CADMIUM CYCLE

Jerome O. Nriagu

Owing to a lack of pertinent data, no attempt has been made to look holistically at the global biogeochemical cycle of cadmium. On the basis of the data reported in the chapters of this volume, I have endeavoured to construct a tentative model of the global cadmium cycle. Such a model can be useful in identifying and evaluating any perturbations in the global cadmium systems attributable to cultural activities. Admittedly, the sources, sinks, and distributions of cadmium in many ecosystems have yet to be properly evaluated, and the cadmium transfer rates between systems are only poorly known. Nevertheless, the phenomenological model described here provides some relative measure of both the quantities of cadmium in the global systems and the importance of the various transport pathways in the cadmium cycle.

1. RESERVOIRS AND FLUX RATES

The applicable mathematical aspects of the global elemental cycle have been summarized by Garrels et al. (1975) and Mackenzie et al. (1979). Basically, a global model of an element consists of persumed reservoirs (or spheres) which may be *active* (available to the biota) or *passive* (unavailable to the biota). The total amount of the element in each reservoir is often referred to as the *burden* (or pool) and is derived simply by multiplying the average concentration by the total mass (or volume) of the reservoir. The metabolism and reactions of the element within the various reservoirs generally lead to the evolution of the concentrations of the various phases of the element with time; such stochastic processes will not be addressed in this report.

Table 1 shows the principal reservoirs and burdens for cadmium in the earth's crust. It should be noted that the active pools, notably the atmosphere, soils, lakes, rivers, and ocean waters, are subjected to large inputs of cadmium from pollutant sources. The atmospheric cadmium burden is particularly small and is therefore the most likely to be influenced by wastes from human activities (see below). Until the technology needed to profitably extract the cadmium from fossil fuels is developed, the environmental impacts of the cadmium burden in fossil fuel deposits must remain a matter of some concern. Whether the enlargement of the cadmium burden in a given reservoir gives rise to a net buildup of cadmium in the living biomasses is still conjectural at this point. It has been suggested (see Elias et al., 1975; Nriagu, 1979b) that the widespread contamination of the environment with lead has resulted in a 10- to 100-fold increase in the lead contents (or body burdens) of most living flora and biota (including human beings). There is nothing so far to suggest that an analogous buildup of cadmium has not occurred.

The exchange of cadmium between reservoirs usually occurs along established pathways involving stream, ice, and groundwater flows, atmospheric transport and deposition, volcanism, uplift, subaqueous weathering, and sedimentation, as well as the various biological pumps. Most global models assume that the total system is at stationary state (i.e., there is not a net production or consumption of the element in the total system) so that the exchange rates (fluxes) for cadmium among the various reservoirs can be derived by solving the appropriate set of simultaneous mass balance equations. Table 2 summarizes the flux rates for cadmium along the principal transport paths linking the various global reservoirs.

Figure 1 depicts the atmospheric segment of the cadmium cycle. Notice that

Table 1. Cadmium Burdens and Residence Times in the Principal Global Reservoirs

Reservoir	Pool Mass (g)	Cadmium Concentration	Total Cadmium in Pool (g)	Residence Time (yr)
Atmosphere	5.1×10^{18} m^3	0.03 ng/m^3	1.5×10^8	7 days
Hydrosphere				
Oceans				
Dissolved	1.4×10^{24}	0.06 µg/kg	8.4×10^{13}	2.1×10^4 (deep sea)
Suspended particulates (total)	1.4×10^{18}	1.0 µg/g	1.4×10^{12}	—
Particulate organic matter	7×10^{16}	4.5 µg/g	3.2×10^{11}	1.3
Fresh waters				
Dissolved	0.32×10^{20}	0.05 µg/kg	1.6×10^9	—
Sediments	6.5×10^{17}	0.16 µg/g	1.0×10^{11}	3.6
Glaciers	1.65×10^{22}	0.005 µg/kg	8.2×10^{10}	—
Groundwater	4×10^{18}	0.1 µg/kg	4×10^8	—

Sediment pore waters	3.2×10^{23}	0.2 µg/kg	6.4×10^{13}	—
Swamps an marshes, biomass	6×10^{15}	0.6 µg/g	3.6×10^{9}	—
Biosphere				
Marine plants	2×10^{14}	2.0 µg/g	4×10^{8}	18 days
Marine animals	3×10^{15}	4.0 µg/g	1.2×10^{10}	
Land plants	2.4×10^{18}	0.3 µg/g	7.2×10^{11}	20 days
Land aminals	2×10^{16}	0.3 µg/g	6×10^{9}	
Freshwater biota	2×10^{15}	3.5 µg/g	7×10^{9}	3.5
Human biomass	4×10^{9} persons	50 mg/person	2×10^{8}	1–40
Terrestrial litter	2.2×10^{18}	0.6 µg	1.3×10^{12}	42
Lithosphere (down to 45 km)	5.7×10^{25}	0.5 µg/g	2.8×10^{19}	10^{9}
Sedimentary rocks	2.5×10^{24}	1.0 µg/g	2.5×10^{18}	—
Shale and clay	1.9×10^{24}	1.3 µg/g	2.47×10^{18}	—
Limestone	0.35×10^{24}	0.08 µg/g	2.8×10^{16}	—
Sandstone	0.3×10^{24}	0.07 µg/g	2.1×10^{16}	—
Soils (to 100 cm)	3.3×10^{20}	0.2 µg/g	6.6×10^{13}	3000
Organic fraction	6.8×10^{18}	0.9 µg/g	6.1×10^{12}	>200

Table 2.   Annual Flux Rates of Cadmium Along Major Pathways of the Global Cycle

Process/Pathway	Material Flux[a] (g/yr)	Average Cadmium Concentration[b] (μg/g)	Annual Cadmium Flux (g)
Rivers			
Dissolved	0.32×10^{20}	0.05 g/kg	1.6×10^9
Particulate load	1.5×10^{16}	1.5	2.25×10^{10}
Biological uptake			
Fresh waters	1.0×10^{15}	2.0	2×10^9
Oceans	6×10^{16}	4.0	2.4×10^{11}
Continents	12×10^{16}	0.3	3.6×10^{10}
Atmospheric emissions			
Anthropogenic sources	—	—	7.3×10^9
Natural sources	—	—	0.8×10^9
Atmospheric fallout			
Lands	—	—	5.7×10^9
Oceans	—	—	2.4×10^9
Waste disposal			
Land, sewage	2.1×10^{13}	23	4.8×10^8
Land, fly ash	2.8×10^{14}	0.8	2.2×10^8
Oceans	2.5×10^{13}	23	5.8×10^8
Fertilizers	9.4×10^{13}	7	6.6×10^8
Litter fall	5.2×10^{16}	0.6	3.1×10^{10}
Accumulation in sediments			
Lithogenous deposition	1.5×10^{16}	1.5	2.2×10^{10}
Hydrogenous precipitation	[c]	[c]	2.0×10^8
Skeletal input	3×10^{15}	1.0	3×10^9
Continental denudation	2.5×10^{16}	0.5	1.2×10^{10}
Human consumption	4×10^9 persons	0.15 mg/day·person	2.2×10^8

[a] Mostly from Nriagu (1979a).

[b] The annual flux rate (shown) is derived as the product of average cadmium concentration and amount of material processed. Average cadmium concentrations are derived mostly from chapters in this volume.

[c] Assuming that the hydrogenous precipitation of manganese is 4×10^{12} g/yr (Elderfield, 1976), that the average cadmium content of marine manganese nodules is 8 μg/g, and that the Mn/Cd ratio in nodules is 2×10^4 (see Cronan, 1976).

the annual anthropogenic flux exceeds the emission from natural sources by almost an order of magnitude. The value of the atmospheric cadmium burden is based on an average residence time for airborne cadmium of 7 days (see Nriagu, 1978). It has also been assumed that 30% of total cadmium emission

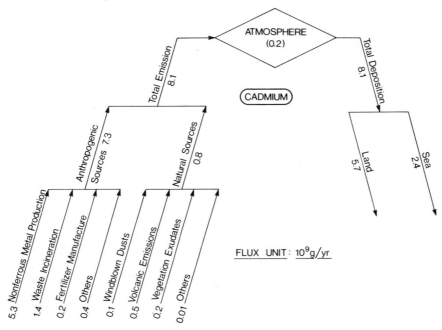

Figure 1. The atmospheric component of the global cadmium cycle.

is deposited in the oceans (e.g., see Nriagu, 1978, 1979a). It is quite significant that the atmospheric influx of cadmium now exces the amount of this element being processed by the biota and flora in many terrestrial ecosystems (e.g., see Yost, 1978).

Some aspects of the marine geochemistry of cadmium have been discussed by Eaton (1976) and Boyle et al. (1976). Present calculations suggest that the oceans currently gain 2.4×10^9 g Cd/year through the atmosphere, at the expense of the lithospheric cadmium pool. Such a gain represents about 20% of all the cadmium burden of marine biota and flora (see Table 1). In addition, the oceans annually gain 7.5×10^9 g Cd via stream runoff, a value derived as the difference between the runoff particulate input and the denudational loss of cadmium from the continents (Table 2). These present-day net gains of cadmium by the oceans stem directly from human influence.

Cadmium inputs to land areas from atmospheric fallout and waste disposal plus fertilizer applications are estimated to be 5.7×10^9 and 1×10^9 g/ year, respectively (Table 2). Since the current uses of cadmium (1.7×10^9 g in 1976) are mostly dissipative, it follows that the soils are currently gaining 9.4×10^9 g Cd each year, mostly because of human activities. The additional cadmium input to soils is comparable to the amount of the metal processed annually by terrestrial animals (Table 2). It should be emphasized that both the burden and the amount of cadmium processed annually by terrestrial biota and flora are relatively small, and thus highly susceptible to perturbations in flux rates engendered by human activities. It is estimated that about 10% of the cadmium in soils is associated with the organic matter (Table 1).

2. RESIDENCE TIMES

For a system at steady state, the residence time (τ) is usually defined as

$$\tau = \frac{A}{dA/dt}$$

where A is the quantity of cadmium in the reservoir at a given time, and dA/dt is the instantaneous rate of addition (or subtraction) of cadmium from the reservoir. Basically, the residence time suggests the time scale of a particular process and should be used in a circumspect manner, considering the uncertainties in estimating A and dA/dt and the fact that the global systems may no longer be at steady state at the present time.

The calculated residence times for cadmium in the various reservoirs are shown in Table 1. The very short residence time precludes an inexorable buildup of the element in the atmosphere. In general, the active cadmium pools (including particulate organic matter, animal and plant biomasses, some agricultural soils, and dissolved phases in fresh water and nearshore marine waters) have relatively short residence times, a reflection of the biological impacts on the dynamics of processes in such reservoirs. Furthermore, any increase in the cadmium burden of an active reservoir will have an immediate impact on the levels and effects of cadmium in the associated biota. From an environmental standpoint, more attention needs to be paid to the cadmium burdens of active pools than those of passive pools (such as the deep sea, sediments, and some soils), where a rapid buildup of cadmium would seem to be unlikely.

The global cycle of cadmium on the 1-year time frame is shown in Figure 2.

2.1. Additional Insight into the Pollution Component

The cadmium burden in the total human biomass is estimated to be 2×10^8 g (Table 1), whereas the intake of cadmium by the entire human population is estimated to be 2.2×10^8 g/year (Table 2). Although the cadmium flow through human beings is relatively insignificant, human beings have become a major macrobiological agent in the present-day biogeochemical cycle of the element. It has been estimated (Table 3) that the all-time amount of cadmium that has

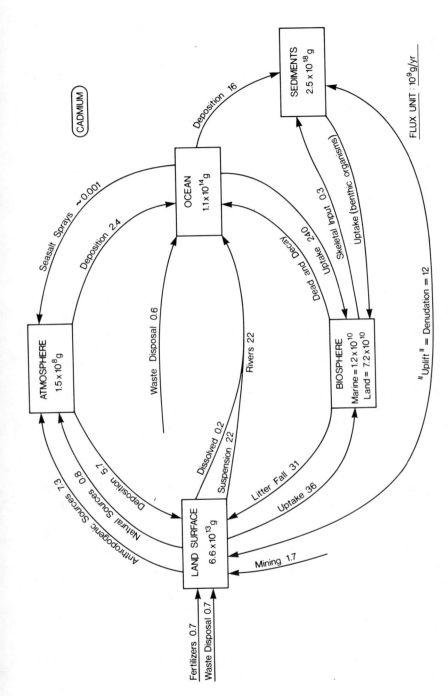

Figure 2. The global cycle of cadmium.

Table 3.   All-Time Production and Anthropogenic Emissions of Cadmium[a]

Period	Cadmium Production ($\times 10^{10}$ g)	Cadmium Emission ($\times 10^9$ g)
Pre-1850	—	63
1850–1900	—	19
1901–1910	—	8.9
1911–1920	0.1	11
1921–1930	0.7	14
1931–1940	2.6	17
1941–1950	4.8	22
1951–1960	8.4	34
1961–1970	14	54
1971–1980	15	74
Total (all-time)	50	316

[a] From Nriagu (1979c).

been released to the atmosphere is about 320×10^9 g. In addition, 500×10^9 g Cd has been produced and dissipated at the earth's surface. The massive redistribution of cadmium apparently has affected the concentration and flux of the metal in many ecosystems. A notable example is the Arctic snowfields, where there has been a 2- to 0-fold increase in the atmospheric deposition of cadmium during the past 300 years (Jaworowski et al., 1975; Weiss et al., 1975).

An impressive record now exists of the profligate contamination of local ecosystems with cadmium. Around smelters, for example, the concentration of cadmium in the atmosphere occasionally exceeds 500 ng/m^3 (Nriagu, this volume, Chapter 3; Muskett et al., 1979), whereas cadmium contents of surficial soils in excess of 500 μg/g are not unusual (John, 1975; Schwartz, 1979). Highly elevated levels of cadmium have also been encountered in sediments, as well as in fauna and flora near smelters (e.g., see Dugdale and Hummel, 1978) and other point sources. The average airborne cadmium level in most cities is 5 to 50 ng/m^3, or a factor of 10 to 100 above the background level (Nriagu, this volume, Chapter 3); the atmospheric depositions of cadmium in many cities currently exceed 2 ng/m^2·year. Also, the average cadmium concentrations in street and household dusts in many urban areas range from 3 to 50 μg/g (Solomon and Hartford, 1976; Harrison, 1979), values that are orders of magnitude higher than the 0.2 μg/g which is typical of uncontaminated soils. There is little doubt that the environmental levels of cadmium in urban areas are highly elevated; the designation of such areas as "hot spots" of cadmium may not be entirely inappropriate.

Although itai-itai disease, which is indicative of overexposure to cadmium, is rare, a large number of people are now being exposed to elevated levels of cadmium via various environmental routes (U.S. Environmental Protection Agency, 1978). Our current understanding of the subclinical effects of such exposure levels is tenuous at best. An overview of the levels and effects of cadmium in human beings is presented in the following section.

REFERENCES

Boyle, E. A., Sclater, F., and Edmond, J. M. (1976). "On the Marine Geochemistry of Cadmium," *Nature* **263**, 42–44.

Cronan, D. S. (1976). "Manganese Nodules and Other Ferro-Manganese Oxide Deposits." In J. P. Riley and R. Chester, Eds., *Chemical Oceanography.* Academic Press, New York, pp. 217–263.

Dugdale, P. J. and Hummel, B. L. (1978). "Cadmium in the Lead Smelter at Belledune: Its Association with Heavy Metals in the Ecosystem." In *Cadmium 77: Proceedings of the 1st International Cadmium Conference.* Metal Bulletin Ltd., London, England, pp. 53–75.

Eaton, A. (1976). "Marine Geochemistry of Cadmium," *Mar. Chem.* **4**, 141–154.

Elderfield, H. (1976). Manganese Fluxes to the Oceans," *Mar. Chem.,* **4**, 103–132.

Elias, R., Hirao, Y., and Patterson, C. C. (1975). "Impact of Present Levels of Aerosol Lead Concentrations on Both Natural Ecosystems and Humans." In *Proceedings of the International Conference on Heavy Metals in the Environment,* Vol. II. Institute of Environmental Studies, University of Toronto, Ont., pp. 257–271.

Garrels, R. M., Mackenzie, F. T., and Hunt, C. (1975). *Chemical Cycles and the Global Environment.* Kaufmann, Inc., Los Altos, Calif., 206 pp.

Harrison, R. M. (1979). "Toxic Metals in Street and Household Dusts," *Sci. Total Environ.* **11**, 89–97.

Jaworowski, Z., et al. (1975). "Stable and Radioactive Pollutants in a Scandinavian Glacier," *Environ. Pollut.* **9**, 305–315.

John, M. K. (1975). "Transfer of Heavy Metals from Soils to Plants." In *Proceedings of the International Conference on Heavy Metals in the Environment,* Vol. II. Institute of Environmental Studies, University of Toronto, Ont., pp. 365–377.

Mackenzie, F. T., Lantzy, R. J., and Paterson, V. (1979). "Global Trace Metal Cycles and Predictions," *J. Int. Assoc. Math. Geol.,* in press.

Muskett, C. J., Roberts, L. H., and Page, B. J. (1979). "Cadmium and Lead Pollution from Secondary Metal Refinery Operations," *Sci. Total Environ.* **11**, 73–87.

Nriagu, J. O. (1978). "Lead in the Atmosphere." In J. O. Nriagu, Ed., *Biogeochemistry of Lead in the Environment,* Part 1A. Elsevier, Amsterdam, pp. 137–183.

Nriagu, J. O. (1979a). "The Global Copper Cycle." In J. O. Nriagu, Ed., *Copper in the Environment,* Part I. Wiley, New York, pp. 43–75.

Nriagu, J. O. (1979b). "Lead in the Atmosphere and Its Effects on Lead in Humans." In R. L. Signhal and J. A. Thomas, Eds., *Lead Toxicity.* Urban and Schwartzenberg, Baltimore, in press.

Nriagu, J. O. (1979c). "Global Inventory of Natural and Anthropogenic Emissions of Trace Metals to the Atmosphere," *Nature* **279**, 409–411.

Schwartz, J. (1979). "Cadmium Puts U.K. Village at Risk," *Nature* **277**, 225.

Solomon, R. L. and Hartford, J. W. (1976). "Lead and Cadmiun in Dusts and Soils in a Small Urban Community," *Environ. Sci. Technol.* **10**, 773–777.

U.S. Environmental Protection Agency (1978). *Health Assessment Document for Cadmium.* EPA, Office of Research and Development, Washington, D.C.

Weiss, H., Bertine, K., Koide, M., and Goldberg, E. D. (1975). "Chemical Composition of Greenland Glacier," *Geochim. Cosmochim. Acta* **39**, 1–10.

Yost, K. J. (1978). "Some Aspects of the Environmental Flow of Cadmium in the United States. In *Cadmium 77: Proceedings of the 1st Cadmium Conference.* Metal Bulletin Ltd., London, pp. 147–166.

CADMIUM IN HUMAN BEINGS

S. Yasumura, D. Vartsky, K. J. Ellis and S. H. Cohn

Cadmium is highly toxic to human beings (Gleason et al., 1969). Acute toxicity of cadmium in human beings was described as early as 1858 (Savot, 1858), although cadmium poisoning as such was not definitively recognized for another 62 years. During World War II the potential of cadmium as a weapon, in the form of cadmium oxide fumes, was considered (Gafafer, 1964). Friberg (1949, 1950) produced in Sweden the first of a series of studies relating specific symptoms and a high incidence of illness and death among industrial workers to cadmium oxide exposure. Prominent among the reactions produced were emphysema of the lungs and renal damage. Anosmia, yellowing of the teeth, and mild liver damage were also observed. Later, other investigators confirmed these findings (Baader, 1951; Hunter, 1954; Buxton, 1956; Bonnell et al., 1959).

Acute cadmium poisoning can result from inhalation of cadmium fumes or dust, or from ingestion of heavily contaminated food or water. Prohibition of uses of cadmium that would bring it into contact with food and drink has successfully reduced the latter form of poisoning to a relatively rare occurrence. Nevertheless, severe gastrointestinal symptoms and several deaths following cadmium ingestion have been reported (Gleason et al., 1969; U.S. Public Health Service, 1962; Stokinger, 1963; Browning, 1961; McKee and Wolf, 1963; Nordberg et al., 1973). The sublethal dose of ingested cadmium has been estimated to be in the range of 326 mg. Acute symptoms following exposure include severe abdominal pain associated with nausea, vomiting, diarrhea, headache, and vertigo. Lethal doses, in the range of 350 mg to 9 g, further induce shock

Table 4.   Lethal Doses of Various Chemical Forms of Cadmium Listed by NIOSH[a]

The number in parentheses is the milligrams of cadmium in the compound.

Substance	Toxic Concentration or Dose*[b]
Cadmium (fume), CdO?	9 mg/m^3 (7.9), human beings (approximate lethal concentration, inhaled)
Cadmium chloride, CdCl$_2$	88 mg/kg (54.0), rat, LD$_{50}$ (oral)
Cadmium fluoborate, Cd(BF$_4$)$_2$	250 mg/kg (98), rat, LD$_{50}$ (oral)
Cadmium fluoride, CdF$_2$	150 mg/kg (112), guinea pig, LD$_{50}$ (oral)
Cadmium fluorosilicate, CdSiF$_6$·6H$_2$O	100 mg/kg, (31), rat (approximate lethal dose, oral); 670 mg/m^3 (208), mouse, LC$_{50}$ (inhaled)
Cadmium lactate, Cd(C$_3$H$_5$O$_3$)$_2$	13.9 mg/kg (5.4), rat, LD$_{50}$ (subcutaneous)
Cadmium oxide, CdO	50 mg/m^3 (43.8), human beings (approximate lethal concentration, inhaled); 72 mg/kg (63), rat, LD$_{50}$ (oral)
Cadmium phosphate, Cd(H$_2$PO$_4$)$_2$·2H$_2$O	650 mg/m^3 (213), mouse (approximate lethal concentration, inhaled)
Cadmium sulfate, CdSO$_4$	27 mg/kg (14.6), dog, LD$_{50}$ (subcutaneous)
Cadmium stearate, Cd[CH$_3$(CH$_2$)$_{16}$—CO$_2$]$_2$	1225 mg/kg (200), rat, LD$_{50}$ (oral)
Cadmium succinate, Cd(CH$_2$—CO$_2$)$_2$	660 mg/kg (325), rat, LD$_{50}$ (oral)

[a] Christeensen (1971).

[b] LD$_{50}$ is the dose expected to kill 50% of the test population through a route of intake other than respiratory. LC$_{50}$ is the calculated concentration that, when administered by the respiratory route, would be expected to kill 50% of the test population during an exposure of 4 hr.

and collapse. Death may occur within 24 hr, or be delayed for 1 to 2 weeks following liver and kidney damage with attendant anuria and uremia. Fortunately, most cases of cadmium consumption have not been fatal.

In contrast to poisoning by cadmium ingestion, acute inhalatory cadmium poisoning is more widespread. Despite the establishment of air quality standards for industrial workrooms, overexposure is still not uncommon. In a large number of cases it is the naive, careless, or untrained worker who is at greatest risk of acute poisoning (Fulkerson and Goeller, 1973). The lethal dose of inhaled cadmium is estimated to be 1900 min mg/m^3 for cadmium oxide fumes or 10,500 min mg/m^3 for cadmium oxide dusts (see Table 4). It is to be emphasized that the lethal dose has been expressed as an air concentration/time-dependent relationship. A 10-min exposure to 190 mg/m^3 cadmium fumes, or less than 8

Table 5.   Cadmium-Related Occupations Considered Potentially Hazardous[a]

Alloy makers	
Aluminum solder makers	Lithopone makers
Cadmium compound collecting bag cleaners	Metalizers
Cadmium compound collecting bag handlers	Paint makers
Cadmium platers	Paint sprayers
Cadmium smelters	Photoelectric cell makers
Cadmium vapor lamp makers	Pigment makers
Cadmium workers	Small arms ammunition makers
Ceramic makers	Smoke bomb makers
Dental amalgam makers	Solderers
Electrical instrument makers	Solder makers
Electroplaters	Storage battery makers
Engravers	Textile printers
Glass makers	Welders, cadmium alloy
Incandescent lamp makers	Welders, cadmium plated object
Lithographers	Zinc refiners

[a] Gafafer (1964).

mg/m^3 for 4 hrs, will result in death. It should also be noted that other investigators have suggested that the lethal dose is considerably lower. Inhalation of high concentrations of cadmium-contaminated air can occur without producing immediate awareness or discomfort to the exposed individual; some time elapses before the severe reactions manifest themselves. These symptoms include chest pain, shortness of breath, coughing, the production of foamy sputum, hemoptysis, and generalized weakness. Death results from asphyxia or massive pulmonary edema.

Until rather recently, the chronic effects of cadmium poisoning of human beings received little attention. With the single exception of cadmium-induced disease in Japan, very few overt diseases in the world have been attributed to long-term cadmium exposure. In Japan, itai-itai disease, a particularly disabling and painful illness marked by osteomalacia and proteinuria, is regarded by most investigators as a form of cadmium poisoning (Friberg et al., 1974; Fulkerson and Goeller, 1973; Tsuchiya, 1976; International Cadmium Conference, 1977). It is the Japanese experience that has raised the level of awareness and concern that chronic toxicity of cadmium is a potentially serious problem in most of the heavily industrialized areas of the world. In the United States about 100,000 people work in cadmium-related industries (Table 5); the group of industrial workers is probably at greatest potential risk.

The increasing concern with the adverse effects of exposure to environmental cadmium has underscored the need for greater knowledge of the metabolism of cadmium in human beings; the physiological effects of cadmium are, at best, incompletely understood. Most of the data available on metabolic effects have been obtained from autopsies and are related to daily intake and excretion by rough estimation. Extrapolation of findings from experimental studies with animals leaves much doubt about the long-term effects of cadmium exposure on human beings. Furthermore, while the majority of animal experiments range from the recording of acute responses to studies requiring several months or years, the biological half-life of cadmium in man is estimated to be of the order of 10 to 40 years (Friberg et al., 1974). The range of this estimate is disturbingly large; there is a very large variance in data on body burden and urinary excretion of cadmium.

Theoretical metabolic models have been proposed by Tsuchiya and Sugita (1971), by Friberg et al. (1974), and in a more sophisticated mode by Kjellström (1977) and Nordberg and Kjellström (1979). These models are useful in that they make it possible to predict the number of years required to reach a critical body burden that produces pathological changes. The shortcomings of existing models, however, emphasize the need for specific kinetic data from human subjects. One study, that of Rahola et al. (1972), was addressed to this problem. Orally administered radioactive cadmium (^{115m}Cd) was used in the study. Unfortunately, an exact figure for the biological half-time of the slow turnover component was not obtained.

3. BODY CADMIUM INTAKE AND DISTRIBUTION

The body accumulates cadmium almost entirely by intestinal and respiratory absorption. Estimates of gastrointestinal absorption range from 0.5 to 12% (Commission of the European Communities, 1974; Friberg et al., 1974), based on a limited number of balance studies. A generally accepted figure is an absorption of 6% (range, 4.7 to 7.0%) of the ingested ^{115m}Cd dose (Rahola et al., 1972). Absorption would be expected to vary widely, depending on the form in which cadmium is bound and on the nature of other constituents of the diet. In animals the protein content of the diet has an influence on cadmium absorption (Suzuki et al., 1969). Also, calcium-deficient diets enhance body cadmium accumulation (Larsson and Piscator, 1971). Since calcium deficiencies are common among elderly persons and pregnant women, the risk of cadmium toxicity may be greater for these groups than for the general population.

In the United States the average diet contains about 50 μg of Cd (Duggan and Corneliussen, 1972). This level of intake would (with the assumed 6% absorption) contribute 3 μg/day to the body burden. Cadmium intake, of course, varies widely. Two of the factors to be considered are the level of industrial pollution to which the individual is exposed and the dietary intake. Assessment of the latter must include consideration of the geophysical nature of the water supply and the soil in which crops are grown; thus fertilization materials and irrigation practices and techniques have to be investigated.

Additionally, there is wide diversity in the basic diets of individuals, in regard to both the basic staple foods used and the caloric intake. In most diets, grain, cereals, and some root plants such as potatoes contribute the largest amounts of cadmium (Fox, 1976). The highest concentration of cadmium is found in oysters (~50 μg/100 g serving, according to Fox, 1976), the brown meat of crabs (Reynolds and Reynolds, 1971), and whale meat (Ishizaki et al., 1970). Liver and kidneys accumulate relatively large amounts of cadmium and are another dietary source. Offsetting this selective concentration, however, is the practice of maintaining short life spans for domestic animals raised for human consumption. Thus calf kidneys contain only 10% of the cadmium level found in wild animals, and about 1% of that in the human adult kidney (Webb, 1975). Despite the high concentrations of cadmium found in certain foods, it appears unlikely that adverse physiological effects will be encountered except by individuals practicing highly unusual dietary practices or ingesting foods raised in very heavily polluted areas. Clearly, it is essential to investigate carefully the dietary history of individuals participating in human studies in which tissue cadmium levels are measured.

A proportionately larger fraction of cadmium is absorbed by the lungs than by the gastrointestinal system. Approximately 25 to 50% of the inhaled dose is absorbed (Friberg et al., 1974). Nonindustrial air concentrations of cadmium are so low that airborne exposure adds very little to the body burden. If a level of 0.025 μg/m^3 is assumed as an average urban air cadmium concentration (National Air Pollution Control Administration, 1968), and an average daily inhaled air volume of 20 m^3 is further assumed, the lungs would be exposed to 0.5 μg Cd/day, with a maximum of 0.25 μg absorbed. However, Menden et al. (1972) estimate that cigarette smokers inhale 2.0 μg for each pack of cigarettes smoked; this amount could add as much as 1.0 μg Cd per pack to the body burden. Finally, it should be emphasized that aveolar absorption rates vary according to the differences in the particulate and chemical forms of airborne cadmium. As would be expected, the smaller the particle and/or the more soluble the chemical form, the greater the absorption.

After gaining entrance into the body, cadmium is sequestered in the liver and kidneys. More than half the total body burden of cadmium is found in these two organs, with the larger fraction in the kidney (see Friberg et al., 1974, p. 59, for a list of references; also Elinder et al., 1976). This organ specificity is more striking if the body burden is expressed as a concentration. The liver has 5 times more cadmium per unit weight than other parts of the body; the kidney concentration is about 50 times that of the rest of the body (Tipton and Cook, 1963). Thus it is the kidney that bears both the largest absolute amount and the highest concentration of cadmium; not unexpectedly, it is the organ of failure in high chronic cadmium exposure. Other organs such as the pancreas and salivary glands accumulate cadmium to a lesser extent (see Cherry, this volume, Part II). The mammary gland forms an effective barrier to cadmium transport into milk (Lucis et al., 1972). This filtering action is particularly fortunate since large quantities of milk are consumed by most children. Similarly, the placenta prevents cadmium from entering the fetal circulation; thus the newborn infant is essentially cadmium-free. From birth on there is an increase in the body burden to an estimated 30 mg Cd in the adult American (Schroeder and Balassa, 1961; Perry et al., 1976). The reported total body burden is lower in Europe and higher in Japan (Friberg et al., 1974).

Unfortunately, the blood and urine concentrations of cadmium are poor indices of the body burden. Although recent exposure to cadmium is reflected by an increase in the blood level (Kjellström, 1977) and cigarette smokers have higher blood levels than nonsmokers (Ulander and Axelson, 1974; Beevers et al., 1976; Zielhius et al., 1977), low blood cadmium levels have been observed in individuals with high body burdens (Friberg et al., 1974). Urine values appear to be related to total body burden on a group-average basis (Kjellström, 1977), but individual variations are so large that the value of urine level as a clinical index of cadmium body burden is questionable. It should be emphasized that there is an extremely high interlaboratory variation in the range of reported values for cadmium in whole blood, plasma, and urine. Probably, methodological problems are involved in these discrepancies.

4. CADMIUM TISSUE CONCENTRATIONS IN HUMAN BEINGS

4.1. Kidney Concentration and Renal Function

Cadmium is accumulated preferentially in the proximal portion of the nephron; thus the kidney cortex concentration is much higher than that in the medullary region. As a rule of thumb, the cadmium concentration of the cortex is about 1.5 times the value for whole kidney (Table 6). In the United States mean levels of cadmium body burdens for individuals at age 50 range from 25 to 50 μg/g kidney cortex (Hammer et al., 1973; Schroeder and Balassa, 1961), but higher and lower levels for "normal" populations have been reported in Japan and Europe, respectively, (Piscator, 1976). Renal tubular damage may occur when the cadmium levels exceed about 200 μg/g wet weight of kidney cortex (Friberg et al., 1974). This figure is an approximation based on animal data and a compilation of renal biopsy and autopsy data taken from industrially exposed workers with insignificant morphological signs of renal damage. It was assumed that these cases were representative of persons with early-stage renal dysfunction

Table 6.   Conversion Factors to Obtain Cadmium Values Normalized to Wet Weight of Tissue

Type of Tissue	Multiplication Factor	Reference
Liver	0.013 × ash weight	Tipton and Cook (1963)
	0.29 × dry weight	Piscator (1971)
Whole kidney	0.011 × ash weight	Tipton and Cook (1963)
	0.21 × dry weight	Piscator (1971)
Kidney cortex[a]	1.5 × whole kidney	Geldmacher et al. (1968)

[a] Slightly different values have been obtained by other investigators. See Friberg et al. (1974, p. 60).

(mild proteinuria), or at the "threshold" level at which damage would be expected. Kidney cadmium levels are much lower after extensive tubular damage has occurred. Presumably, the accumulation of cadmium by the nephron is a saturable process beyond which the cytotoxic effects become manifest. As cell damage progresses, cellular debris, including cadmium, is excreted. In support of this hypothesis are observations of abnormally high urine cadmium levels associated with progressively worsening proteinuria.

Cadmium-induced tubular proteinuria is characterized by the excretion of many low molecular weight proteins (Piscator, 1966a, 1966b). Among these small proteins is β_2-microglobulin, for which a sensitive radioimmunoassay has been developed and made available in commercial kit form (Phadebas β_2-Micro Test Kit). This is not a specific assay for cadmium-induced proteinuria, but appears to correlate well with the degree of tubular damage. Also, β_2-microglobulinuria appears at an earlier stage of nephropathy than do glucosuria and aminoaciduria (Piscator, 1976). Nomiyama et al. (1977) suggest that urine β_2-microglobulin levels might be used as an indicator of cadmium exposure and an early sign of a decrease in the ability of the nephron to reabsorb proteins from the glomerular filtrate.

The mechanisms involved in the development of cadmium-induced renal tubular destruction are incompletely understood. Intracellular cadmium is largely bound to thionein; this complex appears to be biologically inert (Webb, 1975). Piscator (1976) suggests that cadmium bound to liver proteins is slowly released into the blood in the form of metallothionein and taken up by renal tubules from the glomerular filtrate. However, extracellular cadmium thionein is nephrotoxic in animal experiments (Nordberg et al., 1975; Goyer et al., 1977). This would imply that metallothionein is locally synthesized by the cells of the nephron and that the plasma transport of cadmium to the kidney may be in a form other than cadmium thionein. Normally, plasma metallothioein levels are very low, but in red blood cells cadmium is bound to both large and small molecular weight proteins. Nordberg et al. (1971) suggest that the release of metallothionein from red cells is of great importance. There is a need for further research in this area, particularly in human subjects, for whom the existing data are sparse.

Cadmium-induced kidney damage has not been reported in nonindustrially exposed persons, but the production of chronic renal failure by a synergistic effect of cadmium in concert with other etiological factors cannot be ruled out. As previously mentioned, the reported normal renal cortical level of cadmium in adults is about 50 μg/g, with large individual variations that reflect differences in age, geographic area, and smoking habits. Since the placenta is an efficient barrier to cadmium and the fetal kidney is essentially afunctional, in that glomerular filtration does not occur, renal cadmium is absent at birth. There is a rapid, age-related increase in kidney cadmium from birth to age 50; a decrease is seen thereafter (Elinder et al., 1976). The decrease after age 50 may be explained by the fact that environmental cadmium concentrations have increased or that there has been a reduction in food intake and/or cigarette smoking among older individuals. The importance of renal metabolic changes in older individuals is unknown, but the aging kidney has a decreased number of nephrons and a lower glomerular filtration rate (Papper, 1973). Also, nephron shrinkage and an increasing thickening of the basement membranes of Bowman's capsule and the convoluted tubules occur after age 50 (Darmady, 1974). These morphological changes would be consistent with the slight decrease in kidney cadmium concentrations observed in older persons.

As shown by Perry et al. (1961), and by others subsequently, renal cadmium concentrations differ by group, according to geographic area. Africans have lower values than groups of Americans, Swiss, and Indians. The highest renal cadmium levels were measured in Asians. Additional age-related renal cortical cadmium differences by country (data from many investigators) have been summarized by Friberg et al. (1974, p. 64). These differences are thought to be related to eating habits and the level of cadmium contamination that has entered the food chain. Consistent with this hypothesis is the observation that group data from a single location yield a much smaller standard deviation than do data from racial groupings composed of persons from widely separated locations (Perry et al., 1961).

Renal cadmium levels in smokers are approximately double those found in nonsmokers (Elinder et al., 1976). There is a dose-response relationship: the greater the number of cigarette pack-years, the higher the organ cadmium level (Lewis et al., 1977). Pipe and cigar smokers accumulate slightly more cadmium than nonsmokers. Ex-cigarette smokers would be expected to have a body cadmium burden related to their smoking history; because of the long biological half-life of cadmium, renal cadmium loss is minimal in most cases.

4.2. Concentration in Liver and Other Tissues

As mentioned above, liver and kidney cadmium levels represent approximately 50% of the total body burden of this element. The kidney/liver cadmium concentration ratio is approximately 10, but the range of reported values is very large (Friberg et al., 1974). In occupationally exposed workers the ratio is much lower. In one study (Smith et al., 1960) the liver contained the major fraction of the body cadmium 5 to 10 years after exposure had ceased. However, some renal cadmium loss associated with tubular damage may account for some of the markedly low kidney/liver cadmium concentration ratios reported among exposed workers.

Liver cadmium is largely bound to intracellular metallothionein (NATO Conference, 1974). It seems likely that death due to renal failure precludes the development of serious liver impairment. Nevertheless, high liver cadmium levels may be a useful index of body burden, particularly during acute exposure to cadmium compounds.

Other tissues such as the pancreas (Tipton and Cook, 1963; Elinder et al., 1976) concentrate cadmium to a minor extent, but specific organ pathologies have not been correlated with cadmium concentration.

5. PATHOLOGICAL CONDITIONS

5.1. Hypertension

There is some evidence that high blood pressure is related to the body burden of cadmium. However, this thesis is still controversial. Evidence for a causal relation is provided by reports of cadmium-induced hypertension in animals (see Ohanian et al., this volume, Part II), as well as postmortem findings in human subjects with histories of hypertensive disease.

Schroeder (1965) reported a significantly higher concentration of cadmiun in the kidneys of persons dying of hypertension-related diseases than in those dying from other causes. Similar findings were reported by Lener and Bibr (1971). In contrast, Morgan (1969) found no correlation between kidney cadmium levels and hypertension. Østergaard (1977) reported that the renal cadmium concentration was actually higher in normotensive patients than in the hypertensive group. She offered the hypothesis that hypertension-induced renal damage and enhanced cadmium excretion may have resulted in renal loss of cadmium. More recently she suggested (Østergaard, 1978) that a high selenium intake may allow a higher renal cadmium level without the development of hypertension.

The importance of accounting for differences in smoking habits is emphasized by Lewis et al. (1972). When the smoking history was taken into account, no differences in cadmium levels appeared in any tissues, including kidney, liver, and pancreas. There is strong documentation of increased prevalence of arterial hypertension among those who smoke cigarettes. Thus, if there is an increased incidence of hypertension among cigarette smokers, and smoking is known to increase the kidney cadmium content, it seems reasonable to find more renal cadmium in a random group of hypertensive patients when smoking habits have not been factored into the results.

Perry and Schroeder (1955) found high urine cadmium levels in several hypertensive patients. Similarly, slightly greater urinary excretion of cadmium in hypertensive women was noted by McKenzie and Kay (1973). However, only patients with marked renal impairment were excluded from their study; patients with increased cadmium excretion, associated with mild renal damage, were not excluded. In addition, Wester (1973, 1974) found no differences in urine cadmium levels in hypertensive and normotensive subjects.

An association between high plasma cadmium levels and hypertension has also been reported (Glauser et al., 1976; Thind et al., 1976). However, Beevers et al. (1976a, 1976b), Moore et al. (1977), and Wester (1973), placed particular emphasis on controlling differences in smoking habits. Under these conditions they were unable to demonstrate any association between hypertension and blood cadmium concentration.

Occupational exposure to cadmium dusts is not associated with an increased incidence of hypertensive disease (Bonnell et al., 1959; Hammer et al., 1972). Also, hypertension is not observed in patients with itai-itai disease (Perry, 1972). These observations argue against the concept that human hypertension is cadmium induced. However, the hypertension produced in animals appears to occur only when low doses of cadmium are administered (see Ohanian et al., this volume, Part II). Thus the possibility that only a critically low-level exposure to cadmium will cause elevated blood pressure in hypertension-prone individuals cannot be ruled out.

Finally, in the previously mentioned investigation by Schroeder (1965), emphasis was placed on the high kidney Cd/Zn ratios of individuals dying of hypertensive causes. The implication of this observation is that the absolute kidney cadmium level may vary; but as zinc is replaced by cadmium, the zinc-dependent enzymatic processes are compromised, leading to hypertension in some as yet unknown way.

In preliminary studies (unpublished data on patients enrolled in the Brookhaven Laboratory hypertension program headed by Dr. J. Iwai), kidney cadmium was measured by *in vivo* neutron capture prompt-gamma ray activation analysis (Vartsky et al., 1977) in 12 patients with a history of essential hypertension. No differences in the amount of cadmium in the kidney were observed

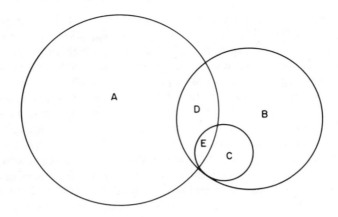

A = HYPERTENSIVE POPULATION

B = INCREASED CADMIUM EXPOSURE

C = OCCUPATIONAL CADMIUM EXPOSURE

D = A ∩ B

E = A ∩ C

Figure 3.   Venn diagram illustrating subject selection criteria. For any study in which the cadmium measurements are independent with two outcomes (namely, normal or elevated kidney cadmium content), the probability of observing increased cadmium in a hypertensive population in a limited number of measurements is proportional to the individual's cadmium exposure. At present, liver and kidney cadmium measurements have been made on subjects taken at random from group C. If, however, one samples from group D, the required number of measurements can be significantly reduced to test the hypertensive hypothesis. The normotensives selected as controls must also come from group B. For an equal probability of detecting elevated cadmium levels in a hypertensive population, the necessary number of measurements is reduced by the ratio of the individual exposure levels. With a portable measurement system (currently under construction), it will be possible to select not only from group A but also from groups B, D, and E. This represents a significant advantage in that a smaller number of measurements will still provide the statistical accuracy necessary to test different hypotheses.

in these patients, as compared to normotensive control subjects, provided that smoking habits were factored into the analyses. However, since hypertension is probably multifactoral in origin (Page, 1949), the possibility of the existence of a cadmium-related hypertensive subgroup is still open to question. A simple theoretical model (Figure 3) provides an experimental design approach to this problem. In essence the model predicts that the incidence of hypertension associated with a modest increase in renal cadmium content should be statistically higher among a chronically cadmium-exposed population where the exposure is well below occupational levels of contact but is above the average national level.

5.2.   Osteomalacia

In bone, cadmium levels are low even in heavily exposed individuals (Piscator, 1977). Nevertheless, osteomalacia has been found in some exposed workers (Adams et al., 1969) and is one of the hallmarks of itai-itai disease. Cadmium-induced osteomalacia is always associated with renal dysfunction, as judged by long-standing proteinuria. Individuals with a history of a low dietary calcium intake are particularly prone to this form of skeletal disease. Renal tubular damage may result in the loss of 1-hydroxylase activity. Under these conditions, 25-hydroxycholecalciferol is not converted to 1,25-dihydroxycholecalciferol, the active form of vitamin D. Since this dihydroxylated vitamin D metabolite promotes protein carrier-mediated transport of calcium across the intestinal membrane, its absence results in a depressed gastrointestinal absorption of calcium (DeLuca, 1976), and secondary hyperparathyroidism might be expected. The skeletal lesions respond to treatment with massive doses of vitamin D, particularly if supplemented with bicarbonate and phosphate (Kazantzis, 1977). It should be pointed out that the pathogenesis of cadmium-induced osteomalacia differs markedly from the sequelae of events leading to the renal osteodystrophy of chronic renal failure origin. In particular, cadmium-induced renal damage is associated with hypoosmolar polyuria and hypophosphatemia, whereas chronic renal failure is characterized by low urine volume with phosphate retention.

5.3.   Pulmonary Effects of Inhaled Cadmium

Exposure to sublethal air concentrations of cadmium may result in anosmia, dyspnea, and emphysema (Gleason et al., 1969). Lung function tests on exposed individuals often reveal deficits compatible with a diagnosis of emphysema. There is evidence of further deterioration with time, even in the absence of further exposure (Bonnell et al., 1959). Unfortunately, smoking histories were not factored into the data in any of the earlier studies. Lack of these data leaves some doubt with respect to the pulmonary effects that are exclusively and causally related to chronic exposure to airborne cadmium. More recently, in an epidemiological survey of workers exposed to cadmium, Lauwerys et al. (1977) selected a control group to match the sex, age, weight, height, smoking habits, and socioeconomic status of the exposed group. A slight but significant reduction in lung function, as assessed by forced vital capacity, forced expiratory volume in 1 sec, and peak expiratory flow rate, was found only in older workers exposed for an average of 28 years. Smith (1977) found similar mild pulmonary effects in a group of workers chronically exposed to high concentrations of airborne cadmium. There is great uncertainty about the exposure dose of cadmium, particularly in the estimation of past exposures in these studies. Nevertheless, on the basis of their study, Lauwerys et al. (1974) speculate that a cadmium oxide dust concentration of 20 $\mu g/m^3$ is probably safe. This level is considerably lower than the current permissible threshold limit values in the United States (American Conference of Governmental and Industrial Hygienists, 1973). It should be noted that the threshold limit values are largely based on possible kidney damage, as assessed by the degree of proteinuria, rather than deficits of lung function. In effect, it has been concluded that the kidneys are affected at a dose which is lower than the exposure level sufficient to cause a detectable decrease in the functional capacity of the lungs.

Nonindustrial airborne cadmium dusts are sufficiently low that they do not pose a major threat to pulmonary function. However, cigarette smokers have an increased body burden of cadmium; it is possible that effects of heavy smoking may be synergistic in promoting respiratory disease in individuals exposed to cadmium in airborne industrial effluents. It is conceivable that persons with a heavy smoking history residing in areas surrounding some industrial complexes are at higher risk with respect to deficits in pulmonary function.

5.4. Neoplastic and Other Changes

There is some evidence that long-term occupational exposure to cadmium is associated with an increased incidence of cancer. However, the supporting data are far from conclusive. Potts (1965) and Kipling and Waterhouse (1967) reported a higher than expected incidence of prostatic carcinoma among workers in cadmium-related industries. Since the number of cases included in these surveys was small, they hesitated to conclude that there was a causal relation to cadmium exposure. More recently, Lemen et al. (1976) reported similar findings, again suggesting the possibility of a causal association with cadmium exposure. Since benign prostatic hypertrophy is virtually universal among older men, it is conceivable that the added burden of cadmium might be one of many toxic substances which could increase the rate of malignant change in this tissue. Habib et al. (1976) observed a small but significant increase in the prostatic cadmium level in benign prostatic hypertrophy, and a marked cadmium increase associated with carcinoma of this tissue. Presumably these patients were not cadmium workers. It would appear that the prostate concentrates cadmium when undergoing hyperplastic or malignant change. However, whether the cadmium accumulation is the result of biochemical changes involved in prostatic growth or is causal in the development of malignancy is unknown. The role of cadmium, if any, in the pathogenesis of prostatic carcinoma remains to be clarified.

To date, there is no evidence of an increased incidence of bronchiogenic carcinoma among cadmium production workers. However, Morgan (1970) reported higher liver and kidney cadmium levels in patients dying from cancer of the lung than in those dying of other forms of cancer. Unfortunately, smoking history was not assessed. Lewis et al. (1972) measured the tissue levels of cadmium in autopsy samples. When smoking habits were factored in, patients with cancer of the bronchus or lung did not have more cadmium in their liver, kidney, or lung tissue than did those dying of other causes.

Other types of neoplastic changes have been observed in experimental animals, but there are no reports of similar cadmium-induced cancers in human beings. There is some evidence, based on the marked increase in the number of chromatid breaks in peripheral leukocytes, that genetic effects may be associated with itai-itai disease (Shiraishi and Yosida, 1972). This type of study has not been extended to include cadmium-exposed workers. Finally, it has been speculated that the lower birth weight of babies born to mothers who smoke cigarettes may be due to an increased body cadmium burden (Webster, 1978). The paucity of data relative to neoplastic, mutagenic, and teratogenic effects of cadmium in human beings may be the result of failure of investigators to look for associated changes, or may simply reflect the relative unimportance of cadmium as an etiological factor in these pathologies. Cadmium is so toxic that the exposure levels necessary to cause neoplastic or genetic effects may well result in death long before overt growth defects appear.

6.   *IN VIVO* MEASUREMENT OF KIDNEY AND LIVER CADMIUM

Vartsky et al. (1977) have built a facility at Brookhaven National Laboratory (BNL) for *in vivo* measurements of human kidney and liver cadmium by neutron capture prompt-gamma ray analysis. The method employs an ^{85}Ci-^{238}Pu-Be neutron source and a gamma ray detection system consisting of two Ge(Li) detectors (Figure 4). A typical gamma ray spectrum is shown in Figure 5. The dose delivered to the liver and left kidney is 666 mrem (detection limit is 1.8 $\mu g/g$ Cd in the liver and 2.5 mg Cd for one kidney).

To date, measurements have been made in 20 normal male volunteers (Ellis et al., 1979). The group consisted of 12 cigarette smokers and 9 nonsmokers with a mean age of 50 years (range, 31 to 62). The data summarized in Table 7 reveal values comparable to those obtained by other investigators from tissue samples at autopsy (Lewis et al., 1972; Elinder et al., 1976). The urine and plasma cadmium levels were determined by atomic absorption, utilizing a graphite furnace and deuterium arc system. Summary data are shown in Table 8. The urine and plasma cadmium values for smokers were slightly but not significantly greater than those observed for nonsmokers. As expected, β_2-microglobulins were within normal limits for both groups.

Prompt-gamma ray neutron activation analysis provides the only currently

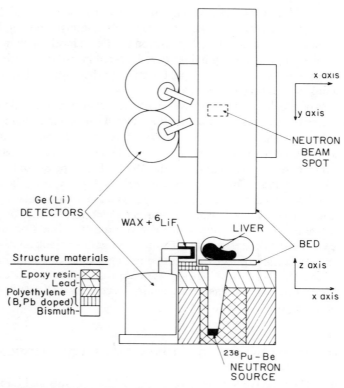

Figure 4.   Facility for *in vivo* liver and kidney cadmium measurements.

available means of measuring *in vivo* levels of liver and kidney cadmium. The radiation hazard of neutron activation is very low; the dose used is less than 5% of the annual maximal permissible dose to a single organ of a radiation worker

Table 7.   In Vivo Measurement of Cadmium

Subjects	Number	Liver[a] (μg/g)	Kidney[a] (mg)	Renal Cd/ Hepatic Cd Ratio
Smokers	12	4.1 (1.6)	5.8 (1.7)	10.6
Nonsmokers	8	2.3 (1.6)	3.1 (2.0)	11.3
Smokers/NonSmokers ratio		1.78	1.87	—
P		0.05	0.02	NS[b]

[a] Values expressed as geometric mean (standard deviation).
[b] Not significant.

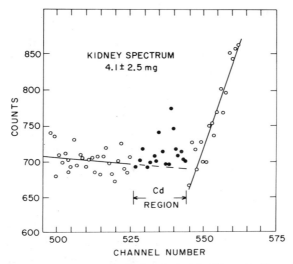

Figure 5. Gamma ray spectrum of cadmium in the left kidney of a 52-year-old male.

set by the ICRP (1959). With this technique, long-term studies with yearly measurements made on selected individuals become possible. At the present time, in addition to BNL, only the group at the University of Birmingham, England (Biggin et al., 1974; McLellan et al., 1975), has the facilities to measure *in vivo* cadmium levels. The Birmingham unit requires a cyclotron for the production of neutrons; accessibility is thus limited. More recently, a small ^{238}Pu-Be portable unit (20 Ci), built by Harvey et al. (1976), appears to be suitable for industrial screening purposes. This unit, however, lacks the sensitivity of either the cy-

Table 8. Plasma and Urine Cadmium and β_2-Microglobulin Levels[a]

Subjects	Number	Plasma Cadmium (μg/l)	P	Urine Cadmium (μg/l)	P
Smokers	12	2.26 (1.67)	0.07	2.42 (1.68)	0.06
Nonsmokers	8	1.49 (1.55)		1.44 (1.77)	

Subjects	Number	Plasma β_2-microglobulin (mg/l)	P	Urine β_2-microglobulin (μg/l)	P
Smokers	12	1.52 (1.39)	NS[b]	46.2 (6.0)	NS
Nonsmokers	6	1.74 (1.10)		28.1 (4.7)	

[a] All values expressed as geometric mean (standard deviation).
[b] Not significant.

clotron unit or the much larger ^{238}Pu-Be BNL facility. At the present time, BNL is building a trailer-mounted unit which will be used in the field throughout the United States for epidemiological studies. The sensitivity of this unit is expected to be better than 2.5 mg/kidney and 1.8 μg/g liver, the limits of detection of the Brookhaven permanently based system. The relative expense and the sophisticated expertise required to mount such a program may limit the availability of the technique to small-scale research-oriented needs, for the near future at least. Nevertheless, it is anticipated that critically important data, particularly on industrially exposed workers, will provide a better basis for the setting or revision of standards for industrial and environmental cadmium pollution.

7. RESEARCH NEEDS

Although animal studies have provided valuable information with respect to the effects of cadmium, there are insufficient data on cadmium metabolism in human beings. For example, at present only very rough estimates of the critical renal cadmium threshold concentrations associated with irreversible nephron damage are available. A better assessment of the dose-effect relationship is of paramount importance. With the advent of *in vivo* prompt-gamma ray neutron activation analysis, it is possible to obtain such data *in vivo* in the near future.

There is also a clear need to improve the technology for the measurement of low tissue concentrations of cadmium in view of the wide interlaboratory variation in reported values.

Furthermore, a better estimate is needed of the biological half-time of cadmium in human beings, particularly in conjunction with factors such as age, sex, weight, dietary habits, smoking habits, and state of health. With an increasing understanding of these facets of cadmium metabolism in man, it will be possible to predict more accurately the pathophysiological consequences of cadmium pollution. From these considerations standards for workroom air, food, and water can be revised to provide an adequate safety margin.

ACKNOWLEDGMENT

The section on cadmium in human beings has been authored under Contract EY-76-C-02-0016 with the U.S. Department of Energy. Accordingly, the U.S. Government retains a nonexclusive, royalty-free license to publish or reproduce the published form of this portion of the contribution, or allow others to do so, for U.S. Government purposes.

REFERENCES

Adams, R. G., Harrison, J. F., and Scott, P. (1969). "The Development of Cadmium-Induced Proteinuria, Impaired Renal Function, and Osteomalacia in Alkaline Battery Workers," *Quart. J. Med.* **38**, 425–443.

American Conference of Governmental and Industrial Hygienists, (1973). *Threshold Limit Values for Chemical Substances and Physical Agents in the Workroom Environment with Intended Changes for 1973.* Cincinnati, Ohio, pp. 12, 36.

Baader, E. W. (1951). "Die Chronische Kadmiumvergiftung," *Deut. Med. Wochenschr.* **76**, 484–487.

Beevers, D. G., Erskine, E., Robertson, M., Beattie, A. D., Goldberg, A., Campbell, B. C., Moore, M. R., and Hawthorne, V. M. (1976a). "Blood Lead and Hypertension," *Lancet* **2**, 1–3.

Beevers, D. G., Goldberg, A., Campbell, B. C., Moore, M. R., and Hawthorne, V. M. (1976b). *Lancet* **2**, 1222–1224.

Biggin, H. C., Chen, N. S., Ettinger, K. V., Fremlin, J. H., Morgan, W. D., Nowotny, R., Chamberlain, M. J., and Harvey, T. C. (1974). "Cadmium by *in vivo* Neutron Activation Analysis," *J. Radioanal. Chem.* **19**, 207–214.

Bonnell, J. A., Kazantzis, G., and King, E. (1959). "A Followup Study of Men Exposed to Cadmium Oxide Fume," *Br. J. Ind. Med.* **16**, 135–145.

Browning, E. (1961). *Toxicity of Industrial Metals.* Butterworths, London.

Buxton, R. St. J. (1956). "Respiratory Function in Men Casting Cadmium Alloys. II: The Estimation of the Total Lung Volume, Its Subdivisions and the Mixing Coefficient," *Br. J. Ind. Med.* **13**, 36–40.

Christeensen, H. E., Ed. (1971). *Toxic Substances: Annual List 1971.* U.S. Department of Health, Education, and Welfare, National Institute for Occupational Safety and Health, Rockville, Md.

Commission of the European Communities (1974). "Problems of the Contamination of Man and His Environment by Cadmium and Mercury." In *Proceedings of the International Symposium, Luxembourg, July 1973.*

Darmady, E. M. (1974). "The Aging Kidney and Transplantation," *Kidney Int.* **5**, 307 (abstract).

DeLuca, H. F. (1976). "Recent Advances in our Understanding of the Vitamin D Endocrine System," *J. Lab. Clin. Med.* **87**, 7–26.

Duggan, R. E. and P. E. Corneliussen (1972). "Dietary Intake of Pesticide Chemicals in the United States (III), June 1968–April 1970," *Pestic. Monit. J.* **5**, 331–341.

Elinder, C. G., Lind, B., Kjellström, T., Linnman, L., and Friberg, L. (1976). Cadmium in Kidney Cortex, Liver, and Pancreas from Swedish Autopsies: Estimation of Biological Half-Time in Kidney Cortex, Considering Calorie Intake and Smoking Habits, *Arch. Environ. Health* **31**, 292–302.

Ellis, K. J., Vartsky, D., Zanzi, I., Cohn, S. H., Yasumura, S. (1979). "Cadmium: In vivo Measurement in Smokers and Nonsmokers," *Science* **205**, pp. 323–325.

Fox, M. R. S. (1976). "Cadmium Metabolism—A Review of Aspects Pertinent to Evaluating Dietary Cadmium Intake by Man." In A. S. Prasad and D. Oberleas, Eds., *Trace Elements in Human Health and Disease,* Vol. II: *Essential and Toxic Elements.* Academic Press, New York, pp. 401–416.

Friberg, L. (1949). "Proteinuria and Emphysema among Workers Exposed to Cadmium and Nickel Dust in a Storage Battery Plant." In *Proceedings of the 9th International Congress on Industrial Medicine, 1948,* pp. 641–644.

Friberg, L. (1950). "Health Hazards in the Manufacture of Alkaline Accumulators with Special Reference to Chronic Cadmium Poisoning," *Acta Med. Scand.* **138,** 1–124.

Friberg, L., Piscator, M., Nordberg, G., and Kjellström, T. (1974). *Cadmium in the Environment,* 2nd ed. CRC Press, Cleveland, Ohio.

Fulkerson, W. and Goeller, H. E. (Eds.). (1973) Cadmium, the Dissipated Element. ORNL NSF EP 21.

Gafafer, W. M. (1964). *Occupational Diseases, a Guide to Their Recognition.* U.S. Department of Health, Education and Welfare, Public Health Service, Washington, D.C., p. 103.

Geldmacher, V., Mallinckrodt, M., and Opitz, O. (1968). "Zur Diagnostik der Cadmiumvergiftung der normale Cadmiumgehalt menschlicher Organe und Korperflüssigkeiten," *Arbeitsmed. Sozialmed. Arbeitshyg.* **3,** 276.

Glauser, S. C., Bello, C. T., and Glauser, E. M. (1976). "Blood Cadmium Levels in Normotensive and Untreated Hypertensive Humans," *Lancet* **1,** 717–718.

Gleason, M. N., Gosselin, R. S., Hodge, H. C., and Smith, R. P. (1969). *Clinical Toxicology of Commercial Products: Acute Poisoning (Home and Farm).* Williams and Wilkins, Baltimore.

Goyer, R. A., Cherian, M. G., and Richardson, L. D. (1977). "Renal Effects of Cadmium." In *Proceedings of the International Cadmium Conference.* Metal Bulletin Ltd., London, pp. 183–186.

Habib, F. K., Hammond, G. L., Lee, I. R., Dawson, J. B., Mason, M. K., Smith, P. H., and Stitch, S. R. (1976). "Metal-Androgen Interrelationships in Carcinoma and Hyperplasia of the Human Prostate," *J. Endocrinol.* **71,** 133–141.

Hammer, D. I., Finklea, J. F., Creason, J. P., Sandifer, S. H., Keil, J. E., Priester, L. E., and Stara, J. F. (1972). "Cadmium Exposure and Human Health Effects." In D. D. Hemphill, Ed., *Trace Substances in Environmental Health,* Vol. 5, University of Missouri Press, Columbia, pp. 269–283.

Hammer, D. K., Colucci, A. V., Hasselblad, V., Williams, M. E., and Pinkerton, C. (1973). "Cadmium and Lead in Autopsy Tissues," *J. Occup. Med.* **15,** 956–963.

Harvey, T. C., McLellan, J. S., Thomas, B. J., Dykes, P. W., and Fremlin, J. F. (1976). in *2nd East Kilbride Conference on Progress and Problems of in Vivo Activation Analysis.* Scottish Universities Research and Reactor Center, Glasgow (abstr.).

Hunter, D. (1954). "Cadmium Poisoning," *Arh. Hig. Rada* **5,** 221–224.

International Cadmium Conference (1977). *Proceedings.* Metal Bulletin Ltd., London.

Ishizaki, A., Fukushima, M., and Sakamoto, M. (1970). "Distribution of Cd in Biological Materials. II: Cadmium and Zinc Contents of Foodstuffs," *Jap. J. Hyg.* **25,** 207–222.

Kazantzis, G. (1977). "Some Long Term Effects of Cadmium on the Human Kidney." In *Proceedings of the International Cadmium Conference.* Metal Bulletin Ltd., London, pp. 194–198.

Kipling, M. D. and Waterhouse, J. A. H. (1967). "Cadmium and Prostatic Carcinoma" (letter), *Lancet* **1,** 730–731.

Kjellström, T. (1977). "Accumulation and Renal Effects of Cadmium in Man." Doctoral Thesis, Department of Environmental Hygiene, Karolinska Institute, Stockholm.

Larsson, S.-E. and Piscator, M. (1971). "Effect of Cadmium on Skeletal Tissue in Normal and Calcium-Deficient Rats," *Isr. J. Med. Sci.* **7,** 495–498.

Lauwerys, R. R., Buchet, J., Roels, H. A., Brouwers, J., and Stanescu, D. (1974). "Epidemiological Survey of Workers Exposed to Cadmium," *Arch. Environ. Health* **28,** 145–148.

Lauwerys, R. R., Stanescu, D., Roels, H., and Buchet, J. P. (1977). "Effects of Cadmium on the Lungs," In *Proceedings of the International Cadmium Conference.* Metal Bulletin Ltd., London, pp. 201–204.

Lemen, R. A., Lee, J. S., Wagoner, J. K., and Blejer, H. P. (1976). "Cancer Mortality among Cadmium Production Workers," *Ann. N.Y. Acad. Sci.* **271,** 273–279.

Lener, J. and Bibr, B. (1971). "Cadmium and Hypertension," *Lancet* **1,** 970.

Lewis, G. P., Jusko, W. J., and Coughlin, L. (1972). "Cadmium Accumulation in Man: Influence of Smoking, Occupation, Alcoholic Habit and Disease," *J. Chronic Dis.* **25,** 717–726.

Lewis, G. P., Jusko, W. J., Coughlin, L. I., and Hartz, S. (1972). "Contribution of Cigarette Smoking to Cadmium Accumulation in Man," *Lancet* **1,** 291–292.

Lucis, O. J., Lucis, R., and Shaikh, Z. A. (1972). "Cadmium and Zinc in Pregnancy and Lactation," *Arch. Environ. Health* **25,** 14–22.

McKee, J. E. and Wolf, H. W. (Eds.) (1963). *Water Quality Criteria,* 2nd ed. Resources Agency of California, State Water Quality Control Board, pp. 148–150.

McKenzie, J. M. and Kay, D. L. (1973). "Urinary Excretion of Cadmium, Zinc and Copper in Normotensive and Hypertensive Women," *N.Z. Med. J.* **78,** 68–70.

McLellan, J. S., Thomas, B. J., Fremlin, J. H., and Harvey, T. C. (1975). "Cadmium—Its *in vivo* Detection in Man," *Phys. Med. Biol.* **20,** 88–95.

Menden, E. E., Elia, V. J., Michael, L. W., and Petering, H. G. (1972). "Distribution of Cadmium and Nickel of Tobacco during Cigarette Smoking," *Environ. Sci. Technol.* **6,** 830–832.

Moore, M. R., Goldberg, A., Campbell, B. C., and Beevers, D. G. (1977). "Cadmium, Lead and Hypertension." In *Proceedings of the International Cadmium Conference.* Metal Bulletin Ltd., London, pp. 198–201.

Morgan, J. M. (1969). "Tissue Cadmium Concentrations in Man," *Arch. Intern. Med.* **123,** 405–408.

Morgan, J. M. (1970). "Cadmium and Zinc Abnormalities in Bronchiogenic Carcinoma," *Cancer* **25**, 1394-1398.

National Air Pollution Control Administration (1968). *Air Quality Data from the National Air Surveillance Networks and Contributing State and Local Networks,* 1966 Ed. Air Quality and Emission Data Program, APTD-68-9, U.S. Department of Health, Education, and Welfare, Public Health Service, Consumer Protection and Environmental Health Service, Durham, N.C.

NATO Conference (1974). *Ecotoxicity of Heavy Metals and Organo-Halogen Compounds.* Proceedings of conference at Mont Gabriel, Canada. Plenum Press, New York.

Nomiyama, K., Nomiyama, H., Yotoriyama, M., and Taguchi, T. (1977). Some Recent Studies on the Renal Effects of Cadmium. In *Proceedings of International Cadmium Conference.* Metal Bulletin, Ltd., London, pp. 186-194.

Nordberg, G. F. and Kjellström, T. (1978). *Environ. Health Prospect.,* in press.

Nordberg, G. F., Piscator, M., and Nordberg, M. (1971). "On the Distribution of Cadmium in Blood," *Acta Pharmacol Toxicol.* **30**, 289-295.

Nordberg, G., Slorach, S., and Stenstrom, T. (1973). "Cadmium Poisoning Caused by a Cooled Soft-Drink Machine," *Lakartidningen* **70**, 601-604.

Nordberg, G. F., Goyer, R. A., and Nordberg, M. (1975). "Comparative Toxicity of Cadmium-Metallothionein and Cadmium Chloride on Mouse Kidney," *Arch. Pathol.* **99**, 192-197.

Østergaard, K. (1977). "The Concentration of Cadmium in Renal Tissue from Smokers and Non-smokers," *Acta Med. Scand.* **202**, 193-195.

Østergaard, K. (1978). *Acta Med. Scand.,* in press.

Page, I. H. (1949). "Pathogenesis of Arterial Hypertension," *J. Am. Med. Assoc.* **140**, 451-458.

Papper, S. (1973). "The Effects of Age in Reducing Renal Function," *Geriatrics* **28**, 83-87.

Perry, H. M., Jr. (1972). "Cardiovascular Diseases Related to Geochemical Environment: Hypertension and the Geochemical Environment," *Ann. N.Y. Acad. Sci.* **199**, 202-216.

Perry, H. M., Jr., Thind, G. S., and Perry, E. F. (1976). "The Biology of Cadmium," *Med. Clin. North Am.* **60**, 759-769.

Perry, H. M., Jr., and Schroeder, H. A. (1955). "Concentration of Trace Metals in Urine and Treated and Untreated Hypertensive Patients Compared with Normal Subjects." *J. Lab. Clin. Med.* **46**, 936.

Perry, H. M., Jr., Tipton, I. H., Schroeder, H. A., Steiner, R. L., and Cook, M. J. (1961). "Variation in the Concentration of Cadmium in Human Kidney as a Function of Age and Geographic Origin," *J. Chronic Dis.* **14**, 259-271.

Piscator, M. (1966a). "Proteinuria in Chronic Cadmium Poisoning. III: Electrophoretic and Immunoelectrophoretic Studies on Urinary Proteins from Cadmium Workers, with Special Reference to the Excretion of Low Molecular Weight Proteins," *Arch. Environ. Health* **12**, 335-344.

Piscator, M. (1966b). "Proteinuria in Chronic Cadmium Poisoning. IV: Gel Filtration and Ion-Exchange Chromatography of Urinary Proteins from Cadmium Workers," *Arch. Environ. Health* **12**, 345–359.

Piscator, M. (1971). In L. Friberg, M. Piscator, and G. Nordberg, Eds., *Cadmium in the Environment*. CRC Press, Cleveland, Ohio.

Piscator, M. (1976). "The Chronic Toxicity of Cadmium." In A. S. Prasad and D. Oberleas, Eds., *Trace Elements in Human Health and Disease*, Vol. II: *Essential and Toxic Elements*. Academic Press, New York, pp. 432–441.

Potts, C. L. (1965). "Cadmium Proteinuria—the Health of Battery Workers Exposed to Cadmium Oxide Dust," *Ann. Occup. Hyg.* **8**, 55.

Rahola, T., Aaran, R.-K., and Miettinen, J. K. (1972). "Half-Time Studies of Mercury and Cadmium by Whole Body Counting." In *Assessment of Radioactive Contamination in Man*. Symposium on the Assessment of Radioactive Organ and Body Burdens, Stockholm, Nov. 22–26, 1971. International Atomic Energy Agency, Vienna, pp. 553–562.

Reynolds, C. V. and Reynolds, E. B. (1971). "Cadmium in Crabs and Crabmeat," *J. Assoc. Public Anal.* **9**, 112.

Savot, Dr. (1858). "Poisoning by a Powdered Silver-Polishing Agent," *Presse Med. (Bel.)* **10**, 69–70.

Schroeder, H. A. (1965). "Cadmium as a Factor in Hypertension," *J. Chronic Dis.* **18**, 647–656.

Schroeder, H. A. and Balassa, J. J. (1961). "Abnormal Trace Metals in Man: Cadmium," *J. Chronic Dis.* **14**, 236–258.

Smith, J. P., Smith, J. C., and McCall, J. (1960). "Chronic Poisoning from Cadmium Fumes," *J. Pathol. Bacteriol.* **80**, 287–296.

Smith, T. (1977). "Effects of Cadmium on the Lungs." International Cadmium Conference, Proc. Metal Bulletin Ltd., London, pp. 205–206.

Stokinger, H. E. (1963). "The Metals (Excluding Lead)." Chapter XXVII in F. A. Patty, Ed., *Industrial Hygiene and Toxicology,* Vol. II: *Toxicology* (B. W. Fassett and D. D. Irish, Eds.). Interscience, New York.

Suzuki, S., Taguchi, T., and Yokohashi, G. (1969). "Dietary Factors Influencing upon the Retention Rate of Orally Administered $^{115m}CdCl_2$ in Mice with Special Reference to Calcium and Protein Concentrations in Diet," *Ind. Health* **7**, 155.

Tipton, I. H. and Cook, M. J. (1963). "Trace Elements in Human Tissue. II: Adult Subjects from the United States," *Health Phys.* **9**, 103–143.

Tsuchiya, K. (1976). "Epidemiological Studies on Cadmium in the Environment in Japan: Etiology of Itai-Itai Disease," *Fed. Proc.* **35**, 2412–2418.

Tsuchiya, K. and Sugita, M. (1971). "A Mathematical Model for Deriving the Biological Half-Life of a Chemical," *Nord. Hyd. Tidskr.* **53**, 106–110.

Ulander, A. and Axelson, O. (1974). "Measurement of Blood-Cadmium Levels," *Lancet* **I**, 682–683.

U.S. Public Health Service (1962). *Public Health Service Drinking Water Standards.*

Vartsky, D., Ellis, K. J., Chen, N. S., and Cohn, S. H. (1977). "A Facility for *in vivo* Measurement of Kidney and Liver Cadmium by Neutron Capture Prompt Gamma Ray Analysis," *Phys. Med. Biol.* **22,** 1085–1096.

Webb, M. (1975). "Cadmium." *Br. Med. Bull.* **31,** 246–250.

Webster, W. (1978). "Cadmium-Induced Fetal Growth Retardation in the Mouse," *Arch. Environ. Health* **73,** 36–42.

Wester, P. O. (1973). "Trace Elements in Serum and Urine from Hypertensive Patients before and during Treatment with Chlorthalidone," *Acta Med. Scand.* **194,** 505–512.

Wester, P. O. (1974). "Trace Elements in Serum and Urine from Hypertensive Patients Treated for Six Months with Chlorthalidone," *Acta Med. Scand.* **196,** 489–494.

Zielhuis, R. L., Stuik, E. J., Herber, R. F. M., Salle, H. J. A., Verberk, M. M., Posma, F. D., and Jager, J. H. (1977). "Smoking Habits and Levels of Lead and Cadmium in Blood in Urban Women," *Int. Arch. Occup. Environ. Health* **39,** 53–58.

2

PRODUCTION, USES, AND PROPERTIES OF CADMIUM

Jerome O. Nriagu

National Water Research Institute, Canada Center for Inland Waters, Burlington, Ontario, Canada

1. INTRODUCTION

The toxic properties of cadmium seem to have been well known to miners long before the metal was identified as an element. For example, Agricola (1490–1555), who lived and worked in the mining district of Schneeberg, warned workmen to wear "boots of rawhide," to put on gloves "long enough to reach to the elbow," and to don "loose veils over their faces" because "a certain kind of *cadmia* eats away the feet of the workmen and injures their lungs and eyes" (Agricola, 1556). Cadmium as such was discovered in 1817 by F. Strohmeyer, a professor of metallurgy at Gottingen, Germany. Strohmeyer noticed that certain samples of zinc carbonate from Salzgitter, Germany, turned yellow on heating instead of white and guessed that the color was due to the oxide of a previously unknown element. He separated some of the metallic oxide from the zinc carbonate by careful precipitation with hydrogen sulfide and subsequently reduced the sulfide to obtain the first sample of the new element. The new metal was confusingly named cadmium, from *cadmia,* the ancient Greek name for calamine (zinc carbonate), from which it was isolated. At about the same time K. S. L. Hermann of Schonebeck sent a piece of zinc (from Silesian zinc ores), which had been rejected because of the formation of a yellow precipitate when it was dissolved in acid and saturated with hydrogen sulfide, to Strohmeyer, who immediately identified the sample as the sulfide of the same metal that he had just discovered. Shortly thereafter (in 1819) W. Meissner and C. J. B. Karsten confirmed the new element.

The production, uses, and properties of cadmium are summarized in this chapter. From the environmental viewpoint the exploitation of cadmium represents an enigma. It is obtained wholly as a by-product in the production of zinc and lead, and apparently would have been released to the environmnt if it were not so recovered. On the other hand, the current uses of cadmium are almost completely dissipative; less than 5% of the metal consumed is now recovered. This chapter should thus serve to emphasize the great need for effective recovery of cadmium from secondary sources.

2. PROPERTIES

Cadmium is a soft, silvery white, ductile metal with a faint bluish tinge. A selection of the physical properties of cadmium is given in Table 1. Cadmium vapor is very reactive, quickly forming finely divided cadmium oxide in the air.

The atomic weight of 112.40 is derived from a mixture of eight stable isotopes (Table 2). The radioactive isotopes are ^{106}Cd and ^{113}Cd, whereas the artificial ones include mass numbers of 105, 107, 109, 115, and 117 (Table 2).

Cadmium falls between silver (atomic no. 47) and indium (atomic no. 49)

Table 1. Selected Physical Properties of Cadmium

Atomic number	48
Atomic weight	112.40
Boiling point	765°C
Bond strengths in diatomic molecules (cal/mol)	
Cd—Cd	2.7
Cd—Cl	49.4
Cd—F	73
Cd—H	16.5
Cd—In	33
Cd—S	48
Crystal structure, distorted close-packed hexagonal	
a	2.97887
c	5.61765
d(Cd—Cd)	2.979
Density at 20°C (g/cm^3)	8.65
Electrical resistivity (microhm-cm)	
Solid, 0°C	6.83
Liquid, 400°C	33.7
Liquid, 700°C	35.8
Electronegativity function	
Pauling	1.69
Allred-Rochow	1.46
Mulliken	1.4
Electronic work function (eV)	4.22
Index of refraction	
Liquid, 4360 Å	39
Solid, 6300 Å	1.13
Ionization potential (ev)	
I	8.99
II	16.904
III	37.47
Latent heat of fusion (cal/g)	13.2
Latent heat of vaporization (cal/g)	286.4
Linear coefficient of thermal expansion (μin./inch.°C)	
25°C	29.8
321–540°C	150
Magnetic susceptibility (cgs units)	19.8×10^{-6}
Photoelectric work function (eV)	4.01
Resistivity (microhm-cm), 20°C	7.6
Mechanical properties	
Tensile strength (psi)	10,000
Elongation (% in 1 in.)	50
Brinell hardness	21
Modulus of elasticity (psi)	8.0×10^6

Table 1. *Continued*

Melting point	321°C (610°F)
Solidification shrinkage (%)	4.74
Specific heat (cal/g·°C)	
Liquid, 321–700°C	0.0632
Solid, 25°C	0.055
Specific volume, (cm³/g)	
20°C	0.1156
330°C	0.1248
Surface tension (dynes/cm)	
350°C	586
400°C	600
450°C	600
Thermal conductivity (cal/cm²·cm·°C·sec)	
20°C	0.22
358°C	0.105
435°C	0.119
Thermal neutron cross section, 2200 m/s	
Absorption (barns)	2500
Scattering (barns)	7
Resonance absorption integral (barns)	Not known
Valence	+2
Vapor pressure (mm Hg)	
394°C	1.0
578°C	60
711°C	400
767°C	760
Viscosity (dynes/sec·cm², poises)	
349°C (660°F)	0.0144
603°C (1117°F)	0.0110
Volume expansion on melting (%)	5.1

and has zinc and mercury as its neighbors in the periodic table. Cadmium and zinc, however, differ from mercury in that the latter forms particularly strong Hg-C bonds. Like zinc, cadmium is almost always divalent in all stable compounds, and its ion is colorless.

Cadmium is not affected by dry air but oxidized readily in moist air with the formation of a protective coating of oxides. Cadmium resists corrosion in rural atmospheres but is attacked aggressively by the pollutant sulfur dioxide and ammonia in urban and industrial environments.

Cadmium is readily oxidized by steam, and at red heat burns in air to form the brown oxide that so often colors the zinc oxide fumes in smelter plants. It dissolves in most inorganic and some organic acids, nitric acid being the best

Table 2. **Isotopes of Cadmium**[a]

Mass Number	Half-Life	Decay Mode	Natural Abundance (%)
103	10 min	β^+, γ	
104	57 min	β^+, γ	
105	55 min	β^+, γ	
106	Stable	—	1.22
107	6.5 hr	β^+, γ	
108	Stable	—	0.88
109	450 days	γ	
110	Stable	—	12.39
111	Stable	—	12.75
112	Stable	—	24.07
113	Stable	—	12.26
114	Stable	—	28.86
115	53.5 hr	β^-	
116	Stable	—	7.58
117	2.4 hr	β^-	
118	49 min	β^-	
119	2.7 min	β^-	

[a] Data from *Handbook of Chemistry and Physics* (1978–1979).

of the acid solvents. Unlike zinc, cadmium is not amphoteric and hence is not soluble in alkalies. It forms strong complexes with cyanides and ammines; many σ-bonded organometallic compounds of the element are also known.

When placed in neutral or alkaline solutions, both the zinc and the cadmium surfaces show a mosaic of active and passive regions that do not appear to be in true equilibrium with the solution (Aylett, 1973).

The thermochemical properties of cadmium and its compounds are summarized in Tables 3, 4a, and 4b.

3. OCCURRENCE

Cadmium is a rare element; its concentration in the lithosphere ranges from 0.1 to 0.2 $\mu g/g$, making it the sixty-seventh element in order of abundance (Fleischer et al., 1974; Gong, 1975). It is thus rarer than mercury and about $1/350$ as abundant as zinc (whose crustal abundance is 70 $\mu g/g$). Cadmium is a strongly chalcophilic element, that is, concentrated in sulfide minerals, a characteristic in which it follows zinc and mercury and, to a lesser extent, lead and copper. Only a few cadmium minerals are known: greenockite (hexagonal CdS), hawleyite

Table 3.   Thermochemical Data for Solid Cadmium Compounds[a]

Compound	ΔH_f° (kcal/mol)	ΔG_f° (kcal/mol)	S° (cal/mol)
Cd, γ	0	0	12.37
Cd, α	−0.14	−0.14	12.37
CdO	−61.7	−54.6	13.1
Cd(OH)$_2$	−134.0	−113.2	23
CdF$_2$	−167.4	−154.8	18.5
CdCl$_2$	−93.57	−82.21	27.55
CdBr$_2$	−75.57	−70.82	32.8
CdBr·4H$_2$O	−356.73	−298.29	75.6
CdI$_2$	−48.6	−48.13	38.5
Cd(OH)Cl	−119.0	−101.8	21
CdS	−38.7	−37.4	15.5
CdSO$_4$	−223.06	−196.65	29.41
CdSO$_4$·H$_2$O	−296.26	−255.46	36.81
CdSO$_4$·8/3H$_2$O	−413.33	−350.22	54.88
CdSO$_4$·2Cd(OH)$_2$		−429.6	
CdSeO$_3$	−137.5	−119.0	34.0
CdTe	−22.1	−22.0	24
CdCl$_2$·2NH$_3$	−152.0	−106.1	51.0
CdCl$_2$·4NH$_3$	−195.4	−116.2	79.0
CdCl$_2$·6NH$_3$	−237.9	−126.2	109.2
Cd$_3$(PO$_4$)$_2$		−587.1	
Cd$_3$(ASO$_4$)$_2$		−410.2	
CdSb	−3.44	−3.11	22.2
CdCO$_3$	−179.4	−160.0	22.1
CdC$_2$O$_4$·3H$_2$O		−360.4	
CdSiO$_3$	−284.2	−264.2	23.3
Cd(BO$_2$)$_2$		−354.87	

[a] Data from Wagman et al. (1968).

(cubic CdS), cadmoselite (CdSe), monteponite (CdO), otavite (CdCO$_3$), and saukovite or cadmian metacinnabar [(Hg,Cd)S]. All of these minerals are rare and are not commercial sources of the metal.

Cadmium occurs characteristically as isomorphic impurities in, or surface coatings of, other sulfide minerals, especially the zinc sulfides (sphalerite and wurtzite), where concentrations as high as 5% have been reported (Chizhikov, 1966). The common average cadmium cntent of zinc sulfides, however, is 0.02 to 1.4%, with a median value of about 0.3% (Fleischer et al., 1974). Fourteen other metal sulfides have also been reported to contain over 500 μgCd/g; the most important of these include galena (PbS), metacinnabar (HgS), tetrahe-

Table 4a. **Thermochemical Data for Dissolved Cadmium Species**[a]

Ion or Speices	ΔH_f° (kcal/mol)	ΔG_f° (kcal/mol)	S° (cal/mol)
Cd^{2+}	−18.14	−18.54	−17.5
$Cd(OH)^+$		−62.4	
$Cd(OH)_2$	−128.1	−93.73	−22.6
$Cd(OH)_3^-$		−143.6	
$Cd(OH)_4^{2-}$		−181.3	
$HCdO_2^-$		−86.9	
CdF_2	−172.14	−151.8	−24.1
$CdCl^+$	−57.5	−53.63	10.4
$CdCl_2$	−98.04	−81.29	9.5
$CdCl_3^-$	−134.1	−116.4	48.5
$Cd(ClO_4)_2$	−79.96	−22.66	69.5
$Cd(OH)Cl$		−96.5	
$CdBr^+$	−48.0	−46.35	9.5
$CdBr_2$	−76.24	−68.24	21.9
$CdBr_3^-$		−97.4	
CdI^+	−33.7	−33.8	38.5
CdI_2	−44.52	−43.20	35.7
CdI_3^-		−62.0	
CdI_4^{2-}	−81.7	−75.5	78.0
$CdSO_4$	−235.46	−196.51	−12.7
$CdSeO_3$	−161.3	−124.0	−4.6
CdN^+		62.6	
$Cd(N_3)_2$	133.38	147.9	34.1
$Cd(N_3)_3^-$		226.8	
$Cd(N_3)_4^{2-}$		309.5	
$CdNO_2^+$		−30.7	
$Cd(NO_3)_2$	−117.26	−71.76	52.5
$Cd(NH_3)^{2+}$		−28.4	
$Cd(NH_3)_2^{2+}$		−38.0	
$Cd(NH_3)_4^{2+}$	−107.6	−54.1	80.4
$Cd(N_2H_2)_2^{2+}$		39.3	
$Cd(N_2H_2)_3^{2+}$		69.4	
$Cd(N_2H_2)_4^{2+}$		98.5	
$CdP_2O_7^{2-}$		−489.1	
CdC_2O_4	−215.3	−179 6	−6.6
$Cd(C_2O_4)_2^{2-}$		−348.6	
$Cd_2C_2O_4^{2-}$		−205.6	
$Cd(CHO)^+$		−104.2	
$Cd(CHO)_2^0$	−221.56	−186.23	26
$Cd(CH_3CO_2)^+$		−109.2	
$Cd(CHO)_3^-$		−272.8	

Table 4a. *Continued*

Ion or Speices	ΔH_f° (kcal/mol)	ΔG_f° (kcal/mol)	S° (cal/mol)
$Cd(CHO)_4^{2-}$		-356	
$Cd(CH_3CO_2)_2$	-250.46	-195.12	23.9
$Cd(CH_3CO_2)_3^-$		-286.4	
$Cd(CH_3CO_2)_4^{2-}$		-374.6	
$CdCN^+$		15.4	
$Cd(CN)_2$	53.9	63.9	27.5
$Cd(CN)_3^-$		84.8	
$Cd(CN)_4^{2-}$		121.3	
$Cd(NH_2CH_3)^{2+}$		-17.4	
$Cd(NH_2CH_3)_2^{2+}$	-58.7	-15.2	40.0
$Cd(NH_2CH_3)_3^{2+}$		-11.8	
$Cd(NH_2CH_3)_4^{2-}$	-98.9	-7.7	84.5
$Cd(NH_2CH_2COO)^+$		-84.4	
$Cd(NH_2CH_2COO)_2$		-157.2	
$Cd(CNS)^+$		1.8	
$Cd(CNS)_2$	18.40	25.76	51.4
$Cd(CNS)_3^-$		45.3	

[a] Data from Wagman et al. (1968).

drite-tennartite $[(Cu,Zn)_2(SB,As)_4S_{13}]$, and chalcopyrite ($CuFeS_2$). Some secondary zinc silicates and carbonates contain up to 1.2% Cd (see Fleischer et al., 1974).

The average concentrations of cadmium in the major rock types are shown in Table 5. Its concentration in igneous rocks is generally low and shows no clear relation to the concentration of any other major or minor element. The Zn/Cd ratios of igneous rocks vary widely, although the average ratio for mafic rocks appears to be higher than that for granitic rocks (910 vs. 740: Gong, 1975). Also, cadmium tends to be enriched in shales, oceanic and lacustrine sediments, and phosphorites but depleted in red shales, sandstones, and limestones relative to igneous and metamorphic rocks (Gong et al., 1977).

4. RECOVERY

The recovery of cadmium from its host minerals is predicated upon two major factors: it has lower boiling and melting points than zinc, and its oxide is more easily reduced than the other associated metal oxides. The first commercial process (no longer in use) consisted of heating the roasted zinc ore mixed with coal or coke in a retort which had a primary condenser and a secondary condenser

Table 4b. Stability Constants of Selected Organocadmium Complexes[a]

Ligand Progenitor	Ionic Strength, I[b]	Log k_1	Log β_2	Log β_3	Log k_{CdHL}[c]
Formic acid	1.0	1.04	1.23	1.75	
Acetic acid	0	1.93	3.15		
Propanoic acid	1.0	1.19	1.86		
Butanoic acid	1.0	1.25	1.98		
Isovaleric acid	1.0	1.34	2.30	2.5	2.0 (β_4)
Nitroacetic acid	0.4 (18°C)	0.19			
Glycolic acid	0	1.87	2.0		
			($I = 2.0$)		
Lactic acid	1.0	1.30	2.1	2.5	
DL-2-Hydroxybutanoic acid	2.0	1.23	2.15	2.26	2.45 (β_4)
Glyceric acid	2.0	1.60			
D-gluconic acid	1.0	2.10			
3-Hydroxypropanoic acid	2.0	1.15	2.20		
N-Acetylglycine	0.6	1.23			
Hippuric acid	1.0	0.95	1.45		
Ethoxyacetic acid	1.0	1.07	1.69	1.54	
(Ethylthio)acetic acid	1.0	1.27	2.12	2.51	2.72 (β_4)
Oxalic acid	0 (18°C)	3.89			
Malonic acid	1.0	1.92	2.88		0.69
Methylmalonic acid	0.1	2.58			1.27
Ethylmalonic acid	0.1	2.59			1.28
Succinic acid	1.0 (20°C)	2.72	2.79		0.99
			($I = 0$)		
Malsic acid	0.2	2.2	3.6	3.8	
Itaconic acid	1.0	1.72			0.85
Adipic acid	0.1	2.1			
Phthalic acid	0.1	2.5			
Malic acid	0.1 (20°C)	2.36			1.34
Diglycolic acid	0.1	3.21			1.02
Thiodiacetic acid	0.1	3.14	5.57		
Selenodiacetic acid	0.1	2.57	1.82		
Citric acid	0.5	3.15	4.54		
Ditartronic acid	0.1	5.44			3.83
Catechol	0.1 (30°C)	8.2			
Titron	1 0	7.69	13.29		
Nitroso-NW acid	0	3.12			
Nitroso-Schllkopf's acid	0.1	3.26	5.37		
Nitroso-Schaffer's acid	0.1	3.35	6.45		
Nitroso-R acid	0	4.65	6.59		

Table 4b. *Continued*

Ligand Progenitor	Ionic Strength, I^b	Log k_1	Log β_2	Log β_3	Log $k_{CdHL}{}^c$
Eriochrome Black T	0.3 (20°C)	12.74			
Acetylacetone	0	3.83	6.65		
Tropolone	0.1	4.60			
Kojic acid	0.1 (20°C)	4.6			
4-(Phenylthio)benzene-sulfonic acid	0.2	0.63		3.04	
Acetohydroxamic acid	0.1 (20°C)	4.5	7.8		
Thiourea	0	1.5	2.2	2.6	3.1 (β_4)
N-Methylthiourea	0.1	1.42	2.40	2.9	4.1 (β_4)
N-Ethylthiourea	0.1	1.46	2.18	3.5	4.4 (β_4)
N,N-Ethylenethiourea	1.0	1.31	2.1	2.7	3.4 (β_4)
Semicarbazide	0.1 (30°C)	1.26	2.79		
Thiosemicarbazide	1.0	2.57	4.70	5.86	
Glycine	0	4.69	8.40	10.68	
Alanine	1.0	3.80	7.10		
Leucine	1 0	3.84	6.54	8.60	
Phenylalanine	0.37 (20°C)	3.87	6.73		
Aspartic acid	0.1	4.39	7.55		
Glutamic acid	0.1	3.9			
Homoserine	0.1	3.69			
Penicillamine	0.1	10.8			
S-Methylcysteine	0.1	3.77	7.09		
Methionine	0.1	3.67	7.03		
Ethionine	0.1	4.68	9.22		
Histidine	0.1	5.39	9.66		
Tryptophan	0.37 (20°C)	4.47	8.18		
EDMA	0.2	8.48	13.23		
N-Acetyl-L-histidine	0.16	2.70	4.65		
EDDA	0.1	8.99			
EDDG	0.1	8.76			
EHPG	0.1	13.13			
HBED	0.1	17.52			8.11
Iminodiacetic acid	0.1	5.71	10.12		
MIDA	0.1	6.75	12.43		
N-(Cyanomethyl) iminodiacetic acid	0.1 (20°C)	4.48	8.5		
NTA	0.1	9.78	14.39		
HIDA	0.1	7.24	12.31		
N-(2-Methoxyethyl) iminodiacetic acid	0.1	7.53	13.18		

Table 4b. *Continued*

Ligand Progenitor	Ionic Strength, I^b	Log k_1	Log β_2	Log β_3	Log $k_{CdHL}{}^c$
Nitrilotriacetic acid monoamide	0.1 (20°C)	3.82	6.47		
N-(2-Mercaptoethyl) iminodiacetic acid	0.1 (20°C)	16.72	22.07		1.49
Ethylenediamine-N,N-diacetic acid	0.1 (20°C)	10.53	16.59		
HEDTA	0.1	13.1			
EDTA	0.1	16.36			3.1
PDTA	0.1	17.6			
DL-(Propylethylene)-DTA	0.1 (20°C)	18.05			
DL-2,3-Butylene-DTA	0.1 (20°C)	18.8			2.77
CPDTA	0.1 (20°C)	18.25			
CDTA	0.1 (20°C)	19.93			3.0
Trimethylene-DTA	0.1 (20°C)	13.90			3.06
Tetramethylene-DTA	0.1 (20°C)	12.02			3.10
EEDTA	0.1	16.1			
EGTA	0.1	16.5			
TEDTA	0.1	14.28			
DTPA	0.1	19.0			4.17
TTHA	0.1	18.6			8.5
Diglycine	0.1	2.90	5.36		6.4 ($I = 1.0$)
L-Carnosine	0.1	3.19			8.38
Triglycine	0.15	2.70	5 3		6.3 ($I = 0.8$)
Tetraglycine	0.15	2.65	5.2		6.3 ($I = 0.8$)
Anthranilic acid	0	1.83			
N-(8-Quinolyl)glycine	0.1	2.7	5.3		
Picolinic acid	0	4.79	8.25		
Quinaldic acid	0	4.12	6.83		
Dipicolinic acid	0.1 (20°C)	6.75	11.15		
Methylamine	2.1	2.75	4.81	5.94	6.55 (β_4)
p-Toluidine	0.3	0.26			
Metanilic acid	1.0	0.26	0.56		
Ethanolamine	0.1	2.77	4.09	5.6	
2-Mercaptoethylamine	0.15	10.97	19.75		
L-Cysteine methyl ester	0.1	8.89	16.2		
Ethylenediamine	0	5.41	9.91	12.69	
1,7-Diaza-4-thiaheptane	1.0 (30°C)	5.47	8.99		

Table 4b. *Continued*

Ligand Progenitor	Ionic Strength, I^b	Log k_1	Log β_2	Log β_3	Log $k_{CdHL}{}^c$
Diethanolamine	0.1	2.40	4.52		
N-Methylethelene-diamine	1.0	5.47	9.56	11.4	1.5
Diethylenetriamine	0.1 (20°C)	8.4	13.8		
Tetraethylenepentamine	0.1	14.0			
Triethanolamine	0.1	2.70	4.60	5.21	
N,N-Dimethylethylene-diamine	1.0	4.81	8.11	9.41	
N,N,N¹-Trimethyl-ethylene diamine	1.0	4.56	6.73	7.7	0.83
Nitrilotris(2-ethylamine)	0.1 (20°C)	12.3			
Imidazole	0.16	2.80	4.99	6.46	7.53 (β_4)
Histamine	0	4.82	8.22		
L-Histidine methyl ester	0.16	3.98	6.79	8.0	
2-(2-Pyridyl)imidazole	0.1	4.70	8.16	10.74	
Pyridine	0.5	1.34	2.13	2.41	
3-Picoline	0.1 (30°C)	1.34	2.3	2.5	
4-Picoline	0.1 (30°C)	1.51	2.47	2.90	
Pyridoxamine	0.1	4.59			
Aminopyridine	0.1 (30°C)	1.52	2.19	2.88	
2-Picolylamine	0	4.67	8.54	11.1	
2-(Methylaminomethyl) pyridine	0	4.55	8.02	10.7	
2-Pyridylhydrazine	0.1 (20°C)	4.36	8.18		
Oxine	0 (20°C)	7.78			
Sulfoxine	0	7.70	14.2		
2,2¹-Bipyridyl	0.1	4.18	7.7	10.3	
Di-2-picolylamine (DPA)	0.1	6.44	11.74		
1,10-Phenanthroline	0.1	5.8	10.6	14.6	
2-Methyl-1,10-phenanthroline	0.1	5.15	9.65	13.3	
2,9-Dimethyl-1,10-phenanthroline	0.1	4.1	7.4	10.4	

[a] From Martell and Smith (1975).

[b] Data pertain to 25°C unless specific otherwise.

[c] Unless noted otherwise, the constant listed is for the equilibrium reaction $H^+ + ML = MHL$, where L is the ligand.

Table 5.   Cadmium in Principal Rock Types and Geological Materials [a]

	Cadmium Content (μg/g)	
Type of Material	Mean	Range
Igneous rocks		
Ryolites	0.23	0.03–0.57
Granites	0.12	0.02–1.6
Basalts, diabases, gabbros	0.14	0.1–1.0
Eclogites	0.10	0.03–1.6
Ultramafic rocks	0.02	0.001–0.03
Metamorphic rocks		
Gneisses	0.04	0.007–0.26
Schists	0.02	0.005–0.87
Sedimentary rocks		
Carbonate rocks	0.08	0.001–0.5
Sandstones, conglomerates	0.068	0.01–0.41
Shales	1.3	0.02–11
Deep ocean sediments	0.5	0.04–17
Oceanic manganese oxides	8.0	<3.0–21
Phosphorites	25	<10–500
Recent sediments		
Lake sediments	0.91	0.02–6.2
Stream sediments	0.16	0.03–0.4

[a] Based largely on compilations by Fleischer et al. (1974) and Gong (1975).

termed a "prolong." As the roasted ore was reduced, the vaporized zinc was caught in the first condenser, whereas the cadmium, owing to its lower boiling point, was condensed in the prolong as "blue powder" containing 3 to 5% Cd. The blue powder was further upgraded by refuming. This early process was very inefficient, giving only a 25% cadmium recovery.

Today the initial recovery of cadmium during the processing of cadmium-bearing ores follows a wide variety of hydro- and pyrometallurgical procedures. In pyrometallurgical recovery the cadmium is collected as flue dusts from the roasting of zinc ore, sintering of the roasted zinc ore, copper smelting, or lead blast furnace smelting. Because the initial cadmium content of the feed into lead and copper smelters is usually low, the fumes must be upgraded to the bleed-off level (10 to 50% Cd) by repeated refuming in a kiln or reverberatory furnace. In the hydrometallurgical process the ore concentrates are roasted in a single multiple-hearth unit, followed by leaching of the roasted ore and the flue dusts with dilute sulfuric acid to take most of cadmium into solution. The leachate may be used as feed for electrolytic zinc cells or stripped of its cadmium content by using zinc dust to precipitate out the cadmium-copper sludge. Occasionally, the winnings from the pyrometallurgical process are leached with sulfuric acid; some of the impurities are removed by chemical means before precipitating out the cadmium (as a sponge), using zinc dust.

The cadmium-enriched materials may be refined by distillation either batchwise (in horizontal or Belgian retort plants) or continuously (in plants using vertical retorts or blast furnaces and employing the engineering principles of rectification). Alternatively, the refinery involves processing in electrolytic or electromotive plants. In the former the cadmium is recovered by electrolyzing purified solutions; in the latter the cadmium (in the form of sponge) is displaced from purified solutions by zinc. The sponge may then be briquetted and purified further by vacuum distillation. About 40% of the primary cadmium produced in 1969 was derived from electrolytic zinc plants (Fleischer et al., 1974).

Recovery processes for cadmium are described in detail by Chizhikov (1966) and Howe (1967).

5. BY-PRODUCT–COPRODUCT RELATIONSHIPS

Primary cadmium is recovered entirely as a by-product from residues obtained during the smelting of zinc, lead, zinc-lead, zinc-copper, and complex ores; the Zn/Cd ratios in representative ores are shown in Table 6. At present, secondary recovery of cadmium is relatively small. The total supply of cadmium thus depends largely on activities in the zinc industry and, to a much lesser extent, the lead and copper industries. Cadmium production now constitutes an integral part of the zinc industry and accounts for about 4 to 10% of the income generated from zinc sales. However, from the overall standpoint of zinc producers, cadmium recovery is inadequate to materially affect the marketing or production policy for the entire industry. There is therefore little incentive to search for ores with higher concentrations of cadmium. Hence future supplies of cadmium must depend on better recovery from the present (as well as secondary) sources, and the discovery and development of other cadmium-bearing ores. The by-product relationship also means that the price elasticity of cadmium is small and that changes in cadmium price are determined chiefly by the supply and price of zinc.

6. PRODUCTION

The historical pattern of cadmium supply has tended to reflect not only the amount of zinc produced but also the method of zinc production. Up to the end

Table 6.   Cadmium/Zinc Ratios in Selected Ores and Ore Concentrates[a]

Ore	Cd/Zn (%)
Upper Silesia deposits	0.23
Zinc ore, Belgian factories	0.24
Joplin deposits (U.S.A.)	0.62
Copper-zinc deposits (average)	0.30
Lead-zinc deposits (average)	0.40
Canadian electrolytic	0.51
Allied (Canada)	0.21
Mitsubishi (Japan)	0.40
American Zinc Company	0.73
Espanola del Zinc (Mexico)	0.40
New Jersey Zinc Company	0.42
St. Joe Minerals (Pennsylvania)	0.39
St. Joe Minerals (Pennsylvania)	0.48
St. Joe Minerals (Pennsylvania)	0.65
Broken Hill (Australia)	0.41
Broken Hill (Australia)	0.38
Broken Hill (Australia)	0.41

[a] From the compilations in Chizhikov (1966) and Sargent and Metz (1975).

of the nineteenth century, pyrometallurgical processing of the blue powder was the only means of obtaining cadmium from zinc ores. After the discovery of cadmium in 1817, this method was first used to produce cadmium on an industrial scale in 1829 in Upper Silesia. Although the production of cadmium from zinc operations in the Rhine area commenced in 1859, the yearly production never exceeded 100 kg until 1871 (Chizhikov, 1966) because no commercial uses for the metal had been found. After 1871, when cadmium and its compounds began to be used as paint pigments and for other purposes, output of the metal expanded rapidly. The yearly production in 1871 and 1872 were 710 and 1871 kg, respectively, but by 1910 had exceeded 43×10^3 kg with about 90% of the supply coming from German mines in Upper Silesia (Chizhikov, 1966).

Cadmium was first produced in the United States in 1906 from the zinc-plant blue powder, and in 1910 the American Smelting and Refining Company initiated the recovery of cadmium from lead-smelter flue dusts. By the outbreak of World War I the yearly production of cadmium stood at 80×10^3 kg and by 1918 had risen to 165×10^3 kg. Because the German supply was cut off during World War I, the production of cadmium in the United States was stepped up and by 1919 had reached 45×10^3 kg, making the country the world's leading producer of the metal, a position retained until about 1975.

A sharp rise in the amounts of cadmium produced followed the introduction of the hydrometallurgical method of zinc production in 1917 and the development, in 1919, of the Udylite process for electroplating steel and iron with cadmium. In the period between the world wars, worldwide cadmium production increased from 84×10^3 kg in 1919 (after the sharp drop from the 165×10^3 kg output in 1918) to 3.6×10^6 kg in 1938, about a 45-fold increase. During the same period production in the United States increased from 45×10^3 kg in 1919 to 1.85×10^6 kg in 1938 and accounted for over 50% of all the cadmium output. Cadmium recovery came into effect in several zinc producing countries during the interval: Australia in 1922, France in 1925, Belgium and Canada in 1928, Norway and Japan in 1931.

Between 1940 and 1944 worldwide cadmium production increased from 4.02 $\times 10^6$ to 4.92×10^6 kg in response to the great demand for the metal for military purposes. After World War II cadmium production increased at a fairly steady rate of about 5%/year and stood at 11.5×10^6 kg in 1960. There was a noticeable drop in growth rate in the early 1960s, but between 1964 and 1969 cadmium production increased by about 50% to the all-time high figure of 17.6×10^6 kg, attained in 1969. Since 1970 the pressure on the industry brought about by the concern regarding cadmium in the environment and its effects on human health has resulted in a continuing downward trend in the worldwide supply of this element. The production figure in 1977 was 17.1×10^6 kg (*Minerals Yearbooks*).

The historical trend in cadmium production is shown in Table 7.

7. SUPPLY PATTERN

The generic affinity of cadmium with zinc implies that world resources of cadmium are determined by zinc resources and cadmium supply is a function of zinc production. In general, however, cadmium production as reported seems to have fluctuated more than that of zinc, and the annual statistics for cadmium production show a slower rate of growth than those for zinc between 1960 and 1978. However, the ratios of production of the two metals have remained fairly constant since 1946 (Table 8), implying that there is no real decline in cadmium recovery and that improvements in methods of recovery are being offset by lower cadmium contents in zinc ores from some new mines. In Canada, for example, the zinc mines that have come into production in the past decade commonly have lower average cadmium concentrations than the older mines (*Canadian Minerals Yearbooks*).

The historical trends in the supply of cadmium from different regions of the

Table 7. Historical Trends in Regional Production of Cadmium[a]

Year	Canada	U.S.A.	South America	Europe	USSR	Asia	Africa	Global Total
1907		6		33				39
1908		3.4		33				36.4
1909		2.3		37				39.3
1910		2.0		41				43
1911		11.0		42.6				53.6
1912		22.2		42.6				64.8
1913		25		37				62
1914		41		39				80
1915		41		37				78
1916		61		58				119
1917		94		78				172
1918		58		107				165
1919		45		39				84
1920		59		22				81
1921		31		23				54
1922		60		3		37		100
1923		83		1		125		209
1924		59		5		162		226
1925		228		5		182		415
1926		368		71		163		602
1927		488		32		157		677
1928	223	851		52		174		1,300
1929	351	1126		104		202		1,783
1930	207	1260		136		233		1,836
1931	147	477		226		202		1,052

Table 7. *Continued*

Year	Canada	U.S.A.	South America	Europe	USSR	Asia	Africa	Global Total
1932	30	363		333		165		891
1933	112	1033		404		165		1,714
1934	133	1260	(385)[b]	471		175		2,039
1935	263	1577	(560)	572		225		2,637
1936	357	1648	(535)	772		276		3,053
1937	338	1812	(620)	1094		211		3,455
1938	317	1702	(762)	1211		262		3,492
1939	426	2001	(816)	1451		255		4,133
1940	412	2792	(816)	640		175		4,019
1941	568	3147	(907)	750		195	3	4,663
1942	521	3322	(854)	544		268	27	4,682
1943	357	3809	(802)	531		294	23	5,014
1944	239	3834	(701)	471		357	22	4,923
1945	293	3598	(1053)	212		268	18	4,389
1946	364	2812	(717)	279		239	17	3,711
1947	326	3632	(778)	376		221	26	4,581
1948	348	3439	(905)	674		313	18	4,792
1949	384	3640	(820)	669		317	25	5,035
1950	385	4024	(689)	966	68	389	30	5,862
1951	602	3681	(893)	1296	82	353	24	6,038
1952	430	3805	(734)	1311	91	457	21	6,115
1953	507	4392	(954)	1394	91	510	32	6,926

1954	493	4271	(592)	1599	213	570	63	7,209
1955	870	4424	(1295)	1892	310	650	166	8,312
1956	1061	4815	(858)	1782	360	682	277	8,977
1957	1074	4785	(759)	1872	476	795	413	9,415
1958	835	4388	64	1934	1300	796	507	9,824
1959	980	3902	64	2216	1500	837	475	9,974
1960	1069	4618	84	2273	1700	872	531	11,147
1961	1008	4747	157	2671	1500	1040	551	11,674
1962	1182	5052	149	2383	1590	1242	324	11,922
1963	1123	4491	408	2485	1680	1506	410	12,103
1964	1258	4744	420	2717	1770	1717	485	13,111
1965	430	4386	284	2716	1900	2004	177	11,897
1966	772	4745	311	2538	2050	2282	293	12,991
1967	934	3946	318	2821	2200	2527	439	13,185
1968	942	4831	374	3057	2200	2771	499	14,674
1969	963	5736	378	4100	2310	3588	498	17,573
1970	837	4293	454	4736	2360	3381	561	16,622
1971	711	3597	362	4411	2400	3471	468	15,420
1972	1020	2760	396	4577	2450	3993	468	16,664
1973	1400	3402	410	4963	2500	4077	398	17,150
1974	1242	3023	710	5277	2585	4023	410	17,270
1975	1191	1989	746	4923	2630	3479	390	15,348
1976	1290	2046	762	6131	2900	3787	400	17,316

[a] Compiled from *Minerals Yearbooks*. Quantities listed are in metric tons.
[b] Mine production exported as ore concentrates.

53

Table 8.   World Mine Production of Zinc and Cadmium[a]

Year	Cadmium ($\times 10^6$ kg)	Zinc ($\times 10^9$ kg)	Cd/Zn Production %
	Metal Production		
1946–1950 (avg)	4.98	1.86	0.27
1955–1959 (avg)	9.44	3.04	0.31
1960	11.1	3.26	0.34
1961	11.7	3.48	0.34
1962	11.9	3.58	0.33
1963	12.1	3.69	0.33
1964	13.1	4.03	0.32
1965	11.9	4.31	0.28
1966	13.0	4.50	0.29
1967	13.2	4.83	0.27
1968	14.7	4.97	0.30
1969	17.6	5.34	0.33
1970	16.6	5.46	0.30
1971	15.4	5.52	0.28
1972	16.7	5.64	0.30
1973	17.2	5.71	0.30
1974	17.3	5.79	0.30
1975	15.3	5.56	0.28
1977	17.1	5.90	0.29

[a] Compiled from *Minerals Yearbooks*.

world are indicated in Table 7. Until about 1917, Germany was the leading producer of cadmium, but with the cession of Upper Silesia to Poland after World War II the German output was drastically reduced. For over 50 years after the end of World War I the United States was the leading producer of cadmium by a wide margin, at times accounting for over two thirds of the total world output (Table 7). The U.S. production, however, is not solely from domestic mines. It has been estimated that since 1940 about 60% of the primary cadmium produced in the United States has been of foreign origin, obtained as concentrates chiefly from Mexico, Canada, Peru, and Zaire (formerly Belgium Congo). In the recent past the portion of world smelter output of cadmium coming from the United States has shown a steady decline from about 50% in 1954–1958 through 35% in 1964–1968 to about 15% during 1974–1977.

Over the years, considerable quantitites of cadmium have been recovered from the zinc mines of Canada, Australia, Mexico, and Peru. Most of the cadmium from Mexico, Peru, and Australia is exported to the United States as concentrates. Japan, Norway, Italy, and France have also produced sizable quantities of cadmium from domestic ores. Belgium and the United Kingdom are significant producers of refined cadmium from mostly imported materials. South-West Africa, though not a major zinc producer, is a leading producer of cadmium by virtue of the high cadmium tenor in the ores mined. Cadmium-rich, zinc-copper concentrates from South-West Africa are exported principally to the United States, United Kingdom, Belgium, and France, where they are processed to recover the metal. It is estimated that Poland and the USSR have also been relatively important cadmium producers.

Between 1964 and 1977 world production of cadmium increased by about 30%, but there were some important shifts in the major producing areas. Thus cadmium production in Canada and Japan increased substantially with the growth in zinc production but declined in the United States as a result of the closing of old zinc smelters (Table 7). Since about 1974 the world output has remained fairly constant at about 17.2×10^6 kg/year. In 1977 Japan, the USSR, the United States, and Germany were the largest cadmium producers with 16, 16, 12, and 8%, respectively, of the world total.

8. CONSUMPTION

The first major commercial application of cadmium in paint pigments occurred well over 60 years after the element was discovered. Other earliest uses of cadmium included low-melting alloys, electroplating, glass making, photography, as salts in dentistry, dying, and calico printing, and as chemical reagents (Mentch and Lansche, 1958). Sharp increases in cadmium consumption occurred around 1919 and again in the 1930s, associated with the development of cadmium electroplating and the expansion in the use of bearing alloys containing cadmium. In general, accurate worldwide consumption figures for cadmium are lacking. It is known, however, that during both World War II and the Korean War the Korean War (1950–1951) the demand for cadmium exceeded the supply, (1950–1951) the demand for cadmium exceeded the supply, and efforts were made either to curtail the consumption or to find additional sources of the metal. When demand tends to exceed supply in peace times, the resulting increases in cadmium prices are likely to discourage some applications and thus restore a balance. The net effect is a cyclic pattern of rapid reversals in trends of consumption, industrial stocks, and foreign trade (Figure 1).

Since 1974 the Western World's supplies of cadmium have exceeded the demand, and the industrial stocks have soared to high levels. Industrial stocks in the United States stood at 2.4×10^6 kg in 1975 but declined to 1.9×10^6 kg in 1977 (*Minerals Yearbook*). Current consumption has been dampened by the poor economic situation and by pressures from environmentalists.

Since the end of World War I, the United States has remained the largest user

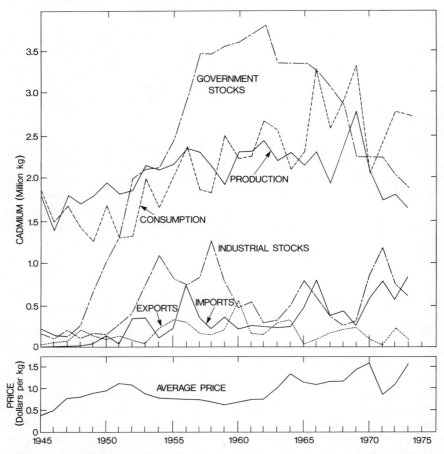

Figure 1. Trends in supply-demand and average price of cadmium in the United States since World War II. (From *Minerals Yearbook*, 1973.)

of cadmium. Cadmium consumption in the country increased from about 45,400 kg in 1919 to about 1.4×10^6 kg in 1924, and by 1941 the demand exceeded 3.5×10^6 kg. The consumption between 1942 and 1945 averaged 3.7×10^6 kg/year and during the Korean War reached 4.3×10^6 kg/year. Between 1951 and 1966 apparent consumption of cadmium in the United States increased at the rate of about 5%/year. Up to 1966 the United States accounted for 50 to 80% of all the cadmium consumed. Since 1966, the peak year, however, cadmium consumption in the United States has continued to decline. In 1977 U.S. cadmium consumption stood at 4.1×10^6 kg. As noted previously, a substantial portion of the cadmium consumed in the United States is imported; in 1976 the principal exporters of the metal to the United States included Canada, 29%; Australia, 12%; Mexico, 11%; Yugoslavia, 9%; and Belgium, 8%. The supply-demand relationships for cadmium in the United States from 1966 to 1975 are shown in Table 9. It should be noted that electroplating accounted for about 50% of all the cadmium consumption in the United States during 1966–1975 and that plastic stabilizers and batteries combined represented about 40% of the U.S. cadmium consumption. The flow of cadmium through the U.S. economy is depicted in Figure 2.

Between 1964 and 1974 cadmium consumption in Europe, which rose by nearly 6%/year, represented about 50% of all the cadmium consumed in the Western World (Table 10). Of the western European countries, the rise in cadmium consumption was highest in France (80% increase from 1965 to 1974) with most of the growth being attributed to battery and pigment output. West German consumption rose by about 40% during the period, mostly because of sharp increases in the use of cadmium in plating, batteries, and chemicals. The pattern of consumption in the United Kingdom remained fairly steady during the period.

Between 1960 and 1969 cadmium consumption in Japan grew by about 18%/year. The rapid growth can be attributed to the production of small cells used in calculators, shavers, cameras, cassette recorders, and other products. Cadmium consumption in Japan declined dramatically in 1969–1970, however, following the incidence of itai-itai in the Jintsu River basin.

Five countries—France, Germany, Japan, the United Kingdom, and the United States—currently account for over 80% of all the cadmium consumed in the Western World. In 1974 the apparent per capita consumption was about 33 g in West Germany, 28 g in France, 27 g in the United Kingdom, 22 g in the United States, and 9 g in Japan (Stubbs, 1978). The Japanese consumption peaked at about 16 g in 1969–1970. Furthermore the consumption statistics show that the use of cadmium in plating in the five countries declined from about 41% to about 30% of total consumption from 1965 to 1974. For these five countries the use of cadmium compounds, including pigments, increased from about 30 to 40%, whereas consumption in batteries rose from 10 to 14% during 1965–1974 (Stubbs, 1978).

It should be stressed that most uses of cadmium are completely dissipative; less than 5% of all the cadmium currently consumed is recycled. The low rate of secondary recovery has undesirable consequences in terms of environmental dispersion and the future global resource of cadmium. On the other hand, if the cadmium were not commercially separated from the host zinc and lead ores, it would be dispersed in the environment on a much wider scale as flue particles and other industrial waste products.

Table 9. Cadmium Supply-Demand Relationships in the United States during 1966–1975[a]

Item	1966	1967	1968	1969	1970	1971	1972	1973	1974	1975
World production										
United States	2,350	1,900	2,100	2,325	1,778	1,777	2,390	2,840	1,900	1,200
Rest of world	11,972	12,635	14,453	17,067	16,450	15,344	16,110	15,960	16,900	16,300
Total	14,322	14,535	16,553	19,392	18,228	17,121	18,500	18,800	18,800	17,500
Components of U.S. supply										
Domestic mines	2,350	1,900	2,100	2,325	1,778	1,777	2,390	2,840	1,900	1,200
Shipments of government stockpile excesses	184	516	404	1,378	3	1	479	385	1,006	
Imports, metal	1,679	794	964	539	1,246	1,750	1,211	1,946	1,985	2,618
Imports, content of ore and flue dust	2,898	2,069	3,028	3,655	2,957	2,188	1,762	874	1,433	993
Industry stocks, Jan. 1	2,042	1,751	1,139	924	1,127	2,391	2,649	1,662	1,326	1,569
Total U.S. supply	9,153	7,030	7,635	8,821	7,111	8,107	8,491	7,707	7,650	6,380
Distribution of U.S. supply										
Industry stocks, Dec. 31	1,751	1,139	924	1,127	2,391	2,649	1,662	1,326	1,569	2,841
Exports	190	346	265	542	187	33	509	153	31	198
Demand	7,212	5,545	6,446	7,152	4,533	5,425	6,320	6,228	6,050	3,341
U.S. demand pattern										
Transportation	913	850	990	1,100	885	1,060	1,250	1,250	1,100	600
Coating and plating	3,150	2,450	2,925	3,260	1,380	1,650	1,900	1,745	1,800	900
Batteries	250	200	200	250	150	180	450	620	600	450
Paints	1,000	555	675	725	648	775	950	1,120	1,100	600
Plastics and synthetic products	1,196	1,108	1,261	1,350	1,193	1,425	1,250	995	1,000	550
Other	703	382	395	467	277	335	520	498	450	241
Total U.S. primary demand	7,212	5,545	6,446	7,152	4,533	5,425	6,320	6,228	6,050	3,341

[a] From Bureau of Mines (1977). Quantities are in short tons.

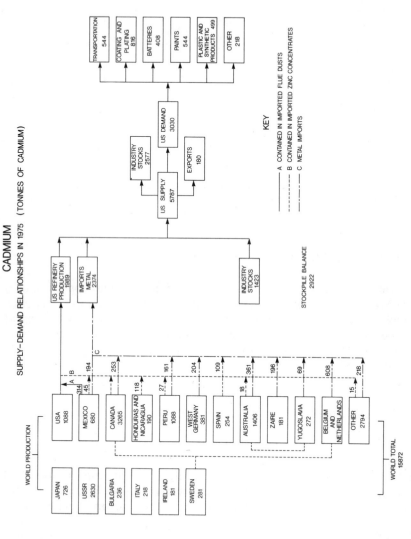

Figure 2. Flow of cadmium (metric tons) in the U.S. economy during 1975. (From Bureau of Mines, 1977.)

Table 10.   Cadmium Consumption in Europe and the
Western World during 1964–1975[a]

| Year | Cadmium Consumption ($\times 10^6$ kg) | |
	Western World	Europe
1964	9.73	4.29
1965	10.4	4.57
1966	13.3	5.16
1967	11.9	4.88
1968	13.3	5.31
1969	15.7	6.36
1970	11.8	5.82
1971	12.4	5.57
1972	14.5	6.30
1973	14.9	7.17
1974	13.8	7.60
1975	9.01	4.97

[a] From Stubbs (1978).

9.   USES OF CADMIUM AND ITS COMPOUNDS

An inventory of the quantities of cadmium used in its various applications by
the five principal consumers of the metal is given in Table 11. If this table is used
as a guide to the relative levels of cadmium use in other countries, it would appear
that about 90% of all the cadmium now consumed goes into four use categories:
plating (about 35%), pigments (about 25%), stabilizers (about 15%), and batteries (about 15%).

9.1.   Electroplating

Although the first patent for the electroplating of cadmium was issued in England in the 1840s, the plating of metals on a commercial scale was not accomplished until about 1919. Basically, coating steel, iron, copper, brass, and
other objects with cadmium confers resistance to corrosion. Properties of cadmium metal that have led to its widespread use in electroplating include (a) low
contact resistance, (b) ready deposition on intricately shaped objects, (c) good
corrosion resistance to alkali and seawater, (d) good solderability of plated parts,
(e) high ductility so that plated parts can be formed or stamped, (f) a coating
life that is a straight-line function of coating thickness, and (g) the ability to
protect steel, to which it is anodic, through sacrificial corrosion.

Table 11. End Uses of Cadmium in Western World Countries[a]

Quantities are in metric tons.

End Use	U.S.A.			Federal Republic of Germany			France			U.K.			Japan		
	1965	1970	1974	1965	1970	1974	1965	1970	1974	1965	1970	1974	1965	1970	1974
Batteries	279	136	544	102	220	302	114	250	300	117	110[b]	180[b]	51	176	320
Pigments	1160	587	997	280	440	660	239	350	650	272	335	416	185	444	252
Stabilizers	385	1081	906	120	190	290	68	60	100	136	168	207	161	341	173
Plating	2447	2052	2718	210[b]	270[b]	300[b]	346	300	350	493	490	457	168	135	457
Alloys	[c]	[c]	[c]	11	30	34	28	50	50	178	133	96	42	169	99
Others	408	251	441	696[d]	650[d]	430[d]	11	18	4	74	77	85	180	220	78
Total	4679	4107	5605	1419	1800	2016	806	1028	1454	1270	1313	1441	787	1485	927

[a] From Stubbs (1978).

[b] Estimated

[c] Included in "Others."

[d] Includes solders.

61

Practically all cadmium plating is done by electrodeposition, although the cadmium may also be applied by vacuum deposition, dipping, or spraying. Cadmium is relatively expensive and is usually preferred in more sheltered applications where a thin coating is sufficient for protection and its bright appearance is an advantage. Cadmium plating is used extensively in the automobile and aircraft industries to coat nuts, bolts, screws, springs, fasteners, washers, nails, rivets, carburetor and magnet parts, body finishes, and so on. Other common uses include casket hardware, builders' hardware, radio and television parts, marine hardware, wire screen, tools, stampings, screw machine parts, electronic apparatus, and household appliances. Cadmium is employed in the packing industry for articles that come into contact with greases containing oleic and stearic acids, but cannot be used in articles that come into contact with food because of its toxic properties.

Table 11 indicates the 1965–1974 trends in the use of cadmium for electroplating in Japan, the United Kingdom, West Germany, France, and the United States.

9.2. Pigments

Cadmium compounds are extensively used as pigments in a wide variety of products, especially plastics, ceramics, paints, coated fabrics, textiles, rubber, glass, enamels, and printing inks. Advantageous features of cadmium pigments include (a) high indices of refraction, which give the appearance of richer, deeper colors, (b) the purity of shade and the brilliance essential for coloring today's plastics, (c) excellent heat resistance, up to 600°C, (d) good resistance to H_2S and alkali, (e) good hiding power and easy dispersability, and (f) good light resistance and, in some environments, excellent weathering resistance. It is clear from these unique properties why cadmium compounds are the choice pigments in many applications, and there are presently no common or easily available substitutes for some uses. Cadmium pigments are acid sensitive, however, and thus generally have poor outside weathering properties. Furthermore, they possess poor tinting strength and are relatively expensive and toxic.

Cadmium pigments range from primrose (CdS + ZnS) through golden yellow (CdS) to red and maroon (CdS + CdSe or CdS + HgS). About 80% of all the cadmium pigment sales in Europe goes to the plastic industry; the corresponding figures for Japan and the United States are 60 to 80% and 75%, respectively. Ceramics generally account for about 10% of the cadmium pigment sales, with the remaining 10% going to paints and coatings. In addition to their traditional use as artists' colors, cadmium pigments have also been used as traffic paints, in high quality industrial finishes such as car paints, and for the glass enamel red label on the familiar Coca-Cola bottles. The consumption of cadmium pigments in France, Japan, West Germany, the United Kingdom, and the United States from 1965 to 1974 is shown in Table 11.

9.3. Plastic Stabilizers

Since cadmium-based stabilizers were introduced into the industry in the early 1940s, they have found widespread use in retarding discoloration due to the breakdown of polyvinyl chloride (PVC) resin during molding operations. It is commonly assumed that cadmium salts stabilize the PVC by an esterification reaction with unstable allylic chlorine atoms in the polymer. The cadmium chloride formed during the reaction, however, accelerates the decomposition of the PVC, which is usually overcome by the synergistic exchange reaction between cadmium and the added barium salts. In addition, epoxy compounds and organic phosphite esters are incorporated in stabilizer formulations to mask any harmful excess chlorine. The cadmium-barium-(zinc)-epoxy organophosphites presently dominate the flexible PVC market and are used extensively for calendering and plastisols.

Advantageous features of cadmium-based stabilizers include (a) stability against heat and light, (b) absorption and neutralization of hydrogen chloride evolved by the resin during and after processing, (c) antioxidant protection, (d) ultraviolet screening, (e) disruption of double bonds and displacement of labile chlorine atoms, (f) neutralization or inactivation of resin impurities and stabilizer degradation products, and (g) nonstaining, nonmigrating characteristics and lack of detrimental effects on the mechanical and electrical properties of the PVC. Because of the potential adverse health effects, however, cadmium-based stabilizers are not recommended for food-grade flexible PVC.

Stabilizers now account for about 15% of all the cadmium consumed. Between 1965 and 1974 the consumption of cadmium by the plastic industries registered substantial increases in western Europe and the United States (Table 11) but declined sharply in Japan following the "decadmium" movement in the PVC stabilizer field, which began around 1970.

9.4. Batteries

The nickel-cadmium battery was invented in Sweden by Waldemar Jungner and Karl Berg around 1900, but came into worldwide commercial use only after World War II. The basic charge-discharge reactions for nickel-cadmium and the less common silver-cadmium batteries are as follows:

$$2NiOOH + Cd + 2H_2O \underset{\text{recharge}}{\overset{\text{discharge}}{\rightleftharpoons}} 2Ni(OH)_2 + Cd(OH)_2$$

$$Ag_2O_2 + Cd + 2H_2O \underset{recharge}{\overset{discharge}{\rightleftarrows}} 2Ag + 2Cd(OH)_2$$

Outstanding features of the cell electrochemistry include (a) the reactions are perfectly reversible, (b) the cell does not gas when charged properly or when discharged, (c) the active materials do not react with the electrolyte when the battery is an open circuit, and (d) the specific gravity does not change during the complete cycle.

Nickel-cadmium alkaline batteries are used as the sealed cells in radios, alarm systems, emergency lighting, pacemakers, calculators, motor starters, walkie-talkies, portable appliances, and tools. Vented cell types are much larger and are used in buses, diesel locomotives, aircraft, spacecraft, military applications, and standby power and lighting. Advantages of the nickel-cadmium battery include (a) long life, up to 3000 cycles of charge and discharge, (b) lack of damage from overcharging, reverse charging, or short circuits, (c) high power density and delivery of maximum current with minimum voltage drop, (d) low rates of self-discharge, (e) quick charging and high discharge rate without damage to cell, (f) ease of maintenance, (g) wide range of operating temperatures, from -55 to $75°C$, and (h) easy recovery of the cadmium from the dead cells. The batteries, however, are relatively expensive.

The manufacture of nickel-cadmium storage batteries is the fastest growing segment of the cadmium industry (see Table 11). In 1972 a total of 16.3 million nickel-cadmium batteries (11.7 million sealed and 4.6 million vented) was produced in the United States alone with a shipment value of $47.6 million (Sargent and Metz, 1975).

9.5.   Miscellaneous Uses

Miscellaneous uses of cadmium represent about 10% of the present total worldwide consumption. Of the residual applications, alloys of cadmium account for 5 to 10% of the total global demand for the metal. Small quantities of cadmium find multifarious applications in chemical, electrical, and electronic industries, as well as in nuclear engineering.

Alloys

Cadmium alloys with most metals (Table 12) to form compounds with a wide variety of properties and commercial applications. *Fusible alloys* (with very low melting points) are used in fire protection devices, for bending pipes and thin sections, for foundry patterns, spotting fixtures, proofcasting molds, gun chambers, and fusible cores, for soldering and sealing, for safety plugs for compressed gas cylinders, and for holding irregular pieces for machining. *Solders*

Table 12.   Cadmium Alloy Systems[a,b]

Alloy	Description
Ag–Cd	
α	Cd in Ag: 42·8 at. % at 440°C.
β	AgCd, b.c.c. (>450°C), $a = 3·332$ (E).[b]
γ	Ag_5Cd_8, γ-brass structure, $a = 10·002$ (E).
ϵ	$AgCd_3$, h.c.p. (E).
Al–Cd	Cd in Al: 0·14 at.% at 649°C.
	Al in Cd: very low.
Au–Cd	
α_1	Au_3Cd, f.c. tetragonal, $a = 4·116$, $c = 4·137$.
β	AuCd, b.c.c., >100°C, $a = 3·312$ (E).
ϵ'	$AuCd_3$, γ-brass structure (?) (E?).
Ba–Cd	BaCd, b.c.c., $a = 4·215$.
	$BaCd_{11}$, b.c. tetragonal, $a = 12·02$, $c = 7·74$.
Be–Cd	No evidence for compounds or solubility.
Bi–Cd	Eutectic at 146°C (60 wt. % Bi); very limited mutual solubility.
Ca–Cd	CaCd, b.c.c., $a = 3·83$.
	$CaCd_2$, hexagonal $MgZn_2$ structure, $a = 5·99$, $c = 9·65$.
Co–Cd	Co_5Cd_{12}, γ-brass structure (?) (E).
Cr–Cd	No compounds; Cr insoluble in Cd at 650°C.
Cs–Cd	$CsCd_{13}$, f.c.c. $NaZn_{13}$ structure, $a = 13·92$.
Cu–Cd	Cu_2Cd, hexagonal, $MgNi_2$ structure, $a = 5·00$, $c = 16·17$.
	Cu_5Cd_8, γ-brass structure, $a = 9·615$ (E).
	$CuCd_3$, hexagonal, $a = 8·11$, $c = 8·76$ (E).
Fe–Cd	No compounds; very slight solubility.
Hg–Cd	ω-phase: tetragonal, $a = 3·964$, $c = 2·849$ (37 at. % Hg).
K–Cd	KCd_{13}, cubic $NaZn_{13}$ structure, $a = 13·80$.
La–Cd	LaCd, b.c.c., $a = 3·905$.
	$LaCd_2$, hexagonal CdI_2 structure, $a = 5·075$, $c = 3·458$.
	$LaCd_{11}$, cubic, $BaHg_{11}$ structure, $a = 9·339$.
Li–Cd	Li_3Cd, f.c.c. (statistically distributed), $a = 4·259$.
	LiCd, cubic NaTl structure, $a = 6·700$.
	$LiCd_3$, h.c.p. (statistically distributed), $a = 3·086$, $c = 4·899$.
Mg–Cd	Mg_3Cd, hexagonal, Ni_3Sn structure, $a = 6·31$, $c = 5·08$.
	MgCd, orthorhombic (ordered), $a = 5·005$, $b = 3·222$, $c = 5·270$.
	$MgCd_3$, hexagonal, Ni_3Sn structure, $a = 6·23$, $c = 5·04$.
Mn–Cd	No compounds, very slight mutual solubility at 25°C.
Ni–Cd	Ni_5Cd_{21}, γ-brass structure, $a = 9·781$ (E).
Pd–Cd	No compounds; eutectic (82·6 wt. % Pb at 248°C).
	Pb in Cd: ~0; Cd in Pb: 0·5 at. % at 50°C.
Pt–CD	$PtCd_2$, AlB_2 structure.
	Pt_5Cd_{21}, γ-brass structure, $a = 9·88$ (E).
Rb–Cd	$RbCd_{13}$, cubic $NaZn_{13}$ structure, $a = 13·91$.

Table 12.   *Continued*

Alloy	Description
Sn–Cd	Cd in Sn: $\sim$1 at. %.
	β-phase, hexagonal (4·89 at. % Cd at 176°C); $a = 3\cdot2328$, $c = 3\cdot0023$.
Sr–Cd	SrCd, b.c.c., $a = 4\cdot011$.
	SrCd$_{11}$, b.c. tetragonal, $a = 12\cdot02$, $c = 7\cdot69$.
Ti–Cd	Ti$_2$Cd, tetragonal, MoSi$_2$ structure, $a = 2\cdot865$, $c = 13\cdot42$.
	TiCd, tetragonal, PbO structure, $a = 2\cdot904$, $c = 8\cdot954$.
Zn–Cd	No compounds formed; mutual solubility very small at 25°C;
	eutectic (17·4 at. % Zn at 226°C) and solid solutions.

[a] From Aylett (1973).

[b] E = electron compound.

are used extensively for soldering aluminum. *Brazing alloys* are commonly used to join ferrous and nonferrous metals in a strong, leakproof, and corrosion resistant manner. *Bearing alloys* are more heat resistant and can run at higher speeds than tin or lead bearings. *Other alloys* have been used in the production of automobile radiators, as a means to improve the fatigue resistance of lead cable sheathing, and as counterelectrode material for selenium rectifiers. Low-cadmium copper (about 1% cadmium) is employed in (*a*) the manufacture of trolley and telephone wires because of the improved tensile strength imparted by cadmium and (*b*) automobile radiator finstock, replacing the low-silver copper formerly used.

Chemicals

The uses of the various cadmium compounds are summarized in Table 13. It is interesting to note that cadmium compounds are extensively used in smoke alarms because of their unique photohemical properties. It has been estimated that General Electric sold over 4 million smoke detectors in the United States alone in 1977 (Radtke, 1978).

Nuclear Engineering

The isotope [113]Cd (thermal neutron cross section, 200,000 barns) is used as a shield against neutrons. Its use for control rods in nuclear reactors dates back to the first atomic pile constructed by Enrico Fermi in 1942, but it has been superseded in this application by Cd-In-Ag alloy elements. The use of graphite microspheres containing cadmium in extinguishing fires involving fissionable or radioactive materials has recently been proposed (see Radtke, 1978).

Electrical and Electronic Uses

The applications of cadmium in the electrical and electronics industries are extremely wide and include the following.

Table 13. Uses of Cadmium Compounds

Compound	Uses
Cadmium sulfate, $CdSO_4$	Electrolyte in Weston cell
	Antiseptics and astringents
	Starting material in cadmium production
	Fungicide and bactericide
	Lubricant
Cadmium nitrate, $Cd(NO_3)_2$	Coloring porcelain and glass
	Photographic flash powder
	Laboratory reagent
	Nickel-cadmium batteries
Cadmium chloride, $CdCl_2$	Photography
	Dyeing and calico printing
	Testing for pyridine bases
	Pyrotechnics
	Copying papers
Cadmium bromide, $CdBr_2$	Photography
	Process engraving
	Lithography
	Epoxy resins
Cadmium iodide, CdI_2	Marme reagent
	Photography
	Lithography
	Process engraving
Cadmium oxide, CdO	Plating baths
	Catalyst
	Resistor in electric furnaces
	Electrical contacts
	Nonchalking pigment
	Coating for luminescent colors
	Lubricant
	Ascaricide
Cadmium hydroxide, $Cd(OH)_2$	Manufacture of other cadmium salts
	Cadmium-nickel storage batteries
Cadmium sulfide, CdS	Extreme pressure lubricant
	Radiation detection devices
	Pigment manufacture (wide application)
	Solar energy cells
	Fireworks (to color flame blue)
	Infrared windows
	Photosensitive elements
Cadmium selenide, $CdSe$	Pigments (wide variety)
	Phosphors and luminescent materials
	Snooperscopes, sniperscopes, and metascopes
Cadmium tungstate, $CdWO_4$	Luminous pigment

Table 13. *Continued*

Compound	Uses
Cadmium arsenides, antimonide, and telluride	Alloys and semiconductors Phosphors and luminescent materials
Cadmium salicylate, $Cd(C_7H_5O_2)_2 \cdot H_2O$	External antiseptic
Cadmium acetate	Laboratory reagent Porcelains and pottery (to give iridescent effects)
Cadmium carbonate, $CdCO_3$	Manufacture of cadmium salts Laboratory reagent Catalyst
Organocadmium compounds	Catalysts Fungicides and anthelmintics

Phosphors and Luminescent Materials. Applicable compounds include cadmium tungstate (white light without activation), silicate (yellow-pink color), and borate (pink color). Mixtures of cadmium sulfide and zinc sulfide doped with copper and silver give colors over the entire spectrum from deep blue for ZnS to deep red for CdS. Cadmium phosphors are used for direct conversion of the energies of invisible particles (cathode rays, X-ray photons, alpha particles, ultraviolet photons) into visible light. Familiar applications include X-ray fluorescent screens, cathode-ray tubes, television tubes, radar equipment, fluorescent lamps, self-luminscent watch and instrument dials, theatrical black magic, interior decorations, phosphorescent tapes and markers, and localized luminescence during blackouts.

Contact Materials. Copper-cadmium alloy strip is used for electrical contacts where good fatigue and wear resistance are essential. It finds extensive application for heavy duty relays and switches where arcing may occur, such as automobile distributor contacts.

Solar Cells. Although the photovoltaic effect of cadmium sulfide was first reported by D. C. Reynolds in 1954, the processes involved are still not clear. It has been suggested that most of the photovoltaic response stems, not from the CdS, but from a surface microlayer of copper-sulfur compound. The efficiency of cadmium sulfide solar cells is about 5% but can be increased to 8% by doping the CdS with other metal sulfides. Cadmium telluride has also been investigated as a possible component of solar cells.

Photocells. Cadmium sulfide cells are commonly used in photographic exposure meters and in photoelectric smoke alarms. Further research in this field will probably lead to other applications.

10. DATA SOURCES

The preceding sections on the production, consumption, and uses of cadmium are based on the following reports: Mentch and Lansche (1958), Chizhikov (1966), Howe (1967), Heindl (1970), Organization for Economic Cooperation and Development (1975), Sargent and Metz (1975), Bussing (1978), Dickenson (1978), Hewitt (1978), Katoh (1978), Murnane (1978), Radtke (1978), Raede (1978), Stubbs (1978), Wienhenkel (1978), and the *Minerals Yearbooks* (1930–1977).

REFERENCES

Agricola, G. (1556). *De Re Metallica.* Basle. English translation by H. C. Hoover and L. H. Hoover, Dover, New York, 1950.

Aylett, B. J. (1973). "Cadmium." In J. C. Bailar, H. J. Emeleus, R. Nyholm, and A. F. Trotman-Dickenson, Eds., *Comprehensive Inorganic Chemistry.* Pergamon, Oxford, pp. 254–272.

Bureau of Mines (1977). *Minerals in the U.S. Economy: Ten-Year Supply-Demand Profiles for Mineral and Fuel Commodities.* U.S. Department of the Interior, Washington, D.C.

Bussing, J. (1978). "Use of Cadmium Compounds as Stabilizers for PVC." In *Cadmium 77: Proceedings of the 1st International Cadmium Conference.* Metal Bulletin Ltd., London, England, pp. 21–25.

Canadian Minerals Yearbook (1969–77). Department of Energy, Mines and Resources, Ottawa, Ont.

Chizhikov, D. M. (1966). *Cadmium.* Pergamon, Oxford, 263 pp.

Dickenson, J. (1978). "Cadmium Pigments in the United States." *In Cadmium 77: Proceedings of the 1st International Cadmium Conference.* Metal Bulletin Ltd., London, England, pp. 16–18.

Fleischer, M., Sarofim, A. F., Fassett, D. W., Hammond, P., Shacklette, H. T., Nisbet, I. C., and Epstein, S. (1974). *Environ. Health Perspec.* **7,** 253–323.

Gong, H. (1975). "The Geochemistry of Cadmium." Thesis, Department of Geosciences, Pennsylvania State University, 114 pp.

Gong, H., Rose, A. W., and Suhr, N. H. (1977). "The Geochemistry of Cadmium in Some Sedimentary Rocks, "*Geochim. Cosmochim. Acta* **41,** 1687–1692.

Handbook of Chemistry and Physics, 59th ed. (1978–1979). CRC Press, West Palm Beach, Fla.

Heindl, R. A. (1970). "Cadmium." In *Mineral Facts and Problems.* Bur. Mines Bull. 650, U.S. Department of the Interior, Washington, D.C.

Hewitt, K. N. (1978). "Miscellaneous Uses of Cadmium." In *Cadmium 77: Proceedings of the 1st International Cadmium Conference.* Metal Bulletin Ltd., London, England, p. 33–37.

Howe, H. E. (1967). "Cadmium and Cadmium Alloys." In *Encyclopedia of Chemical Technology*, Vol. 3. Wiley, New York, pp. 884–899.

Katoh, F. (1978). "The Current Status of Cadmium Containing PVC Stabilizers in Japan." In *Cadmium 77: Proceedings of the 1st International Cadmium Conference*. Metal Bulletin Ltd., London, England, pp. 25–28.

Martell, A. E. and Smith, R. M. (1975). *Critical Stability Constants*, Vols. 1–3, Plenum, New York.

Mentch, R. L. and Lansche, A. M. (1958). *Cadmium, a Materials Survey*. National Technical Information Service, Rep. AD 680-443, Springfield, Va.

Minerals Yearbooks (1930–77). U.S. Bureau of Mines, Government Printing Office, Washington, D.C.

Murnane, R. (1978). "Cadmium Plating—the Process, Applications, and Probable Trends." In *Cadmium 77: Proceedings of the 1st International Cadmium Conference*. Metal Bulletin Ltd., London, England, pp. 31–33.

Organization for Economic Cooperation and Development (1975). *Cadmium and the Environment: Toxicity, Economy and Control*. Environmental Directorate, OECD, Paris, 88 pp.

Radtke, S. F. (1978). "New Technologies and Responsibilities in Cadmium Research. In *Cadmium 77: Proceedings of the 1st International Cadmium Conference*. Metal Bulletin Ltd., London, England, p. 37–42.

Raede, D. (1978). "The Production and Use of Cadmium Pigments in Europe." In *Cadmium 77: Proceedings of the 1st International Cadmium Conference*. Metal Bulletin Ltd., London, England, p. 13–14.

Sargent, D. H. and Metz, J. R. (1975). *Technical and Microeconomic Analysis of Cadmium and Its Compounds*. U.S. Environmental Protection Agency, Office of Toxic Substances, Rep. EPA 560/3-75-005, Washington, D.C.

Stubbs, R. L. (1978). "Cadmium—the Metal of Benign Neglect." In *Cadmium 77: Proceedings of the 1st International Cadmium Conference*. Metal Bulletin Ltd., London, England p. 7–12.

Wagman, D. D. et al. (1968). *Selected Values of Chemical Thermodynamic Properties*. National Bureau of Standards (USA) Technical Notes 270-3.

Weinhenkel, H. J. (1978). "Consumption, Regulations, Future Trends and Development of Cadmium Pigments." In *Cadmium 77: Proceedings of the 1st International Cadmium Conference*. Metal Bulletin Ltd., London, England, p. 14–16

3

CADMIUM IN THE ATMOSPHERE AND IN PRECIPITATION

Jerome O. Nriagu

National Water Research Institute, Canada Centre for Inland Waters, Burlington, Ontario, Canada

1. INTRODUCTION

A large quantity of data has recently become available on the temporal and spatial distribution of trace metals in the air. Ambient monitoring of airborne metals under various meteorological conditions is needed to develop feasible strategies for maintaining healthy ambient air quality standards. Airborne metal concentrations can also provide important clues regarding the global atmospheric circulation patterns. Previous reviews have considered the atmospheric cycles of lead (Nriagu, 1978) and copper (Nriagu, 1979). This chapter focuses on cadmium in the atmosphere, especially the principal sources, concentrations, physicochemical characteristics, and removal processes of cadmium-bearing aerosols.

2. SOURCES OF CADMIUM IN THE ATMOSPHERE

Cadmium in the atmosphere comes from a wide variety of natural and pollutant sources. Worldwide annual emissions of cadmium from natural sources are estimated to total 843×10^6 g (Table 1). About 25% of the natural cadmium emissions is derived from vegetation (as exudates, slouch, etc.), whereas about 12%, 62%, and 2% are attributable to airborne soil particles, volcanogenic aerosols, and forest fires, respectively (Table 1). Although the degassing of crustal rocks has been suggested as an important source of metals in the atmosphere (Goldberg, 1976; Chernyak and Nussinov, 1976), the release of cadmium from such a source remains to be quantified.

The principal sources of pollutant cadmium in the atmosphere are given in Table 2. Large cadmium emissions are associated with zinc and copper production; these two sources together account for about 60% of all the anthropogenic cadmium emissions. It is estimated that 76% of all the anthropogenic cadmium emissions comes from the nonferrous metal industries. About 22% of the global emission is from inadvertent sources, whereas the emissions associated with the iron and steel industries and industrial applications of cadmium are relatively minor, comprising only about 1%, in each case, of the total pollutant cadmium emissions (Table 2).

The total annual flux of cadmium to the atmosphere is about 80×10^5 kg, 90% of which is derived from anthropogenic sources. This ratio of pollutant to natural emission rates is consistent with the ratio of cadmium concentrations observed in modern and ancient ice samples in the Greenland glacier (Weiss et al., 1975). My value of the global cadmium emission agrees quite well with the figure of 7.3×10^6 kg/year reported by Heindryckx et al. (1975).

There is considerable disparity in the reported intensities of cadmium emissions from the various anthropogenic sources. This point is highlighted in the various estimates of cadmium emissions in the United States (Tables 3 and 4). The widely quoted estimates by Davis and Associates (1970) of 2.3×10^6 kg/year and by Goeller et al. (1973) of 1.3 to 2.2×10^6 kg/year are, in fact, incomplete insofar as they do not include the cadmium emissions associated with Cu, Pb, Zn, and Fe/steel industries. Fleischer et al. (1974) estimated the release of cadmium to the atmosphere during the production and disposal of cadmium

Table 1. Global Emissions of Cadmium from Natural Sources

Source	Global Production $(10^9$ kg/year)[a]	Emission Factor $(\mu g/g)$	Cadmium Emission $(\times 10^6$ g$)$
Windblown dusts	500	0.2[b]	100
Forest fires	36	0.32[c]	12
Volcanogenic particles	25	2.0[d]	520
Vegetation	75	2.8[e]	210
Seasalt sprays	1000	0.00004[f]	1.0
Total			843

[a] See Nriagu (1978).

[b] This value is the average crustal abundance of cadmium given by Fleischer et al. (1974).

[c] Assuming an average ash content of 4% for vegetation and a cadmium concentration in the ash of 8 $\mu g/g$ (see Curtin et al., 1974).

[d] Assuming the average crustal abundance of 0.2 $\mu g/g$ and a 260-fold enrichment of cadmium in the volcanic particles (e.g., see Buat-Menard and Arnold, 1978; Duce et al., 1976).

[e] Assuming the average ash content of plant exudates to be 11% and the cadmium content of the ash to be 25 $\mu g/g$ (see Curtin et al., 1974).

[f] Assuming the mean cadmium concentration in ocean water to be 0.006 $\mu g/kg$ (Bewers et al., 1976) and a 200-fold enrichment of cadmium at the water-air interface.

products to be 1.68×10^6 kg in 1968. The U. S. Environmental Protection Agency (EPA) (1975b) report estimates that 34% of total pollutant cadmium released into the air can be attributed to the primary nonferrous metal industry, with 16% accountable to the industrial conversion and use of cadmium and the remaining 50% to inadvertent sources. By contrast, the estimates by Goldberg (1973) and in the EPA (1975a, 1978) reports show that about 60% of the total cadmium emission in the United States comes from the primary nonferrous metal industry (see Table 3). These discrepancies in the emission intensities reported stem primarily from assumptions regarding the efficiency and industrial applications of effective air pollution abatement practices. The estimates presented in Table 2 assume some measure of particulate emission controls in other countries which are less stringent than those considered in the EPA (1975b) report. At any rate the considerable variance in the emission estimates emphasizes the need for a greater effort to obtain better information.

Table 5 gives an inventory of cadmium emissions from anthropogenic sources in Canada. About 80% of the total cadmium emission is attributable to the primary nonferrous industry. The higher per capita emission of cadmium in Canada as compared to the United States reflects the fact that Canada is one of the leading producers of base metals in the Western World.

Table 2. Worldwide Anthropogenic Emissions of Cadmium during 1975–1976

Source	Annual Handling $(10^9 \text{ kg})^a$	Emission Factor $(g/10^3 \text{ kg}$ of material produced or used)	Cadmitted Emitted per Year $(\times 10^3 \text{ kg})$
Zinc mining operations	5.6	0.5^b	2.8
Primary nonferrous metal production			
Zinc operations	5.6	500^b	2800
Copper smelting	7.9	200^b	1580
Lead smelting	4.6	50^b	230
Cadmium extraction	$(1.7 \times 10^7 \text{ kg})$	6500^b	111
Secondary nonferrous metals	c	c	595
Iron and steel production	1300	0.055	72
Industrial applications			
Electroplating shops	—	$(1000)^d$	8.5
Pigment, plastic, etc.	—	—	38^e
Alloys and batteries	—	—	6.4^e
Inadvertent sources			
Coal combustion	3100^f	0.02^g	62
Oil combustion	$2800 \quad f$	0.001^h	2.8
Wood combustion	640	3.0	200
Waste incineration	150^i	9.0^j	1350
Rubber tire wear	—	—	10^l
Phosphate fertilizers	118	1.0^k	118
Total			7186

[a] Unless noted otherwise; the annual handling figures are from the U.S. Bureau of Mines *Minerals Yearbooks* or the *Canadian Minerals Yearbooks.*

[b] These values represent 50% of the emission factors obtained by Anderson (1973), which were essentially employed in the U.S. EPA (1975a) report.

[c] Worldwide emissions from this source assumed to be 4 times (equivalent to the relative cadmium consumption figures) the emission figures for the United States given in the U.S. EPA (1975a) report.

[d] Assuming that electroplating accounts for 50% of all the cadmium consumed (see U.S. EPA, 1975b) and that the emission factor is 1.0 g/1.0 kg Cd electroplated.

[e] The worldwide figures are assumed to be 4 times the emission rates from these sources reported for the United States (U.S. EPA, 1975a).

[f] From the National Academy of Sciences (1975) report.

[g] Jacko and Neuendorf (1977).

[h] Estimated by the method of Bertine and Goldberg (1971).

[i] Assuming that the refuse incinerated globally is 5 times the quantity processed in the United States: which is 30×10^9 kg/year (National Academy of Sciences, 1975).

[j] Yost (1978).

[k] Anderson (1973).

[l] Assuming the worldwide emissions from this source to be 3 times the figure for the United States given in the U.S. EPA (1975a) report.

Table 3. Comparison of Cadmium Emission Estimates from Various Published Sources for the United States. Emission Estimates are in tons (10^3 kg) per year.

Source of Emission	Davis and Associates (1970)	Goeller et al. (1973)	Duncan et al. (1973)	U.S. EPA (1971)	U.S. EPA (1976)	U.S. EPA 1975b	U.S. EPA (1978)	U.S. EPA (1975a)
Mining (Zn + Pb + Cu)	<1	<1	<1	<1	<1	<1	<1	<1
Primary metals								
Zinc	—	—	619	644	500	102	529	584
Lead	1050	1050	55	163	65	—	48	148
Copper	—	—	388	234	110	—	108	212
Cadmium	—	—	—	60	50	—	43	54
Secondary metals								
Steel scrap	1000	<110	1000	78	400	10.5	104	71
Zinc	—	?	—	2	2 }	2.2	1	19
Copper	125	?	125	65	70 }	—	38	59
Manufacturing								
Pigments	11	11	11	7	11	9.5	9	6
Stabilizers	3	3	3	4	3	2.7	3	3
Miscellaneous	<2	<2	<2	<1	<2	<1	<1	<2
Incineration	95	95	95	48	150	16	131	44
Fossil fuel combustion	—	145–1100	—	198	250	130	59.2	179
Sewage sludge incineration	—	—	—	138	12	20	<1	125
Motor Oil	1	1	—	—	<1	—	<1	—
Rubber tires	6	6	—	6	6	5.2	5	5
Gasoline	—	—	—	—	—	50	5	—
Forest and agricultural burning	—	—	—	—	50	—	13	—
Other	—	—	—	—	<2	—	4	—
Total	2294	1425–2380	2305	1650	<1688	300	1100.2	1511

Table 4. Summary of U.S. Cadmium Flow, Dissipation, and Emissions during 1968–1972.[a] Quantities are in 10^3 kg per year.

Source	Commercial Flow	Dissipations in End Products	Airborne Emissions	Waterborne Effluents	Land-Destined Wastes
In domestic zinc ores	2,250				
Losses in Beneficiation			0.2		250
In domestic zinc concentrates	2,000				
In imported concentrates	600				
Total to zinc Smelters	2,600				
Losses in zinc Smelting			102	7	
In zinc for galvanizing		160			
Corrosion in galvanized products					40
Losses in scrap processing			0.4		12
In ZnO for rubber		15			
Rubber tire wear			5.2		
Net from zinc smelting	2,300				
In domestic flue dusts	700				
In imported flue dusts	400				
Domestic cadmium metal production	3,400				
Cadmium metal from GSA stockpile	500				
Cadmium metal imports	1,700				
Total cadmium metal supply	5,600				
Cadmium metal to electroplaters	3,100				
losses in electroplating				7	77
In electroplated products		3000			
Losses in scrap processing			10		318
Cadmium metal to pigments	700				
Losses in processing			9.5	0.8	16.5
In plastics		675			
Losses in incineration			6		26
Cadmium metal to batteries	230				
Losses in processing			0.7	0.3	9
In batteries		220			
Cadmium metal to alloys and other uses	390				
Losses in processing			2.3		

Table 4. *Continued*

Source	Commercial Flow	Dissipations in End Products	Airborne Emissions	Waterborne Effluents	Land-Destined Wastes
In alloys, etc.		390			
Losses in scrap processing			2.2		20
From phosphate fertilizers					100
From phosphate detergents				10	
Collected in sewage sludge			20		250
From coal combustion			80		370
From oil combustion			50		
From lubricating oils			0.8		
Grand totals	27,640	5630	302	25	1532

[a] U.S. EPA (1975b).

Table 4 also gives the relative amounts of cadmium dissipated in the air, water, and land ecosystems in the United States. Of the total cadmium released to the environment through human activities only 17% is airborne, and only 1.4% is discharged directly into natural waters. Thus most (>80%) of the anthropogenic cadmium is wasted on land; the fate and ecological impacts of the cadmium so dissipated are essentially unknown.

3. DISTRIBUTION OF CADMIUM IN AMBIENT AIR

Over 90% of the total global emission of cadmium comes from such point sources as smelters and utility and metal processing plants, as well as incinerators. The contribution of each point source to the atmospheric cadmium burden depends on the quantity and properties of the cadmium-containing aerosols, the location and height of the emitter, the topography of the adjacent areas, and the prevailing meteorological conditions. The cadmium level in ambient air are also strongly influenced by the land-use spatial arrangements, the wind entrainment of local soil particles, and the aging history and deposition of the atmospheric aerosols. The widespread temporospatial variations found in cadmium concentrations in air should therefore not be surprising.

The actual measurement of suspended aerosols covers a wide variety of conditions, with time resolutions of less than 1 sec to over a week and time intervals that vary from a few seconds to several years (Table 6). The measurement situations clearly determine the applicability of the data to the establishment of air quality criteria or the origin of the airborne particles.

Table 5.   Inventory of Cadmium Emissions in Canada during 1972[a]

Sector	Annual Cadmium Emission ($\times 10^3$ kg)	Percent of Total
Industry		
Primary copper and nickel production	397	78.1
Primary lead production	6.6	1.3
Primary zinc production	1.2	0.2
Primary iron and steel production	1.9	0.4
Specialty copper products	4.0	0.8
Iron and steel foundries	3.3	0.6
Nonferrous alloys	0.8	0.2
Miscellaneous sources	0.6	0.1
Fuel combustion—stationary sources		
Power generation	12	2.4
Industrial and commercial	68	13.4
Domestic	6.8	1.3
Transportation		
Motor vehicles	0.2	0.04
Rain transport	0.8	0.2
Shipping	1.7	0.3
Tire wear	0.1	0.002
Solid waste incineration	3.5	0.7
Pesticide applications	0.03	—
Total	508.5	

[a] From Air Pollution Control Directorate (1976).

3.1.   Remote Locations

On the assumption (*a*) that 50% of the anthropogenic cadmium emission per year is exported outside the area of release and subsequently redistributed uniformly throughout the northern hemisphere to a height of 10 km and (*b*) that the atmospheric residence time for cadmium aerosols is 10 days, the expected concentration of cadmium at relatively remote locations would be about 0.04 ng/m^3. This crude estimate, in fact, falls in the range of concentrations observed at very remote locations (see Table 7). One of the lowest cadmium concentrations found anywhere is the 0.006 ng/m^3 at the Chacaltya Mountain in Bolivia

Table 6.   Measurement Situations and Time Scales Associated with Aerosols[a]

Measurement Situation	Resolution	Interval
Compliance monitoring	24 hr	Years
Episode identification	1 hr	Days
Epidemiological measurements (community and personal)	24 hr	Months
Laboratory chambers (clinical exposure and smog chamber)	1–15 min	Hours
Model development and evaluation	1 hr	Months
Aerial measurements (plumes)	1 sec	Minutes
Aerosols transport	3–6 hr	Months
Source identification	1 hr	Days
Ecological deposition	1 week	Years
Deposition parameters	<1 sec	Hours

[a] Compilation by Wilson (1977).

(Adams et al., 1977). At the other extreme the highest cadmium level to be reported at a remote location is the 3000 ng/m^3 above the hot vent of Mount Etna, Sicily (Buat-Menard and Arnold, 1978). In general, the cadmium level at remote locations is less than 0.4 ng/m^3, and quite often below 0.1 ng/m^3. It is not difficult to surmise that the study of cadmium in air at remote locations has been hampered by the insensitivity of the analytical methods used.

Table 7.   Airborne Cadmium Concentrations at Remote Locations

Location	Cadmium Concentration (ng/m^3)	Enrich-ment Factor	Reference
Cape of Desire (Novaya Zemlya)	0.28	70,000	Egorov et al. (1970)
Dickson Island	0.39	20,000	Egorov et al. (1970)
Jungfraujoch, Switzerland	0.014	4,000	Dams and De Jonge (1976)
Chacaltaya Mt., Bolivia	0.006		Adams et al. (1977)
South Pole	<0.015	<7,400	Maenhaut and Zoller (1976)
North Atlantic (30°N)	0.003–0.62	730	Duce et al. (1975a)
North Atlantic (20°N)	0.2	—	Muller and Beilke (1977)
South Indian Ocean	0.35	4,700	Egorov et al. (1970)
Mount Etna			
Volcanic plume	92	2,000	Buat-Menard and Arnold (1978)
Above hot vent	30,000	3×10^6	Buat-Menard and Arnold (1978)
Northern Norway (marine air)	0.12	1,200	Rahn (1976)
Bermuda coast	0.19 (1973)	550	Duce et al. (1976)
	0.13 (1974)	440	Duce et al. (1976)

A characteristic feature of cadmium in air at remote locations is its enormous enrichment in atmospheric particulates (see Table 7). This enrichment is commonly quantified in terms of an *enrichment factor*, EF:

$$EF = \frac{(C_{Cd}/C_{Al})_{air}}{(C_{Cd}/C_{Al})_{crust}}$$

where C is the average concentration of cadmium or aluminum in the aerosol or crustal rock. The general applicability of the EF concept in environmental studies is discussed in detail by Rahn (1976).

The EF values at remote locations range from 700 to 3×10^6 (Table 7), indicating that a large fraction of the cadmium-rich aerosols is derived from sources other than local soils or bedrock. In a survey of the literature data, Rahn (1976) found the geometric mean enrichment factor for cadmium to be 940 for urban aerosols, 2000 for remote continental aerosols, and 5000 for remote marine aerosols. The high EFs suggest some volatilization process in the generation of cadmium-rich aerosols. For example, the EF values for cadmium in particulates from high-temperature processes include $>10^5$ in fly ash from municipal incinerators (Greenberg et al., 1978), 40 to over 1000 in fly ash from coal-fired power plants (Gordon et al., 1973; Natusch, 1978), and 2×10^3 to 3×10^6 in volcanogenic aerosols (Buat-Menard and Arnold, 1978). The relatively low boiling points (or high volatility) of cadmium oxides may be significant in the preferential release of cadmium to the atmosphere. The anomalously high EF values at remote locations, however, provide no clues as to the fraction of the cadmium-rich aerosols that is derived from natural processes (volcanism, biological mobilization, and rock degassing) as compared to anthropogenic combustion processes.

There is evidence that the concentrations of cadmium in air at the most remote areas of the earth have increased in recent times. Jaworowski et al. (1975) showed that the average cadmium concentration in Storbreen Glacier (Norway) was 0.68 µg/kg around 1200 A.D., 0.81 µg/kg around 1600, 1.2 µg/kg around 1800, 2.4 µg/kg between 1954 and 1964, and 12.0 µg/kg between 1965 and 1972. In other words, there has been a 15-fold increase in the atmospheric fallout of cadmium on the glacier since 1600. In another study, Weiss et al. (1975) found the cadmium contents of pre-1900 Greenland ice layers to be 11 ng/kg, compared to 640 ng/kg in the post-1966 ice, or a 59-fold increase in the rate of cadmium deposition. Undoubtedly, these increases in atmospheric fallout of cadmium reflect the increased anthropogenic release of the metal into the atmosphere.

3.2. Rural and Urban Areas

In recent years, numerous synoptic studies and long-term monitoring programs have included measurements of cadmium concentrations in atmsopheric particulates at urban and rural locations. On the basis of these data, it is estimated that the cadmium levels in rural areas are generally <1.0 ng/m^3, whereas in urban areas the common range is 1 to 50 ng/m^3. The ratio of cadmium in urban air to cadmium in rural air usually varies from 10 to 50.

Since 1957 the concentrations of cadmium in ambient air have been measured at numerous National Air Surveillance Network (NASN) stations in the United States. Owing to changes in sampling strategy and analytical methodology, the quality of the data is not always the same. In this connection it should be noted that the cadmium data for 1969 are anomalously high for no apparent reason. The following discussion is restricted to the post-1970 data, which seem to be the most reliable.

Between 1970 and 1974 the atmospheric cadmium level at none of the 46 nonurban NASN sampling stations exceeded 0.3 ng/m^3, which is the discrimination limit of the emission spectrographic method of analysis (NASN, 1976). This observation is in accord with normal concentrations of <1.0 ng/m^3 in rural areas of Sweden (Friberg et al., 1971). By contrast, the average cadmium concentrations in air at semirural stations in the United Kingdom were 4 ng/m^3 in 1975 and <3 ng/m^3 in 1976 (Cawse, 1976, 1977). The mean atmospheric cadmium concentration at two rural locations (Botrange and Oostende) in Belgium was 7 ng/m^3 between 1972 and 1975 (Kretzschmar et al., 1977). Muller and Beilke (1977) found the cadmium level in air at two rural sites in Germany to be <2 ng/m^3.

Table 8 shows the urban areas in the United States with elevated atmospheric levels of cadmium between 1970 and 1974. The mean cadmium concentrations at these NASN monitoring stations were 16 ng/m^3 in 1970, 14 ng/m^3 in 1971, 13 ng/m^3 in 1972, 6.7 ng/m^3 in 1973, and 7.6 ng/m^3 in 1974. These values for the selected areas are much higher than the average data for all the urban areas in the United States—3, 4, 2, 1, and 2 ng/m^3 for 1970, 1971, 1972, 1973, and 1974, respectively (NASN, 1976). The average data for all the stations are biased by the high values recorded occasionally or at particular stations (see Table 9); in fact, over 70% of all the measurements in the urban areas were below the analytical discrimination limit (see NASN, 1976).

The mean cadmium level in ambient air in many towns and cities in Ontario was 6 ng/m^3 during 1971 (Barton et al., 1975). The highest ambient cadmium levels were recorded at stations located at fire halls in the communities of Bramalea and Peterborough, where 17 and 15 ng/m^3, respectively, were obtained (Barton et al., 1975).

Between May 1972 and April 1975 the mean atmospheric cadmium concentration in Belgian cities was 19 ng/m^3, the mean 98 percentile being 104 ng/m^3 (Kretzschmar et al., 1977). Just and Kelus (1971) found the range in mean yearly concentrations of cadmium in air in 10 Polish cities to be 2 to 50 ng/m^3. Between 1969 and 1971 Nagata et al. (1972) observed that the levels of cadmium in air in different parts of Tokyo varied from 10 to 53 ng/m^3 and

Table 8. Representative Data for Urban Areas in the United States with Elevated Levels of Airborne Cadmium[a]

Location	Sampling Year	Cadmium Concentration (ng/m^3)
Tucson, Ariz.	1970	6.5
	1971	25
	1972	<8.0
	1973	<6.0
	1974	<11
Denver, Colo.	1970	10
	1971	10
	1972–74	<10
Waterbury, Conn.	1970	25
	1971	4.8
	1972	17
	1973	<8.0
	1974	14
Ashland, Ky.	1970	11
	1971	15
	1972	9.3
	1973	<7.0
	1974	10
St. Paul, Minn.	1970	6.0
	1971	<0.2
	1972	8.6
	1973–74	<7.0
Jersey City, N.J.	1970	8.1
	1971	5.6
	1972	12
	1973–74	<7.0
Newark, N.J.	1970	12
	1971	<22
	1972	16
	1973	9.0
	1974	5.2
Allentown, Pa.	1970	8.1

Table 8. *Continued*

Location	Sampling Year	Cadmium Concentration (ng/m³)
	1971	7.5
	1972	12
	1973	8.0
	1974	13
Bethlehem, Pa.	1970	14
	1971	19
	1972	18
	1973	6.8
	1974	13
Philadelphia, Pa.	1970	12
	1971	7.8
	1972	5.7
	1973	5.0
	1974	<7.0
El Paso, Tex.	1970	62
	1971	47
	1972	44
	1973	21
	1974	24
New York City, N.Y.[b]	1969	9.7
	1972	6.0
	1973	7.0
	1974	6.0
	1975	4.0
Elizabeth, N.J.[b]	972	10
	1973	9.0
	1974	0.2
Bridgeport, Conn.[b]	1972	9.0
	1973	5.0
	1974	0.2

[a] Data from National Air Surveillance Network (1976) unless noted otherwise.
[b] Compiled by Lioy et al. (1978).

Table 9. Cumulative Frequency Distribution (concentrations in μg/l) of Airborne Cadmium in Urban Areas in the United States[a]

Year	Number of Observations	Min.	10	30	50	70	90	95	99	Max.	Arithmetic Mean	Arithmetic Standard Deviation	Geometric Mean	Geometric Standard Deviation
1970	797	LD[b]	LD	LD	LD	LD	0.008	0.012	0.027	0.099	0.003	0.007	LD	4.40
1971	717	LD	LD	LD	LD	LD	0.010	0.017	0.059	0.295	0.004	0.016	LD	5.21
1972	708	LD	LD	LD	LD	LD	0.007	0.011	0.034	0.112	0.002	0.007	LD	4.97
1973	559	LD	LD	LD	LD	LD	LD	0.007	0.012	0.032	0.001	0.003	LD	4.89
1974	594	LD	LD	LD	LD	LD	0.006	0.009	0.022	0.077	0.002	0.005	LD	4.90

[a] National Air Surveillance Network (1976).

[b] LD denotes below the analytical discrimination limit.

recorded a maximum 24-hr average value of 530 ng/m^3. The atmospheric concentrations of cadmium found in urban areas in many parts of the world are summarized in Table 10.

As expected, towns with large metal industries have higher average concentrations of cadmium in the air. The highest annual average cadmium concentrations (mean, 35 ng/m^3; range, 5 to 715 ng/m^3 between 1972 and 1975) in Belgium were recorded in Liege, where the major metal industries are concentrated (Kretzschmar et al., 1977). A review of the available cadmium data for urban areas of the United States (EPA, 1975a) showed that El Paso, Texas, had the highest 24-hr average recorded (730 ng/m in 1964), the highest quarterly average (150 ng/m^3 in 1969), the highest annual average (120 ng/m^3 in 1964), and the highest percentage of cadmium in airborne particulates (700 μg/g in 1969). Although the actual values may be in error, the consistent top ranking of El Paso in terms of airborne cadmium levels must be related to the large lead smelter located in the area. In the St. Louis, Missouri area, Peden (1977) found that the largest average atmospheric concentrations of cadmium occurred downwind of a zinc smelter, and that the second highest average occurred in the vicinity of both coal-burning steel industries and power plants. In the lead smelting town of Helena, Montana, the average cadmium concentrations in air range from 60 to 290 ng/m^3 (Rupp et al., 1978), with maximum 24-hr values of up to 700 μg/m^3 (EPA, 1972). Hot spots of airborne cadmium with halos of high concentrations that extend for tens of kilometers have generally been associated with smelters, as in the New Lead Belt of southeast Missouri (Dorn et al., 1975, 1976), and other point sources (see below).

As expected, Washington, D.C., which has little industry or major point sources of cadmium aerosols, is characterized by low levels of ambient cadmium, the value during the summer of 1974 being 3.5 ng/m^3 (Kowalczyk et al., 1978). In a recent report, Kowalczyk et al. (1978) showed that most (>90%) of the airborne cadmium in the Washington, D.C., area came from refuse incineration and that the remainder of the suspended cadmium was derived mainly from coal and oil combustion. Greenberg et al. (1978) also reached the conclusion that refuse combustion is a major source of airborne cadmium in many cities with no large smelters.

The effect of automotive emission on ambient cadmium levels in roadside ecosystems is not clear. Creason et al. (1971) found no roadside gradients in cadmium concentrations in the atmosphere or dustfall at four sites in Cincinnati, Ohio, between July and September 1968. Other studies (Lagerwerff and Specht, 1970; Ward et al., 1977), however, have observed gradients in the cadmium contents of roadside soils which tend to correlate with the traffic density. A buildup of cadmium in roadside soils may be expected from the wear of automobile tires and occasionally from the wind removal and erosion of material being hauled along the road.

Table 10. Representative Cadmium Concentrations in Urban and Rural Atmospheres

Locality	Sampling Year	Cadmium Concentration (ng/m^3)	Reference
Osaka (downtown), Japan	—	112	Sugimae and Hasegawa (1973)
Salehard, USSR	1968–69	5.3	Egorov et al. (1970)
Sevastopol, USSR	1968–69	2.3	Egorov et al. (1970)
Petropavlovsk-Kamchatsti, USSR	1968–69	0.6	Egorov et al. (1970)
Magadan, USSR	1968–69	4.2	Egorov et al. (1970)
Tashkent, USSR	1968–69	180	Egorov et al. (1970)
Semipalatinsk, USSR	1968–69	96	Egorov et al. (1970)
Tien Shan, USSR	1968–69	176	Egorov et al. (1970)
Antwerp, Belgium	1972–75	14	Kretzschmar et al. (1977)
Brussels, Belgium	1972–75	13	Kretzschmar et al. (1977)
Charleroi, Belgium	1972–75	15	Kretzschmar et al. (1977)
Liege, Belgium	1972–75	35	Kretzschmar et al. (1977)
Mol, Belgium	1972–75	16	Kretzschmar et al. (1977)
Botrange, Belgium	1972–75	6	Kretzschmar et al. (1977)
Oostende, Belgium	1972–75	8	Kretzschmar et al. (1977)
Ghent, Belgium	1972	9.4	Kretzschmar et al. (1977)
Utrecht			
Urban/industrial area	1970	11	Minderhoud and Boogerd (1975)
Urban area	1970	21	Minderhoud and Boogerd (1975)
Industrial area	1970	16	Minderhoud and Boogerd (1975)
Enschede, Belgium	1970	10	Minderhoud and Boogerd (1975)
Wageningen, Belgium	1972	8.0	Minderhoud and Boogerd (1975)
Witteveen, Belgium	1972	5.0	Minderhoud and Boogerd (1975)
Den Helder, Belgium	1972	4.0	Minderhoud and Boogerd (1975)
Rotterdam, Holland	1971–72	7.7	Evendijk (1974)
Schiedan, Holland	1971–72	5.9	Evendijk (1974)
Maassluis, Holland	1971–72	2.7	Evendijk (1974)
Heidelberg, West Germany	1971	27	Bogen (1973)
Frankfurt, West Germany	1974	6.1	Muller and Beilke (1977)
Corviglia, West Germany	1974	2.1	Muller and Beilke (1977)
Munich, West Germany	1971	36	Schramel et al. (1974)
St. Moritz, Switzerland	1974	1.2	Muller and Bielke (1977)
Paris, France	1971	20	Belot et al. (1971)
Chilton, England	1975	2.7	Cawse (1976, 1977)
Chilton, England	1976	2.7	Cawse (1976, 1977)
Leiston, England	1975	5.5	Cawse (1976, 1977)
Collarfirth, Shetland Is.	1975	<2.0	Cawse (1976, 1977)
	1976	<0.7	Cawse (1976, 1977)
Plynlimon, Wales	1976	<0.9	Cawse (1976, 1977)
	1975	2.8	Cawse (1976, 1977)

Table 10. *Continued*

Locality	Sampling Year	Cadmium Concentration (ng/m³)	Reference
Styrrup, England	1976	<1.0	Cawse (1976, 1977)
	1975	7.2	Cawse (1976, 1977)
Trebanos, Wales	1976	<3.0	Cawse (1976, 1977)
	1975	4.5	Cawse (1976, 1977)
Wraymires, England	1976	<3.0	Cawse (1976, 1977)
	1975	1.7	Cawse (1976, 1977)
Stoke on Trent, England	1976	<3.0	Cawse (1976, 1977)
	1976	3.8	Cawse (1976, 1977)
Leeds, England	1976	3.2	Cawse (1976, 1977)
Sutton, England	1969	0.4	Hamilton (1974)
Stockholm, Sweden	—	5.0	Friberg et al. (1971)
Rural areas, Sweden	—	0.9	Friberg et al. (1971)
Chicago, Ill.	1972–73	4.2	Yost (1978)
Chicago, Illinois	1970	3.0	Henry and Blosser (1971)
Cleveland, Ohio	1971	3.9	King et al. (1976)
Cincinnati, Ohio	1970	2.0	Henry and Blosser (1971)
San Francisco Bay area	1970	3.3	John et al. (1973)
St. Louis, Mo.	1970	5.0	Henry and Blosser (1971)
Washington, D.C.	1970	0.3	Henry and Blosser (1971)
Washington, D.C.	1974	3.5	Greenberg et al. (1978)
Los Angeles, Calif.	1975	7.8	Davidson (1977)
Chedron, Neb.	1973	0.6	Struempler (1975)
Chedron, Neb.	1974	0.3	Struempler (1975)
Boston, Mass.	—	2.0	Gordon et al. (1973)
Gary, Ind.	1972–73	2.6	Yost (1978)
Hammond, Ind.	1972–73	4.1	Yost (1978)
East Chicago, Ind.	1972–73	5.4	Yost (1978)
Porter County, Ind.	1972–73	1.4	Yost (1978)
State of Indiana	1972–73	1.7	Yost (1978)
Brampton, Ont.	1971	9.0	Barton et al. (1975)
Chatham, Ont.	1971	2.0	Barton et al. (1975)
Hamilton, Ont.	1971	8.0	Barton et al. (1975)
London, Ont.	1971	8.0	Barton et al. (1975)
Oshawa, Ont.	1971	7.0	Barton et al. (1975)
Ottawa, Ont.	1971	5.0	Barton et al. (1975)
Sarnia, Ont.	1971	7.0	Barton et al. (1975)
Toronto, Ont.	1971	11	Barton et al. (1975)
Thunder Bay, Ont.	1971	2.5	Barton et al. (1975)
Waterloo, Ont.	1971	4.0	Barton et al. (1975)
Windsor, Ont.	1971	14.0	Barton et al. (1975)

Table 11.   Cadmium Contents of Airborne Particulates at Representative Locations and Aerosols from Pollutant Sources

Location	Cadmium in Particulates (μg/g)[a]
Cape of Desire, USSR	204
Dickson Island, USSR	119
Salehard, USSR	623
Sevastopol, USSR	501
Petropavlovsk-Kamchatski, USSR	70
Magadan, USSR	72
Tien Shan, USSR	1010
North Indian Ocean	1200
South Indian Ocean	52
Atlantic Ocean (20°N)	17
Frankfurt, West Germany	66
Kleiner Feldberg, West Germany	54
Corviglia, West Germany	57
Chacaltaya Mt., Bolivia	1.1
Chilton, England (1975)	108
Leiston, England (1975)	
Collafirts, Shetland Is. (1975)	<78
Plynlimon, Wales (1975)	140
Styrrup, England (1975)	206
Trebanos, Wales (1975)	173
Wraymires, England (1975)	100
Chicago, Ill.	46
East Chicago, Ind.	44
Gary, Ind.	27
Hammond, Ind.	43
Porter County, Ind.	29
State of Indiana	31
Urban areas, U.S.A.	
1970	33
1971	46
1972	24
Walker Branch Watershed, Tenn.	137
Urban areas, Ont.	71
Coal fly ash[b]	0.1–50
Municipal incinerator fly ash[c]	5–200
Stack effluents, open hearth steel furnace[d]	20–600
Stack effluents, copper-zinc smelters[e]	10–9500

[a] The data used in deriving the cadmium contents of ambient aerosols are given in Table 10 and the references therein.
[b] Hillenbrand et al. (1973), Natusch (1978), Furr et al. (1977).
[c] Jacko and Neuendorf (1977), Greenberg et al. (1978).
[d] Lee et al. (1975), Jacko and Neuendorf (1977), Yost (1978).
[e] Hallowell et al. (1973), Jacko and Neuendorf (1977), Chizhikov (1966), Fleischer et al. (1974).

Table 11 shows the cadmium contents of airborne particulates at different locations. The general trend in cadmium enrichment appears to be urban < rural < remote locations. This trend is consistent with the known enrichment of cadmium in the very fine particulates. Notice also that the concentrations of cadmium in ambient particulates are generally in the range of the concentrations observed in aerosols from many industrial point sources.

The air within and surrounding industrial point sources of cadmium, in general, contains elevated levels of the metal. Table 12 shows the ambient cadmium concentrations in different areas of a lead smelting plant. The range observed was 0 to 1169 $\mu g/m^3$; the data occasionally exceed the often accepted standard for cadmium in industrial air of 40 to 50 $\mu g/m^3$. In other interesting studies Lagerwerff and Brower (1975) found the airborne concentrations of cadmium near an inactive smelter with piles of mine chat tailings to vary from 20 to 160 ng/m^3, with a mean value of 73 ng/m^3. In Sweden, levels of 600 and 300 ng/m^3 were recorded at distances of 100 and 500 m, respectively, from a factory that was using Cu-Cd alloys (Friberg et al., 1971). In Japan, airborne cadmium levels of 160 to 500 ng/m^3 occurred with a 500 m radius of two zinc smelters (Friberg et al., 1971).

4. TEMPORAL VARIATIONS IN ATMOSPHERIC CADMIUM LEVELS

Diurnal and day-to-day variations in atmospheric concentrations of cadmium have been observed in urban and rural locations. These changes are determined primarily by variations in source emissions, wind direction, and local ventilation factors. Thus periods of unusually high levels of metals in air are often associated with the development of stagnant anticyclones; Demuynck et al. (1976) recorded a 12-fold increase in airborne cadmium levels during one such incident in Ghent, Belgium. Many studies have demonstrated that a shift in wind trajectory from unpolluted to polluted sectors can engender a substantial buildup of cadmium in air at a given location (e.g., Harrison and Winchester, 1971; Harrison et al., 1971; Heindryckx and Dams, 1974; Duce et al., 1976; Neustadter et al., 1976). For example, Thrane (1978) showed that the airborne cadmium levels in rural areas of Norway could be increased by 5 standard deviations above the average when the air masses came from the populated and industrialized parts of Europe. Duce et al. (1975b) found the concentration of cadmium in air over New York Bight to be 0.52 ng/m^3 when the winds were onshore but 1.4 ng/m^3 when the offshore winds came from New York City. In a recent case study Lioy et al.

Table 12. Cadmium Levels in Air in Different Areas of a Lead Smelter in Belledune, New Brunswick, Canada[a]

	Ambient Cadmium Concentration ($\mu g/m^3$)			
Location	Test 1 Dec. 1974	Test 2 Mar. 1975	Test 3 Jan. 1976	Test 4 Dec. 1976
Sinter Plant Area				
Control room	1.0	11	7.0	—
Ignition platform	1.3	288	49	208
Foreman's office	0.9	11	1.0	4.0
Charge shuttle	113	—	456	745
Tip end	10	83	88	225
M-10 sinter fines bin	23	42	41	121
East bottom floor	52	61	15	96
Head M-8	40	91	81	71
Engineering shop	1.0	3.0	1.0	7.0
Ross chain area	24	42	64	62
Technical office	1.0	3.0	0.1	5.0
Screening and crushing building				
North of cooling drum	18	71	83	89
South of cooling drum	17	36	83	—
Ross rolls	38	70	212	311
Blast Furnace Area				
Control room	1.0	5.0	3.0	—
Foreman's office	3.0	9.0	—	—
Upper slagging	3.0	6.0	8.0	29
Lower slagging	—	65	29	62
Foreman's office (new)	1.0	5.0	2.0	—
Dross plant	0.1	4.0	21	52
Emergency end	6.0	75	21	94
Furnace top	—	—	585	1169
General foreman's office	1.0	12	—	—
Smoking room	—	10	3.0	—
Mechanical room	2.0	15	4.0	5.0
Charge control room	1.0	2.0	3.0	3.0
Sample crusher	242	2.0	—	—
Sinter cross feeders	19	715	178	251
Lead Refinery Area				
Foreman's office	2.9	2.0	5.0	—
Cupel area	0.2	14	21	26
Retort area	0.0	10	21	23
Kettle area east	0.2	4.0	5.0	16

Table 12.   *Continued*

Location	Ambient Cadmium Concentration ($\mu g/m^3$)			
	Test 1 Dec. 1974	Test 2 Mar. 1975	Test 3 Jan. 1976	Test 4 Dec. 1976
Cu fce. area (old)	1.4	12	21	55
General area near office	1.0	12	28	65
Casting area	0.1	3.0	4.0	11
Materials handling office	0.4	40	5.0	3.0
Smoking room	0.8	1.0	1.0	6.0
De-Cu area west	1.7	20	15	117
De-Cu area east	1.3	18	13	81
New lunch room	0.8	2.0	—	—
Bi-refinery	1.2	3.0	6.0	4.0
Materials handling general area	0.6	1.0	3.0	3.0

[a] Dugdale and Hummel (1978).

(1978) used the dispersion normalization technique (Kleinman et al., 1976) and a trajectory analysis to show that an incident of 25-fold elevation in airborne cadmium level (observed on August 4 and 5, 1976) at High Point, New Jersey, could be attributed to a parcel of air which had come all the way from Allentown-Bethlehem, Pennsylvania, located about 100 km upwind.

Pronounced seasonality in airborne cadmium levels has been observed at many locations. Studies in Belgium (Dams, 1974; Kretzchmar et al., 1977) and Great Britain (Cawse, 1976; Salmon et al., 1978) commonly show higher cadmium concentrations in winter than in summer. For example, Cawse (1976) showed that at Styrrup, Nottinghamshire, England, the level of cadmium in air particulates in the first and last quarters of 1975 was 15 times greater than that measured between April and September. The enhanced winter cadmium levels are usually attributed to extra emissions associated with domestic heating and power consumption, which are highest in winter. By contrast, Kleinman et al. (1973) found that the concentrations of cadmium at several locations in New York City exhibited a spring minimum and a summer maximum. The seasonal pattern was attributed to changes in atmospheric dispersion of the atmospheric pollutants. Table 13 shows quarterly average cadmium values in air at several urban areas in the United States. No consistent seasonal trend is apparent in these data, recorded between 1970 and 1974. The discordance in the data of Table 13 may simply reflect the fluctuations in industrial (and power) plant emissions; these are, it will be recalled, by and large the principal sources of cadmium in the atmosphere.

The data bases from the various long-term air quality monitoring programs

Table 13.   Seasonal and Annual Average Cadmium Concentrations in U.S. Cities During 1970[a]

City	Quarterly Average Concentration (ng/m^3)				Annual Average
	1st	2nd	3rd	4th	
Tucson, Ariz.	7.0	<0.2	6.0	13	6.5
Denver, Colo.	7.0	<0.2	7.0	26	10
Bridgeport, Conn.	23	<0.2	11	11	11
Waterbury, Conn.	54	33	7.0	7.0	25
Peoria, Ill.	12	<0.2	7.0	6.7	—
East Chicago, Ind.	30	25	8.0	12	19
Indianapolis, Ind.	<0.2	15	6.0	6.0	6.6
Asland, Ky.	23	8.0	7.0	5.0	11
Detroit, Mich.	<0.2	9.0	5.0	5.0	4.7
St. Paul, Minn.	10	7.0	8.0	<0.2	6.0
St. Louis, Mo.	11	<0.2	—	15	—
Newark, N.J.	12	19	12	8.0	12
Jersey City, N.J.	10	8.0	7.0	8.0	8.1
New York City, N.Y.	<0.2	11	10	8.0	7.1
Cincinnati, Ohio	<0.2	13	6.0	15	8.6
Cleveland, Ohio	7.0	12	10	6.0	8.8
Youngstown, Ohio	5.0	<0.2	9.0	9.0	5.6
Allentown, Pa.	12	6.0	7.0	8.0	8.1
Bethlehem, Pa.	7.0	<0.2	43	5.0	14
Philadelphia, Pa.	7.0	8.0	27	7.0	12
El Paso, Tex.	99	82	—	66	62
Lynchburg, Va.	10	14	14	16	14

[a] Data from National Air Surveillance Network (1976).

are generally inadequate for determining the long-term trends in airborne cadmium levels. This is so because the spectrographic method of analysis is not adequate for determining the low ambient cadmium concentrations prevailing at most of the monitoring stations. Nevertheless, there is evidence to suggest that the ambient levels of cadmium, like those of the other chalcophilic elements, have declined significantly over the last decade. The available data on NASN sites in urban areas of the United States reveal a drop in the median (50th percentile) cadmium value from about 2.5 ng/m^3 in 1965 to 0.4 ng/m^3 in 1970 and thereafter. Between 1965 and 1974 there was a 57% decline in the 90th percentile value of atmospheric cadmium in urban areas of the United States; the decline between 1970 and 1974 alone was 44% (Faoro and McMullen, 1977). Decreases of unknown magnitude are also believed to have occurred in the airborne cadmium levels in nonurban areas of the United States (Faoro and McMullen, 1977) and Britain (Salmon et al., 1978). The downward trend in atmospheric cadmium concentrations is probably related to reduced particulate emissions from industries and power plants and improved incineration and waste burning practices.

5. PROPERTIES OF CADMIUM AEROSOLS

The chemical forms of cadmium in aerosols are essentially unknown. The mechanisms by which cadmium-rich aerosols may be formed following a volatilization process include (a) simple condensation of the volatilized cadmium or its compounds, (b) coprecipitation with other metals or their compounds, (c) preferential uptake by solid or molten particles, and (d) capture in a matrix of water vapor and condensable gases, especially sulfur oxides and nitrogen oxides. Intuitively, then, the airborne cadmium species may be expected to include elemental cadmium, cadmium sulfide, cadmium oxides, and hydroxides, as well as mixed oxides with other metals, particularly copper and zinc. Particulates derived from smelters may also contain cadmium as inclusions in zinc and copper sulfides and oxides.

Dreesen et al. (1977) showed that <35% of the cadmium in fly ash from a coal-fired power plant could be dissolved in dilute (<0.1 M) acids. Their result is in contrast with the observation by Bolter et al. (1975) that 34 to 70% of the cadmium in baghouse dusts from lead smelting activities was soluble in water. Hodge et al. (1978) showed that, on the average, about 80% of the cadmium in southern California aerosols was leachable into seawater. The results of the last two studies are consistent with the common observation that 30 to 95% of the cadmium in atmospheric precipitation samples is in soluble form (see below). In the first instance, fly ash from power plants accounts for only a small fraction of the cadmium released annually into the atmosphere (see Table 2). Also, the cadmium-enriched particles emitted from smelters and combustion sources possess large surface areas, which would enhance the dissolution of such cadmium aerosols in rain droplets with low pH. Natusch and his associates (Linton et al., 1976; Natusch and Wallace, 1974; Natusch, 1978) have established that cadmium is more highly concentrated on the surfaces than in the interior of particles derived from high-temperature combustion or smelting operations. The ratio of surface to bulk concentration of cadmium in coal fly ash has been estimated to be about 30 (Natusch, 1978). Aside from the effect on the physiological extractability of the metal, the accumulation of cadmium on particle surfaces may also account for the apparent high solubility of cadmium aerosols in rainwater.

Physical characterization of cadmium aerosols is largely confined to particle size distribution studies. Before discussing these data, a comment should be made on terminology. Most of the work on cadmium-containing particles involves aerodynamic sizing by means of impactor air centrifuge or other transport behavior of the particles in an applied field. *Aerodynamic sizes* (usually expressed in terms of a sphere of unit density that possesses the same aerodynamic properties as the particle in question) are particularly applicable to environmental studies because the suspension, atmospheric transport, fallout, and washout of the aerosols are determined by the aerodynamic properties. Most often, though, what is reported is the way the mass of an element varies with the particle size, or its mass-size function (MSF). Another common statistic used in particle size description is the *mass-median diameter* (MMD), defined as the diameter below which 50% of the total mass of an element is found.

The reported MMD values must be used in a circumspect manner, considering that recent studies (Whitby, 1977; Willeke and Whitby, 1975, Meszaros, 1977) have shown that the size distribution for most aerosols is often multimodal. Figure 1 summarizes the various mechanisms that contribute to the formation, transformation, and removal of atmospheric aerosols with trimodal size distribution. The implication is clear that the MMDs for cadmium aerosols show significant temporospatial variations. Few studies have addressed the changes in MMD of cadmium aerosols. In fact, evidence that the mass-size function for airborne cadmium has a biomodal distribution has only recently been reported (Peden, 1977).

In his compilation of the published data for rural and urban sites, Rahn (1976) showed that the MMD for cadmium was 2.2 μm. The range in the MMD data was 0.6 to 10 μm, implying that cadmium in air occurs predominantly in the respirable particle size fraction. Particles in the 0.5 to 3 μm size range are commonly assumed to be respirable (Stuart, 1976; Yeh et al., 1976).

A detailed study of the mass-size functions for cadmium aerosols emitted from several industrial and municipal point sources has been made by Jacko and Neuendorf (1977). Their data show that about 50% of the cadmium aerosols emitted from open hearth furnaces with no control devices was <2 μm in size; the respirable figure was increased to about 80% when the particulate emission was controlled by an electrostatic precipitator. Between 60% and 80% of all the cadmium aerosols emitted during zinc smelting operations was shown to have an aerodynamic diameter of <2 μm. About 65% of the cadmium from municipal incinerators with scrubber dust control was found to be contained in the <2 μm fraction. The respirable cadmium fraction from a coal-fired power plant with electrostatic dust control was only about 30% (Jacko and Neuendorf, 1977).

Several studies (e.g., Lee and von Lehmden, 1973; Davison et al., 1974; Lee et al., 1975; Block and Dams, 1976) have determined the distribution of cadmium in size-fractionated fly-ash samples from coal combustion. In general, it is found that 50 to 90% of the cadmium present is associated with particles <5 μm in diameter. Dorn et al. (1976) used an eight-stage Anderson impactor

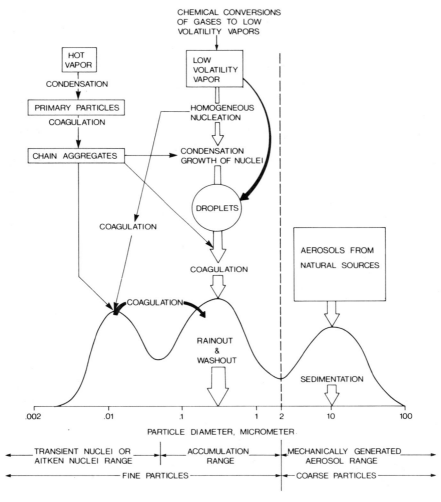

Figure 1.   Schematic illustration of the principal mechanisms of formation of aerosols with multimodel size distribution. (From Whitby, 1977.)

sampler in studying the MSF of cadmium at two locations, one about 800 m and the other about 75 km from the smelter stack in the New Lead Belt of southeast Missouri. Their data show that roughly 88% and 65% of the cadmium aerosols in the test farm (near the smelter) and control farm, respectively, were <5 μm in diameter. In fact, about 78% of the cadmium in air near the smelter was associated with particles <2 μm in diameter.

All the data available thus indicate that the cadmium in airborne particulates occurs predominantly in the respirable size fraction. The size distribution of ambient cadmium particulates is discussed in Chapter 4, this volume, by Davidson.

6. REMOVAL OF CADMIUM FROM THE ATMOSPHERE

The atmosphere has become a major contributor of cadmium to many ecosystems. This point is dramatically illustrated by the data in Table 14, which show that the rate of ombrogenic input of cadmium now exceeds the rate of cadmium assimilation by the living biota in some ecosystems. Another disturbing fact is illustrated by the data in Table 15, namely, the ombrogenic cadmium is mostly in soluble form and hence readily available to living matter. Deposition of anthropogenic cadmium now occurs in the most remote regions of the globe.

Table 14.   Distribution of Cadmium Among Components on the Dune, Marsh, and Flood Plain Ecosystems at a Terrestrial Site in East Chicago, Indiana [a]

Compartment	Cadmium (g/ha)		
	Dune	Marsh	Flood Plain
Atmospheric input (total per year)	8.50	8.50	8.50
Above-ground vegetation (total)	12.43	0.77	2.08
Herbaceous	1.13	0.25	2.08
Shrubs, seedlings, and woody vines	0.44	0.52	0.00
Tree (total)	10.86	—	—
Stem (total)	6.02	—	—
Heartwood	2.01	—	—
Sapwood	1.24	—	—
Bark	2.77	—	—
Branch (total)	4.84	—	—
Wood + bark	3.55	—	—
Leaves	1.29	—	—
Standing dead	1.76	—	—
Roots (total)	21.36	115.93	59.13
Woody	7.62	—	—
Fibrous	13.74		
Litter fall (total per year)	0.15	0.02	0.02
Litter layer (01 + 02)	52.42	No est.	No est.
Mineral soil			
0–2.5 cm	1445.44	826.00	938.50
2.5–14.0 cm	2173.11	2487.00	1933.00

[a] Yost (1978).

Table 15. Concentrations and Deposition of Cadmium in Atmospheric Precipitation at Stations in Ontario[a]

	Cadmium Concentration (μg/l)[b]		Cadmium Fallout Rate (mg/m^2·year)	
Location	Filtered Samples	Total	Soluble	Total
Skead	1.5 (0.1–10)	1.9 (0.4–10)	0.84	1.1
Killarney	0.9 (0.2–8.8)	1.2 (0.3–26)	0.51	0.77
Gore Bay	0.8 (0.1–4.6)	0.8 (0.1–5.3)	0.44	0.44
Jamot	0.9 (0.2–10)	0.9 (0.2–12)	0.51	0.51
Windy Lake	0.6 (0.05–7.0)	0.7 (0.2–7.0)	0.36	0.40
Mount Lake	0.5 (0.05–7.3)	0.7 (0.2–30)	0.36	0.44
Gogama	1.1 (0.1–24)	1.1 (0.3–24)	0.62	0.62
Temagami	0.9 (0.2–4.0)	1.1 (0.1–14)	0.51	0.62
Espanola	0.9 (0.2–8.0)	0.9 (0.2–50)	0.58	0.58
Sudbury, South	1.3 (0.1–7.0)	1.3 (0.2–7.0)	0.80	0.91
Sudbury, North	2.3 (0.5–5.9)	2.8 (1.3–9.1)	1.6	1.9
Sault Ste Marie	0.8 (0.2–36)	0.8 (0.1–36)	0.58	0.58
Chapleau	0.7 (0.2–6.8)	0.7 (0.3–6.8)	0.47	0.47
Wawa	1.0 (0.2–11)	1.1 (0.3–11)	0.55	0.62
Timmins	0.5 (0.03–1.9)	0.7 (0.05–9.5)	0.33	0.44
Kapuskasin	0.7 (0.2–3.0)	0.7 (0.2–3.0)	0.40	0.40
Sparrow Lake	0.5 (0.01–2.4)	0.5 (0.1–3.1)	0.36	0.36
Lake St. Peter	0.6 (0.1–4.1)	0.7 (0.1–8.1)	0.44	0.51
Traverse	0.5 (0.1–2.7)	0.5 (0.1–2.7)	0.33	0.33
Shawanagan	0.4 (0.01–1.4)	0.6 (0.1–2.0)	0.36	0.58
Mattawa	0.8 (0.1–3.1)	0.8 (0.1–3.3)	0.55	0.55
Hearst	0.3 (0.01–1.2)	0.6 (0.05–4.2)	0.18	0.29
Hornepayne	0.5 (0.1–13)	0.6 (0.1–18)	0.29	0.33
Powassan	0.6 (0.1–3.4)	0.6 (0.2–5.0)	0.44	0.44
Cobalt	0.5 (0.1–1.1)	0.7 (0.3–1.3)	0.11	0.15
Marten River	0.5 (0.03–5.6)	0.6 (0.2–5.6)	0.26	0.33
Verner	0.5 (0.1–2.9)	0.5 (0.1–2.7)	0.32	0.32
South Baymouth	0.5 (0.03–2.5)	0.5 (0.1–2.5)	0.29	0.29
Millerd Lake	2.1 (1.0–6.2)	2.2 (1.4–6.6)	1.3	1.4
Wavy Lake	0.7 (0.2–0.9)	0.7 (0.2–1.1)	0.44	0.44
Flack Lake	0.5 (0.2–1.4)	0.8 (0.4–1.7)	0.47	0.69
Kumska Lake	0.8 (0.4–1.2)	0.8 (0.4–1.9)	0.58	0.58
Laundrie Lake	0.4 (0.3–0.8)	0.5 (0.3–1.0)	0.29	0.33
Average, all stations	0.8 (0.01–36)	0.8 (0.01–50)	0.47	0.51

[a] Data from Kramer (1976).
[b] Ranges in concentrations reported are shown in parentheses.

Before discussing the processes and rates of cadmium fallout, a word on the atmospheric residence times of cadmium aerosols is in order. Few studies have specifically addressed this question. By analogy with the work on the general particulate residence times, it is surmized that the residence time for cadmium aerosols in unpolluted troposphere is 2 to 10 days (Junge, 1963; Martell and Poet, 1972). In urban and polluted atmospheres the residence time may be expected to be 0.1 to 4 days for particles 1.0 to 10 μm in diameter and generally >4 days for particles of smaller diameter (Esmen and Corn, 1971; Martell and Poet, 1972; Whelpdale, 1974). Near an industrial point source (<100 km or so) the residence time of particulates usually is less than 2.0 hr (Nriagu, 1978). In one of the few studies of its kind, Hodge et al. (1978) estimated the residence times of cadmium in the atmosphere at La Jolla and Ensenado, California, to be 0.7 and 0.5 days, respectively.

6.1. Cadmium in Rainfall and Snowfall

Rainfall and snowfall play an important part in the removal of cadmium from the atmosphere. Wet removal may occur by incorporation of trace substances into cloud droplets (rainout) or by raindrop scavenging below the clouds (washout). The effectiveness of wet removal is a function of particle diameter, the larger particles being removed more efficiently by both rainout and washout.

In their study of wet removal of heavy metals from the atmosphere, Muller and Beilke (1977) showed that rainout is slightly more effective than washout in urban and polluted areas. In remote areas, however, rainout processes become clearly dominating. Furthermore, they showed that under steady conditions the heavy metal concentrations in rainwater decrease with increasing intensity of the rain. Muller and Beilke also observed higher fluctuations in the cadmium contents of rainwater during a shower than in a continuous rain. Struempler (1976) likewise found a decrease in cadmium concentration with greater duration of precipitation and reported a higher concentration of the element in thunderstorm showers than in slow rainfall. Kramer's (1976) data on "precipitation only" samples serve to emphasize the importance of wind direction (in relation to any point sources) in determining the time trend and concentrations of trace metals in rainwater or snowmelts.

The concentrations of cadmium in rainwater range from <0.1 to over 50 μg/l (Table 16). The ombrogenic levels of cadmium show temporospatial variations that may be related to changes in source contributions and to meteorological variables. Higher concentrations of cadmium have been reported in snowfalls than in rainfalls (see Struempler, 1976). The relative efficiency of snowfall versus rainfall in removing cadmium aerosols from the atmosphere, however, has not been studied systematically.

Table 16. Distribution and Deposition of Cadmium in Atmospheric Fallout

Location	Type of Sample	Cadmium Concentration[a]	Cadmium Fallout Rate (mg/m^2·year)	Reference
London, S.W. 7, England	Bulk precipitation	6 (1–11)	2.47	Harrison et al. (1975)
London, S.W. 2, England	Bulk precipitation	30 (2–108)	2.96	Harrison et al. (1975)
Hyde Park, London, England	Bulk precipitation	6 (1–2)	2.47	Harrison et al. (1975)
Hounslow, London, England	Bulk precipitation	25 (2–72)	7.66	Harrison et al. (1975)
Heathrow, England (air side)	Bulk precipitation	7 (2–14)	2.72	Harrison et al. (1975)
Heathrow: England (land side)	Bulk precipitation	37 (4–90)	4.94	Harrison et al. (1975)
Cambridgeshire, England	Bulk precipitation	32 (5–50)	—	Harrison et al. (1975)
Rural U.K.	Bulk precipitation	<5	<1.0	Cawse (1977)
Switzerland (7 stations)	Bulk precipitation	—	0.6	Imboden et al. (1975)
Gottingen, Germany	Bulk precipitation	0.58 (0.1–1.0)	—	Ruppert (1975)
Ontario (40 stations)	Bulk precipitation	0.8 (0.01–50)	0.51	Kramer (1976)
Chalk River, Ont.	Rainfall	1.4	—	Merritt (1976)
Upper Great Lakes Basin	Bulk precipitation	—	0.62	ACRES (1975)
Lower Great Lakes Basin	Bulk precipitation	1.0	0.55	Kuntz, pers. commun. (1978)
Ogden Dunes, Ind.	Bulk precipitation	0.63	—	Yost (1978)
South Bend, Ind.	Bulk precipitation	1.47	—	Yost (1978)
Gary, Ind.	Bulk precipitation	0.77	—	Yost (1978)
LaPorte, Ind.	Bulk precipitation	1.34	—	Yost (1978)
Pinney Purdue, Ind.	Bulk precipitation	0.83	—	Yost (1978)
Lafayette, Ind.	Bulk precipitation	0.94	—	Yost (1978)
Hammond City Hall, Ind.	Bulk precipitation	7.2	—	Yost (1978)

Table 16. *Continued*

Location	Type of Sample	Cadmium Concentration[a]	Cadmium Fallout Rate (mg/m²·year)	Reference
Mt. Moosilauke, N.H.	Bulk precipitation	0.6	0.88	Schlesinger et al. (1974)
Delaware watersheds	Bulk precipitation	—	0.69	Biggs et al. (1973)
Walker Branch Watershed, Tenn.	Bulk precipitation	—	1.3	Lindberg et al. (1975)
Chedron, Neb.	Rainfall + nowfall	0.31	0.12	Struempler (1976)
Lloyd, N.Y.	Dustfall	—	0.33	Feeley et al. (1976)
Southwest Manhattan, N.Y.	Dustfall	—	2.9	Kleinman et al. (1977)
Midtown Manhattan, N.Y.	Dustfall	—	3.6	Kleinman et al. (1977)
Bronx, N.Y.	Dustfall	—	1.9	Kleinman et al. (1977)
Queens, N.Y.	Dustfall	—	1.9	Pinkerton et al. (1975)
New York Harbor	Dustfall	—	7.2	Kleinman et al. (1977)
Midwest U.S. cities				Hunt et al. (1972)
Residential area	Dustfall	—	>0.5	
Commercial area	Dustfall	—	>0.8	
Industrial area	Dustfall	—	>0.9	
Cincinnati, near roadways	Dustfall	13 ppm	0.76	Creason et al. (1973)
Southeast Missouri				Dorn et al. (1973)
Near smelter	Dustfall	—	52	
75 km from smelter	Dustfall	—	1.1	
East Helena, Mont.				Rupp et al. (1978)
Within 2-km radius of smelter	Dustfall	200–1000 ppm	12–48	
12 km from smelter	Dustfall	200–1000 ppm	0.5–2.3	
Richmond, Calif.	Dustfall	—	0.29	Feeley et al. (1975)
La Jolla, Calif.	Dustfall	—	0.15	Hodge et al. (1978)
Ensenada, Baja, Calif.	Dustfall	—	0.11	Hodge et al. (1978)
Southern coast of California	Dustfall	—	0.18	Davidson (1977)
Northeast United States	Moss specimen	0.4–2.7 ppm	0.19	Groets (1976)
Southern Sweden	Moss specimen	—	0.4–0.7	Ruhling and Tyler (1971)

[a] Concentrations of cadmium in precipitation samples are given in micrograms per liter; the data for dust samples, in parts per million. The reported ranges in concentration are shown in parentheses.

Sophisticated models have now been developed to describe the washout of cadmium and other aerosols from the atmosphere by snowfall (e.g., Sood and Jackson, 1970; Knutson et al., 1976; Koenig, 1977) and rainfall (Hidy, 1973; Slinn, 1977; Dana and Hales, 1977; Kerrigan and Rosinki, 1977; Norment, 1977; Lange and Knox, 1977; Molenkamp, 1977). Such models will not be considered in this review. It is generally true that the data currently available are inadequate to validate even the very simple models (e.g., see Gatz, 1976; Davidson, this volume).

6.2. Cadmium in Bulk Precipitation Samples

Bulk precipitation samples include both wet and dry fallouts; representative cadmium concentrations in these samples are summarized in Table 16. The highest concentrations are usually associated with local sources of cadmium emission. For example, the cadmium concentration in bulk precipitation samples near the smelters at Sudbury is about 3 μg/l, compared to an average of 0.8 μg/l for the entire province of Ontario (Kramer, 1976). The halo effect decreases with distance away from Sudbury (Figure 2).

The ratio of wet to dry deposition of cadmium has been estimated at several locations. Cawse (1976) found the percentages of dry to total (bulk) deposition in rural areas of the United Kingdom to vary between 40 and 90. In Walker Branch Watershed, Tennessee, dry deposition was estimated to contribute about 30% of the total cadmium influx (Lindberg et al., 1975). The ratio of dry to total fallout for cadmium has been shown to be 0.33 in the New York Bight area (Duce et al., 1976). Kramer (1976) showed that dry deposition accounted for 68% of total cadmium deposition at Lively, Ontario, whereas the same process was estimated to be responsible for 53% of the atmospheric cadmium input into southern Lake Michigan (Gatz, 1975). Indeed, Slinn's (1977) models of the washout phenomena predict that the ratio of dry to wet deposition should be close to unity for most aerosols.

A wealth of data has recently become available on the atmospheric input of cadmium into the various ecosystems (Table 16). In addition, considerable insight into the rate of atmospheric fallout of cadmium has come from studies of epiphytes, peats, soils, and lacustrine sediments. The following list represents only some of the highlights of these studies.

1. The rate of ombrogenic deposition of cadmium has increased sharply in recent (since the industrial revolution) times in many parts of the world. The changes

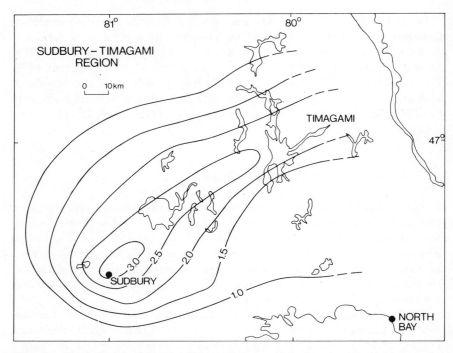

Figure 2.   Distribution of cadmium (μg/l) in atmospheric precipitation around the smelting complex near Sudbury, Ontario.

in rates of cadmium deposition recorded in ice cores have already been noted. Studies of the cadmium contents of dated lacustrine and coastal marine sediments also show that the rates of cadmium loading into these environments have generally increased by 50 to 500% in modern times (e.g., see Bruland et al., 1974; Erlenkeuser et al. 1974; Kemp and Thomas, 1976; Nriagu et al., 1978). Additional documentation of a secular increase in cadmium deposition has come from studies of the cadmium contents of ancient and present-day moss speciments (Gelting and Ponten, 1971; Rasmussen, 1977).

2. Regional differences in atmospheric deposition of cadmium stem primarily from differences in population density, level of industrial activity, and prevailing wind direction. Ruhling and Tyler (1973) observed a tenfold difference in atmospheric fallout of cadmium in southern and northern parts of Scandinavia, which suggests that most of the element came from distant sources in western Europe. In a similar regional study, Grodzinska (1978) found that

Table 17. Lead and Cadmium Levels in Settled Floor Dusts in the Community of Champaign-Urbana, Illinois[a]

Site	Number of Samples	Lead (ppm)	Lead ($\mu g/m^2$)	Cadmium (ppm)	Cadmium ($\mu g/m^2$)
Residential sites					
A. Rugs	7	830	1490	24	43
Nonrugs	13	1780	1020	29	17
B. Rugs	5	420	460	17	20
Nonrugs	10	480	140	25	6
F. Rugs	6	240	350	66	71
Nonrugs	1	290	60	105	22
G. Rugs	11	220	110	14	7
Nonrugs	3	320	70	71	8
Average, 12 sites		600	680	18	25
Nonresidential sites					
Offices, hallways	36	3,380	3,010	9	8
Offices, nonrugs	11	1,450	640	13	7
Classrooms, nonrugs	8	930	160	1	0
Offices, rugs	6	2,320	11,780	1033	2443
Offices, nonrugs	1	2,960	560	1060	200
Public school, rugs/mats	10	730	3,990	29	122
Public school, nonrugs	17	650	200	7	4
Public school, rugs/mats	12	410	2,360	30	115
Public school, nonrugs	26	430	210	19	5
Hospitals, entry areas, rugs	11	620	7,940	20	390
Hospitals, corridors, rooms	11	360	180	22	4
Supermarkets	34	490	100	9	2
Chemical laboratories	35	11,400	3,390	185	57
All floors	254	1,400	2,040	44	110

[a] Solomon and Hartford (1976).

the rates of cadmium accumulation in national parks located in the industrialized southern portion of Poland were higher than the rates in parks in the less developed northern parts of the country. On the other hand, the relatively high rates of cadmium deposition in the northeast portion of the United States have been attributed to the fact that this region lies downwind of the industrial areas of both the Great Lakes Basin and the Middle Atlantic states (Schlesinger et al., 1974; Groets, 1976).

3. Local variations in cadmium deposition depend primarily on land-use spatial arrangements. There is now overwhelming evidence that the accumulation of cadmium is much higher in cities than in the adjacent rural areas (e.g., see Goodman and Roberts, 1971; Groets, 1976; Garty et al., 1977). Also,

enrichment of cadmium in soils and plants has been observed near housing units, highways, railroads, and power plants. It would be jejune to describe all these local studies. The results of an interesting survey (Solomon and Hartford, 1976) of cadmium and lead in dusts and soils in a small urban community are summarized in Table 17. Apparently the local buildup of cadmium and other heavy metals may reach biologically unhealthy levels.

4. In general, the deposition of cadmium decreases exponentially with distance from a single point source such as a smelter (Burkitt et al., 1972; Goodman and Roberts, 1971; John et al., 1975; Ellison et al., 1976) or a regionalized point source such as a city (Groets 1976). Highly significant correlations also exist between the rate of cadmium deposition and the wind direction and speed (Little and Martin, 1972; De Koning, 1974; Ellison et al., 1976). Contamination of soils, vegetation, and organisms with cadmium and other metals is quite prevalent near heavy metal smelters (Little and Martin, 1972; De Koning, 1974; Beavington, 1975; John et al., 1976; Dugdale and Hummel, 1978).

REFERENCES

ACRES (1975). *Atmospheric Loading of the Upper Great Lakes,* Vol. 3, Report submitted to Canada Centre for Inland Waters, by ACRES Consulting Services, Ltd., Toronto, Ont., Appendix 7.

ACRES (1977). *Atmospheric Loading of the Lower Great Lakes and the Great Lakes Drainage Basin.* Prepared by ACRES Consulting Services for Environment Canada, Canada Center for Inland Waters, Burlington, Ont., 75 pp.

Adams, F., Dams, R., Guzman, L., and Winchester, J. W. (1977). "Background Aerosol Composition on Chacaltaya Mountain, Bolivia," *Atmos. Environ.* **11**, 629–634.

Air Pollution Control Directorate (1976). *National Inventory of Sources and Emissions of Cadmium.* APCD, Environmental Protection Service, Int. Rep. APCE-76-2, Environment Canada, Ottawa, Ont.

Anderson, D. (1973). *Emission Factors for Trace Substances.* National Technical Information Service, Rep. PB-230-894, Springfield, Va.

Andren, A. W. and Lindberg, S. E. (1977). "Atmospheric Input and Origin of Selected Elements in Walker Branch Watershed, Oak Ridge, Tennessee," *Water, Air, Soil Pollut.* **8**, 199–215.

Andren, A. W., Elzerman, A. W., and Armstrong, D. E. (1976). "Chemical and Physical Aspects of Surface Organic Microlayers in Freshwater Lakes," *J. Great Lakes Res.* **2** (Suppl. 1), 101–110.

Barton, S. C., Shenfield, L., and Thomas, D. A. (1975). "A Review of Heavy Metal Measurements in Ontario." In *Abstracts, International Conference on Heavy Metals in the Environment, Toronto, Ont.,* pp. C91–C93.

Beavington, F. (1975). "Heavy Metal Contamination of Vegetables and Soil in Domestic Gardens around a Smelting Complex," *Environ. Pollut.* **9,** 211–217.

Belot, Y., Diop, B., and Marini, T. (1971). *Composition Minerale de la Matiere Particulaire en Suspension dans l'Air dans des Zones Urbaines.* Cited in Rahn (1976).

Bertine, K. K. and Goldberg, E. D. (1971). "Fossil Fuel Combustion and the Major Sedimentary Cycle," *Science* **173,** 233–235.

Bewers, J. M., Sundby, S., and Yeats, P. A. (1976). "The Distribution of Trace Metals in the Western North Atlantic off Nova Scotia," *Geochim. Cosmochim. Acta* **40,** 687–696.

Biggs, R. B., Miller, J. C., Otley, M. J., and Shields, C. L. (1973). *Trace Metals in Several Delaware Watersheds.* Final Report, University of Delaware Water Resources Center.

Block, C. and Dams, R. (1976). "Study of Fly Ash Emission during Combustion of Coal," *Environ. Sci. Technol.* **10,** 1011–1017.

Bogen, J. (1973). "Trace Elements in Atmospheric Aerosol in the Heidelberg Area Measured by Instrumental Neutron Activation Analysis," *Atmos. Environ.* **7,** 1117–1125.

Bolter, E., Butz, T., and Arseneau, J. E. (1975). "Mobilization of Heavy Metals by Organic Acids in the Soils of a Lead Mining and Smelting District," *Trace Subst. Environ. Health* **9,** 107–112.

Bruland, K. W., Bertine, K., Koide, M., and Goldberg, E. E. (1974). "History of Metal Pollution in Southern California Coastal Zone," *Environ. Sci. Technol.* **8,** 425–432.

Buat-Menard, P. and Arnold, M. (1978). "The Heavy Metal Chemistry of Atmospheric Particulate Matter Emitted by Mount Etna Volcano," *Geophys. Res. Lett.* **5,** 245–248.

Burkitt, A., Lester, P., and Nickless, G. (1972). "Distribution of Heavy Metals in the Vicinity of an Industrial Complex," *Nature* **238,** 327–328.

Canadian Minerals Yearbooks (1974–76). Publishing Center, Department of Supply and Services, Ottawa, Ont.

Cawse, P. A. (1974–1977). *A Survey of Atmospheric Trace Elements in The U.K.:* Results for 1972–73, 1974, 1975, 1976. AERE Publ. R-7669, R-8038, R-8393, R-8869. H.M. Stationery Office, London.

Chernyak, Yu. B. and Nussinov, M. D. (1976). "Volatilization from Solid Particles of the Regolith," *Nature* **264,** 241.

Chizhikov, D. M. (1966). *Cadmium.* Pergamon Press, London.

Council on Environmental Quality (1975). *Environmental Quality—1975,* Sixth Annual Report. U.S. Government Printing Office, Washington, D.C.

Council on Environmental Quality (1976). *Environmental Quality—1976,* Seventh Annual Report. U.S. Government Printing Office, Washington, D.C.

Creason, J. P. et al. (1971). "Roadside Gradients in Atmospheric Concentrations of Cadmium, Lead and Zinc," *Trace Subst. Environ. Health* **5,** 129–141.

Curtin, G. C., King, H. D., and Mosier, E. L. (1974). "Movement of Elements into the Atmosphere from Coniferous Trees in Subalpine Forests of Colorado and Idaho," *J. Geochem. Explor.* **3**, 245–263.

Dams, R. (1974). *Study of National Air Pollution by Combustion, Part 1: Inorganic Composition of Airborne Particulate Matter.* Cited in Salmon et al. (1978).

Dams, R. and De Jonge, J. (1976). "Chemical Composition of Swiss Aerosols from the Jungfraujoch," *Atmos. Environ.* **10**, 1079–1084.

Dana, M. T. and Hales, J. M. (1977). "Washout Coefficients for Polydisperse Aerosols." In *Precipitation Scavenging (1974)*. National Technical Information Service, CONF-741003, Springfield, Va., pp. 247–257.

Davidson, C. I. (1977). "The Deposition of Trace Metal-Containing Particles in the Los Angeles Area," *Powder Technol.* **18**, 117–126.

Davis, W. E. and Associates (1970). *National Inventory of Sources and Emissions of Cadmium.* National Technical Information Service, Rep. PB 192250, Springfield, Va.

Davison, R. L., Natusch, D. F. S., Wallace, J. R., and Evans, C. A. (1974). "Trace Elements in Fly Ash," *Environ. Sci. Technol.* **8**, 1107–1112.

De Koning, H. W. (1974). "Lead and Cadmium Contamination in the Area Immediately Surrounding a Lead Smelter," *Water, Air, Soil Pollut.* **3**, 63–70.

Demuynck, M., Rahn, K. A., Janssens, M., and Dams, R. (1976). "Chemical Analysis of Airborne Particulate Matter during a Period of Unusually High Pollution," *Atmos. Environ.* **10**, 21–26.

Dorn, C. R., Pierce, J. O., Chase, G. R., and Phillips, P. E. (1975). "Environmental Contamination by Lead, Cadmium, Zinc and Copper in a New Lead-Producing Area," *Environ. Res.* **9**, 159–172.

Dorn, C. R., Pierce, J. O., Phillips, P. E. and Chase, G. R. (1976). "Airborne Pb, Cd, Zn and Cu Concentration by Particle Size near a Pb Smelter," *Atmos. Environ.* **10**, 443–446.

Dreesen, D. R. et al. (1977). "Comparison of Levels of Trace Elements Extracted from Fly Ash and Levels Found in Effluent Waters from a Coal-Fired Power Plant," *Environ. Sci. Technol.* **10**, 1017–1019.

Duce, R. A., Hoffman, G. L., and Zoller, W. H. (1975). "Atmospheric Trace Metals at Remote Northern and Southern Hemisphere Sites: Pollution or Natural?" *Science* **187**, 59–61.

Duce, R. A. et al. (1976). "Trace Metals in the Marine Atmosphere: Sources and Fluxes." In H. L. Windon and R. A. Duce, Eds., *Marine Pollutant Transfer*. D. C. Heath, Lexington, Mass., pp. 77–119.

Dugdale, P. J. and Hummel, B. L. (1978). "Cadmium in the Lead Smelter at Belledune: Its Association with Heavy Metals in the Ecosystem." In *Cadmium 77: Proceedings of the 1st International Cadmium Conference*. Metal Bulletin Ltd., London, pp. 53–75.

Duncan, L. G. et al. (1973). *Selected Characteristics of Hazardous Pollutant Emissions.*

Mitre Corporation, Final Report for Environmental Protection Agency Contract No. 68-01-0438.

Egorov, V. V., Zhigalovskaya, T. N., and Malakhov, S. G. (1970). "Microelement Content of Surface Air above the Continent and the Ocean," *J. Geophys. Res.* **75,** 3650–3656.

Ellison, G., Newham, J., Pinchin, M. J., and Thompson, I. (1976). "Heavy Metal Content of Moss in the Region of Consett (North East England)," *Environ. Pollut.* **11,** 167–174.

Erlenkeuser, H., Suess, E., and Willkomm, H. (1974). "Industrialization Affects Heavy Metal and Carbon Isotope Concentrations in Recent Baltic Sea Sediments," *Geochim. Cosmochim. Acta* **38,** 823–842.

Esmen, N. A. and Corn, M. (1971). "Residence Time of Particles in Urban Air," *Atmos. Environ.* **5,** 571–578.

Evendijk (1974). Cited in Rahn (1976).

Faoro, R. B. and McMullen, T. B. (1977). *National Trends in Trace Metals in Ambient Air, 1965–1974.* U.S. Environmental Protection Agency, Office of Air and Waste Management, Rep. EPA-450/1-77-003, Research Triangle Park, North Carolina, 32 pp.

Feeley, H. W., Volchok, H. L., and Toonkel, L. (1976). "Trace Metals in Atmospheric Deposition." Health and Safety Laboratory, HASL-308, *Environ. Quart.,* Oct. 1, 155–1120.

Fleischer, M. et al. (1974). "Environmental Impact of Cadmium: Review by the Panel on Hazardous Trace Substances," *Environ. Health Perspect.* **7,** 253–323.

Flocchini, R. G. et al. (1976). "Monitoring California's Aerosols by Size and Elemental Composition," *Environ. Sci. Technol.* **10,** 76–82.

Friberg, L., Piscator, M., and Nordberg, G. (1971). *Cadmium in the Environment,* 2nd ed., Chemical Rubber Co., Cleveland, Ohio, pp. 19–22.

Furr, A. K. et al. (1977). "National Survey of Elements and Radioactivity in Fly Ashes," *Environ. Sci. Technol.* **11,** 1194–1201.

Garty, J., Galun, M., Fuchs, C., and Zisapel, N. (1977). "Heavy Metals in the Lichen *Caloplaca aurantia* from Urban, Suburban and Rural Regions in Israel," *Water, Air, Soil Pollut.* **8,** 171–188.

Gatz, D. F. (1975). "Pollutant Aerosol Deposition into Southern Lake Michigan," *Water, Air, Soil Pollut.* **5,** 239–251.

Gatz, D. F. (1976). "Wet Deposition Estimation using Scavenging Ratios," *J. Great Lakes Res.* **2** (Suppl. 1), 21–32.

Gelting, G. and Ponten, A. (1971). *Heavy Metal Pollution in the Uppsala Area.* Internal Report, Department of Plant Biology, Uppsala.

Goeller, H. E., Hise, E. C., and Flora, H. B. (1973). "Societal Flow of Zinc and Cadmium." In W. Fulkerson and H. E. Goeller, Eds., *Cadmium, the Dissipated Element,* Oak Ridge National Laboratory, Publ. ORNL NSF-EP-21, Tennessee.

Goldberg, A. J. (1973). *A Survey of Emissions and Controls of Hazardous and Other Pollutants.* U.S. Environmental Protection Agency, Washington, D.C.

Goldberg, E. D. (1976). "Rock Volatility and Aerosol Composition," *Nature* **260,** 128–129.

Goodman, G. T. and Roberts, T. M. (1971). "Plants and Soils as Indicators of Metals in the Air," *Nature* **231,** 287–292.

Gordon, G. E., Zoller, W. H. and Gladney, E. S. (1973). "Abnormally Enriched Trace Elements in the Atmosphere," *Trace Subst. Environ. Health* **7,** 167–173.

Greenberg, R. R., Zoller, W. H., and Gordon, G. E. (1978). "Composition and Size Distributions of Particles Released in Refuse Incineration," *Environ. Sci. Technol.* **12,** 566–573.

Grodzinska, H. (1978). "Mosses as Bioindicators of Heavy Metal Pollution in Polish National Parks," *Water, Air, Soil Pollut.* **9,** 83–97.

Groets, S. S. (1976). "Regional and Local Variations in Heavy Metal Concentrations of Bryophytes in the Northeast United States," *OIKOS* **27,** 445–456.

Hallowell, J. B., Cherry, R. H., and Smithson, G. R., Jr. (1973). "Trace Metals in Effluents from Metallurgical Operations." In *Cycling and Control of Metals.* U.S. Environmental Protection Agency, National Environmental Research Center, Cincinnati, Ohio, pp. 75–81.

Hamilton, E. I. (1974). "The Chemical Elements and Human Morbidity—Water, Air and Places—a Study of Natural Variability," *Sci. Total Environ.* **3,** 3–85.

Harrison, P. R. and Winchester, J. W. (1971). "Area-wide Distribution of Lead, Copper and Cadmium in Air Particulates from Chicago and Northwest Indiana," *Atmos. Environ.* **5,** 863–880.

Harrison, P. R., Matson, W. R., and Winchester, J. W. (1971). "Time Variations of Lead, Copper and Cadmium Concentrations in Aerosols in Ann Arbor, Michigan," *Atmos. Environ.* **5,** 613–619.

Harrison, R. M., Perry, R., and Wellings, R. A. (1975). "Lead and Cadmium in Precipitation: Their Contribution to Pollution," *J. Air Pollut. Control Assoc.* **25,** 627–630.

Heindryckx, R. and Dams, R. (1974). "Continental, Marine and Anthropogenic Contributions to the Inorganic Composition of the Aerosol of an Industrial Zone," *J. Radioanal. Chem.* **19,** 339–349.

Heindryckx, R. et al. (1975). "Mercury and Cadmium in Belgian Aerosols." In *Problems of the Contamination of Man and His Environment by Mercury and Cadmium.* Commission of European Communities, Luxembourg, pp. 135–148.

Henry, W. M. and Blosser, E. R. (1971). *Identification and Estimation of Ions, Molecules and Compounds in Particulate Matter Collected from Ambient Air.* Battelle, Columbus, Laboratories, Tech. Rep. CPA-70-159, Columbus, Ohio.

Hidy, G. M. (1973). "Removal Processes of Gaseous and Particulate Pollutants." In S. I. Rasool, Ed., *Chemistry of the Lower Atmosphere.* Plenum, New York, pp. 121–176.

Hillenbrand, L. J., Engdahl, R. B., and Barrett, R. E. (1973). *Chemical Composition of Particulate Air Pollutants from Fossil-Fuel Combustion Sources.* National Technical Information Service, PB-219009, Springfield, Va.

Hodge, V., Johnson, S. R. and Goldberg, E. D. (1978). "Influence of Atmospherically Transported Aerosols on Surface Ocean Water Composition," *Geochem. J.* **12,** 7–20.

Hunt, W. F., Pinkerton, C., McNulty, O., and Creason, J. (1972). "A Study of Trace Element Pollution of Air in 77 Midwestern Cities," *Trace Subst. Environ. Health* **4,** 57–63.

Imboden, D. M., Hegi, H. R., and Zobrist, J. (1975). "Atmospheric Loading of Metals in Switzerland." *EAWAG News,* Vol. 4, Swiss Federal Institute of Technology, Dubendorf, Switzerland, pp. 5–7.

Jacko, R. B. and Neuendorf, D. W. (1977). "Trace Metal Particulate Emission Test Results from a Number of Industrial and Municipal Point Sources," *J. Air Pollut. Control Assoc.* **27,** 989–994.

Jaworowski, Z. et al. (1975). "Stable and Radioactive Pollutants in a Scandinavian Glacier," *Environ. Pollut.* **9,** 305–315.

John, M. K., Van Laerhoven, C. J., and Cross, C. H. (1976). "Cadmium, Lead, and Zinc Accumulation in Soils near a Smelter Complex," *Environ. Lett.* **10,** 25–35.

John, W., Kaifer, R., Rahn, K., and Wesolowski, J. J. (1973). "Trace Element Concentrations in Aerosols from San Francisco Bay Area," *Atmos. Environ.* **7,** 107–118.

Junge, C. E. (1963). *Air Chemistry and Radioactivity.* Academic Press, New York, 382 pp.

Just, J. and Kelus, J. (1971). "Cadmium in the Atmosphere of Ten Selected Cities in Poland," *Rocz. Panstw. Zakl. Hig.* **22,** 249; *Chem. Abstr.* **75,** 91050.

Kemp, A. L. W. and Thomas, R. L. (1976). "Impact of Man's Activities on the Chemical Composition in the Sediments of Lakes Ontario: Erie and Huron," *Water, Air, Soil Pollut.* **5,** 469–490.

Kerrigan, T. C. and Rosinski, J. (1977). "Scavenging of Aerosol Particles in Severe Convective Storms." In *Precipitation Scavenging (1974).* National Technical Information Service, CONF. 741003, Springfield, Va., pp. 466–493.

King, R. B. et al. (1976). "Elemental Composition of Airborne Particulates and Source Identification," *J. Air Pollut. Control Assoc.* **26,** 1073–1084.

Kleinman, M. T., Kneip, T. J., and Eisenbud, M. (1973). "Meteorological Influences on Airborne Trace Metals and Suspended Particulates," *Trace Subst. Environ. Health* **7,** 161–166.

Kleinman, M. T., Kneip, T. J., and Eisenbud, M. (1976). "Seasonal Patterns of Airborne Particulate Concentrations in New York City," *Atmos. Environ.* **10:** 9–11.

Kleinman, M. T. et al. (1977). "Fallout of Toxic Metals in New York City." In *Biological Implications of Metals in the Environment.* National Technical Information Service, CONF-750929/RAS, Springfield, Va., pp. 144–152.

Kneip, T. J., Kleinman, M. T., and Eisenbud, M. (1973). "Relative Contribution of Emission Sources to the Total Airborne Particulates in New York City." In *Proceedings of the 3rd International Clean Air Congress, Dusseldorf, Germany.*

Knutson, E. O., Sood, S. K., and Stockham, J. D. (1976). "Aerosol Collection by Snow and Ice Crystals," *Atmos. Environ.* **10**, 395–402.

Koenig, L. R. (1977). "Cloud Glaciation Characteristics Important to Parameterizing Scavenging." In *Precipitation Scavenging (1974)*. National Technical Information Service, CONF-741003, Springfield, Va., pp. 672–694.

Kowalczyk, G. S.: Choquette, C. E., and Gordon, G. E. (1978). "Chemical Element Balances and Identification of Air Pollution Sources in Washington, D.C.," *Atmos. Environ.* **12**, 1143–1153.

Kramer, J. R. (1976). "Fate of Atmospheric Sulfur Dioxide and Related Substances as Indicated by Chemistry of Precipitation." Unpublished Report, Department of Geology, McMaster University, Hemilton, Ont.

Kretzschmar, J. G., Delespaul, I., Rijck, Th.D., and Verduyn, G. (1977). "The Belgian Network for the Determination of Heavy Metals," *Atmos. Environ.* **11**, 263–271.

Lagerwerff, J. V. and Brower, D. L. (1975). "Source Determination of Heavy Metal Contaminants in the Soil of a Mine and Smelter Area," *Trace Subst. Environ. Health* **9**, 207–215.

Lagerwerff, J. V. and Specht, A. W. (1970). "Contamination of Roadside Soil and Vegetation with Cadmium, Nickel, Lead and Zinc," *Environ. Sci. Technol.* **4:** 583.

Lange, R. and Knox, J. B. (1977). "Adaptation of a Three-Dimensional Atmospheric Transport-Diffusion Model to Rainout Assessments." In *Precipitation Scavenging (1974)*. National Technical Information Service, CONF-741003, Springfield, Va., pp. 732–758.

Lee, R. E., Jr., and von Lehmden, D. J. (1973). "Trace Metal Pollution in the Environment," *J. Air Pollut. Control Assoc.* **23**, 853–857.

Lee, R. E., Jr., Goranson, S. S., Enrione, R. E., and Morgan, G. B. (1972). "National Air Surveillance Cascade Impactor Network. II: Size Distribution Measurements of Trace Metal Components," *Environ. Sci. Technol.* **6**, 1025–1030.

Lee, R. E., Jr., Crist, H. L., Riley, A. E., and MacLeod, K. E. (1975). "Concentration and Size of Trace Metal Emissions from a Power Plant, a Steel Plant and a Cotton Gin," *Environ. Sci. Technol.* **9**, 643–647.

Lindberg, S. E., Andren, A. W., Raridon, R. J., and Fulkerson, W. (1975). "Mass Balance of Trace Elements in Walker Branch Watershed: Relation to Coal-Fired Steam Plants," *Environ. Health Perspect.* **12**, 9–18.

Linton, R. W. et al. (1976). "Surface Predominance of Trace Elements in Airborne Particles," *Science* **191**, 852–854.

Lioy, P. J., Wolff, G. T., and Kneip, T. J. (1978). "Toxic Airborne Elements in the New York Metropolitan Area," *J. Air Pollut. Control Assoc.* **28**, 510–512.

Little, P. and Martin, M. H. (1972). "A Survey of Zinc, Lead and Cadmium in Soil and Natural Vegetation around a Smelting Complex:" *Environ. Pollut.* **3**, 241–254.

Maenhaut, W. and Zoller, W. H. (1977). "Determination of the Chemical Composition of the South Pole Aerosol by Instrumental Neutron Activation Analysis," *J. Radioanal. Chem.* **37**, 637–650.

Martell, E. A. and Poet, S. E. (1972). "Residence Times of Natural and Pollutant Aerosols in the Troposphere." In W. E. Brittin, R. West, and R. Williams, Eds., *Air and Water Pollution.* Colorado Associated University Press, Boulder, Colo.: pp. 459–476.

Merritt, W. F. (1976). "Trace Element Content of Precipitation in a Remote Area." In *Measurement, Detection and Control of Environmental Pollution.* Proceedings of Symposium, International Atomic Energy Agency, Vienna, pp. 75–87.

Meszaros, A. (1977). "On the Size Distribution of Atmospheric Aerosol Particles of Different Composition," *Atmos. Environ.* **11**, 1075–1081.

Minderhoud, A. and Boogerd, J. P. (1975). "Determination of Cadmium in Air Particulate Matter by X-Ray Fluorescence." In *Problems of the Contamination of Man and His Environment by Mercury and Cadmium.* Commission of European Communities, Luxembourg, pp. 203–209.

Molenkamp, C. R. (1977). "Numerical Modeling of Precipitation Scavenging by Convective Clouds." In *Precipitation Scavenging (1974).* National Technical Information Service, CONF-741003, Springfield, Va., pp. 769–793.

Muller, J. and Beilke: S. (1977). "Wet Removal of Heavy Metals from the Atmosphere." In *Proceedings of the International Conference on Heavy Metals in the Environment,* Institute for Environmental Studies, University of Toronto, Toronto, Ont., pp. 987–999.

Nagata, R. et al. (1972). *Air Pollution by Heavy Metals Contained in Particulate Matter in Tokyo.* Annual Report, Tokyo Metropolitan Research Institute for Environmental Protection, p. 5.

National Academy of Sciences (1975). *Mineral Resources and the Environment,* and the Supplementary Report. Washington, D.C.

National Air Surveillance Network (1966). *Air Quality Data from the National Air Sampling Networks and Contributing State and Local Networks, 1964–65.* U.S. Department of Health, Education and Welfare, Public Health Service, Cincinnati, Ohio.

National Air Surveillance Network (1976). *Air Quality Data for Metals 1970 through 1974 from the National Air Surveillance Networks.* National Technical Information Service, PB-260905, Springfield, Va., 154 pp.

Natusch, D. F. S. (1978). "Potentially Carcinogenic Species Emitted to the Atmospheres by Fossil-Fueled Power Plants," *Environ. Health Perspect.* **22**, 79–90.

Natusch. D. F. S. and Wallace, J. R. (1974). "Urban Aerosol Toxicity: The Influence of Particle Size," *Science* **186**, 695–699.

Neustadter, H. E., Fordyce, J. S., and King, R. B. (1976). "Elemental Composition of Airborne Particulates and Source Identification: Data Analysis Techniques," *J. Air Pollut. Control Assoc.* **26**, 1079–1084.

Norment, H. G. (1977). "A Precipitation Scavenging Model for Nuclear Fallout Research." In *Precipitation Scavenging* (*1974*). National Technical Information Service, CONF-741003, Springfield, Va., pp. 695–731.

Nriagu, J. O. (1978). "Lead in the Atmosphere." In J. O. Nriagu, Ed., *The Biogeochemistry of Lead in the Environment*. Elsevier, Amsterdam, pp. 137–183.

Nriagu, J. O. (1979). "Copper in the Atmosphere and Precipitation." In J. O. Nriagu, Ed., *Copper in the Environment*. Wiley, New York, Part I, pp. 43–75.

Nriagu, J. O., Kemp, A. L. W., Wong, H. K. T., and Harper, N. (1979). "Sedimentary Record of Heavy Metal Pollution in Lake Erie," *Geochim. Cosmochim. Acta.* **43**, 247–258.

Pattenden, N. J. (1974). *Atmospheric Concentrations and Deposition Rates of Some Trace Elements Measured in the Swansea/Neath/Port Talbot Area*. AERE Publ. R-7729. H. M. Stationary Office: London.

Peden, M. E. (1977). "Flameless Atomic Absorption Determinations of Cadmium, Lead and Manganese in Particle Size Fractionated Aerosols." In *Methods and Standards for Environmental Measurement,* Natl. Bur. Stand. Spec Publ. 464, pp. 367–377.

Pinkerton, C., Lagerwerff, J. V., Creason, J. P., Hinners, T. A., and Hammer, D. I. (1975). "Trace Metals in Dustfall and Soil in New York City." In *Abstracts, International Conference on Heavy Metals in the Environment, Toronto, Ont.,* pp. C205–C207.

Rahn, K. A. (1976). *The Chemical Composition of the Atmospheric Aerosol*. Technical Report, Graduate School of Oceanography, University of Rhode Island, Kingston, 265 pp.

Ranticelli, L. A., Perkins: R. W., Tanner, T. M., and Thomas, C. W. (1971). "Stable Elements of the Atmosphere as Tracers of Precipitation Scavenging." In *Precipitation Scavenging* (*1970*). National Technical Information Service, CONF-700601, Springfield, Va., pp. 99–108.

Rasmussen, L. (1977). "Epiphytic Bryophytes as Indicators of the Changes in the Background Levels of Airborne Metals from 1951–75," *Environ. Pollut.* **14**, 37–44.

Ruhling, A. and Tyler, G. (1971). "Regional Differences in the Deposition of Heavy Metals over Scandinavia," *J. Appl. Ecol.* **8**, 497–507.

Ruhling, A. and Tyler, G. (1973). "Heavy Metal Deposition in Scandinavia," *Water, Air, Soil Pollut.* **2**, 445–455.

Rupp, E. M. et al. (1978). "Composite Hazard Index for Assessing Limiting Exposures to Environmental Pollutants: Application through a Case Study," *Environ. Sci. Technol.* **12**, 802–807.

Ruppert, H. (1975). "Geochemical Investigations on Atmospheric Precipitation in a Medium-Sized City (Gottingen, F.R.G.)," *Water, Air, Soil Pollut.* **4**, 447–460.

Salmon, L., Atkins, D. H. F., Fisher, E. M. R., Healy, C., and Law, D. V. (1978). "Retrospective Trend Analysis of the Content of U.K. Air Particulate Material 1957–1974," *Sci. Total Environ.* **9**, 161–200.

Schlesinger, W. H., Reiners, W. A., and Knopman, D. S. (1974). "Heavy Metal Concentrations and Deposition in Bulk Precipitation in Montane Ecosystems of New Hampshire, U.S.A.," *Environ. Pollut.* **6,** 39–47.

Schramel, P., Samsahl, K., and Pavlu, J. (1974). "Determination of 12 Selected Microelements in Air Particles by Neutron Activation Analysis," *J. Radioanal. Chem.* **19,** 329–337.

Semkin, R. G. and Kramer, J. R. (1976). "Sediment Geochemistry of Sudbury-Area Lakes," *Can. Mineral.* **14,** 73–90.

Slinn, W. G. N. (1977). "Some Approximations for the Wet and Dry Removal of Particles and Gases from the Atmosphere," *Water, Air, Soil Pollut.* **7,** 513–543.

Solomon, R. L. and Hartford, J. W. (1976). "Lead and Cadmium in Dusts and Soils in a Small Urban Community," *Environ. Sci. Technol.* **10,** 773–777.

Sood, S. K. and Jackson, M. R. (1970). "Scavenging by Snow and Ice Crystals." In *Precipitation Scavenging (1970).* National Technical Information Service, CONF-700601, Springfield, Va., pp. 121–136.

Struempler, A. W. (1975). "The Trace Element Composition of Atmospheric Particulates during 1973 and the Summer of 1974 at Chedron, Nebraska," *Environ. Sci. Technol.* **9,** 1164–1168.

Struempler, A. W. (1976). "Trace Metals in Rain and Snow during 1973 at Chedron, Nebraska," *Atmosph. Environ.* **10,** 33–37.

Stuart, B. O. (1976). "Deposition and Clearance of Inhaled Particles," *Environ. Health Perspect.* **16,** 41–53.

Sugimae, A. and Hasegawa, T. (1973). "Emission Spectrographic Determination of Trace Metals in Airborne Particulates Collected on a Membrane Filter," *Jap. Anal.* **22,** 3–9.

Thrane, K. E. (1978). "Background Levels in Air of Lead, Cadmium, Mercury and Some Chlorinated Hydrocarbons Measured in South Norway," *Atmos. Environ.* **12,** 1155–1561.

U.S. Bureau of Mines *Minerals Yearbooks* (1974–1976). U.S. Department of the Interior, Washington, D.C.

U.S. Environmental Protection Agency (1972). *Helena Valley, Montana, Area Environmental Pollution Study.* EPA, Office of Air Programs, Publ. AP-91, Research Triangle Park, North Carolina.

U.S. Environmental Protection Agency (1975a). *Scientific and Technical Assessment Report on Cadmium.* EPA, Office of Research and Development, EPA-600/6-6-75-003, Washington, D.C.

U.S. Environmental Protection Agency (1975b). *Technical and Microanalysis of Cadmium and Its Compounds.* EPA, Office of Toxic Substances, Rep. EPA 560/3-75-005, Washington, D.C.

U.S. Environmental Protection Agency (1976). *Cadmium: Control Strategy Analysis.* EPA, Rep. GCA-TR-75-36-G, Research Triangle Park, North Carolina.

U.S. Environmental Protection Agency (1978). *Sources of Atmospheric Cadmium.* Draft Report under Contract 68-02-2836, EPA, Research Triangle Park, North Carolina.

Ward, N. I., Brooks, R. R., and Roberts, E. (1977). "Heavy-Metal Pollution from Automotive Emissions and Its Effect on Roadside Soils and Pasture Species in New Zealand," *Environ. Sci. Technol.* **11**, 917–920.

Weiss, H., Bertine, K., Koide, M., and Goldberg, E. D. (1975). "Chemical Composition of Greenland Glacier," *Geochim. Cosmochim. Acta* **39**, 1–10.

Whelpdale, D. M. (1974). "Particulate Residence Times," *Water, Air, Soil Pollut.* **3**, 293–300.

Whitby, K. T. (1977). "Physical Characterization of Aerosols." In *Methods and Standards for Environmental Measurement,* Natl. Bur. Stand. Spec. Publ. 464, pp. 165–172.

Willeke, K. and Whitby, K. T. (1975). "Atmospheric Aerosols: Size Distribution Interpretation," *J. Air Pollut. Control Assoc.* **25**, 529–534.

Wilson, W. E. (1977). "Chemical Characterization of Aerosols: Progress and Problems." In *Methods and Standards for Environmental Measurement.* Natl. Bur. Stand. Spec. Publ. 465, pp. 323–325.

Yeh, H. C., Phalen, R. F., and Raabe, O. G. (1976). "Factors Influencing the Deposition of Inhaled Particles," *Environ. Health Perspect.* **15**, 147–156.

Yost, K. J. (1978). "Some Aspects of the Environmental Flow of Cadmium in the United States." In *Cadmium 77: Proceedings of the 1st International Cadmium Conferences.* Metal Bulletin Ltd., London, pp. 147–166.

Zoller, W. H., Gladney, E. S., and Duce, R. A. (1974). "Atmospheric Concentration and Sources of Trace Metals at the South Pole," *Science* **183**, 198–200.

4

DRY DEPOSITION
OF CADMIUM FROM
THE ATMOSPHERE

Cliff I. Davidson

*Departments of Civil Engineering and Engineering and Public
Policy, Carnegie-Mellon University, Pittsburgh, Pennsylvania*

1. INTRODUCTION

In Chapter 3 of this volume Nriagu discussed airborne concentrations of cadmium in urban and remote areas. Measured deposition rates for several locations
were also listed. This chapter focuses on calculations of dry deposition from the
free atmosphere.

First, mathematical models for aerosol deposition are discussed, including
studies by Chamberlain (1966, 1968), Sehmel (1971), and Davidson and Fried-

lander (1978). Cadmium airborne mass-size distributions, which are key inputs to these models, are then presented. The distributions are based on literature data of several investigators. Finally, the size distribution data are used with the models to predict cadmium deposition under different conditions.

2. DEPOSITION MODELS

It is convenient to begin by defining the deposition velocity, v_g, which represents the particle flux onto a surface normalized by the total airborne concentration of particles:

$$v_g \text{ (cm/sec)} = \frac{-J \text{ (downward flux, g/cm}^2\text{·sec)}}{C \text{ (airborne concentration, g/cm}^3)} \tag{1}$$

Several measurements of J and C for various species and surfaces have been made, yielding values for overall deposition velocity. These studies are of limited value, however, if a small fraction of the airborne mass is responsible for the bulk of the deposition. In that case, even a small perturbation in the total airborne concentration may induce large changes in flux. The concept of an aerosol deposition velocity is thus more useful when accompanied by a more rigorous deposition model.

The models to be discussed in this section consider v_g to be a function of aerosol characteristics, state of the atmosphere, and surface conditions. Aerosols are normally characterized in terms of their *aerodynamic diameter,* representing the size of a spherical, unit-density particle whose aerodynamic behavior in the atmosphere is identical to that of the original particle. The deposition velocity curves presented here refer to spherical particles with unit density. Ambient particles of nonunit density and nonspherical shape may be applied to these curves, provided that their size distribution is presented in terms of aerodynamic diameters. Fortunately many ambient particle sampling instruments fractionate the aerosols according to their aerodynamic diameters.

"Atmospheric state" in this context refers to the variation of windspeed with height. Key parameters of the wind profile are the friction velocity, $u*$, and the surface roughness height, z_0. In models where an adiabatic lapse rate is not assumed, the variation of temperature with height may be a model input.

Surface conditions may be parameterized in several ways. The roughness height, z_0, is often used, and attempts have been made to include the size and spacing of roughness elements in models. Only qualitative agreement between theory and experiment has been achieved with these models. In other studies, deposition velocity curves have been determined experimentally for a given surface structure. Generalizations of these surface-specific models to other surfaces are tenuous.

In all of the detailed models to be discussed in this section, the general flux equation relating J and C is invoked:

$$J = -(D + \epsilon)\frac{dC}{dz} - v_s C \qquad (2)$$

where D = Brownian diffusivity (cm^2/sec)
    ϵ = eddy diffusity (cm^2/sec)
    z = height above the surface (cm)
    v_s = sedimentation velocity (cm/sec)
This equation is treated somewhat differently in each of the three models.

2.1. Deposition Model of Chamberlain

In one of the earliest aerosol deposition models, Chamberlain (1966) assumed that $\epsilon \gg D$ and also that the eddy diffusivities for mass and momentum were identical. Using $\epsilon = 0.4u^*z$, he integrated equation 2 between a reference height above the canopy and a hypothetical mass concentration sink, z_v, near the surface. The appropriate value of z_v was determined by experiment.

By performing the integration, an expression for v_g was derived:

$$v_g(z) = \frac{v_s}{1 - (z_v/z)^\alpha} = \frac{-J}{C(z)} \qquad (3)$$

where $\alpha = v_s/(0.4u^*)$ and $C(z_v) = 0$.

Note that the deposition velocity is a function of height above the surface. This is so because the concentration of equation 1 varies with height while flux is assumed constant. It may thus be necessary to specify the height at which concentration measurements are taken in order to apply a deposition model. This is especially true if measurements are obtained very close to a surface, but is relatively unimportant for measurements made at heights of 1 m or greater.

A typical plot of v_g versus particle diameter is shown in Figure 1. This curve was obtained using wind tunnel data for the deposition of monodisperse particles. Note that the deposition velocity is always greater than sedimentation, although v_g approaches v_s for large particles. Also note that the curve increases as particle diameter decreases below about 0.2 μm. This is due to the increase in Brownian diffusivity with decreasing diameter.

2.2. Deposition Model of Sehmel

In the model of Sehmel (1971) the mass concentration sink is assumed to occur at one particle radius from the surface. Several expressions for eddy diffusivity

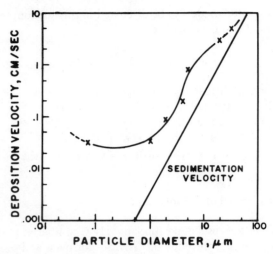

Figure 1.   Deposition velocity, v_g versus particle diameter, d_p, redrawn from Chamberlain (1966). The points shown were determined in wind tunnel experiments employing monodisperse particles depositing on Italian rye grass. The canopy height, z_c, is 6 cm; roughness height, z_0, is 0.65 cm; and friction velocity, u^*, is 70 cm/sec. Reference height for these points is 7.5 cm. Sedimentation velocity is also shown.

are incorporated, depending on the distance above the surface. For example, $\epsilon = 0.4u^*z$ is assumed to hold only at large distances from the surface; other expressions apply at small distances. These expressions for eddy diffusivity are then used in equation 2, which is integrated between the concentration sink and the reference height.

The expressions were determined experimentally by conducting wind tunnel measurements. After collecting a large number of data points, Sehmel generalized his deposition velocity curves for any surfaces and wind conditions where u^* and z_0 are known. Some of his curves include corrections for nonadiabatic lapse rates. Typical smooth surface v_g versus d_p curves are shown in Figure 2, for a reference height of 1 cm.

Sehmel (1971) hypothesized that the controlling "resistance" to particle deposition occurs in a narrow layer just above the surface. Concentration data obtained at relatively large heights above a surface can be used to determine deposition velocity; however, the resulting values of v_g depend on both the surface resistance and a less important diffusional resistance. The latter resistance depends on atmospheric transport of particles down toward the surface, which is poorly understood. A reference height of 1 cm was chosen for the wind tunnel experiments as the smallest practical height where concentration measurements could be obtained. The curves of Figure 2 thus represent reasonable maximum values of v_g for the indicated surface and airflow conditions. Greater reference

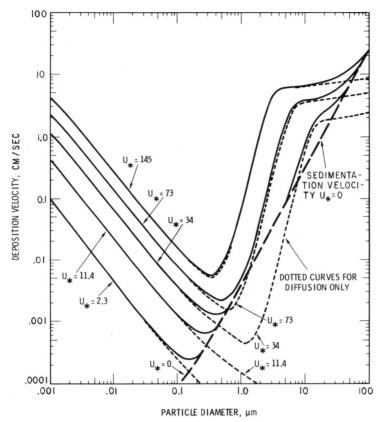

Figure 2. Deposition velocity versus particle diameter (from Sehmel, 1971). Curves are for a smooth surface and for a particle density of 1 g/cm^3. The reference height is 1 cm.

heights will yield smaller deposition velocities because of the increased influence of diffusional resistance.

According to Sehmel's model, the minimum in each of the v_g curves of Figure 2 becomes less pronounced as surface roughness increases. Plots of v_g versus d_p for a rough surface are shown in Figure 3. The reference height is 1 m; hence these curves show the influence of surface resistance and diffusional resistance. From a practical standpoint, Figure 3 can be used to predict the deposition of many species, such as cadmium, whose airborne mass-size distributions are given in the literature. Although some of these literature data were obtained at heights other than 1 m, the deposition velocity is relatively insensitive to height at distances greater than 1 m above a surface.

Figure 3 applies to a neutral (adiabatic) atmosphere. Sehmel and Hodgson

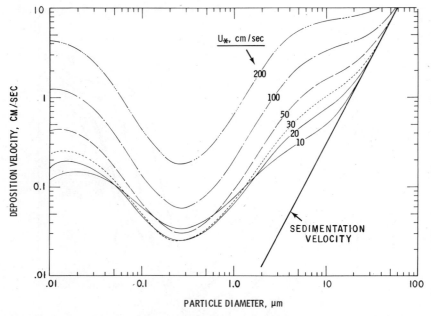

Figure 3. Deposition velocity versus particle diameter for a rough surface (Sehmel and Hodgson, 1978). The curves are for a particle density of 1 g/cm^3, $z_0 = 3$ cm, and a 1-m reference height. An adiabatic atmosphere is assumed.

(1978) have shown that the influence of atmospheric stability on 1-m deposition velocities is minor.

2.3.   Deposition Model of Davidson and Friedlander

Strictly speaking, u^* and z_0 do not uniquely determine v_g for a given particle size. The reason is that small surface irregularities which may not affect momentum transport may alter the mass deposition properties of a surface. For example, Wells and Chamberlain (1967) have shown that filter paper having tiny vertical fibers on the surface is a much better aerosol collector than brass, even though wind profiles above the surfaces are nearly identical. To resolve this ambiguity, a filtration model (Davidson and Friedlander, 1978) has been developed that incorporates details of the surface structure. The filtration model uses wind tunnel data from the literature for aerosol deposition on fiber elements of a filter, for example, cylinders transverse to air flow. A vegetated surface is then modeled as a fiber filter. The resulting equation for filtration applies only within the canopy, whereas equation 2 applies from a reference height above

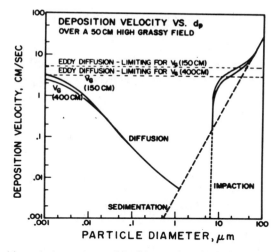

Figure 4. Deposition velocity versus particle diameter for a field of wild oat grass, *Avena fatua* (Davidson and Friedlander, 1978). Reference heights are 1.5 and 4 m above the ground. Values of parameters are $u* = 21.2$ cm/sec, $z_0 = 4$ cm, and $z_c = 50$ cm. Other inputs to the filtration model are fiber diameter = 0.2 cm, fiber spacing = 8600 fibers/m², and windfield zero plane displacement = 25 cm in an adiabatic atmosphere. Particle density is 1 g/cm³.

the vegetation down to a particle sink, z_v, within the canopy. These two equations are solved simultaneously for the two unknowns, J and C.

Typical curves for v_g versus d_p are shown in Figure 4. Since the curves were calculated from filtration theory and the model involves a number of assumptions, experimental verification was necessary. Measurements of ambient condensation nuclei concentration profiles yielded qualitative agreement between theory and experiment; application of this method to predict lead aerosol deposition and subsequent comparison with experiment also yielded qualitative agreement.

It is important to note that only very small differences exist between the 1.5-m and 4-m reference height curves. This shows the insensitivity of v_g to changes in height at relatively large distances from the surface. Also note the "eddy diffusion-limiting" v_g values. These represent the deposition velocities when filtration by the vegetation is very efficient. In this case the overall deposition is determined by the rates at which wind eddies deliver particles to the canopy.

Figure 4 applies to species of vegetation that may be modeled as a collection of cylinders. It is straightforward to apply the method to other vegetation, provided that the windfield and mass transport characteristics are known. For example, Chamberlain (1968) has used the Polhausen formula for laminar boundary layers to calculate diffusional mass transport to the surfaces of smooth, flat leaves. Reasonable agreement between wind tunnel data and theory was achieved.

Although other deposition models are available in the literature, the three models discussed here cover a reasonably wide range of v_g values. For additional information on deposition, the reader is referred to *Atmosphere-Surface Exchange of Particulate and Gaseous Pollutants* (1974).

3. SIZES OF CADMIUM-CONTAINING AEROSOLS

Size distribution measurements of cadmium-containing aerosols have been conducted in several urban and nonurban areas. In general, commercially available cascade impactors (e.g., Rao, 1975) or cyclone samplers (Bernstein et al., 1976) have been used for sampling. These devices are imperfect, but impactor or cyclone sampling is by far the most convenient method of fractionating aerosols by size for subsequent elemental analysis.

Information obtained by using an impactor includes the total mass of an element (or total mass of all particles regardless of chemical composition) in given size ranges. A typical impactor may have several stages. Except for the minimum and maximum aerodynamic particle diameter in each size range, no other information on the distribution of particle sizes on a stage is normally available. Thus the "smooth curve" true size distributions are unknown.

In this chapter all size distributions are plotted as though particles are uniformly distributed in log d_p on each impactor stage. This greatly simplifies the calculation of deposition rates. Data to construct these histograms have been obtained from differential or cumulative distribution plots, or tables in the original references.

Table 1 summarizes several cadmium mass-size distribution measurements obtained from the literature. In general, lower and upper limits for each distribution are not given in the original references. All distributions (except those obtained without a backup filter) are assumed to have a lower limit of 0.05 μm aerodynamic diameter, after Hidy (1973, pp. 254–264). An upper limit of 25 μm is assumed for the nonisokinetically sampled ambient distributions. The source measurement of Lee et al. (1975) is assumed to have an upper limit of 50 μm. These values were chosen based on the distributions of Davidson (1977), whose data show upper limits of 40 μm for isokinetically sampled urban aerosols. It should be recognized that the values of 0.05, 25, 40, and 50 μm are only estimates of the true minimum and maximum aerodynamic diameters.

The distributions summarized in Table 1 are shown in Figures 5 to 23. These distributions are normalized to their respective total airborne cadmium masses. The differences in shapes among the size distributions may be due in part to variations in sampling techniques, as well as differences in the actual size spectra

Table 1. Descriptions, from the Literature, of Cadmium Airborne Mass-Size Distributions

Figure	Reference	Dates of Sampling	Location of Sampling	Type of Sampler	C_T (ng/m^3)	Approx. MMD (μm)
5	Lee et al. (1968)	September 1966 Average of 14 runs, 24 hr each	Downtown Cincinnati, Ohio	Andersen impactor with backup filter, 1.2 m above the ground	80	3.1
6	Lee et al. (1968)	February 1967 Average of 3 runs, 4 days each	Fairfax, Ohio	Same as Figure 5	20	6
7	Lee et al. (1975)	April 1973 One run of 2 hr	Coal-fired electric power plant in Illinois; source sampling at output of electrostatic precipitator	Pilat in-stack impactor with backup filter	100	5
8	U. S. Environmental Protection Agency (1973, 1974)	April 1973 through January 1974 Average of 37 runs, 24 hr each	Kellogg, Idaho, near a lead smelter	Modified Andersen impactor with backup filter	1050	1.5
9	Harrison et al. (1971)	April 1968 Average of 21 runs, 2 hr each	Ann Arbor, Mich.	Modified Andersen impactor with backup filter, on building roof	60	0.5
10	Rahn (1976, p. 98)	April–May 1972 One 3-week run	Liege, Belgium	Andersen impactor with backup filter	118	2

Table 1. *Continued*

Figure	Reference	Dates of Sampling	Location of Sampling	Type of Sampler	C_T (ng/m³)	Approx. MMD (μm)
11	Dorn et al. (1976)	Winter, spring, summer 1972 Average of 3 runs, 27 days each	Southeast Missouri, 800 m from a lead smelter	Andersen impactor, no backup filter, 11.7 m above the ground	24.8	1.3
12	Dorn et al. (1976)	Winter, spring, summer 1972 Average of 3 runs, 14 days each	75 km from the smelter of Figure 11	Same as Figure 11	3.7	3.3
13	Duce et al. (1976)	May–June 1975 One run of 112 hr	Southeast coast of Bermuda	Sierra high-volume impactor with backup filter, 20 m above the ground	0.08	0.7
14	Duce et al. (1976)	July 1975 One run of 79 hr	Same as Figure 13	Same as Figure 13	0.07	0.6
15	Peden (1976)	Summer 1972 Average of 9 runs, avg. 9 days each	Pere Marquette State Park, Illinois, upwind of St. Louis	Andersen impactor with backup filter	1.76	1.28
16	Peden (1976)	Summer 1975 Average of 4 runs, avg. 8 days each	Wood River, Ill., industrial area near St. Louis	Andersen impactor, no backup filter	2.71	1.82

17	Peden (1976)	Collinsville, Ill., industrial area near St. Louis	Summer 1973 Average of 2 runs, avg. 5 days each	Same as Figure 15	1.79	0.95
18	Peden (1976)	Alton, Ill., industrial area near St. Louis	Summer 1975 Average of 4 runs, avg. 8 days each	Same as Figure 16	1.2	1.11
19	Peden (1976)	KMOX radio transmitter, industrial area near St. Louis	Summer 1973 Average of 2 runs, avg. 6 days each	Same as Figure 15	2.04	0.95
20	Peden (1976)	Centreville, Ill., downwind of a zinc smelter near St. Louis	Summer 1972 Average of 3 runs, avg. 10 days each	Same as Figure 15	3.97	1.53
21	Bernstein (1977, p. 87)	New York City, 30th St. and 1st Ave.	April 1976 One 7-day run	Cyclone sampling system with backup filter, on roof of 15-story building	9	1.7
22	Davidson (1977)	Pasadena, Calif.	May 1975 One run of 61 hr	Modified Andersen impactor with backup filter, on roof 4-story building, approx. isokinetic sampling	11.6	2.6
23	Davidson (1977)	Same as Figure 22	July 1975 One run of 62 hr	Same as Figure 22	4.1	3.6

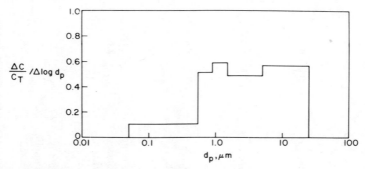

Figure 5. Normalized cadmium mass-size distribution in Cincinnati, Ohio. (Lee et al., 1968.)

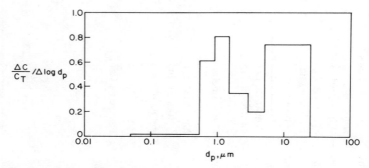

Figure 6. Normalized cadmium mass-size distribution in Fairfax, Ohio. (Lee et al., 1968.)

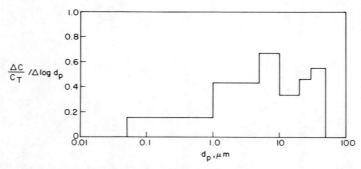

Figure 7. Normalized cadmium mass-size distribution emitted from a coal-fired electric power plant in Illinois. (Lee et al., 1975.)

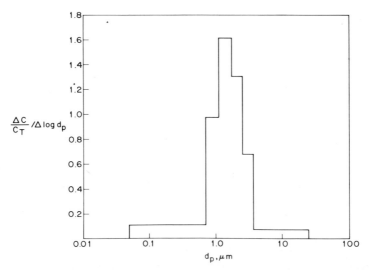

Figure 8. Normalized cadmium mass-size distribution in Kellogg, Idaho, near a lead smelter. (U.S. Environmental Protection Agency, 1973, 1974.)

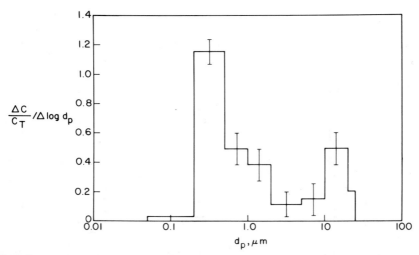

Figure 9. Normalized cadmium mass-size distribution in Ann arbor, Michigan. (Harrison et al., 1971.)

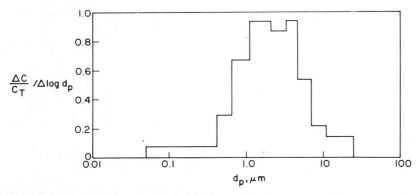

Figure 10.   Normalized cadmium mass-size distribution in Liege, Belgium. (Rahn, 1976.)

at these locations. For several of the distributions, personal communication with the authors was necessary to update or clarify data given in the original references.

Although other cadmium mass-size distributions are discussed in the literature, the histograms of Figures 5 to 23 present a reasonable picture of typical cadmium distributions. These data will now be used with deposition models to calculate cadmium depositions on various surfaces.

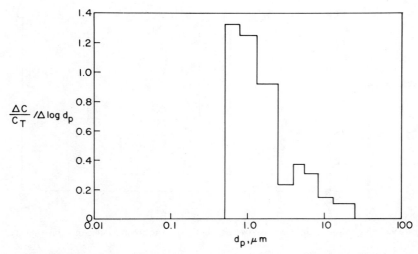

Figure 11.   Normalized cadmium mass-size distribution in southeast Missouri, 800 m from a lead smelter. (Dorn et al., 1976.)

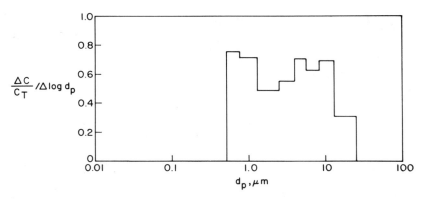

Figure 12. Normalized cadmium mass-size distribution in southeast Missouri, 75 km from the lead smelter of Figure 11. (Dorn et al., 1976.)

4. CALCULATION OF CADMIUM DEPOSITION

As a lower limit for cadmium deposition from the free atmosphere on rough natural surfaces, Davidson (1977) applied sedimentation velocities to the size distributions of Figures 22 and 23. The calculated flux values are $0.448 \pm 10\%$ and $0.224 \pm 25\%$ ng/cm²·day, respectively. The corresponding deposition velocities are $0.45 \pm 11\%$ and $0.63 \pm 29\%$ cm/sec.

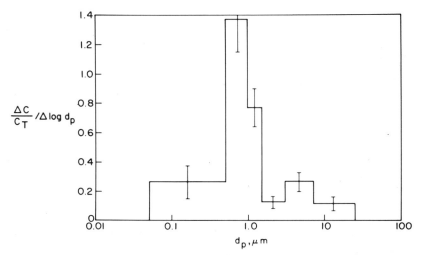

Figure 13. Normalized cadmium mass-size distribution on Bermuda, May–June 1975. (Duce et al., 1976.)

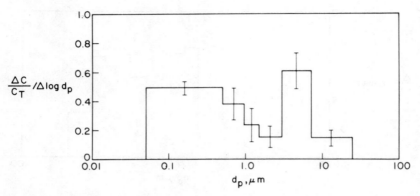

Figure 14. Normalized cadmium mass-size distribution on Bermuda, July 1975. (Duce et al., 1976.)

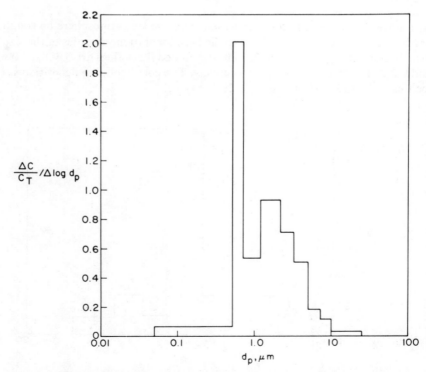

Figure 15. Normalized cadmium mass-size distribution at Pere Marquette State Park, Illinois. (Peden, 1976.)

130

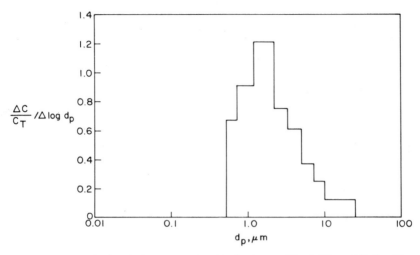

Figure 16. Normalized cadmium mass-size distribution at Wood River, Illinois. (Peden, 1976.)

The latter flux value may be compared with the measured deposition on a smooth, flat Teflon plate exposed simultaneously with the size distribution sampling: $0.238 \pm 43\%$. This value is actually reduced to $0.153 \pm 58\%$ when the estimated component of nonsedimentation deposition is subtracted. Although

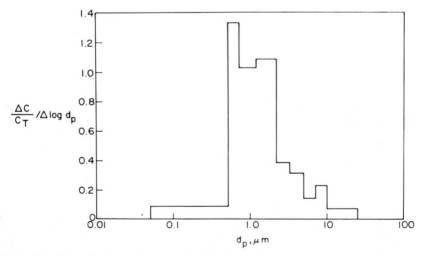

Figure 17. Normalized cadmium mass-size distribution at Collinsville, Illinois. (Peden, 1976.)

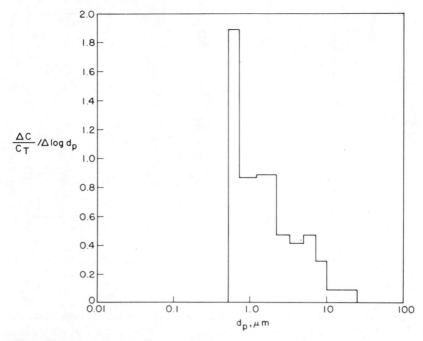

Figure 18.   Normalized cadmium mass-size distribution at Alton, Illinois. (Peden, 1976.)

the uncertainties are large, the qualitative agreement suggests that sedimentation controls cadmium deposition on smooth, flat surfaces in light winds.

Estimates of cadmium deposition on rough surfaces can be obtained by using the deposition velocity curves presented earlier. Note that these curves apply to particle densities of 1 g/cm^3. Since the size distributions are presented in terms of aerodynamic diameters, it is acceptable to use the distribution data directly without a density correction. The only difficulty with this approach concerns the lower end of the size spectra; deposition of very small particles may be controlled by Brownian diffusion, which is independent of density. Thus geometric rather than aerodynamic diameters are needed when characterizing the submicron material. In nearly all instances, however, cadmium deposition is seen to be controlled by supermicron particles—hence this inconsistency may be ignored.

Another problem with using size distribution plots is that the deposition calculations are very sensitive to the largest particle sizes present. Most of the size distributions were obtained nonisokinetically and may underestimate the amount of material in the upper size ranges. Furthermore, the true size range of particles on the top impactor stage is unknown. It also must be remembered that the size plots assume uniform distribution of particle size with respect to log d_p on each

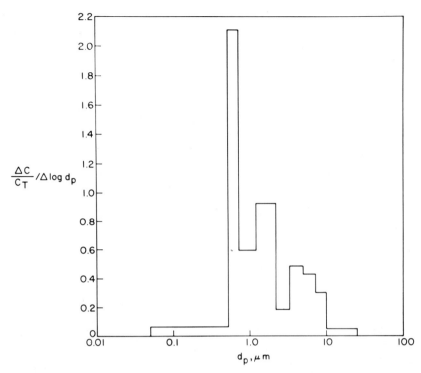

Figure 19. Normalized cadmium mass-size distribution at KMOX radio transmitter, Illinois. (Peden, 1976.)

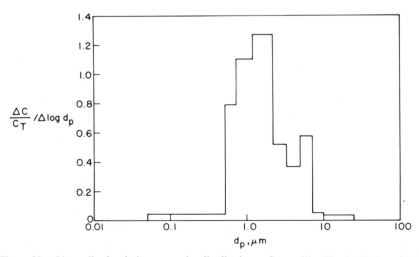

Figure 20. Normalized cadmium mass-size distribution at Centerville, Illinois. (Peden, 1976.)

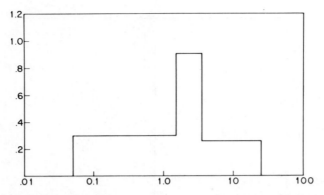

Figure 21. Normalized cadmium mass-size distribution in New York City. (Bernstein, 1977.)

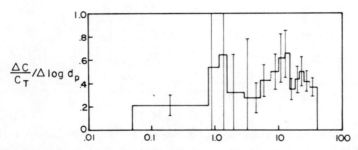

Figure 22. Normalized cadmium mass-size distribution in Pasadena, California, May 1975. (Davidson, 1977.)

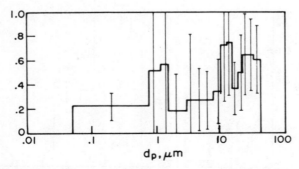

Figure 23. Normalized cadmium mass-size distribution in Pasadena, California, July 1975. (Davidson, 1977.)

134

Table 2. Cadmium Deposition Fluxes (ng/cm²·day), Using Deposition Models of Figures 1, 3, and 4, with Size Distributions of Figures 5 through 23 Fluxes due to sedimentation alone are also given.

| Figure | Model | | | Sedimentation |
	Figure 1	Figure 3	Figure 4	
5	5.6	2.9	8.7	1.1
6	1.8	0.86	2.9	0.36
7	12	8.9	17	6.8
8	13	15	14	2.3
9	2.6	1.5	3.8	0.68
10	3.7	2.9	3.3	0.77
11	0.63	0.46	0.57	0.13
12	0.25	0.14	0.29	0.058
13	0.0016	0.0011	0.0017	0.00030
14	0.0019	0.0012	0.0018	0.00033
15	0.026	0.026	0.020	0.0049
16	0.071	0.059	0.065	0.015
17	0.034	0.029	0.037	0.0074
18	0.028	0.022	0.025	0.0056
19	0.044	0.034	0.037	0.0082
20	0.066	0.064	0.036	0.011
21	0.29	0.18	0.44	0.051
22	1.1	0.67	1.5	0.45
23	0.48	0.30	0.67	0.22

impactor stage, which is only a rough approximation to the true distributions. For these reasons the calculated cadmium deposition rates must be viewed as estimates.

Figures 1, 3, and 4 have been used with each of the 19 distributions to determine cadmium flux and deposition velocity in Tables 2 and 3. In Figure 3 the curve corresponding to $u^* = 20$ cm/sec has been used. In Figure 4 the 1.5-m curve is used. Since Figure 1 applies to a reference height of 7.5 cm, the deposition rates calculated with this figure are upper limits for size distributions measured at greater heights. The overestimate is not severe, however. Sehmel and Hodgson (1978) have shown that the decrease in v_g for a given particle size as reference height increases from 10 cm to 1 m is much less significant than the decrease in v_g between 1 cm and 10 cm.

Calculated cadmium deposition fluxes are shown in Table 2. In addition to applying the three models, the fluxes due only to sedimentation are shown. The geometric mean aerodynamic particle diameter for each size range of Figures 5 to 23 has been used to determine v_g or v_s for that size range. Note that the geometric mean of a size range, d_{p1} to d_{p2}, is equal to $(d_{p1}d_{p2})^{1/2}$.

Table 3.   Deposition Velocities (cm/sec) of Cadmium-Containing Aerosols, Using Equation 1

Fluxes are those of Table 2, while airborne concentrations for each distribution are given in Table 1.

Size Distribution Figure	Model			
	Figure 1	Figure 3	Figure 4	Sedimentation
5	0.81	0.42	1.3	0.16
6	1.0	0.50	1.7	0.21
7	1.4	1.0	1.9	0.79
8	0.15	0.16	0.16	0.025
9	0.49	0.29	0.72	0.13
10	0.36	0.28	0.33	0.075
11	0.29	0.21	0.27	0.062
12	0.79	0.44	0.92	0.18
13	0.23	0.16	0.25	0.043
14	0.34	0.22	0.32	0.059
15	0.17	0.17	0.13	0.032
16	0.30	0.25	0.28	0.065
17	0.22	0.19	0.24	0.048
18	0.27	0.21	0.24	0.054
19	0.25	0.20	0.21	0.047
20	0.19	0.19	0.10	0.033
21	0.37	0.23	0.57	0.066
22	1.1	0.67	1.5	0.45
23	1.4	0.85	1.9	0.63

Deposition rates for the three rough surface models range from 0.0011 to 17 ng/cm²·day. This wide range is due in part to the variation in total cadmium mass concentrations among the distributions. Note that the sedimentation fluxes are always considerably smaller than the corresponding rough surface fluxes. The differences between sedimentation and rough surface fluxes are least significant for Figures 7, 22, and 23 because of the fraction of large particles in these distributions. In all rough surface models, v_g approaches v_s at the upper end of the size spectrum.

Table 3 lists deposition velocities for each of the fluxes of Table 2. The variation in total cadmium mass concentration among the distributions is not a factor in the Table 3 entries. Variations in rough surface deposition velocities from 0.10 to 1.9 cm/sec are due primarily to differences in the shapes of the size distributions. Of course, differences in wind and surface conditions among the three models also play a role.

It is noteworthy that Figures 5 to 23 show several distributions characterized by similar shapes. In many of the distributions there are two distinct modes. Whitby et al. (1972) have shown that urban aerosols generally have bimodal size distributions. The lower mode is attributed to condensation processes, while the upper mode is the result of mechanical separation processes such as grinding and crushing.

It would be desirable to compare the calculations of Tables 2 and 3 with measured cadmium deposition data. For example, ambient cadmium deposition onto artificial collectors and onto vegetation was discussed by Nriagu in Chapter 3 of this volume. It is difficult to compare measured with calculated values, however, because size distribution information and deposition data were not obtained at the same location. Such data would be useful, however, to validate the calculations of this chapter.

It must be remembered that, because of difficulties in sampling and differences among impactors, the distributions of Figures 5 to 23 should be compared only qualitatively. More research is needed to assess impactor sampling errors (see, e.g., Dzubay et al., 1976), and more data on ambient cadmium size distributions would be desirable. These data are needed not only to study source-sink relations for cadmium, but also to investigate health effects.

5. SUMMARY

Models from the literature for aerosol dry deposition can be used with information on sizes of cadmium-containing aerosols to determine dry deposition fluxes of cadmium. This chapter has focused on three types of models for rough surfaces and on 19 size distributions of cadmium-containing aerosols from the literature.

Dry deposition fluxes of cadmium onto rough surfaces range over several orders of magnitude because of large differences in ambient cadmium mass concentrations. Levels in excess of 1000 ng/m^3 have been measured near sources, whereas remote area measurements show concentrations of less than 0.1 ng/m^3. The rough surface dry deposition velocities of cadmium, which are not functions of airborne mass concentration, vary over a factor of 20. These variations are due to differences in the shapes of the size distributions; a sizeable fraction of cadmium mass in large airborne particles will result in a high deposition velocity.

Because of the approximate nature of currently available deposition models and because of difficulties in accurately characterizing ambient cadmium aerosols, the calculations of this chapter must be viewed as estimates. More research, particularly in the area of ambient cadmium sampling, is needed to assess cadmium deposition more accurately.

ACKNOWLEDGMENTS

The author gratefully acknowledges the assistance of the following individuals in obtaining data for this chapter: Drs. G. G. Akland, D. M. Bernstein, P. R. Harrison, R. E. Lee, Jr., J. O. Nriagu, M. E. Peden, P. E. Phillips, K. A. Rahn, and G. A. Sehmel.

REFERENCES

Atmosphere-Surface Exchange of Particulate and Gaseous Pollutants (1974). National Technical Information Service, CONF-740921, Proceedings of a Symposium, Sept. 4–6, 1974, Richland, Wash.

Bernstein, D. M. (1977). "The Influence of Trace Metals in Disperse Aerosols on the Human Body Burden of Trace Metals." Ph.D. Thesis, Department of Environmental Health Sciences, New York University.

Bernstein, D. M., Kleinman, M. T., Kneip, T. J., Chan, T. L., and Lippmann, M. (1976). "A High-Volume Sampler for the Determination of Particle Size Distributions in Ambient Air," *J. Air Pollut. Control Assoc.* **26,** 1069–1072.

Chamberlain, A. C. (1966). "Transport of *Lycopodium* Spores and Other Small Particles to Rough Surfaces," *Proc. R. Soc.* **A296,** 45–70.

Chamberlain, A. C. (1968). "Mass Transfer to Bean Leaves," *Boundary-Layer Meteorol.* **6,** 477–486.

Davidson, C. I. (1977). "The Deposition of Trace Metal-Containing Particles in the Los Angeles Area," *Powder Technol.* **18,** 117–126.

Davidson, C. I. and Friedlander, S. K. (1978). "A Filtration Model for Aerosol Dry Deposition: Application to Trace Metal Deposition from the Atmosphere," *J. Geophys. Res.* **83,** 2343–2352.

Dorn, C. R., Pierce, J. O., Phillips, P. E., and Chase, G. R. (1976). "Airborne Pb, Cd, Zn, and Cu Concentration by Particle Size near a Pb Smelter," *Atmos. Environ.* **10,** 443–446.

Duce, R. A., Ray, B. J., Hoffman, G. L., and Walsh, P. R. (1976). "Trace Metal Concentration as a Function of Particle Size in Marine Aerosols from Bermuda," *Geophys. Res. Lett.* **3,** 339–342.

Dzubay, T. G., Hines, L. E., and Stevens, R. K. (1976). "Particle Bounce Errors in Cascade Impactors," *Atmos. Environ.* **10,** 229–234.

Harrison, P. R., Matson, W. R., and Winchester, J. W. (1971). "Time Variations of Lead, Copper and Cadmium Concentrations in Aerosols in Ann Arbor, Michigan," *Atmos. Environ.* **5,** 613–619.

Hidy, G. M. (1973). *Characterization of Aerosols in California.* Interim Report for Phase I, State of California Air Resources Board.

Lee, R. E., Patterson, R. K., and Wagman, J. (1968). "Particle-Size Distribution of Metal Components in Urban Air," *Environ. Sci. Technol.* **2,** 288–290.

Lee, R. E., Crist, H. L., Riley, A. E., and MacLeod, K. E. (1975). "Concentration and Size of Trace Metal Emissions from a Power Plant, a Steel Plant, and a Cotton Gin," *Environ. Sci. Technol.* **9,** 643–647.

Peden, M. E. (1976). "Flameless Atomic Absorption Determinations of Cadmium, Lead, and Manganese in Particle Size Fractionated Aerosols." In *Proceedings of the 8th IMR Symposium, Sept. 20-24, 1976, Gaithersburg, Md.* Natl. Bur. Stand. Spec. Publ. 464, pp. 367–377.

Rahn, K. E. (1976). *The Chemical Composition of the Atmospheric Aerosol.* Technical Report, Graduate School of Oceanography, University of Rhode Island.

Rao, A. K. (1975). "An Experimental Study of Inertial Impactors." Ph.D. Thesis, Mechanical Engineering Department, University of Minnesota.

Sehmel, G. A. (1971). "Particle Diffusivities and Deposition Velocities over a Horizontal Smooth Surface," *J. Coll. Int. Sci.* **37,** 891–906.

Sehmel, G. A. and Hodgson, W. H. (1978). *A Model for Predicting Dry Deposition of Particles and Gases to Environmental Surfaces.* Battelle Pacific Northwest Laboratories, Rep. PNL-SA-6721.

U. S. Environmental Protection Agency (1973, 1974). Computer printouts of data supplied by G. G. Akland. U. S. Environmental Protection Agency, Environmental Monitoring and Support Laboratory, Research Triangle Park, North Carolina.

Wells, A. C. and Chamberlain, A. C. (1967). "Transport of Small Particles to Vertical Surfaces," *Br. J. Appl. Phys.* **18,** 1793–1799.

Whitby, K. T., Husar, R. B., and Liu, B. Y. H. (1972). "The Aerosol Size Distribution of Los Angeles Smog," *J. Coll. Int. Sci.* **39,** 177–204.

5

CADMIUM IN NATURAL WATERS

John H. Martin

George A. Knauer

A. Russell Flegal

Moss Landing Marine Laboratories, Moss Landing, California

1. INTRODUCTION

Trace elements in natural waters are generally assumed to be in the low microgram per liter range and to be highly variable in their temporal and spatial distributions. However, within the past 2 to 3 years scientists have proved this assumption to be false, at least for marine waters. It is now known that well-defined and similar distribution patterns exist for Cu, Cd, Ni, Mn, and Zn in open-ocean waters and that the levels of these elements fall within the 1 to 1000 ng/l range (Bender and Gagner, 1976; Bender et al., 1977; Boyle and Edmond, 1975; Boyle et al., 1976, 1977; Bruland et al., 1978a, 1978b, 1979; Martin et al., 1976; Moore and Burton, 1976; Sclater et al., 1976; Sugawara, 1978). This revelation came about largely because of the urgings of Patterson (Patterson and Settle, 1976), who has long argued that scientists attempting to measure

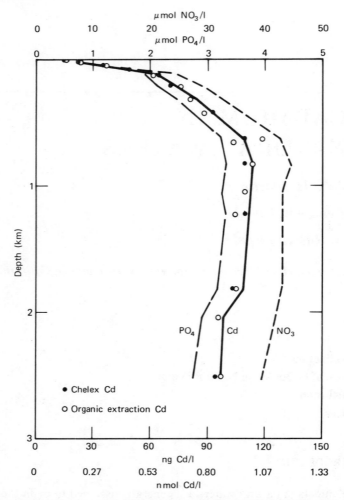

Figure 1.   Depth profiles for cadmium, PO_4 and NO_3 observed off the central California coast (37°05′N, 123°22′W). Cadmium values obtained using extraction and Chelex resin preconcentration techniques are also shown. (From Bruland et al., 1978b.)

lead in natural waters consistently contaminate their samples with gross amounts of this element during both collecting and analytical procedures. Thus, with the exception of one recent report (Schaule and Patterson, 1978), all lead data for marine, and perhaps all natural, waters are invalid.

It also became apparent that contamination problems exist for other elements as well (Brewer and Spencer, 1975). However, in the case of cadmium, these

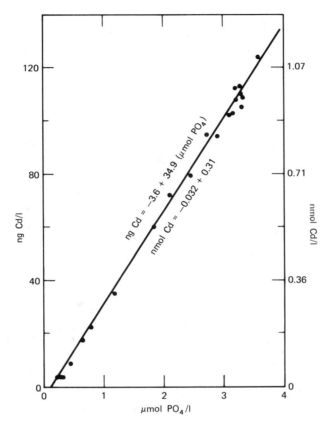

Figure 2. Cadmium versus PO_4 observed off the central California coast; $r = 0.998$. (From Bruland et al., 1978b.)

problems are not severe, and marine scientists have been especially successful in measuring this element.

2. LEVELS OF CADMIUM IN NATURAL WATERS

It is now well established that cadmium is depleted in oceanic surface waters (4 to 5 ng/l) and enriched at mid-depth (up to 125 ng/l) in association with the oxygen minimum (Martin et al., 1976; Boyle et al., 1976; Bender and Gagner, 1976; Bruland et al., 1978b). All of these studies also agree that cadmium is highly correlated with the plant nutrients PO_4 and NO_3 (Figures 1 and 2). The establishment of these oceanic cadmium-nutrient relationships represents a powerful tool for interpreting the biogeochemistry of this element. For example, deviations from the PO_4-predicted cadmium level may indicate the effect of different processes in estuaries and rivers, anthropogenic cadmium enrichment near point sources, or simply bad data in areas where there is no reason for an observed deviation from the expected values.

A few recent studies also suggest general agreement for the amounts of cadmium in fresh waters, that is, about 0.01 to 0.1 μg Cd/l. Examples of river cadmium values include those of Kennedy and Sebetich (1976) for remote California streams (0.01 to 0.03 μg/liter), Trefrey and Presley's (1976) observed concentrations in the Mississippi (0.1 μg/l), and the level reported by Boyle et al. (1976) for the Amazon (0.07 μg/l). Lake values also fall within this same general range. Sugawara (1978) reports an average of 0.02 μg/l for Lake Suwa in Japan; Beamish et al. (1976) were unable to detect cadmium (less than 0.1 μg/l) in 102 Canadian lakes; and Schell and Nevissi (1977), in cooperation with Cutshall, found about 0.02 μg/l in Lake Washington.

3. CONCLUSION

There is general agreement on the distribution of cadmium in natural waters, both fresh and marine. Cadmium is relatively easy to measure, and techniques are now available to study this element in a wide range of environments. With comparable research efforts we should be able to discern the important factors affecting the natural cycling of cadmium, and we are also in a position to fully evaluate human perturbation of these natural cycles.

REFERENCES

Beamish, R. J., Blouw, L. M., and McFarlane, G. A. (1976). *A Fish and Chemical Study of 109 Lakes in the Experimental Lakes Area (ELA), Northwestern Ontario, with Appended Reports on Lake Whitefish Aging Errors and the Northwestern Ontario Bait Fish Industry.* Environment Canada, Fisheries and Marine Service, Tech. Rept. 607, 106 pp.

Bender, M. L. and Gagner, C. L. (1976). "Dissolved Copper, Nickel and Cadmium in the Sargasso Sea," *J. Mar. Res.* **34,** 327–339.

Bender, M. L., Klinkhammer, G. P., and Spencer, D. W. (1977). "Manganese in Seawater and the Marine Manganese Balance," *Deep-Sea Res.* **24,** 799–812.

Boyle, E. and Edmond, J. M. (1975). "Copper in Surface Waters South of New Zealand," *Nature* **253,** 107–109.

Boyle, E. A., Sclater, F., and Edmond, J. M. (1976). "On the Marine Geochemistry of Cadmium," *Nature* **263,** 42–44.

Boyle, E. A., Sclater, F. R., and Edmond, J. M. (1977). "The Distribution of Dissolved Copper in the Pacific," *Earth Planet. Sci. Lett.* **37**, 38–54.

Brewer, P. G. and Spencer, D. W. (1975). "Minor Element Models in Coastal Waters." In T. M. Church, Ed., *Marine Chemistry in the Coastal Environment.* American Chemical Society, Washington, D.C., pp. 80–96.

Bruland, K. W., Knauer, G. A., and Martin, J. H. (1978a). "Zinc in Northeast Pacific Water," *Nature* **271**, 741–743.

Bruland, K. W., Knauer, G. A., and Martin, J. H. (1978b). "Cadmium in the Northeast Pacific Waters," *Limnol. Oceanogr.* **23**, 618–625.

Bruland, K. W., Franks, R. P., Knauer, G. A., and Martin, J. H. (1979). "Sampling and Analytical Methods for the Nanogram per Liter Determination of Copper, Cadmium, Zinc and Nickel in Sea Water," *Anal. Chim. Acta* **105**, 233–245.

Kennedy, V. C. and Sebetich, M. J. (1976) "Trace Elements in Northern California Streams." In *Geological Survey Research 1976*. Washington, D.C., pp. 208–209.

Martin, J. H., Bruland, K. W., and Broenkow, W. W. (1976). "Cadmium Transport in the California Current." In H. L. Windom and R. A. Duce, Eds., *Marine Pollutant Transfer*. D. C. Heath, Lexington, Mass., pp. 159–184.

Moore, R. M. and Burton, J. D. (1976). "Concentrations of Dissolved Copper in the Eastern Atlantic Ocean 23°N to 47°N," *Nature* **264**, 242–243.

Patterson, C. C. and Settle, D. M. (1976). *The Reduction of Orders of Magnitude Errors in Lead Analyses of Biological Materials and Natural Waters by Evaluating and Controlling the Extent and Sources of Industrial Lead Contamination Introduced during Sample Collecting, Handling and Analysis.* Nat. Bur. Stand. Publ. 422, pp. 321–351.

Schaule, B. and Patterson, C. (1978). "The Occurrence of Lead in the Northeast Pacific and the Effects of Anthropogenic Inputs." In M. Branica, Ed., *Lead—Occurrence, Fate and Pollution in the Marine Environment.* Proceedings of an international experts' discussion. Rudger Boskovic Institute, Center for Marine Research, Rovinj, Yugoslavia, October 1977. Pergamon Press, Oxford, England, in press.

Schell, W. R. and Nevissi, A. (1977). "Heavy Metals from Waste Disposal in Central Puget Sound," *Environ. Sci. Technol.* **11**, 887–893.

Sclater, F. R., Boyle, E., and Edmond, J. M. (1976). "On the Marine Geochemistry of Nickel," *Earth Planet. Sci. Lett.* **31**, 119–128.

Sugawara, K. (1978). "Interlaboratory Comparison of the Determination of Mercury and Cadmium in Sea and Fresh Waters," *Deep-Sea Res.* **25**, 323–332.

Trefrey, J. H. and Presley, B. J. (1976). "Heavy Metal Transport from the Mississippi River to the Gulf of Mexico." In H. L. Windom and R. A. Duce, Eds., *Marine Pollutant Transfer*. D. C. Heath, Lexington, Mass., pp. 39–76.

6

DISTRIBUTION AND SPECIATION OF CADMIUM IN NATURAL WATERS

Biserka Raspor

Center for Marine Research, Institute "Rudjer Bošković," Zagreb, Croatia, Jugoslavia

1. INTRODUCTION

Transport of cadmium in the environment may take place through the air or water, and the element reaches human beings through the food chains and by respiration (Nürnberg, 1976). The concentrations of cadmium are usually very low in air, water, and soil from the areas unaffected by human activities in contrast to the levels in polluted areas. The increased concentration is also reflected in the cadmium contents of plants and other biota. Industrial and municipal wastes are usually the main sources of cadmium pollution. A general scheme of pathways of toxic metals, from various environmental compartments to human beings via food chains and respiration, is applicable to cadmium and is presented in Figure 1.

Cadmium present in the aquatic environment is generally taken up into the human body via the gastrointestinal tract from ingesting water and food. It has been shown that 3 to 6% of cadmium is resorbed in the gastrointestinal tract (Nomiyama, 1975). In view of the concentration of cadmium in drinking water, the expert group of the Food and Agriculture Organization/World Health Organization (FAO/WHO) has recommended that cadmium shall not exceed 5 ppb (Nordberg, 1974). In the Federal Republic of Germany the legal tolerance value for cadmium in potable water is 6 $\mu g/l$, and measurement data show that the actual concentration usually found is lower than 1 μg Cd/l. However, the situation can be more serious if the drinking water is processed without special precautions for heavy metal removal from metal-polluted river waters, as is al-

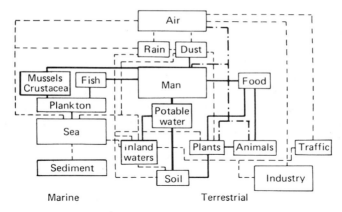

Figure 1.   Scheme of biogeochemical cycle for toxic metals: ——, terrestrial and marine food chain; - - - -, respiration; - - - - -, other pathways. (Reproduced from Nürnberg, 1978.)

ready necessary in various parts of western Europe (Nürnberg, 1976). With increased direct and indirect reuse of wastewater, which is to be processed for drinking water, further information on the levels of various trace metals in water supplies and the extent to which they are removed by conventional treatment processes is clearly desirable (Martin, 1970).

Aquatic organisms represent one type of food for human beings and can thus contribute to their cadmium burden. Direct uptake of heavy metals from surrounding seawater is only one of the possible routes of entry of heavy metals into marine organisms. In this respect the meaning of the "safe concentration" of cadmium consumed by persons in the form of fish food will be discussed.

The "safe concentration" of cadmium is the concentration safeguarding the normal development of marine organisms. In 1972 and 1973 the Water Quality Criteria Committee of the U.S. Academy of Science suggested 0.2 μg Cd/l as the "minimum risk concentration" for marine organisms (Bernhard and Zattera, 1975). Dividing the "safe concentration" by the natural concentration of cadmium in seawater, which is 0.02 μg/l, yields a "safety factor." In the case of cadmium, the "safety factor" is 10, and its meaning is as follows: If the factor is 1, the natural concentration of the pollutant corresponds to the "safe concentration," and no pollution of seawater occurs within this definition. If the factor is higher than 1, as in the case of cadmium, some discharge of the pollutant is tolerable without injuring marine organisms (Bernhard and Zattera, 1975). However, one should bear in mind that, for many metals, mainly the unchelated dissolved species can be taken up by organisms and are able to contribute to toxicity. These species are frequently only a fraction of the total concentration. Nevertheless, there is also a possibility that chelated metals will become resorbed by organisms and in this manner enter the food chain.

A pollutant is more concentrated if it is taken up by marine or aquatic organisms from their food, instead of directly from the water. In the process of accumulation of toxic heavy metals, the physicochemical state of the particular metal is of great importance (Bernhard and Zattera, 1975). It has been known for some time that chelating agents, such as EDTA and humic extracts from soil, usually reduce the toxicity of heavy metals for many aquatic organisms. Of great interest in toxicity tests is the simultaneous determination of the concentrations of the pollutant in seawater, in organisms, and in their food. It has been observed that a correlation exists between the concentrations of a pollutant in seawater and in marine organisms. The higher the concentration of a heavy metal in seawater, the higher becomes the concentration of that metal in marine organisms (like macrophytes and molluscs) collected in their natural habitats. For example, the concentration of cadmium starts to increase west of Morte Point from 0.2 μg/l to reach 4.1 μg/l higher up in the Bristol Channel (Abdullah et al., 1972). Likewise, the cadmium concentration in *Patella vulgata* increases from 8500 μg/kg fresh weight (FW) at Morte Point to 188,500 μg/kg FW at Portished (Peden et al., 1973). *Patella* from unpolluted Devon coastal waters contains only 900 to 2300 μg Cd/kg FW.

To determine the amount of fish food that can be consumed by human beings without exceeding the permissible cadmium intake, it is necessary to know the concentration of the metal in the edible part of fish. According to Stoeppler (1978), the cadmium concentration in the edible part of fish is extremely low, and the table presented by this author indicates that relatively high cadmium concentrations could be detected only in benthic organisms from lower or other trophic levels of the marine food chain (e.g., copepods, mussels, and crustaceans). The following cadmium concentrations were determined:

Organism	Concentration μg Cd/kg FW
Gadus morhua	2
Sepia officinalis	35–55
Portunus sp.	32–74
Mytillus sp.	150–570

According to Bernhard and Zattera (1975), in the edible part of marine fish the typical concentration of cadmium was taken as 400 μg/kg FW (which is significantly higher than the actual concentration in the edible part of marine fish reported (Stoeppler, 1978)). This concentration of 400 μg Cd/kg FW also corresponds to the provisional tolerable weekly intake of cadmium from marine fish for a person with a standard body weight of 70 kg. From these two data the maximum amount of fish that a person may consume without exceeding the provisional tolerance level of cadmium intake can be calculated to be 1 kg marine fish/week (Bernhard and Zattera, 1975). Comparing the resulting maximum amounts of fish that can be consumed in certain countries throughout the world, one notices that, on average, the tolerance level is not surpassed. For example, in Japan the average consumption of fish is 0.6 kg/week per capita; in Portugal, Sweden, and Denmark, 0.4 kg/week per capita; and in Spain, Italy, France, and Greece 0.27, 0.11, 0.16, and 0.19 kg/week per capita, respectively (Bernhard and Zattera, 1975). Naturally, these are average values recorded in certain countries, and the individual fish consumption could be significantly larger, particularly for certain populations, such as fishermen.

In many environmental matrices the cadmium level is generally rather low, lying in the microgram per kilogram or even the nanogram per kilogram range. This will be illustrated in Section 3 by the data on cadmium concentration in the aquatic environment consequently the environmental chemistry of cadmium belongs mainly in the field of trace or even ultra trace chemistry. This requires an appropriate philosophy and style of experimental work if meaningful data are to be obtained.

2. METHODS FOR TRACE METAL ANALYSIS AND TRACE METAL CHEMISTRY

Generally, the requirements for trace metal analysis and chemistry include sensitivity, selectivity, accuracy, precision, ease, rapidity, and reasonable cost (Hume, 1967). The basic requirements of an efficient trace chemical method are listed in a comprehensive manner in Table 1. *Sensitivity* refers to the limiting amount of the substance that is determinable. For environmental trace metal analysis sensitivity down to the parts per billion (ppb) range is required. For seawater and other natural waters the sensitivity range has to be lowered to 1 ng/1 or parts per trillion (ppt). Usually, with respect to minimization of contamination risks and resulting accuracy deficiencies, sensitivity should be high enough to exclude chemical preconcentration steps. *Selectivity* means the discrimination of a particular metal from the others. *Accuracy* indicates the degree to which the true value is approached, that is, the degree of freedom from systematic errors, while *precision* refers to the degree of reproducibility, that is, to the remaining statistical error.

When all the requirements in Table 1 are considered, the number of methods particularly suitable for trace metal determination in the aquatic environment becomes rather limited. There remain only a few techniques that correspond in a comprehensive and satisfactory manner to all the requirements, from having sufficiently high sensitivity (combined with good precision and, in particular, accuracy) to costs that do not necessarily prohibit wide and frequent applications (Nürnberg, 1978a). A list of current analytical methods that meet the challenge of reliable and meaningful ultratrace analysis is as follows:

Table 1.   Required Basic Properties of Trace Chemical Methods[a]

Clear and well-founded fundamental theoretical basis

High sensitivity   Versatile application range

Good precision

Reliable accuracy ◄───────────────────────► Simple chemical sample
                                                             pretreatment

Selectivity ◄───────────────────────► Simultaneous determination of
                                                             ──► substances

Rapidity ◄───────────────────────► Adaptability to automation

Economical cost demands for investment and operation

[a] From Nürnberg (1978a).

1. Differential pulse stripping voltammetry (DPSV).
2. Nonflame atomic absorption spectrometry (NFAAS).
3. Inductively coupled plasma source atomic emission spectrometry (ICPAES).
4. X-ray fluorescence spectrometry (XRFS).
5. Chemical ionization mass spectrometry (CIMS).
6. Spark source mass spectrometry (SSMS).
7. Isotopic dilution mass spectrometry (IDMS).
8. Neutron activation analysis (NAA).

In practice these methods are not equally suitable for quantitative determination of every metal; certain methods have their respective optimal application ranges with respect to the chemical nature of individual metals and the type of matrix in which they occur. Specific methodological properties are required if particular chemical species of a given metal are to be investigated. Frequently the methodological potentialities depend on the type of matrix in a very significant manner. The methods applied for trace metal determinations can be classified into four groups: electrochemical, optical, mass spectroscopic, and radiochemical. The first method in the list given above is electrochemical in nature, the following three belong to the optical spectroscopy branch, the next three are versions of mass spectroscopy, and the last one is in the radiochemical group of analytical methods.

    For study of the most significant toxic trace metals in seawater and inland waters (rivers, lakes, barrages, fountains) the electrochemical approach based on certain advanced models of polarography and voltammetry has become the method of choice (Nürnberg, 1977b). With the increasing interest in high sensitivity for environmental and pollution studies, polarography and voltammetry record to date a rapidly expanding degree of application. This is mainly the result of two methodological developments: (a) wide usage of the stripping approach, also termed inverse voltammetry, and (b) the commercial introduction of differential pulse polarography. Together with the commercial availability of reliable and relatively simple and inexpensive instruments, voltammetry covers the whole range of aquatic trace metal analysis and is specially suitable for trace metal determination in the dissolved state. Since it is the only method among those listed that allows direct determination of the chemical species of dissolved trace metals, its development and application will be discussed further in Section 2.1.

The nonelectrometric alternatives for the analysis of toxic trace metals in environmental samples are not numerous, and they are usually restricted to the determination of the total elemental content and provide no information on speciation, as they are all essentially element-sensitive methods (Nürnberg, 1979). One nonelectrometric method which represents the main alternative for wide-scale application to trace metal determinations in environmental matrices (particularly at the ultratrace level of 1 ppb and below) consists of the nonflame atomic absorption spectroscopy (NFAAS) in its various forms. They have become very popular, to an extent that is not justified for the contemplated application to environmental toxic trace metal chemistry. Here inherent inaccuracy problems arise, particularly in electrothermal AAS with the graphite tube, which is the most frequently applied version of nonflame AAS (Nürnberg, 1979; Stoeppler, 1978). The higher accuracy risks of nonflame AAS stem from several sources, such as instrumental bias, nonspecific adsorption effects, volatization, molecular distillation and preatomization losses, matrix interferences leading to signal suppression, interactions with the graphite tube wall, causing carbide formation, and the fact that rather small sample volumes of the microliter order are used and these are much more sensitive to relatively small contamination effects than the 30 to 50 ml sample volumes common in voltammetry. Moreover, AAS is only a sequential single-element method, although a rather rapid one. Another important aspect for the marine environment is that avoidable interferences by the excess salts constituting the salinity of the sea usually prohibit direct trace metal determination and necessitate prior separation of the trace metals by extraction, which enhances the contamination risk and prolongs the analysis time. Nevertheless, nonflame AAS remains the most important alternative for extended baseline studies and monitoring purposes in the environmental chemistry of trace metals. A definite advantage of AAS in trace metal analysis of biological matrices (marine organisms, detritus, sediments) is the fact that its requirements for prior sample digestion are less stringent with respect to completeness of mineralization of organic matter than those for subsequent voltammetric determinations (Nürnberg and Stoeppler, 1977). Nonflame AAS versions will yield reliable data if accuracy is established by checking a certain fraction of samples in a series from the same matrix by an independent trace analytical method (e.g., voltammetry) to establish the accuracy of the average results of the sample series (Nürnberg, 1979; Stoeppler, 1978).

Another method from the nonelectrometric but optical group is X-ray fluorescence spectrometry (XRFS), which is a multielement method not a relatively expensive one. It can be applied for the investigation of sediments, aquatic organisms, dust particles, and certain organs (i.e., substrates in which accumulation of trace metals has occurred). Usually, however, the much lower concentrations of toxic metals in natural waters and body fluids remain inaccessible to this method (Nürnberg, 1979).

In regard to the group of mass spectroscopic and radiochemical methods it can be concluded that wide-scale application of these methods is limited, in the first place, by the high costs of the required equipment. Moreover, the mass spectrometric approach suffers from the lack of standard reference materials for trace metals in various types of environmental matrix (Nürnberg, 1978c). Neutron activation analysis is in general a versatile multielement method but depends on the availability of a nuclear reactor and special laboratory facilities for handling relatively high amounts of radioactivity. Also, it requires a specially trained staff. Unfortunately, because of low sensitivity, it has no great significance particularly for cadmium.

This short overview of analytical methods applicable to trace metal determinations in environmental samples can be concluded by stating that at present two groups of methods, voltammetry and AAS, remain the predominant approaches for performing the large number of determinations necessary to attain the data from extensive sampling networks and large-scale monitoring programs. Both methods are also suitable, because of their compact equipment and their moderate operation demands, for application in small field laboratories on board research and monitoring vessels or in terrestrial mobile laboratories on vans operating along the coastline (Nürnberg and Stoeppler, 1977).

2.1. Polarographic and Voltammetric Methods

Polarography was developed in 1922 by J. Heyrovský. In general, polarographic and voltammetric methods are based on the remarkably complete, well-established, and detailed theory of electrode processes. In principle, polarography is electrolysis in a small volume of solution, and the principal component of the polarographic system is the test electrode, also termed the indicator or working electrode. According to the type of test electrode, the method is classified as

Table 2.   Typical Sensitivity and Precision Potentialities of Some Polarographic and Voltammetric Methods for Trace Metal Analysis

Method/Electrode[a]	Determination Limit for Cd(II) at RSD ±20%	
	(μg/l)	(mol/l)
A. Direct polarographic measurements by DME		
Linear scan-dc polarography	1000	1×10^{-5}
Phase sensitive ac polarography (only for reversible responses)	20	2×10^{-7}
Differential pulse polarography	5	5×10^{-8}
Square wave polarography (only for reversible responses)	2	2×10^{-8}
B. Inverse voltammetric methods (stripping)		
Conventional ASV/HMDE	0.1	1×10^{-9}
Conventional ASV/MFE	0.01	1×10^{-10}
DPASV/HMDE	0.01	1×10^{-10}
DPASV/MFE	0.001	1×10^{-11}

[a] DME = dropping mercury electrode, HMDE = hanging mercury drop electrode, MFE = mercury film electrode, ASV = anodic stripping voltammetry, DPASV = differential pulse anodic stripping voltammetry.

polarography (when the dropping mercury electrode is used) and as *voltammetry* (when stationary electrodes are used such as the hanging mercury drop electrode, the mercury film electrode, or stationary electrodes from other materials). It should be emphasized that without parallel development of both instrumentation and test electrodes the present range of applications of polarography and voltametry would have not been possible. The following types of test electrodes are available:

1. Dropping mercury electrode (DME).
2. Slowly dropping mercury electrode (SDME), which is intermediate between the DME and the stationary type of mercury electrodes.
3. Hanging mercury drop electrode (HMDE), wax-impregnated graphite electrode (WIG), rotating glassy carbon electrode coated *in situ* with a thin film of mercury (MFE).

The improvements in determination limits when using different types of electrodes and excitation potentials have been summarized for polarographic and voltammetric methods by Nürnberg (1979) and are presented in Table 2. The data in group A of Table 2 for direct polarographic measurements applying the DME show that the determination limits achieved are not high enough or at best

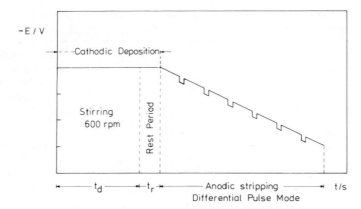

Figure 2.   Principle of DPASV (on HMDE).

just approach the concentration level of cadmium that can usually be found in many types of environmental samples. However, the introduction of the inverse voltammetric approach at stationary test electrodes has led to a significant and sufficient improvement in sensitivity. This increase in sensitivity occurs because cadmium (or any other metal that forms an amalgam) is electrochemically preconcentrated in the working electrode during the deposition step. This kind of preconcentration is not subjected to the frequently intolerable (with respect to reliable ultratrace analysis) contamination risks encountered in chemical preconcentrations. Anodic stripping voltammetry (ASV) is usually applied in connection with stationary electrodes, but examples exist where it has also been combined with the slowly dropping mercury electrode.

The determination limits were increased even further by combining the inverse voltammetric approach with a special type of polarization using a pulse train which is for metals usually applied in the anodic direction so that differential pulse anodic stripping voltammetry (DPASV) results. The effect on the sensitivity increase becomes obvious from the values in group B of Table 2 for DPASV. The principle of DPASV on HMDE is depicted in Figure 2. During the deposition time (t_d) at constant cathodic potential of the stationary test electrode, the solution is stirred with a magnetic Teflon-coated bar to speed up mass transfer toward the test electrode. After the deposition time the stirring is stopped, and a rest period follows. Then the potential is altered in the anodic direction by applying the differential pulse mode. At appropriate potential values an aliquot of the trace metal, deposited as amalgam, is reoxidized and released as ion into the solution. The corresponding current is the recorded signal.

An important practical advantage of DPASV compared with conventional ASV is the saving in overall analysis time, as the deposition time can be restricted to several minutes even at the nanogram per liter level in the ultratrace range.

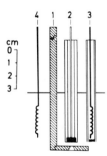

Figure 3.   Schematic representation of the electrode system with efficient mixing of electrolyte. (1) Perforated vibrating Perspex disk; (2) MFE; (3) reference electrode; (4) platinum counter electrode. (Magjer and Branica, 1977.)

The following ways of solution stirring are possible:

1. A Teflon-coated magnetic bar rotating at 600 to 900 rpm, which is usually applied in conjunction with the HMDE.
2. Rotation of the electrode itself at a high and constant speed of 1500 to 3000 rpm, as in the case of the rotating MFE.
3. A vibrator (Magjer and Branica, 1977), where the electrolyte streams toward a stationary working electrode (e.g., MFE). In this case the reproducible transport of the electroactive species from the bulk of the solution toward the electrode, during the deposition time, is achieved by a conically perforated vibrating disk. The applied vibration frequency was 50 Hz, and the vibration amplitude was less than 1 mm. A schematic representation of this electrode system is shown in Figure 3.

Having the appropriate instrumentation and type of working electrode, one may ask, "What type of information can be obtained by applying polarography or voltammetry to trace metal analysis?" The resulting response due to an electrode reaction taking place in a certain potential range is a polarographic wave or a peak-shaped voltammogram. This is, in fact, a current (i)-potential (E) response. The limiting or peak currents, i_{lim} or i_p, are proportional to the concentration of the substance studied in the bulk of the sample solution, while the half-wave, peak, or half-peak potentials ($E_{1/2}$, E_p, $E_{p/2}$) are correlated with the chemical nature and structure of the determined species. From the recorded i-E responses (i.e., polarograms and voltammograms) the following quantities and parameters relevant to problems in marine and aquatic chemistry are obtained (Nürnberg and Valenta, 1975):

1. The bulk concentration (c) of the trace metal can be determined.
2. In addition, the overall diffusion coefficient (D) of the substance studied can be obtained because $i \sim \sqrt{Dc}$.

3. The potential range in which the i-E response of the metal appears may give a superficial indication of the nature of the chemical species.
4. Quantitative information on stability constants, ligand numbers, and thus the structures of the complexes is available via the dependence of $E_{1/2}$, E_p, or $E_{p/2}$ of a reversible electrode reaction on the concentration of the respective ligand. Sometimes these data may also be obtained via the alteration of D with ligand concentration. However, in the case of irreversible electrode reactions involving strong trace metal chelates the coordination chemistry data can be obtained by voltammetrically determining the concentration of trace metal remaining unchelated at a given chelator concentration and yielding a reversible voltametric response. This approach will be outlined in detail in Section 5.2.
5. From the kinetically controlled values of i_{lim} or i_p, the formation and dissociation rates of metal complexes can be determined (Crow, 1969), and conclusions on the operative type of mechanism will emerge.

Further discussion (Section 3) will be focused on cadmium determinations in natural waters, with special emphasis on seawater. For the usual concentration of cadmium in seawater polarographic methods are not sensitive enough, and therefore inverse voltammetric methods have to be applied. This statement can best be illustrated by Table 3, where it can be seen that, for the determination of heavy metals at the concentration levels at which they appear in nature, usually only stationary electrodes are of concern.

Two types of stationary electrodes are used for inverse voltammetric determinations of cadmium in natural water samples, the HMDE and the MFE. Because of the increasing need for improvement of the determination limit in the analysis of trace metals dissolved in natural waters, particularly seawater, the HMDE is being replaced by the MFE. Of the various mercury film electrodes suggested in the literature, the rotating glassy carbon electrode, plated *in situ* with a mercury film typically 200 to 1000 Å thick, is the electrode of choice for the ultratrace application. Details about the construction of this electrode are given in the papers by Sipos et al. (1974) and Valenta et al. (1977a). Appropriately polished glassy carbon is an excellent supporting material, which gives low background current and a wide potential range. Glassy carbon with its low porosity and great hardness is more suitable as electrode material than graphite, which has first to be impreganted with paraffin wax. The quality of the glassy carbon is of the outmost importance for obtaining good results in the sub-parts per billion range (Valenta et al., 1977a). Rotation of the MFE provides effective and reproducible mass transport of the solution toward the electrode.

Several authors have tried to compare these two stationary electrodes, the HMDE and MFE, in regard to their electrochemical behavior and advantages (e.g., Batley and Florence, 1974; Lund and Salberg, 1975; Lund and Onshus, 1976). Recently, there has been a further development of the MFE electrode

Table 3. Comparison of Direct Polarographic and of Inverse Voltammetric Methods[a,b]

Method/ Electrode	Concentration in solution (ppb)	(mol/l)	Concentration in Mercury Electrode (ppb)	Accumu- lation Factor	Signal Height ($\mu A/10^3$ ppb)	RSD (%)	Determination Limit at RSD of 20% ($\mu g/l$)	(mol/l)
LSP/DME	2000	1×10^{-5}	—	—	0.1	±2	50	2×10^{-7}
DPP/DME	1000	5×10^{-6}	—	—	0.5	±3	20	1×10^{-7}
ASV/HMDE	50	2×10^{-7}	5000	100	10	±3	2.5	1×10^{-8}
DPASV/HMDE	2.3	1×10^{-8}	276	120	200	±2	0.01	4×10^{-11}
DPASV/MFE	0.05	2×10^{-10}	25	500	600	±8	0.001	4×10^{-12}

[a] From Valenta et al. (1977a) and Nürnberg (1979).

[b] LSP = linear sweep polarography, DPP = differential pulse polarography, ASV = conventional anodic stripping voltammetry, DPASV = differential pulse anodic stripping voltammetry.

Table 4.   Cadmium Determination Ranges and Precision with MFE

Cadmium Range of Concentration		Precision
(ppb)	(mol/l)	(%)
0.2 to 0.02	2×10^{-9} to 2×10^{-10}	± 2.5
10^{-3} to 2×10^{-4}	1×10^{-11} to 2×10^{-12}	± 20

in order to achieve even higher sensitivity and better performance (Valenta et al., 1977a).

Nürnberg, et al. (1976) have indicated the determination ranges for cadmium and the corresponding precision in seawater and inland water samples by applying the rotating MFE in DPASV mode. The results are summarized in Table 4 and show very good precision, remaining satisfactory even at the limits of the determination. At the parts per trillion level, (i.e., tenths and hundredths of ppb) the question arises, Which results reflect the real concentration, and which pollution? A comprehensive study on various trace analytical aspects, particularly the voltammetric parameters that optimize the performance of DPASV in connection with the MFE, has been carried out by Valenta et al. (1977a).

Sipos et al. (1978) have described the application of subtractive anodic stripping voltammetry with the rotating mercury-coated glassy carbon split disk electrode for the determination of trace metals. The results show that this method has, in conventional ASV, a higher sensitivity due to an improved signal/noise ratio achieved by the applicability of subtractive ASV.

In conclusion, the advantages of polarographic and voltammetric determination of cadmium and of trace metals in general are as follows (Nürnberg, 1976):

1. Polarography and voltammetry, as compared to other nonelectrochemical methods, are based on very well defined physicochemical fundamentals in terms of the kinetics and thermodynamics of electrode processes.
2. As polarography and voltammetry are based on Faraday's law, they combine high sensitivity, emerging from the equivalence of an enormous electric charge of 96,500 coulombs with 1 mol of substance, with good precision and inherent methodological accuracy, as well as the absence of the fundamental standardization problems.
3. Simultaneous determination of several trace metals in one run is usually possible, because polarography and voltammetry are oligosubstance methods, compared to single-element methods like the more popular AAS.
4. The most valuable characteristic of the polarographic and voltammetric approach is its sensitivity to the metal species dissolved in the sample, and

not merely to the total metal content; this is a particularly important aspect for all samples from natural waters (i.e., rivers, lakes, and seawater) where direct measurements without a change in the original matrix is possible. These electrochemical methods are unique compared to other highly sensitive and popular methods of trace metal analysis, such as AAS, emission spectroscopy, and NAA mass spectroscopy, because they are sensitive to the chemical state of the substrate studied.

5. Polarography and voltammetry in the presence of excess salts is an advantage rather than a problem, unlike the application of AAS; the excess salts cause no interference but instead fulfill the necessary supporting electrolyte function.

6. Reliable and versatile instrumentation is now commercially available at relatively modest prices, providing a very satisfactory cost benefit ratio.

7. The current somewhat low analysis rate for voltammetry will significantly improve in the near future because of the increasing trend toward automation and computerization in voltammetry. In this latter area a reliable and sturdy automated device for simultaneous voltammetric analysis by DPASV of several toxic trace metals in drinking water and groundwater has been reported by Valenta et al. (1978). This device makes possible on-line monitoring of drinking water over extended periods without supervision by staff and can be also applied to other types of natural waters, and after some feasible modifications even in the presence of higher amounts of surfactants.

3. CONCENTRATION LEVELS OF DISSOLVED CADMIUM IN NATURAL WATERS

This section focuses on dissolved cadmium levels in the aquatic environment. More importance is attributed to the seawater samples because of the importance of the marine environment for human beings and because of the fact this system is better defined in basic composition around the world. By contrast, rivers and lakes can differ significantly in basic composition, depending on their geographical position, thereby rendering direct comparison more difficult. In fact, complete data on the cadmium distributions in seawater, estuaries, and inland waters (lakes and rivers) should include for unpolluted and polluted waters the total cadmium concentration, as well as the levels of the element in the liquid and solids phases. In the liquid phase, in addition to the total dissolved concentration of metal, the amount bound in the form of labile and inert complexes should also be known. Moreover, the metal distribution between liquid and solid phases (the latter consisting of sediments, inorganic, organic particulate matter, and colloids), is of great significance. In spite of the fact that the oceanic "residence time" of cadmium is about 50,000 years, which means that the element is incorporated to a relatively small extent into marine sediments (Boyle et al., 1976), it is of interest and importance to know how cadmium interacts with the sediments and what the possibilities are of its release, particularly in fresh waters.

Besides the distribution of cadmium between two phases, living organisms that are able to accumulate the metal coexist in the aquatic environment and contribute to the biological part of the cadmium cycle in nature. The question arises which chemical forms of cadmium are more easily taken up by the organisms and can thus enter the food chain to human beings, and by what regulatory mechanism organisms of a given trophic level possibly release cadmium into natural waters.

For the elucidation of such a complex and complete set of intercorrelated data, team work by analytical, physical, and organic chemists, biologists, biochemists, geologists, oceanographers, and ecologists will be necessary. It must be based on intercalibrated and reliable data. It should be emphasized that there are still not enough data on the speciation of cadmium in various phases of the environment and in different links of the food chain, and on the toxicity of certain chemical forms of cadmium for various types of organisms. This lack is due largely to the type of analytical methods used for such determinations. In most cases the methods give only data on the total metal concentration, that is, usually they are not species sensitive. Therefore the results obtained by using various methods will be presented, but most of the discussion will be devoted to one group of methods that in our experience is the most promising for the determination and characterization of different dissolved chemical forms of cadmium in natural waters. This is the polarographic or, more generally, the voltammetric approach, which combines sensitivity and selectivity and has the additional advantage that chemical preconcentration steps are avoided, so that fewer contamination problems and losses due to adsorption or incomplete reactions occur.

First it will be advisable to consider the concentration and distribution of cadmium in a few types of natural waters to gain an appreciation of its occurrence and the variability in levels. Therefore a survey of data from the literature on the cadmium concentrations in the dissolved state in different oceans, seas, and coastal waters will be presented.

3.1.   Vertical Distribution of Dissolved Cadmium Concentration in the Ocean

The data summarized in Table 5 give the range of dissolved cadmium concentrations at different depths of the water column in the open oceans. From these data it is possible to conclude that the cadmium concentration in the open ocean

Table 5. Vertical Concentration Distribution of Dissolved Cadmium in Open Oceans

Location	Depth (m)	Remarks	Analytical Method	Cadmium Concentration Range	References
Atlantic (between 45°N and 18°N Biskaya-Antillen)	0.5–1000	Filtered	DPASV on MFE	6.6–24.8 ng/l	Nürnberg et al., unpublished results
Pacific (8°N and 8°S along equator up to 155°W, north to Hawai)	0.5–1000	Filtered	DPASV on MFE	3.9–92.3 ng/l	Nürnberg et al., unpublished results
	1000		DPASV on MFE	84.8 ng/l	
North Atlantic	50–5000	Filtered	Extraction + AAS	20–150 ng/kg	Eaton (1976)
Northeast Atlantic (between Canary Islands and the Cape Verde Islands, along west African coast)	10–2500	Filtered	Chelating resin + AAS	70–710 ng/kg	Riley and Taylor (1972)
Pacific (south of New Zealand,	3–5271	Not filtered	Coprecipitation + AAS	1–12 ng/kg	Boyle et al. (1976)
east of Japan and	8–5446			0.8–9.6 ng/kg	
in Bering Sea)	5–3711			5.4–10.3 ng/kg	

varies with depth from a few nanograms per liter up to several tens of nanograms per liter. Most of the values in Table 5 are in good agreement with the cadmium concentration given for open ocean by Goldberg (1972) as 20 ng/l.

Natural trace metals levels or baseline concentrations of open ocean waters are often useful in assessing anthropogenic effects on the marine environment. Eaton (1976), for example, suggests 60 ng/l as a baseline value for cadmium in the North Atlantic. In general, dissolved cadmium shows within an order of magnitude an essentially homogeneous distribution in the open ocean, although there are some regional variations. Cadmium always occurs in the dissolved state at the ultratrace level. This low level and the numerous steps involved from sampling to analysis could lead to wrong data due to contamination. Currently, because of technological progress, the sensitivity of instruments has increased significantly, but the problem of how and what to measure remains a personal one.

A detailed description of the sources of contamination and the precautionary steps needed in the analysis of aquatic samples is given in a review by Nürnberg (1977b). According to this author, sampling can be the first and a very important source of error. To avoid or minimize contamination, it is necessary to apply the following sampling procedure, particularly for surface water. The most significant and common contamination source is usually the research vessel itself, releasing continuously ultratrace amounts from its metallic body, its antifouling paints, the exhaust of its engine, and wastewater outlets to the surrounding water. Thus, for water sampling in the surface zone (typically 0.5 to 2 m below the surface), rubber boats should preferentially be used. In open-ocean missions a rubber boat should be lowered from the research ship, and samples taken at a distance of several hundred meters from the ship. Under rougher sea conditions, as in North Sea coastal waters, good results have been obtained with small wooden seagoing cutters, from which water samples were taken 3 m apart with a flask mounted to a specially designed sampling gear made from "fiberglass." For many toxic metals polyethylene has turned out to be the optimal choice as material for sampling flasks and also for storage containers, but mercury requires quartz or glass flasks (Nürnberg, 1977b).

Obviously all sampling containers and all subsequently used laboratory ware have to be thoroughly cleaned. First they are put into hydrochloric acid at 60°C for 6 days and then rinsed for 3 days with distilled water, acidified to pH 2 with HCl (Merck, Suprapur) to avoid contamination errors. Termination of the release of trace metals from the laboratory ware and container walls has to be checked by measurements. On the other hand, after this cleaning procedure the walls will have become very apt to adsorb trace metals, and this will cause errors due to losses. Thus the containers have to be conditioned with seawater for a certain time, after which the adsorption sites are mainly occupied by divalent Ca^{2+} and Mg^{2+} ions. The conditioned containers and laboratory ware are then

Table 6.   Variations in Levels of Dissolved Cadmium Traces in Untreated Seawater Samples

Storage time after Sampling (hr)	Cadmium (μg/l)			
	Sample 1	Sample 2	Sample 3	Sample 4
None	0.025	0.028	0.035	0.030
1	0.025	0.028	0.031	0.031
5	0.026	0.026	0.031	0.027

sealed in polyethylene bags and kept there until usage to avoid contamination by airborne dust particles.

The sampling of uncontaminated deep water requires special precautions and recently designed approaches and devices. The type of Nansen bottles used hitherto, which had to be lowered open through the water column on a steel rope from the research ship, are frequently subjected to severe contamination as they traverse open through the contaminated water cloud expending downwards and sidewards from the ship. The use of newly designed all-plastic sampling devices that go closed to greater water depths is almost mandatory to exclude contamination. There was good experience with new sampling equipment constructed on these principles during a recent oceanic mission through the Atlantic, Caribbean, and western Pacific. The levels of Cu, Cd, Pb, and also, partially, Hg were determined across the most important oceanic currents (e.g., the Gulf Stream, Caribbean Stream, Humboldt Stream, South and North Equatorial Stream in the western Pacific) (Nürnberg et al., unpublished data).

The first step in seawater sample pretreatment, preferentially performed within several hours after sampling, is separating the suspended particulate matter by 0.45 μ filtration. Of course, purified and subsequently conditioned filters must be used for which the trace metal blank values of the particular applied batch have been determined. To exclude contamination a technique has been developed (Mart, 1979) whereby containers are linked immediately to the filtration device, and subsequent filtration is performed under nitrogen pressure without bringing the sampled seawater into contact with the atmosphere, from which contamination due to airborne dust particles might occur (Mart, 1979). It was shown by Mart (1979) that within 6 hr after sampling no alteration in the concentration of dissolved cadmium occurs in unfiltered samples. That is an important factor for field missions, because it provides the possibility of performing the filtration under better controlled conditions in a clean laboratory, possibly a mobile one on a van operating along the coast if the sampling tasks end within the indicated time. An example is given by the data in Table 6, dealing with cadmium determinations in four different samples at different time intervals, within 6 hrs.

After filtration the sample solution is acidified to pH 2 with 1 ml/l of 10 M HCl (Merck, Suprapur). If it is to be stored for a longer time before further treatment, it is kept deep-frozen at $-20°C$. In this manner the determined concentration of dissolved cadmium is to be regarded as the essentially mobile amount of dissolved trace metal at the particular sampling location. After the sample pretreatment follows the determination step. For this purpose all parts of the measuring setup that come into contact with the investigated solution have to be scrupulously cleaned. It is recommended that preparation and measuring steps are performed under a clean bench set up to exclude contamination by air borne dust. Carefully performed preparatory work is of vital significance for successful measurements resulting in meaningful data. Sensitive analytical methods and good control over each step performed, with respect to contamination sources, are prerequisites for obtaining precise data which will provide a reliable basis for more general conclusions.

According to Eaton (1976), the distributions of cadmium are characterized by a greater scatter in the values at shallow depths. The lack of more appreciable differences between surface and deep water values (see Table 5) and the rather small variations in deep water values suggest that dissolved cadmium behaves essentially as an inert constituent in the open ocean. Eaton did not find a correlation between the concentrations of dissolved cadmium and of nutrients in any of the profiles, and he concluded that biological cycling appears to be unimportant as a mechanism controlling the vertical distribution of cadmium.

On the other hand, Boyle et al. (1976) and Bruland et al. (1978) observed the general distribution of cadmium along three detailed vertical profiles in the Pacific (Table 5). From the comparative depth profiles of cadmium and of nutrients like phosphate and silicate, Boyle et al. concluded that cadmium profiles resemble those of phosphate more than those of silicate. The concentration of cadmium increases rapidly with depth to a maximum at about 1000 m and then decreases slightly below this value. The covariance with phosphate is consistent for all stations and suggests that cadmium is regenerated in a shallow cycle, like the labile nutrients, rather than deeper in the ocean, as is silicate. These results are compatible with the following data from other authors. According to Nürnberg et al. (unpublished data), in surface Atlantic open-ocean water the dissolved cadmium concentration amounts to 6.6 ng/l and at 1000 m to 24.8 ng/l; in Pacific open-ocean water 3.9 ng Cd/l was determined in surface samples and 92.3 ng/l at 1000-m depth. Bender and Gagner (1976) reported values for the Sargasso Sea of ≤ 10 ng/l Cd in surface water and 25 ng/l in deep waters. Knauer and Martin (1973) reported surface concentrations below the detection limit (22 ng Cd/l) but, for 1000-m depth near Hawaii, found 112 ng Cd/l. The fact that cadmium is less abundant in surface than in deep waters is consistent with the notion that organisms incorporate this metal in their tissue and transport it to deep waters by the sinking of their detritus and debris.

Table 7. Horizontal Concentration Distribution of Dissolved Cadmium in Ocean Waters

Location	Depth (m)	Remarks	Analytical Method	Cadmium Mean Concentration (ng/l)	References
Atlantic (between 45°N and 18°N)	0.5	Filtered	DPASV on MFE	6.6	Nürnberg et al., unpublished results
	35			8.6	
	200			9.5	
Pacific (8°N and 8°S along equator up to 155°W, north to Hawaii	0.5	Filtered	DPASV on MFE	3.9	Nürnberg et al., unpublished results
	35			14.2	
	200			52.3	
Pacific Ocean, transect Hawaii-Monterey	Surface	Filtered 0.8 μ	Chelex-100 + AAS	20	Knauer and Martin (1973)
South Atlantic	Surface	Filtered	Chelex-100 + AAS	70	Chester and Stoner (1974)
Indian Ocean	Surface	Filtered	Chelex-100 + AAS	70	

Table 8. Cadmium Concentrations in Surface Waters of Nearshore Areas and in Closed Seas

Location	Average Cadmium Concentration (ng/l)	Analytical Method	Remark	References
Ligurian and Tyrrhenian coast	14	DPASV on MFE	Filtered	Mart (1979); Nürnberg et al. (1977)
North Sea coastal zone off East and North Frisian Islands and German Bight	20	DPASV on MFE	Filtered	Mart (1979); Nürnberg et al. (1976)
German Wadden Sea	12	DPASV on MFE	Filtered	Mart (1979); Nürnberg et al. (1976)
Northeast Atlantic	80	Chelex-100 + AAS	Filtered	Chester and Stoner (1974)
Northeast Atlantic	40[a]	Extraction + AAS	Filtered 0.2 μ	Preston et al. (1972)
South African Coast	90	Chelex-100 + AAS	Filtered	Chester and Stoner (1974)
China Sea	80	Chelex-100 + AAS	Filtered	Chester and Stoner (1974)
Atlantic Shelf		Chelex-100 + AAS	Filtered	Windom and Smith (1972)
North	110			
South	60			
Atlantic near Scotland	10[a]	Extraction + AAS	Filtered 0.2 μ	Preston et al. (1972)
Irish Sea	55 ng/kg	Extraction + spectrographically	Filtered	Mullin and Riley (1954)
Irish Sea	107 ng/kg	Extraction + spectrographically	Filtered, seasonal variation	Mullin and Riley (1956)

Location	Filtration	Method	Concentration	Reference
Irish Sea, eastern	Filtered 0.2 μ	Extraction + AAS	<10[a]	Preston et al. (1972)
Irish Sea, western	Filtered 0.2 μ	Extraction + AAS	40[a]	Preston et al. (1972)
English Channel	Seasonal variation	Extraction + spectrographically	116 ng/kg	Mullin and Riley (1956)
English Channel	Not filtered	ASV on HMDE	180	Smith and Redmond (1971)
English Channel	Filtered (one sample)	Extraction + AAS	60	Preston et al. (1972)
North Sea	Filtered	Extraction + AAS	410	Preston et al. (1972)
Jervis Bay	Filtered	ASV on MFE	190–1060	Florence (1972)
North Adriatic	Filtered	ASV on MFE	100	Branica (1976)
Monterey Bay	Filtered 0.8 μ	Chelex-100 + AAS	150	Knauer and Martin (1973)
Gulf of Maine	Not filtered	Extraction + AAS	230 ng/kg	Eaton (1976)
South Japan Coast	Filtered	Chelex-100 + AAS	130	Chester and Stoner (1974)
Sea of Japan	Filtered	Chelex-100 + AAS	110	
Java sea	Filtered	Chelex-100 + AAS	60	
Baltic Sea	Not filtered	ASV on HMDE	170	Kremling (1973)
Baltic Sea	Filtered	ASV on HMDE	220	Kremling (1973)
Baltic Sea	Filtered	ASV on HMDE	230	Bruegmann (1974)
Baltic Sea	Filtered	ASV on HMDE	330	Bruegmann (1974)
Baltic Sea	Filtered	ASV on HMDE	290	Bruegmann (1977)

[a] Geometric mean.

3.2. Horizontal Distribution of Cadmium Concentration in Oceanic Water

The mean cadmium concentrations in various open-ocean waters listed in Table 7 refer to surface waters. According to Chester and Stoner (1974), open-ocean waters are those more than 400 km from land masses. The horizontal distribution of the cadmium concentrations in the open oceans, as determined by several authors, is presented in Table 7. It is noticeable that these values range from about 4 to 70 ng/l, which still indicates a natural concentration level of cadmium in seawater media. On the basis of these data it is possible to draw the conclusion that there is no evidence of open-ocean pollution by cadmium.

3.3. Cadmium Concentration in Surface Waters near the Shore and in Closed Seas

Chester and Stoner (1974) collected nearshore water samples within 400 km of the adjacent land masses. In Table 8 the results of the analysis of various surface water samples taken, according to this definition, near the shore are summarized. The cadmium values in the table show discrepancies that may result from systematic analytical differences between laboratories or from contamination burdens due to lack of adequate caution in treating the samples; however, the variability may be real and reflect the existing state of pollution in particular areas of sea. The reasons for these discrepancies have to be established very carefully, as cases exist where, for the same part of the sea, different results were obtained. In this connection the question arises of what is the seasonal variation in dissolved cadmium concentration in relation to nutrient concentration, mixing of the water masses, and other factors. To answer this question with certainty, continous observations in certain areas, with repeated measurements, will be necessary, taking into account all precautions in sampling and pretreating procedures recommended by Nürnberg (1977b). For this reason the data summarized in Table 8 should serve only for a rough orientation in regard to the concentrations of cadmium that can be expected in certain areas of coastal and closed seawaters, indicating which parts are already highly polluted and where, consequently, certain regulatory actions will be needed to maintain marine life.

Results on the cadmium levels in seawater samples taken along the Ligurian and Tyrrhenian coast are given in Table 9. They are divided into four groups, according to the cadmium level determined. The average values from 0.01 to 0.02 ppb Cd coincide with the natural concentration of cadmium in seawater (Bernhard and Zattera, 1975). The category "high" coincides with the shipping routes in the direction of big ports (like Genoa and, to a lesser degree, La Spezia). It was also observed that the category "elevated" correlates with very clear

Table 9.   Levels of Dissolved Cadmium (μg/kg)(n_{total} = 225) in Ligurian and Tyrrhenian Coastal Waters [a]

Metal	Cadmium Level			
	Low	Elevated	High	Average
Cd	0.005–0.009	0.021–0.05	0.051–0.452	0.01–0.02
	(n = 59)	(n = 33)	(n = 9)	(n = 121)

[a] n = number of sampling stations. Data from Mart (1979).

waters, while "low" represents areas rich in algae and/or particulate matter. Obviously, these particulate matter and algae act as efficient scavengers for dissolved heavy metals by chemisorption and uptake, respectively.

On the basis of these data, it is obvious that at such low metal concentrations appropriate and careful handling of samples is a vital prerequisite for obtaining reliable and consistent data. It is also important that the samples do not change with time during storage before they are analyzed. The results summarized in Table 10 illustrate that comparing cadmium levels determined in various saline and lake water samples immediately after sampling and 75 days later reveals excellent agreement. This indicates that the stability of the filtered and acidified samples is satisfactory over extended periods. Evidently the treatment described

Table 10.   Long-Term Stability of Stored Sea and Fresh Water Samples (Summer 1975)[a]

Cd (ppb)			
At Sampling Day	After 75 Days	Difference (%)	Location
0.047	0.049	+ 4.2	
0.032	0.027	−15.6	
0.024	0.022	− 8.3	
0.017	0.017	0	
0.011	0.010	− 9.1	Ligurian Coast
0.041	0.046	+12.0	near
0.017	0.014	−17.6	Fiascherino
0.046	0.044	− 4.4	
0.022	0.021	− 4.5	
0.022	0.022	0	
—	—	—	
0.005	0.003	−34.0	
0.007	0.003	−57.0	Lake Constance
—	—	—	

[a] From Nürnberg et al. (1976).

Table 11.   Typical Ranges of Levels of Dissolved Cadmium in the Rhine River and Three Italian Rivers

Geographical Region	Cadmium (μg/l)
Lake Constance and upper Rhine	0.05–0.24
Torre	0.004
Adige	0.007
Po near Persepino	0.022

above minimizes the problems of contamination and losses due to adsorption quite satisfactorily.

The concentration of dissolved cadmium was determined in the estuary of the Weser River (Valenta et al., 1977b). The samples were taken in the streaming direction of the river close to Brake. An increase in detected cadmium concentration was observed in the range from 0.026 to 0.082 ppb, as a result of urban pollution upstream of the sampling points.

3.4.   Voltammetric Determination of Cadmium Concentration in Freshwater Samples

To illustrate the potentialities of the voltammetric method for freshwater analysis, some results of direct cadmium concentration determinations are presented. In this case no foreign salt or buffer addition was necessary.

Determination of the dissolved cadmium concentration in Lake Constance near Überlingen showed that, because of the significant presence of green algae, the cadmium level was very low, ranging from 0.004 to 0.010 μg/l. These particularly low levels of cadmium in Lake Constance required the full sensitivity potentialities of DPASV.

In Table 11 (Nürnberg, 1978a) the determined ranges of dissolved cadmium in various river waters are presented.

In this section, as the tables indicate, two analytical methods are dominant for determination of trace levels of cadmium in natural waters. One of them, electrothermal AAS, is an indirect method; the other, voltammetry, allows direct measurements. An example of the parallel application of both methods to the determination of cadmium in water samples is given by Duinker and Kramer (1977). In filtered seawater samples the cadmium content was measured at the natural pH of 8.1 with AAS after previous extraction, and directly with DPASV at the HMDE. An average concentration of 0.23 μg/l for cadmium in North Sea water samples of 32‰ salinity has been determined with the voltammetric approach. Extraction of the cadmium from the same sample, followed by AAS measurement, yielded 0.11 μgCd/1. The concentration of cadmium determined with AAS, after previous extraction, is lower. According to the characteristics of the analytical method, however, the same or a higher amount of determined cadmium is to be expected, in comparison with the value determined by voltammetry. The finding of a lower concentration with AAS seems unrealistic and therefore may indicate certain losses of cadmium during the extraction step.

Certainly, it is difficult to consistently achieve the same accuracy with the two independent methods. On the other hand, such comparative measurement is, in general, a practical solution of how to approach the accuracy problem in trace metal determinations, as long as standard reference materials for the matrix type under study are not available. When two independent methods are applied, it is necessary that the obtained results coincide within 20 to 30% in the ultra trace range (Nürnberg and Valenta, 1975). Usually, the method providing the more rapid determination should be chosen as the routine one, while the other method serves for checking and standardization purposes. According to the particular matrix type, either flameless AAS or various modes of voltammetry are to be applied as routine or checking methods, respectively. For the dissolved cadmium contents in all types of natural waters, however, DPASV has become the method of choice also for routine applications.

4. CADMIUM SPECIATION BY INORGANIC LIGANDS IN NATURAL WATERS

Cadmium belongs to group IIb of elements in the periodic system, and in aqueous solution it has the stable 2+ oxidation state. Cadmium is a d^{10}-acceptor and for this reason has an interesting solution chemistry that is fairly well known. Like zinc and mercury, it belongs to the oxyphilic and sulfophilic group of elements.

In natural waters metals may exist bound to particles and in the colloidal or dissolved state. In general, suspended particles are considered to be those greater than 1 μ in size, while colloidal particles fall in the range between 1 μ and 1 m/μ.

Metal ions in general and cadmium in particular can exist as different chemical species in natural waters. Different chemical species can be distinguished in solution and the solid phase. The dissolved metal content may be present in the following forms:

1. As hydrated ions.
2. Complexed with inorganic ligands, like chloro, carbonato, and hydroxo complexes, that is, more or less labile complexes.

3. Chelated in rather stable and in inert complexes with organic ligands such as amino, fulvic, humic, and nucleic acids, proteins, thionines, and chelators of anthropogenic origin (e.g., EDTA, NTA).

Moreover, metal ions can be associated with colloidal particles or occur adsorbed at or incorporated in suspended particulate matter or sediments. The suspended and colloidal particles may consist of individual or mixed hydroxides, oxides, silicates, sulfides, or other compounds, or they may consist of clay, silica, or organic matter to which metals are bound by adsorption, ion exchange, or surface complexation.

When investigating the speciation of metals, the following questions arise:

1. Which is the most abundant chemical form of a metal ion in the natural water type, with respect to the chemical composition of the water, under study, and is this chemical form also the most toxic one for the aquatic organisms?
2. What charge does the chemical species have, and what are their adsorption properties?
3. What is the size of the existing chemical forms of the metal?

The last item of information, together with the charge, can give some indication of the ability of the species to penetrate membranes, particularly in the uptake of metals at the phytoplanktonic trophic level of the marine food chain. In addition to answer to any of these questions of general significance, it is important to have reliable data on the total cadmium concentration in the natural waters under investigation which can vary in both location of sampling and chemical composition.

4.1. Seawater Media

Determination of the total dissolved and the suspended-particulate-matter bound concentrations of trace metals has been discussed in the preceding section. On the basis of accurate and precise data regarding the dissolved cadmium concentration in seawater, one can proceed to observe what is known about the speciation of cadmium in seawater and which experimental methods are able to study this speciation.

The macrocomponent composition of seawater is remarkably constant, and the waters in different parts of the open oceans differ very little in this respect. Table 12 highlights the composition of seawater according to Sverdrup et al. (1942).

The general significance of speciation stems from the fact that the chemical behavior of an element in the marine environment and the rate at which an element participates in the chemical processes depend on its chemical form. Speciation influences the participation of an element in geochemical processes

Table 12.   Macrocomponent Composition of Seawater

Ion	Amount Present in Ocean (mol/l)
Na^+	0.47
K^+	1.0×10^{-2}
Ca^{2+}	1.0×10^{-2}
Mg^{2+}	5.4×10^{-2}
F^-	7×10^{-5}
Cl^-	0.55
SO_4^{2-}	3.8×10^{-2}
pH	7.89
Alkalinity (equiv/l)	2.3×10^{-3}

such as precipitation-dissolution and adsorption-desorption (Kester et al., 1975). Chemical speciation and the reactions by which species are transformed from one form to another have to be elucidated in order to understand the mechanisms by which pollutants affect the marine environment. Thus chemical speciation is generally significant in relation to geochemical and also biochemical (see below) processes in the marine environment.

Evidence exists in the literature that the availability of metals to organisms depends on their speciation. The degree of complex formation and its nature, as well as the oxidation state and the degree of hydration of the metal species, are important factors in determining the ecological significance of a metal (Duinker and Kramer, 1977). According to Provasoli (1963) and Bernhard and Zattera (1969), phytoplankton has a tendency to take up organic complexed trace metals. However, because of cation exchange Gutknecht (1963) and Bryan (1969) reported the preferential uptake of the ionic species by certain algae. Although evidence for both types of chemical species is available, no general conclusion can yet be drawn about the relative effects of organic and inorganic ligands in the uptake of metal complexes in seawater (see Phillips, Chapter 12).

The need for a better understanding of the transport and biological interactions of heavy metals in natural waters has made the study of the chemical speciation of these metals an area of high priority in marine trace chemical research. There are two basic approaches to identifying the chemical species of minor constituents. The preferable and direct one is to design methods that can distinguish different chemical forms present in seawater; the other, more indirect approach is to develop models that are applicable to the marine environment, based on knowledge of the chemistry of the element in question (Kester et al., 1975). This theoretical approach will be discussed at the end of this section.

Speciation Based on Size Differentiation

One factor encountered in defining chemical species in marine systems is separation of the species into the dissolved and particulate phases.

Filtration. The most common approach to the separation of the two phases is filtration. Particle size distinction can be a significant factor in chemical speciation, because it is likely that some chemical processes are surface area dependent and the chemical reactivity of some solid phases is particle size dependent (Kester et al., 1975).

According to the most common convention, the dissolved and particulate phases are separated by passing the solution through a 0.45 μ pore filter. Metal adsorbed on fine mineral material or organic detritus, in addition to that bound within planktonic material or coprecipitated with the hydrated oxides of iron and manganese is removed in large proportions by filtration (Abdullah et al., 1976). Since the size of colloidal particles covers the range from 10^{-3} to $1\ \mu$, filtering the water through 0.45 μ results in incomplete separation of colloids. The filtrate will be not only a real solution but also a mixture, containing colloids up to 0.45 μ in size. In a real solution the size of the particles is usually less than $10^{-3}\ \mu$. For the separation procedure the next question is, How regular in size are the pores of the filter? The fact that in filtered solution colloids of certain size can still be present has to be borne in mind when water samples taken from different areas and depths are compared. The content and the structure of colloidal particles may be completely different for various regions.

In future studies possible improvements over filtration could be obtained by using centrifugation over a range of accelerations, employing density gradient separation. Another approach to metal speciation involves the use of molecular size discriminating techniques such as ultrafiltration and dialysis. These techniques suffer severely from contamination problems, however, and have not yet been successfully applied to heavy metal speciation studies.

Ultrafiltration. Comparison of the filter pore sizes with the species sizes indicates that free metal ions and small complexes will pass through the Amicon PM 10 filter, but metal ions associated with humic substances and colloids will not pass through (Guy and Chakrabarti, 1975). The results indicate that in this separation procedure adsorption is a serious problem. This method is not selective and sensitive enough for low molecular weight complexes. Cadmium complexes with chloro, carbonato, and similar inorganic ligands will pass through the filter together with the hydrated or free metal ions. If the method is used with caution in regard to the adsorption effects, however, it should be possible to distinguish between simple complexes with inorganic ligands and chelates with natural organic ligands.

Dialysis. If the size of species is compared with the size of the membrane pores, it is evident that free metal ions and small complexes can pass through the dialysis membrane. The results of metal complex dialysis reveal another problem, that is, usually a negative charge is associated with the membranes. Under such conditions negative species like metal complexes experience a smaller effective pore size because of the repulsion of similar charges. For this reason a longer time is required to attain equilibrium (Guy and Chakrabarti, 1975).

One of the main problems in ecological studies is to establish the relationship between the chemical form of a given metal and its availability to biological systems (Kremling, 1976). It has been mentioned that AAS is not a species-specific method. To use AAS for metal speciation studies a preliminary separation of the species is necessary. For example, separation based on solvent extraction, coprecipitation, ion exchange, or size fractionation can be applied.

At present the only method allowing direct measurement of the metal concentrations in aquatic samples and potentially capable of distinguishing between the "free" and the complexed part of dissolved metal species is the voltammetric approach.

Speciation Based on Charge and Chemical Form Differentiation

The remarkably high performance and the fundamental properties of stripping voltammetry in its advanced versions lead to striking potentialities in the ultratrace analysis of metal species, with respect to sensitivity, precision, reproducibility, resolution, and simultaneous determination. Another important aspect of this method is speciation (Nürnberg and Valenta, 1975). Applying this method, one can get the answers to three of the questions previously stated in this section that is, those concerning the most abundant chemical form of the metal, its charge, and its adsorption properties. Experiments with aquatic organisms, on the other hand, can give information about the toxicity of a particular chemical form.

A particular domain of advanced voltammetry is determination of the concentrations of toxic trace metals dissolved in seas and inland waters and of their speciation. The potential ability of voltammetry to discriminate between different species on the basis of changes in their electrochemical behavior as a result of changes in experimental parameters is discussed subsequently.

Metals in the form of inorganic complexes (such as chloro, hydroxo, carbonato, bicarbonato, and hydroxychloro complexes) and metals in the form of chelates with dissolved organic matter (such as amino acids, humic materials, or extracellular metabolites) can easily be observed electrochemically by their reduction at the mercury electrode (Abdullah et al., 1976; Nürnberg et al., 1976). Experiments have shown that copper and zinc chelated by organic compounds such as EDTA, alanine, humic, and fulvic acids (Mee, unpublished data) are reduced at the DME and of course can be also plated on solid electrodes such as graphite electrodes coated with a mercury film.

At the natural concentration level of cadmium in seawater, ASV or, preferably DPASV, has to be applied. The electroactive part of cadmium, which equals the sum of the hydrated and labile complexed cadmium, will be determined. Lowering the pH by acidifying the sample below 3.5 releases adsorbed metal from either colloidal hydrated oxides or mineral matter (Abdullah et al. 1976). At pH 2 it is to be expected that even more inert complexes of metals are decomposed and that frequently the total concentration of metal can be determined (but see below). Thus the increase in peak current with decreasing pH does not necessarily mean an increase in concentration (Duinker and Kramer, 1977).

It has to be established by the standard addition method whether the concentration of a metal changes with pH. Differences between the peak heights of acidified and nonacidified samples are not necessarily indicative of the amount of metals in "nonlabile" complexes. It has been suggested that the change in peak heights with pH may be due to the change in activities of the individual metals and not to a concentration change of the electroactive species (Kremling, 1976). However, opposite effects are also possible, as reported by Bubić, et al. (1973). The peak height of electroactive cadmium is greater at pH 8, and for this reason the sensitivity of the method is also higher. The calibration curves, obtained by the standard addition method for cadmium, gave the same concentration of this metal at various pH values. According to these authors, the change in the peak heights of cadmium with pH can be ascribed to the change in the ionic form of the elements and not to a change in concentration.

If the water contains higher amounts of dissolved and suspended organic material, acidification will usually not be sufficient to destroy all inert complexes or chelates; hence the metal will not be completely released and be determined. In this case ultraviolet (UV) irradiation has to be applied to the sample. The acidified sample (pH 1) should be irradiated for a few hours before determination of the total amount of metal. A detailed description for contamination-free UV irradiation has been given by Mart (1979).

Metal bound in organometallic compounds, where certain metals form a central part of the structure (e.g., porphyrin), and metal adsorbed at colloidal matter and/or occluded into the colloids have to be liberated to become reducible species (Abdullah et al., 1976). For water samples two procedures to release the bound metal from the dissolved organic matter or colloidal and mineral material are acid digestion and UV irradiation. Both treatments are rigorous and destructive, causing sufficient decomposition of the organic material. The concentration of the metal obtained after the sample has been acidified and UV irradiated will be higher than the concentration determined at the natural pH of seawater or inland waters.

The results of Bruegmann (1974) concerning the determination of different

Table 13.   Labile and Ultraviolet-Labile Cadmium Contents in Seawater

Depth	Labile Content (μg/l)		UV-Labile Content (μg/l)	
(m)	Filtered	Unfiltered	Filtered	Unfiltered
0.5	0.18	0.25	0.22	0.56
10	0.09	0.11	0.15	0.80
30	0.08	0.13	0.42	0.68
58	0.05	0.13	0.14	0.23

species of cadmium in Baltic seawater samples show that in a sample at natural pH, 40% of this metal was present in labile forms. After acidification of the sample an increase in cadmium concentration was recorded, indicating that this part of the cadmium was not electroactive at pH 8. After UV irradiation an increase in cadmium concentration was again obtained, and the determined concentration corresponded to the total dissolved concentration of the element. To summarize, 40% of the total cadmium was present in the form of labile and hydrated forms at pH 8, 10% occurred as inert complexes, and 50% existed as very inert complexes or was adsorbed or occluded by particles. Such a distribution of cadmium is easy to understand, as it is known that the water samples taken from Warnow Estuary were particularly rich in dissolved and suspended organic and inorganic material (humic acids, organic amines, phosphates, acids, etc.). The same author applied this procedure to another sample of seawater, taken from the area of permanent upwelling in front of northwest Africa, where the water is rich in nutrients and the biological production is high. A comparison of the cadmium concentrations determined in filtered and unfiltered samples, taken at depths from 0.5 to 58 m, without and after UV irradiation, is shown in Table 13. It is interesting to notice that in this case appreciably higher cadmium concentrations were found in the surface layer, even without UV irradiation. From the results it is obvious that higher concentrations of cadmium were obtained in samples that were UV irradiated. The results in Table 13 can also serve as an interesting example of the importance of filtration of water samples taken at regions rich in particulate matter or biological production. In trace metal studies it has to be borne in mind that the concept of "total metal," based on data obtained by filtering the water samples through a 0.45 μ pore size filter, refers to the dissolved part of metal in chemical species that can pass through this pore size. In regions rich in particulate matter it should certainly be interesting to consider the entire water samples, as particles larger than 0.45 μ participate in adsorption-desorption reactions and consequently could exert a strong influence on the level and distribution of the dissolved part of trace metals (Mart, 1979).

Abdullah et al. (1976) investigated the behavior of the chelating ion exchange resin Chelex-100 in the collection of several trace metals in seawater, including cadmium. As the detection method, DPASV at the MFE was applied. Their results show that, when the original seawater sample was applied to the resin, without any prior chemical treatment, all the reducible species of the metal were retained by the resin. The same was observed after the pH of the sample was decreased to 5.4. Acid digestion of the original samples caused an increase in cadmium concentration. Since centrifugation at 48,000 G reduced the amount of metal liberated after the acid digestion of seawater, it seems very probable that the additional fraction of the metal released by decrease of the pH is associated with the colloidal and fine particulate matter. Furthermore, the data suggest that complete separation of the dissolved cadmium complexes with inorganic and organic ligands can be achieved by the use of chelating resins. In other words, colloidal species of metals, metals bound to organic material or colloidal species, or metals adsorbed on colloidal species will remain unaffected by the chelating resin. Although this fraction is certainly important for the geochemical cycle of trace metals in the seawater link of the cycle, its availability to filter feeding organisms by enzymatic attack remains an open question, while release of metal from organic detritus is usually achieved by bacterial activities and oxidative processes.

In conclusion, the physicochemical treatments of natural water samples previously mentioned (such as filtration, percolation through Chelex-100, UV irradiation, and acid digestion) in connection with ASV for the subsequent determination, can provide information on the concentration and distribution of a certain metal between various chemical forms and phases existing in natural waters. Batley and Florence (1976a, 1976b) classified metal species in natural waters arbitrarily into four groups. This classification is based on the ability of a chelating resin, like Chelex-100, to remove metal from labile and particulate-matter-bound metal fractions, and on behavioral rather than chemical differences. The following conclusions can be drawn:

1. Filtration through a 0.45 μ membrane filter removes particulate-matter-bound metal, but colloids of smaller size can pass through and can be present in the filtrate.
2. Percolation of the filtered sample through a column of Chelex-100 results in the removal of the free metal ion, labile organic complexes (e.g., citrate, glycinate), labile inorganic complexes (e.g., carbonato, chloro, hydroxo, sulfato), and nonlabile inorganic complexes which are dissociated by Chelex-100.
3. Acidification of the sample to pH 2 releases the metal adsorbed onto the colloidal and mineral material and the metal contained in many but not all kinds of the dissolved nonlabile organic chelates.
4. Ultraviolet irradiation or acid digestion of the filtered sample releases metal bound or occluded with suspended particles, such as clays, silicates, or organic

Table 14.   Cadmium Speciation Determination in Seawater Sample Taken near Wattamolla Beach[a]

Metal Species	Cadmium	
	(μg/l)	(%)
Free metal + labile organic and inorganic complexes dissociable by Chelex-100 and determinable by ASV	0.03	11
Labile organic complexes and labile metal adsorbed on organic colloids nondissociable by Chelex-100 and determinable by ASV	0.21	75
Nonlabile organic complexes and nonlabile metal adsorbed on organic species nondissociable by Chelex-100 and nondeterminable by ASV	0.03	11
Nonlabile inorganic complexes and nonlabile metal adsorbed on inorganic species nondissociable by Chelex-100 and nondeterminable by ASV	0.02	7

[a] Data from Batley and Florence (1976a, 1976b). The estimated error ±10 to 15%.

detritus, and metal complexes in rather inert organic chelates (see also Mart, 1978).

Batley and Florence (1976a, 1976b) have applied their speciation scheme to the analysis of cadmium in filtered surface seawater, taken near Wattamolla Beach, an area relatively free from urban pollution. Based on their experimental results, the concentrations of various chemical species of cadmium are listed in Table 14. It follows that in this Australian coastal area 86% of the total dissolved cadmium corresponded to complexes with organic material. This seemed to the authors to indicate a remarkably high concentration of dissolved organic material in the coastal waters investigated. From these results Batley and Florence (1976b) concluded that a larger fraction of labile cadmium was present in filtered seawater as labile organic complexes and labile metal adsorbed on organic colloids. Unfortunately, however, the authors did not mention any figure on the content of organic material in their samples. Nevertheless there remain two possibilities to explain the experimental results:

1. Percolating the filtered solution through the column removes Ca and Mg cations, causing another distribution of the complexing organic material, so that the amount that was previously bound to calcium and magnesium becomes available for cadmium complexation.
2. A certain contamination with the organic material occurs from the chelating column, causing artifacts with respect to the real complexation of cadmium by organic material.

As mentioned before, the results of Bruegmann (1974) indicated that 60% of total cadmium concentration was present in an inert state, adsorbed to or occluded into materials of higher molecular weight, but in that case it was explicitly stated that the samples were rich in dissolved and suspended organic material. Moreover, these samples were taken from an estuary and not from the open sea. Thus the statement of Batley and Florence (1976) that the distribution of dissolved cadmium presented in Table 14 is a distribution in a typical seawater sample appears to be questionable. Whilst investigating the chelating properties of the anthropogenic ligand NTA in genuine seawater sample, at the natural concentration level of cadmium, our own results (Raspor et al., 1978a) has unequivocally shown that no measurable chelate formation with the dissolved ligand NTA occurs up to 10^{-4} M NTA, because of the existing strong competition of the present alkali earth cations for the ligand. This finding is of general and typical significance for the concentration requirements for the formation of rather stable and inert cadmium chelates with organic material in seawater and will be discussed in more detail in the next section.

The chapter up to now has focused on how information on the total dissolved concentration of cadmium and its overall distribution between labile and bound parts was obtained by applying certain physicochemical pretreatment procedures to the sample, followed by an appropriate determination method. Now the possibilities of obtaining information about the formation of labile complexes (e.g., chloro, carbonato, sulfato) with a particular metal will be discussed. There are also new and efficient approaches to obtain quantitative and detailed information about the formation of metal chelates with organic materials in natural waters in a more immediate manner. This will be treated in the subsequent section.

The formation of labile complexes of cadmium with inorganic ligands in seawater will first be considered; the most abundant of all are chloride ions. Chlorides, as well as hydroxyl and hydrogen ions, occur in all natural waters and may be regarded as one of the most mobile and abundant labile complexing agents with respect to heavy metals. Seawater contains 0.5657 M Cl^-, and its pH value lies between 8.1 and 8.2. Therefore chloride ions could be a more important factor in the distribution of heavy metals than they have been considered up to now (Hahne and Kroontje, 1973). In spite of the fact that the data obtained by Vanderzee and Dawson (1953) are not directly applicable to seawater, their observation is important, that during the potentiometric determination of consecutive cadmium-chloro stability constants the interaction with Cd(II) increased when the concentration of chlorides was increased.

The degree of cadmium interaction with chloride ions can be followed polarographically from the shift of the reversible half-wave potential ($E_{1/2}$) of Cd(II) reduction with the concentration of the ligand. Determination of the relationship between the half-wave potential shift ($\Delta E_{1/2}$) of a reversible (!) trace metal reduction and the ligand concentration is the common and usual approach for the voltammetric determination of the stability constant and ligand

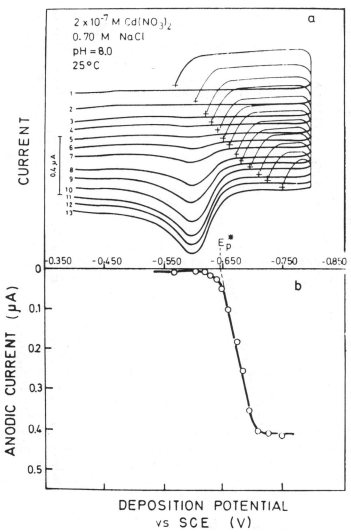

Figure 4. (a) Voltammograms recorded for 2×10^{-7} M Cd concentration in 0.7 M NaCl solution and pH 8, at different deposition potentials. (b) The polarogram constructed by plotting anodic peak current versus deposition potential. (Branica et al., 1977.)

number of labile complexes with sufficient mobile complex equilibria (Crow, 1969; Nürnberg and Valenta, 1975). In this manner voltammetry offers an efficient and reliable way to characterize and identify labile metal complexes in the various types of natural waters. Barić and Branica (1967) obtained from $E_{1/2}$ shifts the value of 1.43 for log K_{CdCl^+} at the ionic strength of seawater, using a dropping amalgam electrode. At the same time these authors confirmed that no significant hydrolysis of cadmium occurs at a pH lower than 9. The cadmium concentration was 10^{-4} M in this study, higher than its natural value in seawater. Application of the HMDE in connection with ASV made it possible to decrease the adjusted cadmium concentration from the 10^{-4} M level to 5×10^{-9} M (Bubić and Branica, 1973). In that study an ASV procedure was employed to determine the ionic state of cadmium at a concentration similar to that in seawater.

The necessary relationship between the shift of the reversible half-wave potential (or, in this case, an equivalent potential value) and the ligand concentration (c_{Cl^-}) was established in the following manner. A series of anodic stripping voltammograms was recorded, referring to systematically altered deposition potentials but always the same deposition time (t_d) (Figure 4a) (Branica et al., 1977). During the deposition time the solution was continuously stirred. Subsequently 15 sec of quiescent time was allowed before a rapid sweep of potential in the anodic direction started. From the series of voltammograms recorded at different deposition potentials the dependence of the anodic peak current on the deposition potential was obtained, as shown in Figure 4b. This dependence yielded a curve very similar in shape to the corresponding classical dc polarographic wave (Figure 5). From that curve the potential (E_p^+) was obtained by extrapolating the slope of the wave to zero current (Figure 4b).

The E_p^+ potentials, determined at different free ligand concentrations, (Cl^- in this case), can be used to evaluate the conditional stability constants of the complexes formed. In this manner the common polarographic approach relevant for labile complexes, based originally on the shift of reversible half-wave potentials ($\Delta E_{1/2}$) as a function of the free ligand concentration, becomes applicable also at the very low natural concentration levels of cadmium. An experimental and theoretical correlation has been found between the classical polarographic half-wave potential ($E_{1/2}$) and the "critical potential" (E_p^+). For a reversible electrode process E_p^+ may be shifted toward more negative potentials up to 100 mV, in comparison with $E_{1/2}$ of the corresponding reversible dc polarographic wave (Figure 5). Although E_p^+ differs in value from the corresponding cathodic $E_{1/2}$, the significance of its shift with the free ligand concentration is the same, and under invariant experimental conditions E_p^+ can therefore be used in the evaluation of stability constants and ligand numbers of labile metal complexes according to the procedure for consecutive complex series of a given ligand proposed by DeFord and Hume (1951). In Figure 5 both polarograms i.e., the classical one and that resulting from the outlined ASV approach, were determined at the same cadmium concentration of 1×10^{-4} M. It is evident that E_p^+ is 35 mV more negative than $E_{1/2}$ of the corresponding classical dc polarographic wave. On the basis of the required experimental evidence the potential E_p^+ does not change with variation in the cadmium concen-

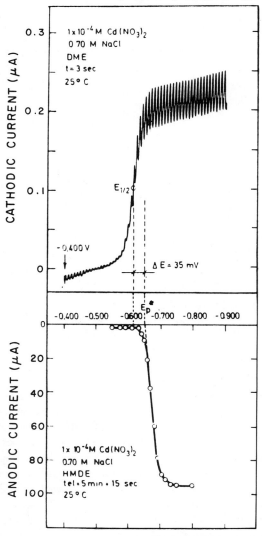

Figure 5. Comparison of cadmium reduction dc polarogram at $1 \times 10^-$ M concentration with the polarogram-like curve obtained by plotting anodic peak current versus deposition potential. (Branica et al., 1977.)

tration but depends on the type and ionic strength of the electrolyte used and on the concentration and type of particular ligand forming labile complexes.

As a model solution for investigating complex formation of cadmium with chloride ions, a mixture of $NaClO_4$ and $NaCl$ having a constant ionic strength

IONIC STATE OF CADMIUM IN SEA WATER

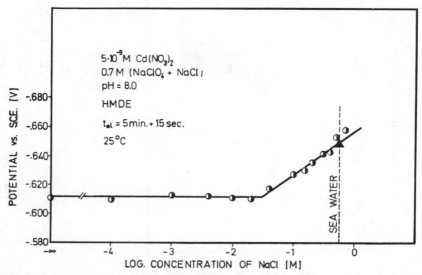

Figure 6. Critical apparent potential (E_p^*) dependence on various chloride concentrations at 5 × 10⁻⁹ M Cd concentration and 0.7 M ionic strength (NaClO₄ + NaCl). (Bubić and Branica, 1973.)

of 0.7 M was used (Bubić and Branica, 1973). From the series of voltammetric measurements, as shown in Figure 4a and 4b, E_p^+ potentials were obtained as a function of the free concentration of chlorides. A graphical presentation of this relationship is given in Figure 6. The chloride concentration corresponding to real seawater is indicated. The authors have determined that E_p^+ does not change with increasing chloride concentration up to 3 × 10⁻² M. At higher concentrations than that, E_p^+ linearly shifts toward negative values. The straight line relationship has a slope of 30 mV per decade of ligand concentration. This corresponds to the theoretically predicted change in potential when the CdCl⁺ species is the predominant one in the solution investigated. The fact that the last two points (Figure 6) are not on the theoretical line indicates the possible existence of CdCl₂⁰ species at elevated chloride concentrations. The existence of a dichloro complex of cadmium was experimentally proved in model solutions of 2.0 M ionic strength. The result of these studies indicated that at chloride concentrations higher than 2.6 × 10⁻¹ M the predominant species in the solution is CdCl₂⁰. Bubić and Branica (1973) reported a value of 1.52 for log K_{CdCl^+} for the conditional stability constant of CdCl⁺ at 0.7 M ionic strength.

The voltammetric procedure described for the characterization and identification of the fraction of cadmium with complexed chloride ions in seawater or analogous model media, and the determination of the respective conditional

Table 15. Stability Constants of Cadmium Carbonato and Hydrogen Carbonato Complexes in 0.7 M Perchlorate Solution[a]

Metal—Ligand	Concentration Level ($10^{-6}\,M$)	
	β_1	β_2
(a) Cd—CO$_3^0$	$(4.67 \pm 0.38)10^2$	$(4.46 \pm 0.36)10^4$
(b)	$(7.98 \pm 0.67)10^2$	$(1.21 \pm 0.10)10^5$
(a) Cd-HCO$_3^+$	1.84 ± 1.25	$(2.28 \pm 3.52)10^1$
(b)	1.54 ± 1.46	24.92 ± 4.00

[a] Data of Sipos et al. (1979a.)

stability constants, can be applied to any other type of labile complex that occurs in sea water or other natural waters and causes a shift of $E_{1/2}$ or E_p^+ in a reversible electrode process with the corresponding change in free ligand concentration. In addition to chloride ions, sea water contains, the following anions, carbonato, hydrogen carbonato, hydroxo and sulfato, as major constituents which form with trace heavy metals more or less labile complexes. The question is, How much do they contribute to the speciation of the dissolved concentration of cadmium in seawater?

A review on the stability constants of carbonato-complexes of cadmium (Bilinski et al., 1976) indicates that none of the values hitherto reported is directly applicable to seawater media.

A more recent voltammetric study by Sipos et al. (1979a) on the stability constants and ligand numbers of carbonato and hydrogen carbonato complexes of cadmium in seawater has settled this problem.

As was mentioned before, the dependence of $E_{1/2}$ and E_p^+ potentials of a reversible electrode process on ligand concentration can be used for the evaluation of stability constants and ligand numbers of labile metal complexes. In the same manner the peak potential (E_p) or half-peak potential ($E_{p/2}$) obtained in sweep voltammetry can be used to obtain the same information. Sweep voltammetric measurements at the HMDE were performed in this study, using the shift in E_p as indicator for labile complex formation. The measurements here considered refer to $10^{-6}\,M$ cadmium concentration. More recent investigations at the 10^{-9} M cadmium level have confirmed the findings reported here. In this context it is to be emphasized that in a fundamental sense the actual cadmium concentration is not significant for this voltammetric determination of the conditional stability constants of consecutive labile complexes and of their ligand numbers. This is valid provided the basic prerequisite holds that the ligand concentration is much larger than that of the trace metal.

The stability constants of carbonato and hydrogen carbonato complexes of cadmium as determined for seawater are listed in Table 15. The stability constant of cadmium-carbonato complexes has been calculated on the basis of 34 experimental points determined in solutions containing more than $10^{-2}\ M\ CO_3^{2-}$. The stability constants of cadmium-hydrogen carbonato complexes have been calculated on the basis of 27 experimental points measured in solutions containing $0.2 \times 0.7\ M$ $NaHCO_3$ and less than $10^{-2}\ M\ CO_3^{2-}$. The stability constants in Table 15 determined by Sipos et al. (1979a) are given in relation to (a) the "ionic medium" pH scale, where the activity coefficient of hydrogen ions approaches zero when H^+ activity approaches zero in the pure ionic medium; and (b) the operational "infinitive dilution" activity scale, where pH is defined by a standard dilute buffer solution. Although the infinitive dilution scale has no fundamental thermodynamic foundation, it is widely used because of its convenience in practice.

In a similar manner the speciation of lead by carbonate in seawater was recently elucidated (Sipos et al., 1979a). Another recent study has revealed that, if the formation of coulombic ion pairs between the major cationic and anionic constituents of seawater is also taken into account, further refinement is achieved in the conditional stability constants values and in the resulting distribution of Cd(II) and Pb(II) over the major labile complex species existing in seawater (Sipos et al., 1979b). This finding will have general significance for quantitative data on the speciation of trace metals by labile complex species in natural waters.

Theoretical Calculation of Cadmium Speciation with Inorganic Ligands in Seawater

In establishing models of trace metal speciation in natural waters it is necessary to take into account the complex network of multiple interactions which results from the fact that any cation can interact with any ligand, yielding a complex. The resultant set of nonlinear equations can then be solved for the concentration of the free, uncomplexed reactant, provided that the following information is available:

1. The total concentration of the various reacting species.
2. The composition and the conditional stability constant of each relevant complex considered.

With the aid of computer programs the equilibrium concentration and subsequently the percentage distribution of each trace metal species can be calculated. If data on the stability constants of all relevant species are available, it is easy to calculate the concentration or percentage distribution of the total dissolved metal content in seawater or inland waters. Models used to compute theoretically the speciation of trace metals under natural conditions, such as those constructed by Sillén (1961), Garrels and Thompson (1962), Kester and Pytkowicz (1969), Stumm and Morgan (1970), Zirino and Yamamoto (1972), and Morel and Morgan (1972), are valuable aids in expanding our capability to describe the

Table 16.   Model Distribution of Dissolved Cadmium in Seawater

References	Cd^{2+} (%)	$CdCl^+$ (%)	$CdCl_2^0$ (%)	$CdCl_3^-$ (%)
Zirino and Yamamoto (1972)	2.5	40	50	6
Ahrland (1975)	3	34	51	12
Dyrssen and Wedborg (1973)	1.8	29	38	28

chemistry of aquatic systems (Elder, 1975). However, ultimately they require confirmation by experimental data before they can be accepted as reflecting completely the reality.

The speciation calculations are performed under the assumption that equilibrium conditions exist in the solution. Although natural waters, particularly the sea, are quite dynamic systems, usually this assumption will be correct for the rather mobile and thus rapidly adjusting equilibria of the considered labile complexes with inorganic ligands. Certainly, the most serious problem arises with respect to the proper choice of stability constant values. If different values for the stability constants are introduced into the program, a different distribution of the species will necessarily result, even if the total concentration of the macrocomponents (i.e., the salinity) remains constant. As an example, calculation of the cadmium distribution between various complexes with chloride ions is presented in Table 16. The calculation refers to 25°C, 1 atm pressure, 35‰ salinity, and pH 8, using different sets of cadmium-chloro stability constant values. Usually, under given conditions the metal is present in one or two heavily predominant forms, with all other forms occurring only in rather insignificant quantities (Elder, 1975); the same can be observed for cadmium speciation in Table 16. In spite of the different percentage distributions in Table 16 it is obvious that chloride complexes completely dominate cadmium chemistry in seawater, with $CdCl_2^0$ as the major species. There are two types of stability constants that can be used to calculate the speciation of the total metal concentration in a particular solution: *thermodynamic* and *operational*. In coordination chemistry the latter is also termed *conditional* stability constant. Its value always refers only to the given medium conditions with respect to composition, ionic strength, pH, and so on.

In the paper by Zirino and Yamamoto (1972) a pH-dependent model for the speciation of cadmium in seawater was constructed with available and estimated thermodynamic stability constants and individual ion activity coefficients. This model was used to calculate the degree of interaction between the trace metal ions and the anions Cl^-, SO_4^{2-}, CO_3^{2-}, HCO_3^-, and OH^-, as a function of pH.

Polynuclear and mixed-ligand complexes were not included in the model. All calculations were made for seawater with a chlorinity of 19‰, at 25°C and 1 atm pressure. Equilibrium was assumed to exist. The concentrations of the uncomplexed metal ion and each complexed species were calculated as a percentage of the total metal concentration using available and estimated individual activity coefficients and stability constants. The danger in such an approach lies in the necessary estimation of the values for stability constants and activity coefficients. In the paper mentioned, individual activity coefficients for the metal ions and their complexes were calculated from the Davies modification of the Debye-Hückel expression (Davies, 1962). The Davies equation is based on the assumption that the charge (z) of the ion is the only factor that has to be considered, according to the expression

$$\log \gamma_{ion} = -0.51 z^2 \left(\frac{I^{1/2}}{1 + I^{1/2}} - 0.3I \right) \tag{1}$$

Using the Davies equation, Zirino and Yamamoto (1972) obtained the following values of the activity coefficients γ, refering to seawater with an effective ionic strength of 0.67 M (Kester and Pytkowicz, 1969):

$$\gamma_0 = 1.0 \quad \text{for neutral species}$$
$$\gamma_1 = 0.74 \quad \text{for monovalent ions}$$
$$\gamma_2 = 0.34 \quad \text{for divalent ions}$$

The assumption about the validity of the Davies equation may approach truth if the ionic strength (I) is low. The different effects of the various ions on the water structure make it unlikely, however, that such an electrostatic equation will be valid at higher ionic strengths. The validity of the assumption that free ion activity coefficients are a function of the ionic strength only, and hence are independent of the composition of the solution, is questionable, especially at higher ionic strengths (Mantoura et al., 1978). In addition, there is a problem with the uncharged species, because they cannot be treated in the same way as the charged ones. In accordance with a semitheoretical equation, such species will have lower activity coefficients (Reardon and Langmuir, 1976):

$$\log \gamma_{IP} = -BI \tag{2}$$

in which $B = 0.1$, 0.3, and 0.5 for 1:1, 1:2, and 2:2 ion pairs, respectively (Mantoura et al., 1978). Table 16 illustrates the fact that differences in the calculated speciation of cadmium originate from the use of different stability constant values. In contrast to errors produced in this way, those introduced by the use of incorrect estimates of activity coefficients are of comparatively minor importance (Mantoura et al., 1978).

The purpose of the paper by Watson et al. (1975) was to extend the calculations, using the equation of Reilly, Wood, and Robinson to obtain the activity coefficients of cadmium chloride and sulfate complexes in artificial seawater. The artificial seawater consisted of four salts: $NaCl$, KCl, $MgCl_2$, and $MgSO_4$. The resulting solution had 35‰ salinity and a 0.7231 M ionic strength. The interactions of cadmium with the macrocomponents of artificial seawater of this composition were the only interactions taken into account in the calculation. The low value of 0.235 for $\gamma_\pm$ of $CdCl_2$ in that artificial seawater at 25°C reflects the pronounced tendencies for complex formation between Cd^{2+} and chloride ions; $\gamma_\pm$ for $CdSO_4$ in that artificial seawater equals 0.057 at 25°C. The authors concluded that the calculated activity coefficients for cadmium ions in their artificial seawater are not to be regarded as a good representation of the activity coefficients in actual seawater, because cadmium displays strong interactions with some of the ions excluded from the treated artificial seawater, for example, CO_3^{2-}, HCO_3^-, and OH^-.

Therefore, for practical applications, it seems more reasonable to use operational or conditional stability constants which deal with the equilibrium concentration ratio of complexes and reactants in the given medium. They are determined at a defined ionic strength with respect to quantity and composition of the solution and are valid and applicable only at that ionic strength. Certainly, it is unlikely that these conditions can always be fulfilled completely and precisely. The data for labile cadmium complexes are not always available at 0.7 ionic strength and because of application of different measuring techniques there is a certain divergence of the values. If operational conditional stability constants are known for a certain range of ionic strength, the approach described in the paper by Raspor et al. (1978a) can be applied to obtain operational or conditional stability constants at defined ionic strength. For example, the operational or conditional stability constant of the $CdCl^+$ complex (β_1) at 0.7 M ionic strength was obtained by graphical interpolation of literature data representing the logarithmic values, at intervals of 10, of β versus the square root of the ionic strength (see Figure 7). From Figure 7 the degree of divergence of the respective determined values as a result of various measuring techniques with different reproducibility and precision becomes evident. The best way would be, of course, to depict such relationships only for data that were determined by the same method; however, this is hardly possible at present on the basis of the data available from the literature.

To obtain reasonable precision in an equilibrium model it is necessary to acquire stability constant values for species of primary significance in seawater, for example, for trace metal complexes with the ligands OH^-, Cl^-, CO_3^{2-}, HCO_3^-, $B(OH)_4^-$, SO_4^{2-}, and F^-, including mixed ligand species [e.g., $(ClOH)^{2-}$]. These measurements must have been obtained in a self-consistent manner for each minor element and under conditions of solution composition, temperature, and pressure characteristic for the marine system in question. With respect to these fundamental aspects, a properly designed equilibrium model

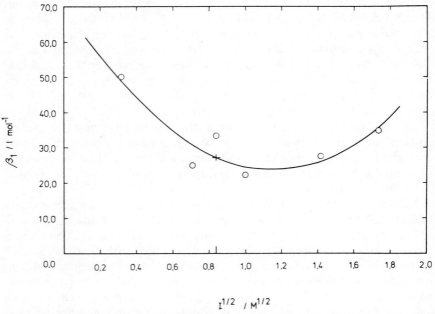

Figure 7.   Dependence of β_{CdCl^+} on the square root of the ionic strength and its evaluation at 0.7 *M* by interpolation of literature data.

provides a useful reference for comparison with experimentally determined operational data on the species considered (Kester et al., 1975).

The polarographic and voltammetric approach, with its special potentialities for reliable and accurate determinations at the ultratrace level, makes it possible as well to check the validity of theoretical equilibrium models obtained by computing the speciation of a certain metal with critically selected sets of stability constants. If the composition of seawater with respect to its macrocomponents is known, as well as the operational or conditional stability constants of all probable labile trace metal complexes, it is possible to calculate according to the general form of the DeFord-Hume equation (Sipos et al., 1979a):

$$\Delta E = \frac{-RT}{nF} \ln \left(1 + \sum_{i=1}^{N} \sum_{j=1}^{M_i} \beta_{Me(Li)j} [L_i]^j \right) \tag{3}$$

the theoretical shift of the reversible half-wave potential $(E_{1/2})$ or of the equivalent potential shift (E_p) or (E_p^+) with the corresponding free ligand concentration. The theoretical shift of these potentials can then be compared with the experimental one. In the case of discrepancies between the theoretical and experimental shifts, the theoretical model has to be amended. The fundamental relations in polarography and voltammetry allow a direct comparison of the

Table 17. Composition of Average River Water[a]

Species	Amount Present in River Water (mol/kg solution)
Na^+	2.83×10^{-4}
K^+	5.9×10^{-5}
Mg^{2+}	1.69×10^{-4}
Ca^{2+}	3.74×10^{-4}
Cl^-	2.20×10^{-4}
SO_4^{2-}	1.17×10^{-4}
A_c	9.54×10^{-4}
C_T	1.02×10^{-3}
Cu	10^{-8}
Zn	10^{-8}
Cd	10^{-8}
Mn	10^{-8}
Hg	10^{-9}
Ni	10^{-8}
Co	10^{-9}
Humic acids	10^{-6}

[a] Livingstone (1963).

theoretical and experimental data and in the case of discrepancies permit relevant corrections of the assumed model.

4.2. Freshwater Media

The same procedure for determining the formation of labile inorganic complexes in freshwater media is also applicable to solutions with smaller amounts of excess salts of different composition, like rivers or lakes. To compare saline and freshwater media, a river water can serve as an example. In contrast to seawater, it is far less possible to generalize the behavior of river waters, because this freshwater type is very variable in composition. The average river water composition (Livingstone, 1963) presented in Table 17 is more a model of soft water than a reality.

Bilinski et al. (1976) determined that the logarithm of the stability constant of the cadmium-carbonato complex, $\log K_{CdCO_3^0}$, at 0.1 M ionic strength, adjusted with KNO_3, is around 3.5. This value indicates rather weak complex formation of cadmium with carbonate ions. The measurements were performed by differential pulse polarography at the HMDE at the 10^{-6} to $4 \times 10^{-7}\ M$ concentration level of cadmium. They are consistent with the results of a study performed by Gardiner (1974a) in 0.001 M KNO$_3$ applying a cadmium-specific electrode, if the effect of different ionic strength is taken into account. As would be expected, an increase in the stability constant value could be observed with the decrease in ionic strength.

The measurements by Gardiner (1974a) were performed with an Orion 94/48 A cadmium-specific electrode. According to the general theory for the measured electrode potential and for ion-specific electrodes particularly, their response corresponds to the activity of free uncomplexed metal ions. Gardiner performed the measurements in very dilute solutions (like river water) in order to avoid interference by other ions present. The electrode changes its characteristics in the presence of copper, so that this source of interference has to be removed before measuring river water samples. One disadvantage of the application of ion-specific electrodes in the investigation of solutions of higher ionic strength is the interference of the other macrocomponent cations. Another disadvantage is the limited applicable concentration range; ion-specific electrodes are applicable only in the concentration range in which a linear relationship between the concentration (or activity) of the investigated ion and the potential exists. According to Gardiner, for the cadmium-specific electrode such a linear relationship exists down to 0.5 mg Cd/l. This limit is insufficient for the actual concentration of cadmium in natural waters. Certainly, the idea of monitoring the free, uncomplexed part of cadmium in fresh waters is of interest and could yield very important information on the concentration of the biologically and biochemically active fraction of dissolved cadmium. To reach a realistic level of determination, however, a substantial improvement in the selectivity and sensitivity of the electrodes would be necessary. From the work of Gardiner (1974a) the following stability constants of labile cadmium complexes are cited, as well as the experimental conditions with respect to total dissolved cadmium concentration [Cd$_T$] and ionic strength:

K_{CdCl^+}: 48 ± 8 M^{-1}, $K_{CdSO_4^0}$: 220 ± 10 M^{-1}, at 20°C, [Cd$_T$]1.0 mg/l; 0.01 M ionic strength

log $K_{CdCO_3^0}$: 4.02 ± 0.04 at 20°C, [Cd$_T$]0.9 mg/l; 0.01 M ionic strength

log K_{CdOH^+}: −9.06 ± 0.08 at 20°C, [Cd$_T$]0.10 mg/l; 0.002 M ionic strength

5. CADMIUM SPECIATION IN NATURAL WATERS IN THE PRESENCE OF ORGANIC LIGANDS

The speciation of metal ions in seawater is considered primarily under aerobic conditions. Under anoxic conditions the formation of sulfide complexes becomes especially important. The major metallic constituents—Na, Mg, Ca, and K—are present mainly as simple hydrated ions but also to a certain extent as complexes of ligands coordinating via oxygen, primarily sulfate, hydroxide, bicarbonate, and carbonate (Ahrland, 1975). In a recent paper by Sipos et al. (1979b) ion pair formation between cations and anions of the major salinity components of seawater was taken into account. The model of seawater obtained therefore gives a more realistic picture of the actual distribution of the major component species in the concentrated multielectrolyte medium.

The aim of the paper by Sipos et al. (1979b), besides introducing a more complete model of seawater, was primarily to investigate how ion pairing of the macrocomponents affects the speciation of trace metals. In general it is possible to say that trace metals that form various complexes with ligands, like hydroxide, sulfate, bicarbonate, and carbonate, will be more affected than the chloro complexes, because the tendency of these anionic ligands to ion pairing with the major cations of seawater is more significant than that with chloride ions.

The ligands present in seawater that are able to form complexes with the various metal ions may be divided into two groups. The first group consists of inorganic anions which are mainly constituents of seawater, like Cl^-, SO_4^{2-}, Br^-, F^-, CO_3^{2-}, and BO_3^-. The second group consists of organic ligands, usually at rather low and often widely varying concentrations. They stem predominantly from the biological activity in the euphotic zone of the sea. In spite of their low concentrations in the dissolved state some of them can be significant ligands because of their high affinity for certain metals.

Metals can be divided into (a) hard and (b) soft metals, characterized by their differing affinities to various ligands. Metal ions of class (a) strongly prefer ligands coordinating via the light donor atoms, fluoride, oxygen, and nitrogen, whereas metals of class (b) are acceptors, which prefer ligands coordinating via the heavier donor atoms of each group (Ahrland, 1975). Complex formation with hard type ligands is mainly due to electrostatic interactions, and the most typical acceptors for these ligands are therefore small multivalent cations. With the soft type ligands metals coordinate via bonds of essentially covalent character. The complexes formed with soft metal acceptors are generally stronger the larger is the radius of the metal ion involved and the lower is its charge. The soft character of a metal increases as the oxidation state decreases.

Although Cd^{2+} is mainly an acceptor of soft character, it shows some tendencies toward the hard metals group. Thus Cd^{2+} is an acceptor of mildly soft character with a chemistry between that of a rather markedly hard acceptor (Zn^{2+}) and that of a very markedly soft acceptor (Hg^{2+}).

All the metal ions, being macrocomponents of seawater, are typically hard acceptors. They prefer hard oxygen donors, like water, sulfate, and carbonate, instead of soft halide donors (e.g., chloride and bromide ions). Because of the low concentration of hard ligands in seawater, some of the highly abundant alkali and alkaline earth ions are mainly present as hydrated ions (Sipos et al., 1979b), and the others as ion pairs with SO_4^{2-}, CO_3^{2-}, and OH^- ions.

For a ligand to be an effective complexing agent at a concentration of 10^{-8} M, however, the conditional stability constant of its complex with cadmium would have to be significantly larger than $10^8 M^{-1}$. This requirement excludes all naturally occurring ligands present at this trace level, except ligands of biological origin, such as free cystein, thionin, nucleic acids, and proteins and their degradation products. Synthetic chelating agents like EDTA could also be important in coastal waters since highly stable chelates are formed. But, because of the competition between seawater macrocomponents, particularly Ca and Mg ions, and trace heavy metal ions for the available ligand content, only concentrations of EDTA $\geq 10^{-7} M$ are likely to become significant (Raspor et al., 1977).

The organic matter found in natural waters includes compounds such as amino acids, polysaccharides, amino sugars, fatty acids, organic phosphorus compounds, aromatic compounds containing —OH and —COOH functional groups, nucleic acids, and porphyrines which contain donor atoms suitable for complex formation (Stumm and Morgan, 1970). Two other important metal complexing agents found in natural waters are humic and fulvic acids. With the free metal ions and the alkaline earth ions they form humates and fulvates of varying stability according to the chemical nature of the metal ion in question and are more soluble in fresh water than in seawater. The latter exerts salting-out tendencies on organic substances with larger hydrocarbon moieties and chains, such as humic material. The most widely distributed functional groups in humic substances that have been shown to participate in the complexing of metals are carboxyl (—COOH), phenolic (—OH), and possibly sterically favorably positioned keto ($=C=O$) and amino (—NH$_2$) groups. At the moment it is difficult to say whether metal humates and fulvates are complexes or chelates, and for practical purposes this is immaterial. What is important is that humic substances have the capacity for binding substantial amounts of metals and that they can exert considerable control over the supply and availability of nutrient elements to plants and animals in soils and soft inland waters (Schnitzer and Khan, 1972).

A better understanding of the chemistry of metal humate and metal fulvate complexes would facilitate the development of more efficient methods of extraction and purification of humic and fulvic acids. It would also provide useful information on the role of humic substances in soil-forming processes, soil structure formation, nutrient ability, and, especially, the mobilization, transport, and immobilization of nonhazardous and toxic heavy metals at the micro- and trace levels in the aquatic environment. Thus the synthesis and properties of complexes of humic substances with metal ions should be of concern to all who are interested in the preservation of the environment (Schnitzer and Kahn, 1972).

Humic substances are widely distributed in nature and arise from the chemical and biological degradation of plant material and from synthetic activities of bacteria. The concentration range of dissolved organic material in natural waters is, according to Stumm and Morgan (1970), 0.1 to 10 mg/l. The upper concentration value is reached in lakes, rivers, and estuaries with substantial biological activity. The lower concentration is more typical for unpolluted and biologically nonproductive fresh waters and for seawater (Gamble and Schnitzer, 1973).

Christman and Ghassemi (1966) carried out chemical degradation studies of organic material from which they concluded that the colored organic material found in water is chemically similar to soil humic material. However, they concluded also that samples of organic matter from different inland surface waters have significantly different molecular size distributions.

5.1. Experimental Evidence of Metal Complexation by Organic Matter in Natural Waters

The significance of organic matter in determining the speciation of trace metal ions has been a subject of continued controversy among the investigators of fresh and marine waters. One view is that the high concentrations of Ca(II) and Mg(II) present in most aquatic systems compete with the trace metals for the low concentration of dissolved organics (Stumm and Morgan, 1970). From another point of view, numerous analyses have indicated that the larger part of trace metals found in natural waters is in the organic fraction. Generally speaking, it should be emphasized that, taking into account differences in chemical behavior of various trace metals and in the composition of natural waters in terms of various degrees of hardness both arguments are correct. For example, according to Allen (1973), cadmium and copper were titrated with humic acid under the same conditions, in the same model solution. The formation of metal-humate complex was recorded via the decrease in the peak current of the uncomplexed amount of metal with increasing concentration of humic acid. Such behavior indicated nonlabile complex formation with both cadmium and copper, but under the same conditions the extent of copper complexation was higher than that of cadmium. The behavior of cadmium is very different from that of copper, which is invariably highly complexed in polluted and unpolluted fresh waters. At the high concentration of 20 mg humic acid/l, according to Allen, about 37% of uncomplexed cadmium is left, whereas copper is already completely complexed at a level of 2 mg humic acid/l.

Up to now, the interactions of metal ions with the humic material present in fresh water and seawater have not been adequately investigated (Mantoura, et al., 1978). There are two principal reasons:

1. The difficulty of isolating and characterizing the humic materials.
2. Until recently, the lack of methods suitable for examination of the interactions between metals and polydentate ligands under conditions of pH and concentration approximating those of natural waters. For this reason some chemists interested in trace metal speciation in natural waters have disregarded the formation of organic complexes in their model systems (see Zirino and Yamamoto, 1972; Dyrssen and Wedborg, 1973). Others have based their models on synthetic complexing agents, such as EDTA, NTA, salicylic acid, and citric acid. The stability constants of metal complexes with synthetic complexing agents may differ considerably from those of complexes with humic materials, which are regarded as important organic metal binding agents in natural waters. As will be discussed later, however, these days the synthetic chelating agents, like EDTA and NTA, are not only model substances, but are also to be expected as pollutants, especially in fresh waters, particularly because of their use in detergents and washing powders. Nevertheless, studies with ligands like EDTA and NTA, which form well-defined and inert chelates with metal ions, can give general information (a) about the reaction path of chelate formation in a complicated medium seawater, (b) on the operative mechanism of this reaction and (c) shed light and provide guidelines for the estimation of the amount of ligand required to chelate a certain percentage of the particular trace metal concentration (Raspor et al., 1978a, 1978b).

To investigate the complexing abilities of fractionated parent humic material with di- and trivalent metal ions, Rashid (1971) used humic material extracted from lagoonal sediments. Figure 8 illustrates the marked differences noticed in the metal binding capacities of those humic acids for di- and trivalent metals, as determined by Rashid. On average, the quantity of divalent metal complexes was 3 to 4 times higher than that of trivalent metals. Among the various divalent metal ions, the differences in the complexed amounts were not large, yet there appears to exist a certain correlation between the complexed amount of a metal and its ionic size. Using three grades of gels, Rashid obtained from the parent humic acid solution, four molecular weight fractions in the following ranges: less than 700, 700 to 10,000, 10,000 to 100,000, and larger than 100,000. The lowest molecular weight fraction (less than 700) appeared to include the most efficient components in complexing the metals in Rashid's study: Al, Fe, Cu, Co, and Ni. The molecular weight fraction below 700 complexed 1.5 times more metal than the parent material, or 2 to 6 times more than any other molecular weight fraction. As the molecular weight increases, complexing ability decreases except for the last fraction (molecular weight above 100,000), for which again some increase in metal binding capacity was noticed. The fraction of molecular weight between 10,000 and 100,000 has the lowest metal binding capacity.

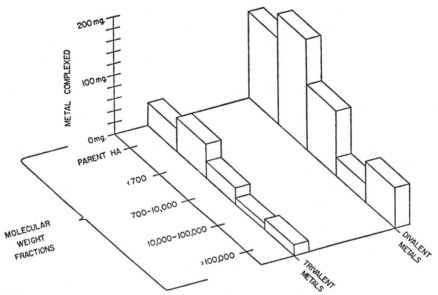

Figure 8. Comparison of the average metal holding capacities of the parent and fractionated samples of marine humic acids in relation to their molecular weight distribution (Rashid, 1971.)

Chau et al. (1970) concluded on the basis of ultrafiltration studies that most of the metal complexing compounds in lake water fall into an arbitrary molecular weight range between 1000 and 10,000.

The molecular weight of organic material and the valence of the metal are important factors governing the metal holding capacity. Appreciable quantities of these compounds of terrestrial origin enter marine basins by means of river water in the mixing zones.

The speciation studies of Musani et al. (1977, 1978) deal with the interaction of some toxic metals and humic acid of terrestrial and marine sediment origin in seawater. The interaction of ^{65}Zn and ^{210}Pb with humic acid (isolated from marine sediments) was investigated by high-voltage paper electrophoresis in seawater and in 0.55 M NaCl. The seawater originated from the Ligurian Sea and was filtered through a 0.45 μ membrane filter. The experiments were performed in 100%, 30% and 10% seawater to simulate estuarine conditions as well. Upon the addition of humic acid to seawater (100%) the radionuclides partially changed their electrophoretic mobility. The chelating effect of humic acid increases as the seawater is diluted and is greater in 0.55 M NaCl than in seawater. However, the concentration of humic acid needed to chelate the radionuclides investigated is above the natural level of dissolved humic material in the open ocean. In waters of lower salinity like estuarine waters, chelates of humic acid may, however, contribute to the distribution of a particular trace metal among the various dissolved chemical species. In general, the results with humic acid of marine sediment origin confirm the previous findings of Musani et al. (1977) that the chelation of Pb(II) with dissolved humic acid of terrestrial as well as of marine sediment origin in seawater is negligibly small, depending on the adjusted humic acid concentration, although the humic acid from the marine sediments has a higher chelating efficiency (Musani et al., 1978). The measurements also reflect the competition of Pb(II) with the alkaline earth ions present in substantial excess, particularly with Ca(II), for the chelation sites.

From the experimental evidence previously mentioned there emerge the following general conclusions:

1. There is a certain molecular weight fraction of humic acid that is more effective as a complexing agent than other fractions, and from the experimental results it seems to be the rather low molecular weight fraction. Unfortunately, detailed data on the structure, functional groups, and coordination abilities of this fraction are still lacking.
2. The actual degree to which metal-humate complex formation affects the speciation distribution of a particular metal in natural waters depends predominantly on the following parameters:
    A. Type of metal and its concentration.
    B. Molecular weight of humic material.
    C. Conditional stability constant of metal-humate and corresponding required concentration level of dissolved humic materials.
    D. Type of natural water, particularly with regard to alkaline earth ion levels, as Ca(II) and Mg(II) are strong competitors for complexing sites in polymeric humic material.

Experimental evidence indicates that cadmium tends to stay in natural water in the ionic form (predominantly as labile chloro complexes) rather than becoming bound with the organic ligands. In soft water media a certain degree of complexation is achieved at lower dissolved concentrations of humic acid, because of less pronounced competition of Ca and Mg ions, as compared with seawater containing comparatively high levels of these alkaline earth ions.

5.2. Synthetic Chelating Agents

During recent years concern has grown over the presence of heavy metals in water. Chelating agents of natural and pollution origin play an important role in solubilization, transport, and reactions of heavy metals in water. Chelation may hinder the removal of metals from water, increase the corrosion of metal surfaces, and affect the oxidation state of metals in natural waters. Chelating agents may have a strong effect on aquatic biota. Sometimes they serve to keep microamounts of essential metal ions in solution and make them available to algae. This role of chelating agents in algal growth in nature can result in eutrophication (Kunkel and Manahan, 1973). The formation of very inert complexes of radionuclides may inhibit uptake by marine organisms until the radionuclides are sufficiently diluted in the seawater masses. For characterization of the physicochemical state of radionuclides in the marine environment it is important to know, as well, the stability of the chelates formed with synthetic chelating agents. The findings by Musani-Marazović and Pučar (1973) and Kozjak et al. (1977) are also important with respect to the prognosis of the fate of cadmium entering the marine environment by waste disposal.

Synthetic chelating agents are important for investigation because usually they form more stable chelates with metal ions than do many of the naturally occurring ligands. A most significant example is the stabilities of their chelates with Cd, Zn, Pb, Ni, and other trace metals, which are several orders of magnitude higher than the stability of the corresponding metal-humate. Even for the very stable humates of copper the conditional stability constants are comparable in order to those with common synthetic ligands. This is the fundamental reason why, in the open sea, metal-humate formation usually remains insignificant for most trace metals, with the possible exception of copper and mercury.

As synthetic chelating materials are pollutants for natural waters, it is also important from the environmental viewpoint to know how they influence the physicochemical state of trace metals in general and cadmium in particular.

It is of fundamental interest and significance to determine in which chemical forms trace metals occur in seawater and other natural water types in the presence of organic ligands and which components of seawater and other natural waters influence metal-chelate formation. In earlier polarographic studies (Raspor and Branica, 1973, 1975a, 1975b), at a higher Cd(II) concentration of about 10^{-4} M in chloride solutions of high ionic strength and with a pH around 8 (i.e., similar to seawater), general information was gained on the physicochemical behavior, the electrochemical reduction, the stability, and the dissociation kinetics of Cd(II) chelates in such aqueous media. As pointed out below, these studies have since been extended to the realistic concentration level of dissolved cadmium in seawater and inland waters.

Nitrilotriacetic Acid (NTA) as Ligand

In some countries the trisodium salt of NTA is being considered, or even already used, as a substitute for polyphosphates in commercial detergents. Around 1970 NTA had been banned owing to the detection of some possibly harmful biological effects (Mottola, 1974). It now appears, however, that the replacement of tripolyphosphates in synthetic detergents by NTA, as practiced, for example, in Canada, does not constitute an environmental hazard (Gardiner, 1976). At summer temperature, NTA is biodegradable in the activated sludge process (Thom, 1971); the degree of degradation appears to fall at low temperatures and with increasing concentrations of NTA. Forsberg and Lindquist (1967) reported that bacteria which utilize NTA as both carbon and nitrogen sources at concentrations of 25 to 1000 mg/l have been isolated from Swedish rivers polluted with sewage. The bacteria have been tentatively identified as *Pseudomonas*. A temperature of 25°C gave optimal growth for these microorganisms. Wong et al. (1973) found that the natural bacterial flora required a long period of adaptation before gaining the ability to utilize NTA at a relative low rate. Their results showed that glycine and acetic acid were the metabolic products of bacterial degradation of NTA. Thompson and Duthie (1968) also concluded that NTA was metabolized to glycine by the release of two acetic acid groups. Glycine was further metabolized, the nitrogen being converted to inorganic nitrogen (see Mottola, 1974).

At any rate, NTA remains a pollutant of inland, estuarine, and coastal waters, as it is an unnatural substance affecting the biochemical cycle of heavy metals. In the review by Mottola (1974) a summary is given of the analytical methods appearing in the literature for the determination of NTA. The broad spectrum of methods supports the statement that the substitution of NTA in detergents stimulated significant scientific interest from the environmental point of view. The bacterial degradation rates of various chelates have been investigated by Chau and Shiomi (1972). According to these authors, 60 days were required for the degradation of Cd(II)-NTA chelate.

Björndal et al. (1972) showed that 6 mg NTA/l inhibits the adsorption of trace metals on solids during treatment. An increasing concentration of NTA mobilizes increasing concentrations of metals from lake sediments, and this effect was measurable even at levels as low as 1.3 mg NTA/1 (Chau and Shiomi, 1972). NTA caused a 15% increase of cadmium in the solution phase after flocculation of the adsorbing solids. The release of metals from lake sediments by chelating agents was also demonstrated by Zitko and Carson (1972) to be highest at low calcium concentrations.

Theoretical Model for Cadmium-NTA Speciation.   In the same manner as the speciation of cadmium with inorganic ligands was calculated, as discussed previously, it is possible to calculate the distribution of cadmium among various chemical forms in the presence of organic ligands in natural waters if provided all the necessary concentration and stability constant data are known. Stumm and Morgan (1970) have considered a hypothetical system consisting of nine metal and ligand ions, each one present at a known analytical concentration. A pH of 8.0, a temperature of 25°C, and a constant ionic strength of 0.5 M were assumed. The purpose of such a model was to consider a system that contains,

Table 18.   Model Distribution of Dissolved Cadmium in Seawater in the Presence of Inorganic and Organic Ligands[a]

Cadmium Total	Free	SO_4 2.5×10^{-2} M	CO_3 2×10^{-4} M	OH 1×10^{-6} M	F 6×10^{-5} M	PO_4 2×10^{-6} M
10^{-9} M	6.9×10^{-10} M	1.3×10^{-10} M	4×10^{-13} M	$.3 \times 10^{-11}$ M	not considered	1.6×10^{-18} M
100%	69%	13%	0.04%	1.3%		<0.00%
	NTA^b 10^{-6} M	Noc^b 10^{-7} M	Gly^b 3×10^{-8} M	Sal^b 10^{-7} M		Cit^b 3.2×10^{-7} M
	2×10^{-10} M	1.6×10^{-22} M	$7.9 \times 10^{-15} M$	7.9×10^{-17} M		10^{-13} M
	20%	<0.00%	<0.00%	<0.00%		0.01%

[a] Data from Stumm and Morgan (1970). [b] NTA = nitrilotriacetate, Noc = nocardamine, Gly = glycine, Sal = salicylate, Cit = citrate.

in addition to the major and minor ions typically found in seawater, a variety of organic substances, all present at low concentrations ($C \leq 10^{-6}$ M). In Table 18 the molar concentration values of the total level of cadmium and the various ligands considered are given, as well as the concentration distribution of cadmium among various complex forms, expressed as molar concentrations and percentages. As Stumm and Morgan mentioned, this system has some but not all of the chemical properties of seawater. As can be noticed, according to the proposed composition of the model seawater, Cd(II) was, besides, the free hydrated form, predominantly found as NTA chelate. This result can be easily explained with the fact that chloride ions, which, especially in the seawater distribution of cadmium, play a very important role, were not taken into account.

If the speciation of Cd(II) in seawater, in the presence of various organic ligands, among them NTA, is calculated taking into account chloride ions as well, a completely different distribution of cadmium species will be obtained, as was also experimentally confirmed (Raspor et al., 1978a). Moreover, according to Stumm and Morgan (1970) and our own observations (Raspor et al. 1978a), at the 10^{-6} M concentration level and under equilibrium conditions NTA is predominantly present in the form of Ca-NTA chelate.

Stability Constant Determination of Cd(II)-NTA Chelate in Natural Waters. In earlier polarographic studies (Raspor and Branica, 1975a) various information on the physicochemical behavior of cadmium at the rather high concentration level of 2×10^{-4} M in the presence of NTA was obtained in model solutions similar to seawater. In a further study using the more sensitive dif-

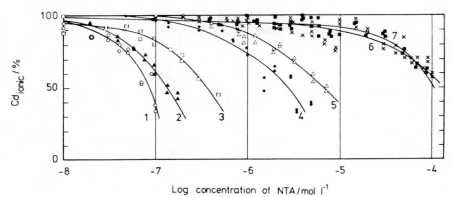

Figure 9. Titration curves of ionic Cd(II) as a function of the logarithm of the NTA concentration in different media of ionic strength 0.7 M (if necessary, adjusted by appropriate additions of $NaClO_4$) and pH 7.3 to 8.3 (see Table 19). (1) 0.7 M $NaClO_4$; (2) 0.1 M NaCl; (3) 0.59 M NaCl; (4) 0.01 M $CaCl_2$; (5) 0.0536 M $MgCl_2$; (6) artificial seawater; (7) Adriatic seawater. The value of $-\log c_{NTA}$ at $Cd_{ionic} = 50\%$ corresponds to the respective value of the logarithm of the operational conditional stability constant ($\log K'_{Cd-NTA}$). Full curves obtained by polynomial curve fitting with polynomials up to the fifth grade. Reproducibility of experimental points: (1) ±44%; (2) and (3) ±10%; (4) ±8%; (5) ±7%; (6) ± 24%; (7) ±15%.

ferential pulse polarographic mode, the concentration of Cd(II) could be decreased to 10^{-7} M, and cadmium chelation with NTA was studied under these concentration conditions, which are closer to the reality in natural waters (Raspor et al., 1978a). Very recently the findings of this investigation have been confirmed in detail by similar studies with Cd, Zn, and Pb at rather realistic levels, that is, 3×10^{-9} M, 3×10^{-8} M, and 3×10^{-8} M, respectively. These measurements became possible by the application of the even more sensitive method of differential pulse anodic stripping voltammetry (Raspor et al., 1979a), the principle of which has been explained in Section 2.

The aim of these investigations was to establish a basis of judgments and conclusions on the interactions of potential pollutants of the NTA type in particular and of chelating organics of natural origin in general with the dissolved amount of toxic trace metals in complicated multielectrolyte seawater and in inland waters. The seawater medium was systematically developed with respect to its major components by appropriate model solutions. In this iterative approach, which is applicable in general to inert complexes (contrary to the labile ones previously discussed) a 10^{-7} M and later a 3×10^{-9} M total dissolved concentration of cadmium was titrated with increasing concentrations of NTA initially in certain model solutions containing major salinity components, then in artificial seawater, and finally in genuine seawater. The decrease in the concentration of unchelated cadmium was monitored via the corresponding reversible voltammetric response. As shown in Figure 9, a series of titration curves was recorded in the various media mentioned, and on this basis the operational and common conditional stability constant values of Cd(II)-NTA in the studied media could be evaluated. Furthermore, the curves in Figure 9 show the NTA concentration required to achieve a certain percentage of chelation of the total Cd(II) concentration.

These systematic investigations revealed the importance of calcium and magnesium, which compete with such trace metals as cadmium for the ligand NTA, in regard to the degree of Cd(II)-NTA chelate formed at a given NTA concentration in the particular medium under study. It is obvious from Figure 9 that at the 10^{-7} M concentration of total cadmium the most important salinity constituents of seawater which influence the degree of Cd(II)-NTA formation are Cl, Ca, and Mg ions. When the concentration of Cd(II) is decreased to the 3×10^{-9} M level, macroconstituents that affect Cd(II)-NTA chelate formation are, in addition, carbonate, bicarbonate, sulfate, and Na ions, as well the presence of other trace metals, like zinc. In seawater, at the 10^{-7} M Cd(II) level the following network of involved equilibria is operative, because of the competition of these macrocomponents for the toxic trace metal and the chelating ligand, respectively (Raspor et al., 1978a):

$$
\begin{array}{c}
\textbf{Me-NTA}^- \\[4pt]
\uparrow \\
K_{\text{Ca-NTA}} = 5.75 \times 10^5 \quad \begin{array}{l} K_{\text{Mg-NTA}} = 5.75 \times 10^4 \\ + \text{Me}^{2+}\ (\text{Ca}^{2+},\ \text{Mg}^{2+}) \end{array} \\[6pt]
\text{CdCl}_2^0 \underset{\beta_2\,=\,62}{\overset{\text{Cl}^-}{\rightleftharpoons}} \text{CdCl}^+ \underset{\beta_1\,=\,27.3}{\overset{\text{Cl}^-}{\rightleftharpoons}} \text{Cd}^{2+} + \text{NTA}^{3-} \rightleftharpoons \text{Cd-NTA}^- \qquad (4) \\[6pt]
\downarrow \\
+\text{H}^+ \quad K_3 = 2.9 \times 10^9 \\[6pt]
\textbf{H-NTA}^{2-}
\end{array}
$$

To obtain a common value for the operational conditional stability constant of the Cd(II)-NTA chelate, all these equilibria of side reactions have to be taken into account. In Table 19 the values of the operational and common conditional stability constants in the different media studied, determined at 10^{-7} M total Cd(II) concentration, are compared. It is obvious that in the same medium there is an appreciable numerical difference between the operational $K'_{\text{Cd-NTA}}$ and the common conditional stability constant $K_{\text{Cd-NTA}}$. The lower the $K'_{\text{Cd-NTA}}$ value, the more pronounced are side reaction effects, as, for example, in artificial and real seawater, where all the side effects act cumulatively. The fact that the values of the common conditional stability constants ($K_{\text{Cd-NTA}}$) have to coincide for the various media, within the experimental error constitutes an internal consistency check of the applied method and indicates that all relevant side reactions of significance were taken into account. It is to be emphasized that a concentration of 10^{-5} M NTA is a minimum for the onset of cadmium chelation in

Table 19.   Operational and Common Conditional Stability Constants of Cd(II)-NTA Determined in Model Solutions and Real Seawater Sample

Electrolyte (M)	pH	log $K'_{Cd\text{-}NTA^-}$	log $K_{Cd\text{-}NTA^-}$
0.7 NaClO$_4$	7.35	7.05	9.60
0.59 NaCl[a]	7.83	6.46	9.90
0.01 CaCl$_2$[a]	7.38	5.55	9.53
0.0536 MgCl$_2$[a]	7.46	5.16	9.33
Artificial seawater	8.29	3.96	9.44
Adriatic seawater	8.01	3.96	9.44

[a] Ionic strength of 0.7 M adjusted by appropriate addition of NaClO$_4$.

seawater, while the chelation of 50% of 10^{-7} M Cd(II) requires 3×10^{-4} M NTA in this medium. As a general conclusion, it turns out that the pollution of natural waters with high salt loads containing larger amounts of alkaline earth cations will inevitably lower the amounts of heavy metals existing in complexes with organic matter.

The same measuring procedure is of course applicable to soft water or freshwater media. In the case of zinc, the formation of EDTA and NTA chelates was determined by DPASV in Ontario lake water, which has, at 0.007 M, a 100 times lower ionic strength than seawater and a correspondingly lower content of Ca and Mg ions (Raspor et al., 1979b).

Ethylenediaminetetraacetic Acid (EDTA) as Ligand

For a long time small quantities of EDTA have been used in detergent preparations to chelate trace metals which catalyze the decomposition of the added perborate. In that way, EDTA is present in detergents and therefore appears in primary and secondary sewage and river waters. Although EDTA is not regarded as toxic to human beings, it is resistant to biodegradation (Bunch and Ettinger, 1967). EDTA is also used to modify the availability of trace metals to plants and animals, and it finds application as well in the treatment of metal poisoning and kidney stones in human beings. In a paper by Gardiner (1976) data on the variation in the concentration of EDTA in domestic sewage during the course of the day are given; these data are presented in Table 20.

The use of EDTA in washing powders leads to a concentration of 0.1 to 0.2 mg EDTA/l in domestic sewage effluents, but an additional concentration of up to 1 mg EDTA/l could come from industrial sources (Gardiner, 1975). At this level EDTA can increase the mobility of trace metals in the aquatic environment, but can also reduce metal toxicity to aquatic life. Thus the tendency of EDTA to keep metal ions in solution leads to increased trace metal mobility in water courses and a reduced rate of trace metal removal by adsorption and precipitation in water reservoirs. According to Gardiner (1975), there are few

Table 20.   Variation in EDTA Concentration in Secondary Domestic Sewage during the Course of a Day[a]

Time (hr):	08.45	09.55	11.15	12.35	14.15	15.40	17.00
EDTA concentration ($\mu g/l$):	210	210	190	150	110	180	180

[a] Data from Gardiner (1976).

observations on the release of adsorbed metals by EDTA. When Cd(II) was added to a sample of Stevenage domestic sewage effluent, which contained 0.1 to 0.2 mg EDTA/l, cadmium was first adsorbed on the suspended solids to an extent expected from studies on systems containing chelating agents, but after a few hours almost all the cadmium had returned to the solution phase (Gardiner, 1974b). This is obviously due to a slow exchange of the EDTA ligand between the calcium excess and cadmium, a tendency promoted by the significantly higher stability of Cd(II)-EDTA chelate.

A similar observation was made in seawater by Maljković and Branica (1971). Applying square wave polarography, the authors observed that the rate of Cd(II)-EDTA chelate formation was reduced in seawater, and they concluded that Ca and chloride ions were the constituents of seawater that exerted a predominant influence on the degree of Cd(II)-EDTA chelate formation. The authors showed that the rate of Cd(II)-EDTA chelate formation in seawater is lower than in a corresponding solution of sodium chloride containing Ca ions. This effect was an interesting starting point for further investigations to obtain more detailed and quantitative kinetic data by which the mechanism could be explained (Raspor et al., 1977). This reaction mechanism has general significance and can be applied to similar chelate formation reactions occurring between trace metals and various organic ligands in natural waters. In Table 21 the resulting values of the conditional operative rate constant of Cd(II)-EDTA chelate formation in various media are summarized, and from these values it is seen that the reaction proceeds more rapidly in sodium chloride solution of the same concentration as that in seawater. In magnesium chloride solution of the same concentration as that in sea water, the reaction is somewhat slower, and k_f becomes measurable by the applied voltammetric method. Chelate formation becomes even slower in 0.01 M $CaCl_2$, in artificial and genuine seawater. Within the experimental error the values of k_f evaluated in artificial and real seawater samples are the same. The small difference between the k_f values for 0.01 M $CaCl_2$ and seawater indicates that Ca^{2+} is the most important competitor. These results indicate also that the rather slow rate of Cd(II)-EDTA chelate formation observed experimentally is not due to some unknown component of natural seawater, because in the case of model solution and artificial

Table 21.   Rate Constant k_f of Cd(II)-EDTA Formation Determined in Model Solutions and Real Seawater Sample

Electrolyte (M)	Reactant Concentration (M)	k_f (mol/1.sec)
0.59 NaCl	5×10^{-7}	Unmeasurably fast
0.0536 MgCl$_2$	5×10^{-7}	2.8×10^3
0.01 CaCl$_2$	5×10^{-7}	4.2×10^2
Artificial sea water	5×10^{-7}	3.3×10^2
Real seawater	5×10^{-7}	$(3.6 \pm 0.5) \times 10^2$

seawater the exact composition is known. In seawater the alkaline earth cations are predominantly present in the free hydrated ionic state. Their concentration is several orders of magnitude higher than the investigated concentration range of Cd(II) and EDTA ($5 \times 10^{-7} M$), and their chelation rate is obviously fast. If a small amount of EDTA (between 10^{-5} and $10^{-7} M$) is added to the solution, Ca^{2+} and Mg^{2+} will act as efficient and rapid scavengers, and the corresponding chelates with EDTA will be formed. Nevertheless, because of the significantly higher stability of Cd(II)-EDTA a relatively slow exchange of the ligand will occur until the equilibrium concentration of Cd(II)-EDTA has been reached. Thus in seawater the following kinetic overall scheme represents the formation of cadmium chelates with organic ligands by the ligand exchange mechanism usually operative:

$$CdCl_2$$
$$\Updownarrow$$
$$CdCl^+$$
$$\Updownarrow$$
$$[Ca(II)\text{-}EDTA]^{2-} + Cd^{2+} \xrightarrow{k_f} [Cd(II)\cdot EDTA]^{2-} + \begin{matrix} Ca^{2+} \\ Mg^{2+} \end{matrix} \qquad (5)$$
$$\Updownarrow$$
$$[Mg(II)\cdot EDTA]^{2-}$$

The scheme reflects the fact that both alkaline earth cations, Ca and Mg, play a role in regard to the rate of Cd(II)-EDTA chelate formation. The rate determining step of EDTA exchange, however, occurs between Ca(II)-EDTA chelate and Cd(II), because the equilibrium between the two alkaline earth-EDTA chelates remains practically adjusted and the Ca(II)-EDTA equilibrium concentration is restored immediately at the expense of Mg(II)-EDTA (Raspor et al., 1977, 1979a).

The same measuring procedure was applied to the determination of the operational rate constants of chelate formation of other trace metals, like lead and zinc, with EDTA in seawater (Raspor et al., 1979a). It should be emphasized that the type of mechanism, which consists of ligand exchange between first-formed alkaline earth chelates and the trace metals, can be generally presumed for the chelation of heavy trace metals by natural organic ligands in seawater and fresh waters of sufficient hardness, as long as the basic assumption that the heavy metal chelate is more stable than the corresponding alkaline earth chelate is valid.

According to Gardiner (1976), in a system containing the alkali and alkaline earth cations only at concentrations to be expected in fresh water, with no complexing ligands present except EDTA and with a pH between 5 and 9, consideration of conditional stability constant values leads to the conclusion that almost all of the EDTA will be associated with calcium. A small fraction will be associated with magnesium. The amounts of protonated and free EDTA and of its complexes with alkali cations remain negligible. In a natural water there will be, of course, competition, not only between the trace metal ions and calcium but also between EDTA and other ligands present, for the available trace metal ion. Both factors tend to reduce the extent of trace metal-EDTA chelate formation. This tendency will be more pronounced the larger are the conditional stability constants of trace metal chelates with other organic and inorganic ligands and the higher are the concentrations of these dissolved ligands in the natural water type in question.

5.3.   Natural Chelating Agents

The stability constants for various metal ions (but not cadmium) with fulvic acid at 0.1 ionic strength have been summarized by Schnitzer and Khan (1972). Because metal-fulvate stability constants are considerably lower than those for metal-EDTA chelates, the authors conclude that metals complexed with fulvic acid will be more readily available to plant roots, microorganisms, and small animals than those chelated by EDTA or similar chelators. They also demonstrate that log $K_{Me\text{-}fulv}$ decreases linearly as ionic strength increases from 0 to 0.15 M. The slope of the lines is steeper for metals that form relatively strong complexes with fulvic acid than for those yielding weaker complexes (Schnitzer and Hansen, 1970).

The study of Mantoura et al. (1978) was undertaken to obtain information about the extent to which metals are bound to organic matter in natural waters. First, a gel filtration was performed in order to determine the conditional stability constants of the complexes formed between humic material and the metals studied under realistic conditions of pH and free metal ion concentration. Second, these stability constant values were used in a model for multimetal and multi-

Table 22. Overall Stability Constants (log K_0) for Humic Compounds with Ca, Mg, Mn, Co, Ni, Cu, Zn, Cd, and Hg at 0.02 M Ionic Strength 20°C and pH 8.0 together with the Values found by Previous Workers.

Humic Compound		log K_0									References
Source	Sample	Ca	Mg	Mn	Co	Ni	Cu	Zn	Cd	Hg	
Peat	FA	3.65	3.81	4.17	4.51	4.98	7.85	4.83	4.57	18.3	Mantoura, et al (1978)
	HA						8.29				
	FP$_2$						8.40				
	FP$_4$						8.30				
	FP$_6$						8.27				
	FP$_8$						8.30				
	FP$_{12}$						8.28				
Lakes	CEL$_1$	3.95	4.00	4.85	4.83	5.14	9.83	5.14	4.57	19.4	
	CEL$_2$	3.73	3.67	4.30	4.75	5.27	8.42	5.05	4.70	20.1	
	CEL$_3$	4.09	3.74		4.90		9.35	5.31		18.4	
	BAL	3.56	3.26		4.67		9.30	5.25		19.3	
Rivers	DEE						9.48	5.36		19.7	
	CON						9.59	5.41		21.1	
Lochs	ET$_1$	3.65	3.50		4.29	5.31	8.89		4.95	20.6	
	ET$_2$	3.27	3.41		4.75	5.19	10.21		4.87	20.6	

										Reference
Sediment	ET_3	4.65	4.09			4.91	11.37	5.87		21.3	
	ET_4		3.92				10.43	4.99			
	ET_5						10.14				
	ET_6						9.91				
Seawater	IR_1	3.60	3.50	4.45	4.83	5.41	8.89	5.27		18.1	
	IR_2	4.12	3.98	4.51	4.79	5.51	9.71	5.31	4.69	18.0	
Soil	FA[a]	3.4	2.2	3.7	4.2	4.2	4.0	3.7			Schnitzer and Khan (1972)
	FA[b]					4.35		3.04	5.08		Cheam and Gamble (1974)
	HA[c]						7.8				Ernst et al. (1975)
	FA[d]				6.6						Malcolm (1972)
	HA									5.2	Strohal and Huljev (1971)

[a] Ionic strength 0.1 M, pH 5.0.
[b] Ionic strength 0.1 M, pH 4–5.
[c] Ionic strength 0.1 M, pH 6.8.
[d] Ionic strength 0.1 M, pH 6.0.

ligand interaction to evaluate the extent to which metals are bound by organics in various types of natural waters.

The overall conditional stability constants (K_0) (expressed on a concentration basis) are given in Table 22, together with a selection of data obtained by other workers using different methods. The concentration of the metal in the effluents was determined by AAS. For the purpose of comparison, Table 22 includes stability constant data previously determined by various other authors. The data refer to metal complex formation with fulvic and humic acids derived from soils. There do not appear to be many data on metal-humic complex formation with humic acids isolated from aquatic media. In general, it can be concluded that the stabilities of the humic complexes of the various metals follow the well-known Irving-Williams order of chelate stability:

$$Mg < Ca < Cd \approx Mn < Co < Zn \approx Ni < Cu < Hg$$

The stability constants determined by Mantoura et al. (1978) were further applied in conjunction with the HALTAFALL program of Ingri et al. (1967) in order to compute a speciation model for fresh water and seawater, taking into account metal-humic acid interaction.

Surface water from the Irish Sea was used in developing a model for seawater. The concentration of the major ions was evaluated from the salinity. The macrocomponent composition of the Irish Sea water sample is in very good agreement with the typical average seawater sample proposed by Sverdrup et al. (1942) (see Table 12). Marine humic compounds were isolated from the surface waters of the Irish Sea.

Lake Celyn water was used as the basis for a speciation model of fresh water, and the constituents were analytically determined and tabulated (Table 23). The macrocomponent composition is close to that quoted by Livingstone (1963) for average river water (see Table 17). The respective metal-humate stability constants for Lake Celyn water are tabulated in Table 22.

The results of the calculation performed by Mantoura et al. (1978) of the distribution of the species existing at equilibrium in the Irish Sea water (IR 1) are presented in Table 24. Of the whole series of metals investigated by these authors, for this review only relevant ones (i.e., Cd, Ca, and Mg), were selected. From the distribution in Table 24 it is evident that the speciation is dominated by inorganic ligands even in the presence of humic material. Because of their relatively high concentrations calcium and magnesium together bind 99.5% of the total humic ligand, despite the relatively low stability constants of their humic complexes. However, less than 0.1% of cadmium was found in the form complexed with humic material.

Lake Celyn water (CEL 1) was used by Mantoura et al. (1978) as a basis for the study of the influence of humic material on the speciation of metals in fresh waters. The model was derived for an analyzed sample, the composition of which

Table 23. Composition of Lake Celyn Water Used in Construction of Models [a]

Species	Amount Present in Lake Celyn Water (mol/l)
Na^+	8.3×10^{-4}
K^+	1.9×10^{-5}
Mg^{2+}	1.0×10^{-6}
Ca^{2+}	2.6×10^{-5}
Cl^-	8.4×10^{-5}
SO_4^{2-}	3.9×10^{-6}
A_c	5.8×10^{-4}
C_T	—
Cu	1.0×10^{-8}
Zn	1.8×10^{-7}
Cd	3.6×10^{-8}
Mn	3.5×10^{-7}
Hg	3.5×10^{-10}
Ni	2.0×10^{-8}
Co	1.7×10^{-8}
Humic acids [b]	1.2×10^{-6}
pH	6.2

[a] Data from Mantoura et al. (1978).
[b] Calculated assuming the molecular weight of 5×10^3.

is presented in Table 23. Again, of the series of metals investigated by Mantoura et al., only Cd, Mg, and Ca are considered here and are shown in Table 25a. According to these data, in lake water only 2.7% of cadmium is bound by humic acids; the rest occurs predominantly as free hydrated ion (92.8%). Contrary to the situation in saline media, the humic substances occur mainly in the uncomplexed state. On the whole, in the freshwater model given by Mantoura et al. trace metals account for only 3.5 mol % of the complexing capacity of the humic acid (Table 25b). In contrast even though they form weaker complexes (see Table 22), calcium and magnesium bind 21.4 mol % of the humic material, because they are present in an excess of at least two orders of magnitude in comparison with the concentration of humic substances (see Table 23). In this model of fresh water, therefore, 75.2 mol % of the humic ligand exists uncomplexed.

An attempt has been made by Mantoura et al. (1978) to examine the changes

Table 24. Model Distribution of Mg, Ca. Cd, and Humic Ligand in Seawater (mol %) at 25°C, pH 8.0, and 1 atm Pressure[a]

Metal	M^{2+}_{free}	MF^+	MCl^+	MCl_2^0	MCl_3^-	MCl_4^{2-}	$M(OH)^+$	$M(OH)_2^0$	MSO_4^0	MCO_3^0	M-Hum	Total Humic Bound with Metal
Mg^{2+}	87.4	1.9					*		10.4	0.3	*	84.3
Ca^{2+}	90.6	*						*	9.2	0.2	*	15.2
Cd^{2+}	0.9	*	16.2	45.9	30.2	*	*		0.3	6.4	*	

* Speciation < 0.1%.
[a] Data from Mantoura et al. (1978).

Table 25a. Speciation of Mg, Ca, and Cd in Lake Celyn Water (mol %), Using Analytical Data Given in Table 23[a]

Component	Free Ion	Chloro	Sulfato	Bicarbonato	Carbonato	Hydroxy	Humic
Magnesium	98.2	—	0.5	0.5	—	—	0.8
Calcium	97.8	—	0.6	0.9	—	—	0.7
Cadmium	92.8	0.7	0.5	3.0	0.2	0.1	2.7

Table 25b. Speciation of Humic Ligand in Lake Celyn Water (mol %)

Component	Free	Mg	Ca	Cu	Zn	Mn	Ni	Cd
Humic	75.2	6.4	15.0	0.8	1.0	1.5	0.1	>0.1

[a] Data from Mantoura et al. (1978).

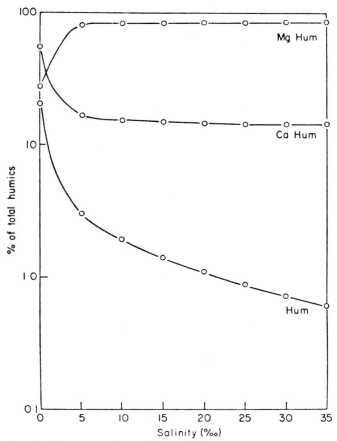

Figure 10.   Calculated distribution of humic material as a function of salinity along a model estuary. (Mantoura et al., 1978.)

in speciation that occur in an estuary, where river water mixes with the sea. In this model it was assumed that the river water has the composition of world-average river water (see Table 17), and the sea that of average seawater (see Table 12). Because of increasing competition from higher concentrations of calcium, magnesium, and various anions the degree of humic acid-trace metal interaction decreases as salinity increases. Although humic acids in fresh water exist predominantly uncomplexed, they become bound to calcium and magnesium up to an extent of 99% in the earliest stages of estuarine dilution, at about 1‰ of salinity (Figure 10). Under natural conditions humic acids could ultimately be removed from the solution, probably by adsorption on colloidal hydroxides of iron and aluminum.

Table 26. Composition of Samples and Determined Distribution of Cadmium[a]

| Sample No. | Type of Sample | Point of Collection | pH Value | Calculated Proportion (%) as | | | | | | $E_0 - E$ (mV) | Observed Proportion as Cd^{2+} (%) |
				$CdOH^+$	$CdCO_3$	$CdCl^+$	$CdSO_4$	Cadmium Humic Complex	Cd^{2+}		
1	Fresh ground water	WPRL	7.2	1.4	3.9	1.8	0.6	0	92	1.5	89
2	Filtered, settled sewage	WPRL	7.7	1.8	9.0	5.3	3.1	39	41	12.3	38
3	Filtered poor quality percolating filter effluent	WPRL Filter 30	8.0	3.2	15	5.2	2.5	38	35	14.4	32
4	Filtered good quality percolating filter effluent	WPRL Filter 1	8.0	3.6	12	6.2	3.0	37	38	15.9	29

5	Filtered river water	Stevenage Brook WPRL	7.7	2.8	6.1	4.6	7.2	24	55	6.9	63
6	Filtered river water	River Ivel Stotfold, England	8.1	6.5	21	2.6	2.6	9.3	59	7.8	54
7	Filtered river water	Pix Brook Arlesey, England	8.1	4.9	16	6.0	4.3	24	44	13.8	35
8	Filtered river water	River Hiz Arlesey, England	8.1	5.7	18	3.8	4.1	16	52	8.6	51
9	Filtered river water	River Lee Luton, England	8.0	4.9	12	10	5.1	12	56	5.5	65
10	Filtered river water	River Lee Luton, England	7.9	3.6	9.7	9.2	7.7	20	51	8.1	53
11	Filtered river water	River Trent Swarkeston, England	7.7	2.6	3.9	3.5	7.2	24	58	7.9	54

[a] Data of Gardiner (1974a). WPRL = Water Pollution Research Laboratory.

217

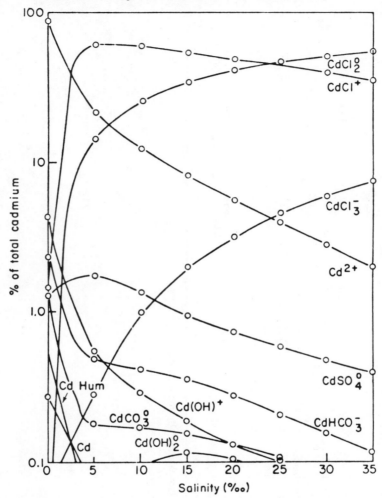

Figure 11. Calculated distribution of Cd(II) as a function of salinity along a model estuary. (Mantoura et al., 1978.)

There is a rapid decrease in the fraction of trace metals, including cadmium, bound to the humic components as salinity increases. This drop is principally the result of calcium and magnesium competition for the humic ligand. Despite the relatively low stability constants of alkaline earth complexes, they are sufficiently abundant to utilize more than 99% of the complexing capacity of humic ligand (Mantoura et al. 1978); see Figure 10. In Figure 11 the speciation of cadmium is graphically presented as a function of the salinity gradient along the model estuary (Mantoura et al., 1978). It is obvious that chloro complexes are particularly important for cadmium speciation; cadmium is present in that form to an extent of about 90%. The proportion of cadmium bound as chloro complexes increases with salinity, as the chloride concentration rises as well.

On the basis of the results presented by Mantoura et al. (1978), of direct experimental evidence considering the interaction of trace metals with humic material (Musani et al., 1977, 1978) in open ocean or seawater, and of experimental data for the chelation of cadmium and lead by the synthetic chelating ligand NTA (Raspor et al. 1978a, 1978b), it is possible to conclude that, at the natural concentration level of dissolved organic matter in the open sea, cadmium will remain predominantly in the labile inorganic complex form. This is due to pronounced competition from alkaline earth cations for the organic ligand and to the high chlorinity of seawater, causing cadmium to be present predominantly as labile chloro complexes. In coastal waters, which can be loaded with higher amounts of organic material, and in low-salinity waters, like estuaries, rivers, and lakes, the distribution of cadmium species will be different.

Gardiner (1974a) investigated the distribution of cadmium in samples of river water and sewage effluents. Data on the source, type, and composition of sewage and river water samples are presented in Table 26, as well as the calculated fractions of cadmium present as free and complexed ions. The results of the calculation were compared with direct determinations obtained by adding 0.1 to 2 mg Cd/l to the sample and measuring with a cadmium-selective electrode the percentage of free cadmium ion. The ionic strength of the samples ranged from 0.008 to 0.017 M. The fact that good agreement was obtained, for most of the samples, between the calculated and experientally determined percentages of free cadmium indicates that no appreciable formation of complexes takes place other than those mentioned in Table 26. In the unpolluted water (sample 1) approximately 90% of cadmium was present as free ion, Cd^{2+}. Predominant complexation in sewage and sewage effluent (samples 2 to 4) resulted from the presence of humic material. The extent of complexation increased with pH (7.2 to 8.0) and with the amount of sewage effluent in the water (sample 9 and 10). The ratio of complexed to uncomplexed cadmium depends only on the stability constant and the concentration of ligand, but not on the total concentration of cadmium, provided that the ligand remains in excess. The extent of complexation with carbonato, that is, with ligands that form single defined complexes, is the same at the lower and higher cadmium concentration. As humic material is a mixture of substances having various molecular weights. it is possible to assume that there will be a difference in complexation between very low concentrations of metal and higher metal concentrations. On the other hand, in seawater, where alkaline earth cations can saturate the active sites of humic substances, such differences are not to be expected.

As a general summary of the discussed experimental evidence and proposed models, the following are the main factors affecting the extent of cadmium interaction with humic material, under equilibrium conditions, in various natural aquatic systems:

1. The stability constant value (K) of the cadmium-humic acid interaction.
2. The concentration of major cations, like Ca(II) and Mg(II), that compete with cadmium for humic acid, and the concentration of major anions, like Cl^- and SO_4^{2-}, that compete for cadmium with humic acid.
3. The pH, which provides a measure of the competition of carbonato and hydrogen carbonato ions for cadmium, and at the same time determines the existing reactive form of humic acid.

6. ADSORPTION EFFECTS OF NATURAL ORGANICS IN VOLTAMMETRIC MEASUREMENTS

The adsorption effects of natural organic substances are a critical parameter in voltammetric measurements. Therefore several authors have investigated the experimental conditions under which trace metal determinations can be performed in the presence of organics.

A voltammetric approach for the determination of cadmium in the presence of humic acid was given by Zur and Ariel (1977). The authors investigated the conditions under which the most reliable determinations of total cadmium concentration in the presence of humic acid could be performed. The measurements were carried out in a model solution, consisting of adjusted concentrations of Cd^{2+}, Cl^-, CO_3^{2-}, Na^+, NH_4^+, NH_3, and humic acid. Commercial humic acid was used for the experiments. The determinations were performed on the glassy carbon rotating disk electrode with the mercury film. The competing equilibrium reactions vary from one pH to another and also depend on the composition of the buffer solution. Optimal conditions for the determination of cadmium in the presence of humic acid were assessed. Three ranges of pH were investigated: slightly basic (8.4 to 8.6), acidic (2.0 to 2.4), and slightly acidic (4.2 to 4.4). In solutions containing Cd(II), humic acid, and ammonia or hydrogen carbonate buffer, adjusted to pH 4.6 to 4.7 by addition of acetate buffer, a good linearity for the cadmium concentration dependence was obtained. It permitted the determination of cadmium in the concentration range from 10^{-7} to 10^{-9} M, by the method of standard addition, in the presence of humic acid, at the preformed mercury film electrode and at the continuously renewed mercury film electrode. The results indicate that the formation of cadmium-humic acid complexes occurs throughout the pH range studied and is affected considerably by the presence of additional inorganic ligands in the sample solution. Acetate buffer is recommended as the preferable medium for the determination of total cadmium in humic acid-containing solutions.

In slightly basic media, Zur and Ariel (1977) recorded nonlabile cadmium-humate formation, but the shift of the peak potential (E_p) in the cathodic direction indicated also a certain labile complex formation. No adsorption effects in that pH region were observed, contrary to the situation in acidic media, where progressive decrease in cadmium peak height, accompanied by peak broadening, indicated adsorption effects. In voltammetry, pronounced adsorption effects could cause serious problems in the reliable analysis of natural waters rich in surface-active substances. In such cases a useful method is to destroy the organic matter previously by UV irradiation, as was carried out by Mart (1979) to reliably exclude contamination.

Brezonik et al. (1976) have investigated the adsorption effects of surface-active substances in metal analysis by ASV. Adsorption effects are potentially more serious for ASV than for polarographic determinations, since the stationary electrode applied in ASV allows the relatively slow sorption process to reach equilibrium. Adsorption effects may depend on pH, as well as the metal-organic complex formation, so that it has to be established whether the decrease in metal peak height results from complex formation with the organic ligand or from the adsorption effects of that organic substance. Adsorption of surface-active substances can affect both diagnostic parameters (i_p and E_p) used in ASV. Thus E_p may shift to more positive values if adsorbed molecule coats the electrode and renders the metal oxidation more irreversible because of inhibition of the electrode reaction. Adsorption affects the peak current in two ways: (*a*) by preventing metal deposition and (*b*) by decreasing the reversibility of the metal oxidation reaction.

A variety of model organic compounds, representative of the types of organic substances occurring in natural waters, were investigated and found to adsorb onto the HMDE in ASV analysis of trace metals. Brezonik et al. (1976) have tested the following compounds: gelatin and alkaline phosphatase as representatives of soluble proteins, Triton X-100 as a nonionic surfactant, and a variety of polysaccharides, including agar, alginic acid, and starch. Gelatin, a well-known maxima suppressor in polarography, was used as a model of the colloidal proteinaceous matter present in natural waters as the result of microbial activities. Varying the concentration of gelatin in neutral and acidic media produced the following effects on the voltammogram of cadmium. There was no effect in neutral solution up to a concentration of 20 mg gelatin/l; however, when the pH was decreased to 3.0, the cadmium peak current was depressed by more than 50%, and its peak potential was shifted by about 56 mV to more positive values. An even larger depression of the cadmium peak was noticed in the presence of the nonionic surfactant Triton X-100. As model compounds of the polysaccharides that can be excreted into natural water by plankton, agar, starch, alginic acid, and polygalacturonic acid were investigated. Freshly prepared agar solutions affected cadmium analysis less than did solutions aged for 1 or more days.

Table 27. Effects of Some Adsorbents on ASV Diagnostic Parameters of Cadmium[a]

Adsorbent	Concentration Range (ppm)	Neutral Conditions[b] (pH 7)			Acidic Conditions[b] (pH 3)	
		Metal	i_p	E_p	i_p	E_p
Agar	1–10	Cd	D up to 40%	No	NC or I (but control)	NC
Alginic acid	4–40	Cd	D up to 50%	NC	I to control	NC
Alkaline phosphatase	1–36	Cd	D up to 15%	NC	D up to 90%	+
Camphor	1–100	Cd	D up to 17%	NC	D up to 25%	+
Gelatin	0.1–20	Cd	NC	NC	D up to 55%	+
Polygalacturonic acid	2–100	Cd	D up to 23%	+	I (but control)	− (Not to control)
Starch	1–10	Cd	D up to 15%	NC	NC	NC
Triton X-100	1–40	Cd	D up to 40%	+	D up to 65%	+

[a] Data from Brezonik et al. (1976).
[b] D = decrease, I = increase, NC = no change, + = anodic shift, − = cathodic shift.

No change in E_p was observed with starch. Ernst et al. (1975) showed that glycin up to 4.8×10^{-4} M has no effect on the E_p of cadmium. In Table 27 the various organic materials investigated by Brezonik et al. (1976) are summarized in regard to the effect on the voltammogram of cadmium at two different pH values.

For a number of substances the influences of adsorption and complex formation on ASV responses are not easily discriminated. For example, the effects of alginic acid on ASV and ion-selective electrode measurements of cadmium are compared in Figure 12. The results indicate a more rapid decrease of the response at low alginic acid concentration than at higher levels. According to Brezonik et al. (1976), the shape of the curves in Figure 12 indicates adsorption reaction, but other results indicate that complex formation with alginic acid may have caused the decrease in i_p (Musani et al., 1977).

Many proteins have both surface-active characteristics and complexation capacity as a result of their numerous amino, carboxyl, and sulfhydryl groups. A decrease in i_p upon protein addition thus could indicate formation of a nonreducible metal-protein complex, rather than adsorption phenomenon.

The foregoing data demonstrate that significant, and in some cases severe, changes can be produced in anodic stripping voltammograms by adsorption phenomena at model adsorbent concentrations that are realistic for natural waters. What remains to be discussed are the implications of these changes in regard to the usefulness of ASV for *in situ* analysis and the question of whether there exist any means whereby the effects of adsorption can be eliminated or compensated for. We have seen that adsorption occurs under neutral and low pH conditions and that, indeed, some adsorbents inhibit electrochemical reactions more strongly at low than at neutral pH. Clearly the difference in i_p between an untreated and an acidified sample does not necessarily define the amount of metal present as nonreducible (presumably organic) or "nonlabile" complexes, nor should the absolute value of i_p for a natural water sample be used to extrapolate concentration values from a standard curve obtained with pure laboratory standard solutions.

A commonly accepted and usually mandatory approach to avoid interference problems is the method of standard additions. If the added metal reacts rapidly with all ligands, so that the fraction of nonreducible metal complexes remains unchanged, the value obtained by the method of additions will permit an estimate of the total dissolved metal content in the sample that is not trapped in rather inert chelates. If, however, the added metal reaches equilibrium with organic ligands very slowly, the value obtained by standard additions will approximate the sum of the free metal ion plus reducible complexes. Evidence that the attainment of equilibrium can be slow for some metal-macromolecular substance interactions is shown in Figure 13, in which the voltammetric responses (i_p) of several metals (Zn, Cd, Pb, and Cu) added to a solution containing humic acid

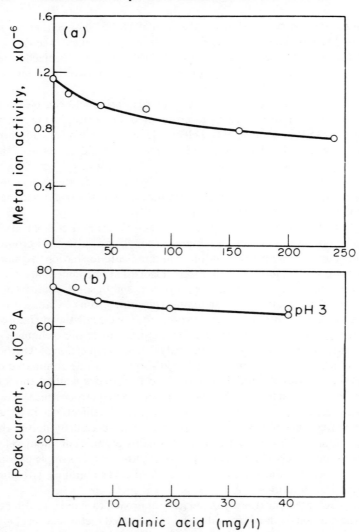

Figure 12.   Effect of alginic acid on cadmium (*a*) Ionic activity (*b*) Peak current (i_p). (Brezonik et al., 1976.)

are seen to decrease as the solution aged (Brezonik et al., 1976). In acid media the method of standard additions yields correct results because the added metal behaves similar to that originally present, that is, the effect of adsorption on the reduction of Cd^{2+} at the mercury electrode is not a function of Cd^{2+} concentration.

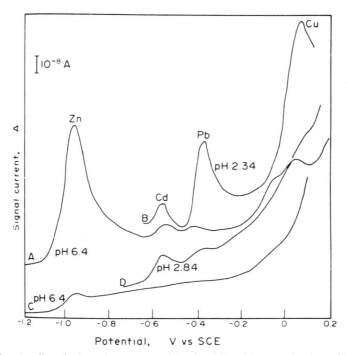

Figure 13.   Anodic stripping voltamograms of Zn, Cu, Cd, and Pb: *A* and *B*, immediately after humic acid addition; *C* and *D*, after 3 days incubation. (Brezonik et al., 1976.)

The method of standard additions will give correct results for the total metal level if it can be assumed that acidification releases all complexed metal (as is unlikely for metal-macromolecule interactions) or that the added metal behaves the same (i.e., is complexed to the same extent) as the metal originally present. As previously mentioned, the organic matter can, however, be destroyed by preceding UV irradiation according to the procedure of Mart (1979).

Obviously, adsorption effects can be an obstacle in determining the total concentrations of trace metals present in natural waters. On the other hand, such effects can be applied usefully for the polarographic or voltammetric determination of surface-active substances. Possibilities for the application of a sensitive electroanalytical method for the determination of traces of surfactants in seawater have been reported by Zvonarić et al. (1973). The analytical method is based on the supression of the polarographic maximum of oxygen due to adsorption of surfactants at the DME-solution interface. This simple method can be applied for the sensitive determination of the global low level of surfactants in seawater. The amount of surfactant in seawater is evaluated in arbitrary relative units by comparison with calibration curves, obtained for artificial seawater and the nonionic surfactant Triton X-100. However, the adsorption analysis is inherently unspecific, because all surface-active agents have qualitatively similar effects. Since in the case of seawater the individual constituents, as well as their percentages, are usually unknown, the result obtained is only a measure in arbitrary Triton X-100 units of the global amount of surfactant in the sample. The Triton X-100 equivalent value can serve, however, as an empirical relative measure of the particular surfactant level. For a more exact elucidation of its significance, the results obtained by this method should be compared with the independently determined content of dissolved organic matter. Since the majority of the organic substances present in seawater are surface active, it might be expected that this method could be of use for the rapid detection of organic pollution levels in the investigated regions of the sea. The measured suppression of the oxygen maximum in offshore samples of the North Adriatic (international waters), according to Zvonarić et al. (1973), is equivalent to the effect of 0.2 to 5 mg of Triton X-100.

Surface-active substances in natural waters can be of natural or pollutive origin. For the preservation of aquatic life it is necessary to be able to control precisely the concentration of synthetic surfactants that reach the waters through industrial and domestic wastewaters. The Kalousek commutator technique can be used for the determination of surface-active substances in aquatic solutions by measuring the adsorption effects of organic molecules at the mercury electrode surface. The adsorption effects of various types of organic substances, either occurring naturally in seawater or added to the sample, have been extensively described in a paper by Kozarac et al. (1976). The technique itself was reported earlier in more detail by Ćosović and Branica (1973). The aim of the work by Kozarac et al. (1976) was to determine the applicability of the method for the estimation of the degree of seawater pollution by surface-active substances. Adsorption isotherms of an anionic detergent and two cationic detergents, lecithin and lauric acid, at the HMDE were used as calibration curves. The adsorption effects at the HMDE were measured in seawater samples from a polluted harbor area. The concentration of surface-active substances in the sample was expressed in equivalents of the anionic detergent, sodium lauryl sulfate, and was compared with the results of the spectrophotometric determination for anionic detergents, based on methylene blue.

In unpolluted seawater the quantity of naturally occurring surfactant substances is, in most cases, below the detection limit of the method described. No samples from an unpolluted seawater area showed any measurable effects.

In view of the fact that samples of polluted seawater contain a mixture of organic substances, it is of great interest to study by polarography the adsorption effects of various mixtures at very low concentrations.

LIST OF ABBREVIATIONS AND SYMBOLS

AAS	Atomic absorption spectroscopy
ASV	Anodic stripping voltammetry
dc polar-ography	Direct current polarography
DME	Dropping mercury electrode
DPASV	Differential pulse anodic stripping voltammetry
DPP	Differential pulse polarography
EDTA	Ethylenediaminetetraacetic acid
HMDE	Hanging mercury drop electrode
MFE	Mercury film electrode (electrode with the thin film of mercury on the glassy carbon carrier
NTA	Nitrilotriacetic acid
SDME	Slowly dropping mercury electrode
WIG	Wax-impregnated graphite electrode
$E_{1/2}$	Half-wave potential in dc polarography
E_p	Potential at the maximum (peak) of the cathodic or anodic current
E_p^+	Critical apparent potential obtained by interpolation from the plot of anodic peak current versus deposition potential
$E_{p/2}$	Half-peak potential
$i_{\lim}$	Limiting current
i_p	Peak current

ACKNOWLEDGMENTS

This chapter has been written within the framework of the joint research project "Environmental Research in Aquatic Systems" of the Institute of Chemistry, Institute for Applied Physical Chemistry, Nuclear Research Centre (KFA), Juelich, and the Center for Marine Research, Institute "Rudjer Bosković," Zagreb. Financial support from the Nuclear Research Centre while the author was attached to Institute for Applied Physical Chemistry is gratefully acknowledged.

Critical discussion of the manuscript and suggestions made by Professor H. W. Nürnberg, collaboration with Dr. K. Schroeder (Central Library, Juelich) in collecting the necessary literature data and with Mrs. Klingner-Thielsch for typing the manuscript are acknowledged with appreciation.

REFERENCES

Abdullah, M. I., Royle, L. G., and Morris, A. W. (1972). "Heavy Metal Concentration in Coastal Waters," *Science* **235**, 158–160.

Abdullah, M. I., El-Rayis, O. A., and Riley, J. P. (1976). "Reassessment of Chelating Ion-Exchange Resins for Trace Metal Analysis of Sea Water," *Anal. Chim. Acta* **84**, 363–368.

Ahrland, S. (1975). "Metal Complexes Present in Sea Water." In E. D. Goldberg, Ed., *The Nature of Sea Water.* Dahlem Konferenzen, Berlin, pp. 219–243.

Allen, H. E. (1973). "Studies of Trace Metals in Aquatic Environments Using Anodic Stripping Voltammetry." In *Modern Analytical Polarography,* Workshop Manual by PAR Corporation, New Jersey, pp. VIII-1-14.

Barić, A. and Branica, M. (1967). "Polarography of Sea Water. I: Ionic State of Cadmium and Zinc in Sea Water," *J. Pologr. Soc.* **13**, 4–8.

Batley, G. E. and Florence, T. M. (1974). "An Evaluation and Comparison of Some Techniques of Anodic Stripping Voltammetry," *J. Electroanal. Chem.* **55**, 23–43.

Batley, G. E. and Florence, T. M. (1976a). "A Novel Scheme for the Classification of Heavy Metal Species in Natural Waters," *Anal. Lett.* **9**, 379–388.

Batley, G. E. and Florence, T. M. (1976b). "Determination of the Chemical Forms of Dissolved Cadmium, Lead, and Copper in Sea Water," *Mar. Chem.* **4**, 347–363.

Bender, M. L. and Gagner, C. (1976). "Dissolved Copper, Nickel and Cadmium in the Sargasso Sea," *J. Mar. Res.* **34**, 327–339.

Bernhard, M. and Zattera, A. (1969). "A Comparison between the Uptake of Radioactive and Stable Zn by a Marine Unicellular Algae." In *2nd National Symposium on Radioecology, Ann Arbor.*

Bernhard, M. and Zattera, A. (1975). "Major Pollutants in the Marine Environment." In E. A. Pearson and E. DeFraja Frangipane, Eds., *Marine Pollutants and Marine Waste Disposal.* Proceedings of the 2nd International Congress, San Remo, Pergamon Press, London, pp. 195–300.

Bilinski, H., Huston, R., and Stumm, W. (1976). "Determination of the Stability Constants of Some Hydroxo and Carbonato Complexes of Pb(II), Cu(II), Cd(II), and Zn(II) in Dilute Solutions by Anodic Stripping Voltammetry and Differential Pulse Polarography," *Anal. Chim. Acta* **84**, 157–164.

Björndal, H., Bouveng, H. O., Solyom, P., and Werner, J. (1972). "NTA in Sewage Treatment. III: Biochemical Stability of Some Metal Chelates," *Vatten* **28**, 5–16.

Boyle, E. A., Sclater, F., and Edmond, J. M. (1976). "On the Marine Geochemistry of Cadmium," *Nature* **263**, 42–44.

Branica, M. (1976). "Development of Methods for Rapid Detection of Trace Metals in Sea Water." In S. P. Meyers, Ed., *Proceedings of the International Symposium on Marine Pollution Research, Baton Rouge,* pp. 114–119.

Branica, M., Novak, D. M., and Bubić, S. (1977). "Application of Anodic Stripping Voltammetry to Determination of the State of Complexation of Traces of Metal Ions at Low Concentration Levels," *Croat. Chem. Acta* **49**, 539–547.

Brezonik, P. L., Brauner, P. A., and Stumm, W. (1976). "Trace Metal Analysis by Anodic Stripping Voltammetry: Effect of Sorption by Natural and Model Organic Compounds," *Water Res.* **10**, 605–612.

Bruegmann, L. (1974). "The Determination of Trace Metals in Sea Water Applying a Stationary Mercury Electrode," *Acta Hydrochim. Hydrobiol.* **2**, 123–138.

Bruegmann, L. (1977). "Distribution of Some Heavy Metals in the Baltic Sea—A Survey," *Acta Hydrochim. Hydrobiol.* **5**, 3–21.

Bruland, K. W., Knauer, G. A., and Martin, J. H. (1978). "Cadmium in Northeast Pacific Waters," *Limnol. Oceanogr.* **23**, 618–625.

Bryan, G. W. (1969). "The Adsorption of Zinc and Other Metals by the Brown Seaweed *Laminaria digitata*," *J. Mar. Biol. Assoc. U.K.* **49**, 225–243.

Buat-Menard, P. and Chesselet, R. (1978). "Variable Control of the Atmospheric Flux on the Trace Element Chemistry of Oceanic Suspended Matter," *Earth Planet. Sci. Lett.,* in press.

Bubić, S. and Branica, M. (1973). "Voltammetric Characterization of the Ionic State of Cadmium Present in Sea Water," *Thalassia Jugosl* **9**, 47–53.

Bubić, S., Sipos, L. and Branica, M. (1973). "Comparison of Different Electroanalytical Techniques for the Determination of Heavy Metals in Sea Water," *Thalassia Jugosl.* **9**, 55–63.

Bunch, R. I. and Ettinger, M. B. (1967). "Biodegradability of Potential Organic Substitutes for Phosphates." In *Proceedings of the 22nd Industrial Waste Conference,* Vol. 1, Purdue Univ. Eng. Ext. Ser. 129, pp. 393–396.

Chau, Y. K. and Shiomi, M. T. (1972). "Complexing Properties of Nitrilotriacetic Acid in the Lake Environment," *Water, Air, Soil Pollut.* **1**, 149–164.

Chau, Y. K. and Chawla, V. K., Nicholson, H. F., and Vollenweider, R. A. (1970). *Proc Great Lakes Res. Conf.* **13**, 659.

Cheam, V. and Gamble, D. S. (1974). "Metal-Fulvic Acid Chelation Equilibrium in Aqueous $NaNO_3$ Solution," *Can. J. Soil Sci.* **54**, 413–417.

Chester, R. and Stoner, J. H. (1974). "Distribution of Zinc, Nickel, Mangenese, Cadmium, Copper and Iron in Some Surface Waters from the World Ocean," *Mar. Chem.* **2**, 17–32.

Christman, R. F. and Ghassemi, M. (1966). *J. Am. Water Works Assoc.* **58** 723.

Ćosović, B. and Branica, M. (1973). "Study of the Adsorption of Organic Substances at a Mercury Electrode by the Kalousek Commutator," *J. Electroanal. Chem.* **46**, 63–69.

Crow, D. R. (1969). *Polarography of Metal Complexes.* Academic Press, London.

Davies, C. W. (1962). *Ion Association.* Butterworths, London, 190 pp.

DeFord, D. D. and Hume, D. N. (1951). "The Determination of Consecutive Formation Constants of Complex Ions from Polarographic Data," *J. Am. Chem. Soc.* **73**, 5321–5325.

Duinker, J. C. and Kramer, C. J. M. (1977). "An Experimental Study on the Speciation of Dissolved Zinc, Cadmium, Lead and Copper in River Rhine and North Sea Water, by Differential Pulse Anodic Stripping Voltammetry," *Mar. Chem.* **5,** 207–228.

Dulka, J. J. and Risby, T. H. (1976). "Ultratrace Metals in Some Environmental and Biological Systems," *Anal. Chem.* **48,** 640A–653A.

Dyrssen, D. and Wedborg, M. (1973). "Equilibrium Calculations of the Speciation of Elements in Sea Water." In E. D. Goldberg, Ed., *The Sea,* Vol. 5. Wiley-Interscience, New York, pp. 181–195.

Eaton, A. (1976). "Marine Geochemistry of Cadmium," *Mar. Chem.* **4,** 141–154.

Elder, J. F. (1975). "Complexation Side Reactions Involving Trace Metals in Natural Water Systems," *Limnol. Oceanogr.* **20,** 96–102.

Ernst, R., Allen, H. E., and Mancy, K. H. (1975). "Characteristics of Trace Metal Species and Measurements of Trace Metal Stability Constants by Electrochemical Techniques," *Water Res.* **9,** 969–979.

Florence, T. M. (1972). "Determination of Trace Metals in Marine Samples by Anodic Stripping Voltammetry," *J. Electroanal. Chem.* **35,** 237–245.

Forsberg, C. and Lindquist, G. (1967). "Experimental Studies on Bacterial Degradation of Nitrilotriacetate, NTA," *Vatten* **23,** 265–277.

Friberg, L., Piscator, H., and Nordberg, G. F. (1971). *Cadmium in the Environment,* 1st Ed., CRC Press, Cleveland, Ohio.

Friberg, L., Piscator, M., Nordberg, G. F., and Kjellström, T. (1974). *Cadmium in the Environment,* 2nd ed., CRC Press, Cleveland, Ohio.

Gamble, D. S. and Schnitzer, M. (1973). "The Chemistry of Fulvic Acid and Its Reactions with Metal Ions." In P. C. Singer, Ed., *Trace Metals and Metal-Organic Interactions in Natural Waters.* Ann Arbor Science Publishers, Ann Arbor, Mich., pp. 265–302.

Gardiner, J. (1974a). "Chemistry of Cadmium in Natural Water. I: Study of Cadmium Complex Formation Using the Cadmium Specific-Ion Electrode," *Water Res.* **8,** 23–30.

Gardiner, J. (1974b). "Chemistry of Cadmium in Natural Water. II: The Adsorption of Cadmium on River Muds and Naturally Occurring Solids," *Water Res.* **8,** 157–164.

Gardiner, J. (1975). "The Complexation of Trace Metals in Natural Waters." In *Proceedings of the 1st International Conference on Heavy Metals in the Environment,* Vol. 1, Toronto, Ont., pp. 303–317.

Gardiner, J. (1976). "Complexation of Trace Metals by Ethylenediaminetetraacetic Acid (EDTA) in Natural Waters," *Water Res.* **10,** 507–514.

Garrels, R. M. and Thompson, M. E. (1962). "A Chemical Model for Sea Water at 25°C and One Atmosphere Pressure," *Am. J. Sci.* **260,** 57–66.

Goldberg, E. D. (1965). "Minor Elements in Sea Water." In J. P. Riley and G. Skirrow, Eds., *Chemical Oceanography,* Vol. 1. Academic Press, London and New York, pp. 163–196.

Goldberg, E. D., Ed. (1972). *Baseline Studies of Pollutants in the Marine Environment and Research Recommendations.* IDOE Baseline Conference, May 24–26, 1972, New York.

Gutknecht, J. (1963). "Zn-65 Uptake by Benthic Marine Algae," *Limnol. Oceanogr.* **8,** 31–38.

Guy, R. D. and Chakrabarti, C. L. (1975). "Analytical Techniques for Speciation of Trace Metals." In *Proceedings of the 1st International Conference on Toronto,* pp. 275–294, *Heavy Metals in the Environment,* Vol. 1.

Hahne, H. C. H. and Kroontje, W. (1973). "Significance of pH and Chloride Concentration on Behaviour of Heavy Metal Pollutants: Mercury(II), Cadmium(II), Zinc(II) and Lead(II)," *J. Environ. Qual.* **2,** 444–450.

Hume, D. N. (1967). "Analysis of Water for Trace Metals: Present Capabilities and Limitations." In R. F. Gould, Ed., *Advances in Chemistry,* Series 67. American Chemical Society, Washington D.C., pp. 30–44.

Ingri, N., Kakolowicz, W., Sillén, L. G., and Warnquist, B. (1967). "High Speed Computers as a Supplement to Graphical Methods. V: HALTAFALL, a General Program for Calculating the Composition of Equilibrium Mixutres," *Talanta* **12,** 1261–1286; errata, *ibid.* **15,** xi–xii (1968).

Kester, D. R. and Pytokwicz, R. M. (1969). "Sodium, Magnesium and Calcium Sulfate Ion Pairs in Sea Water at 25°C," *Limnol. Oceanogr.* **14,** 686–692.

Kester, D. R., Ahrland, S., Beasley, T. M., Bernhard, M., Branica, M., Campbell, I. D., Eichhorn, G. L., Kraus, K. A., Kremling, K., Millero, F. J., Nürnberg, H. W., Piro, A., Pytkowicz, R. M., Steffan, I., and Stumm, W. (1975). "Chemical Speciation in Sea Water." In E. D. Goldberg, Ed., *The Nature of Sea Water.* Dahlem-Konferenzen, Berlin, pp. 17–41.

Kjellström, T., Lind, B., Linnman, L., and Nordberg, G. F. (1974). "Comparative Study on Methods for Cadmium Analysis of Grain with Application of Pollution Evaluation," *Environ. Res.* **8,** 92–106.

Knauer, G. and Martin, J. (1973). "Seasonal Variations of Cadmium, Copper, Mangenese, Lead and Zinc in Water and Phytoplankton in Monterey Bay, California," *Limnol. Oceanogr.* **18,** 597–604.

Kozarac, Z., Ćosović, B., and Branica, M. (1976). "Estimation of Surfactant Activity of Polluted Sea Water by Kalousek Commutator Technique," *J. Electroanal. Chem.* **68,** 75–83.

Kozjak, B., Marinić, Z., Konrad, Z., Musani-Marazović, Lj., and Pučar, Z. (1977). "Electrophoretic Investigations of the Complexing of Cadmium and Zinc with EDTA," *J. Chromatogr.* **132,** 323–334.

Kremling, K. (1973). "Voltammetric Measurements of the Distribution of Zinc, Cadmium, Lead and Copper in the Baltic Sea," *Kiel. Meeresforsch.* **29,** 77–84.

Kremling, K. (1976). "Determination of Trace Metals." In K. Grasshoff, Ed., *Methods of Sea Water Analysis.* Verlag Chemie, Weinhei, New York, pp. 183–191.

Kunkel, R. and Manahan, S. E. (1973). "Atomic Absorption Analysis of Strong Heavy Metal Chelating Agents in Water and Waste Water," *Anal. Chem.* **45,** 1465–1468.

Livingstone, D. A. (1963). *Chemical Composition of Rivers and Lakes.* U.S. Geol. Surv. Pap. 440G.

Lund, W. and Onshus, D. (1976). "The Determination of Copper, Lead and Cadmium in Sea Water by Differential Pulse Anodic Stripping Voltammetry," *Anal. Chim. Acta* **86,** 109–122.

Lund, W. and Salberg, M. (1975). "Anodic Stripping Voltammetry with the Florence Mercury Film Electrode: Determination of Copper, Lead and Cadmium in Sea Water," *Anal. Chim. Acta* **76,** 131–141.

Magjer, T. and Branica, M. (1977). "A New Electrode System with Efficient Mixing of Electrolyte," *Croat. Chem. Acta* **49,** L1–L5.

Malcolm, E. L. (1972). "Comparison of Conditional Stability Constants of North Carolina Humic and Fulvic Acids with Co(II) and Fe(III)." In E. D. Nelson, *Environmental Framework of Coastal Plain Estuaries.* Geological Society of Washington.

Maljković, D. and Branica, M. (1971). "Polarography of Sea Water. II: Complex Formation of Cadmium with EDTA," *Limnol. Oceanogr.* **16,** 779–785.

Mantoura, R. F. C., Dickson, A., and Riley, J. P. (1978). "The Complexation of Metals with Humic Materials in Natural Waters," *Estuarine Coastal Mar. Sci.* **6,** 387–408.

Mart, L. (1979). "Determination and Comparison of Toxic Trace Metal Level in North Atlantic Mediterranean Coastal Waters." Ph.D. Thesis, RWTH Aachen, West Germany.

Martin, J. H. (1970). "Possible Transport of Trace Metals in a Moulted Copepool Exoskeletons," *Limnol. Oceanogr.* **15,** 756–61.

Meites, L., Campbell, B. H., and Zuman, P. (1977). "Development and Publication of New Polarographic and Related Methods of Analysis," *Talanta* **24,** 709–724.

Morel, F. and Morgan, J. (1972). "A Numerical Method for Computing Equilibria in Aqueous Chemical Systems," *Environ. Sci. Technol.* **6,** 58–67.

Mottola, H. A. (1974). "Nitrilotriacetic Acid as a Chelating Agent: Applications, Toxicology and Bio-Environmental Impact," *Toxicol. Environ. Chem. Rev.* **2,** 99–161.

Mullin, J. B. and Riley, J. P. (1954). "Cadmium in Sea Water," *Nature* **174,** 42.

Mullin, J. B. and Riley, J. P. (1956). "The Occurrence of Cadmium in Seawater and in Marine Organisms and Sediments," *J. Mar. Res.* **15,** 103–122.

Musani, Lj., Valenta, P., Nürnberg, H. W., Konrad, Z., and Branica, M. (1977). "Interactions of Pb-210 with Some Natural Organic Materials in Sea Water." In M. Branica, Ed., *Lead—Occurrence, Fate and Pollution in the Marine Environment.* Proceedings of an international experts' discussion, Center for Marine Research, Rovinj, Yugoslavia. Pergamon Press, London, in press.

Musani, Lj., Valenta, P., Nürnberg, H. W., Konrad, Z., and Branica, M., (1978). *Interaction of Some Toxic Metals and Humic Acid of Marine Sediment Origin in Sea Water.* XXVI Congress and Plenary Assembly of ICSEM, Antalya.

Musani-Marazović, Lj. and Pučar, Z. (1973). "The Interaction of ^{109}Cd and EDTA in Sea Water and Sodium Chloride Solutions (Electrophoretic Investigations),"

Thalassia Jugosl. **9**, 101–111.

Nguyen, V. D., Valenta, P., and Nürnberg, H. W. (1978). "Voltammetry of Atmospheric Pollutants. I: The Determination of Toxic Trace Metals in Rain Water and Snow by Differential Pulse Stripping Voltammetry," *Sci. Total Environ.,* in press.

Nomiyama, K. (1975). "Toxicity of Cadmium—Mechanism and Diagnosis." In P. A. Krenkel, Ed., *Heavy Metals in the Aquatic Environment.* Pergamon Press, Oxford, pp. 15–23.

Nordberg, G. F. (1974). "Health Hazards of Environmental Cadmium Pollution," *AMBIO* **3**, 55–66.

Nürnberg, H. W. (1976). *Polarography and Voltammetry in Environmental Research and Surveillance of Toxic Metals.* International Symposium on Industrial Electrochemistry, Madras.

Nürnberg, H. W. (1977a). "Potentialities and Application of Advanced Polarographic and Voltammetric Methods in Environmental Research and Surveillance of Toxic Metals," *Electrochim. Acta,* **22**, 935–949.

Nürnberg, H. W. (1977b). "Potentialities of the Voltammetric Approach in Trace Metal Chemistry of Sea Water." Content of two lectures presented at the Cours Internationaux Post-Universitaires, University of Gent, Aug. 15–20.

Nürnberg, H. W. (1978a). "Potentialities and Applications of Advanced Polarographic and Voltammetric Methods in Aquatic and Marine Trace Metal Chemistry." In P. Ahlberg and L. D. Sundelöf, *Structure and Dynamics in Chemistry.* Proceedings of a symposium, Uppsala, 1977, pp. 270–307.

Nürnberg, H. W. (1978b). "Studies on Toxic Trace Metals in the Environment by Advanced Polarographic Methods," *Proc. Anal. Div. Chem. Soc.* (*London*) **15**, 275–283.

Nürnberg, H. W. (1979). "Polarography and Voltammetry in Studies of Toxic Metals in Man and His Environment," *Sci. Total Environ.* **12**, 35–60.

Nürnberg, H. W. and Stoeppler, M. (1977). Lecons et Séminires de "Systèmes Marins," Quatrième Ecole Européenne d'Eté d'Environment E4, Louvainla Neuve, in press.

Nürnberg, H. W. and Valenta, P. (1975). "Polarography and Voltammetry in Marine Chemistry." In E. D. Goldberg, Ed., *The Nature of Sea Water.* Dahlem-Konferenzen, Berlin, pp. 87–136.

Nürnberg, H. W., Stoeppler, M., and Valenta, P. (1975). "On the Accuracy and Reliability of Trace Metal Determinations in Environmental Matrix Types by Advanced Polarographic and Spectroscopic Techniques," *Thalassia Jugosl.* **11**, 85–100.

Nürnberg, H. W., Valenta, P., Mart, L., Raspor, B., and Sipos, L. (1976). "Application of Polarography and Voltammetry to Marine and Aquatic Chemistry," *Z. Anal. Chem.* **282**, 357–367.

Nürnberg, H. W., Mart, L., and Valenta, P. (1977). "Concentrations of Cd, Pb and Cu in Ligurian and Tyrrhenian Coastal Waters." In *Rapports et Proces-Verbaux,* vol. 24, Fasc. 8., pp. 25–29. XXV Congress and Plenary Assembly ICSEM, Split, 1976.

Peden, J. D., Crothers, J. H., Waterfall, C. E., and Beasley, J. (1973). "Heavy Metals in Somerset Marine Organisms," *Mar. Pollut. Bull.,* **4,** 7–9.

Preston, A., Jefferies, D. F., Dutton, J. W. R., Harvey, B. R., and Steele, A. K. (1972). "British Isles Coastal Water Concentrations of Selected Heavy Metals in Sea Water, Suspended Matter and Biological Indicators," *Environ. Pollut.* **3,** 69–82.

Provasoli, L. (1963). "Organic Regulation of Phytoplankton Fertility." In M. Hill, Ed., *The Sea,* Vol. 2. Interscience, New York, pp. 165–210.

Rashid, M. A. (1971). "Role of Humic Acids of Marine Origin and Their Different Molecular Weight Fractions in Complexing Di- and Trivalent Metals," *Soil Sci.* **111,** 298–306.

Raspor, B. and Branica, M. (1973). "Polarographic Study of the Cadmium-Ethylene-diaminetetraacetate Chelate at pH about 8," *J. Electroanal. Chem.* **45,** 79–88.

Raspor, B. and Branica, M. (1975a). "Polarographic Reduction of Cd(II)-Nitrilotriacetic Acid Chelate in Chloride Solutions of pH about 8," *J. Electroanal. Chem.* **59,** 99–109.

Raspor, B. and Branica, M. (1975b). "Comparison of the Polarographic Reduction of Cd-NTA, -EDTA and -DTPA Chelates in Chelates in Chloride Solutions," *J. Electroanal. Chem.* **60,** 335–339.

Raspor, B., Valenta, P., Nürnberg, H. W., and Branica, M. (1977). "Polarographic Studies on the Kinetics and Mechanism of Cd(II)-Chelate Formation with EDTA in Sea Water." In *Rapports et Procès-Verbaux,* Vol. 24, Fasc. 8., pp. 89–91. XXV Congress and Plenary Assembly of ICSEM, Split, 1976; *Thalassia Jugosl., 13* (1/2) 79–91.

Raspor, B., Valenta, P., Nürnberg, H. W., and Branica, M. (1978a). "The Chelation of Cadmium with NTA in Sea Water as a Model for The Typical Behavior of Trace Heavy Metal Chelates in Natural Waters," *Sci. Total Environ.* **9,** 87–109.

Raspor, B., Nürnberg, H. W., Valenta, P., and Branica, M. (1978b). "The Chelation of Lead by Organic Ligands in Sea Water." In M. Branica, Ed., *Lead—Occurrence, Fate and Pollution in the Marine Environment.* Proceedings of an international experts' discussion, Center for Marine Research, Rovinj, Yugoslavia, Pergamon Press, London, in press.

Raspor, B., Nürnberg, H. W., Valenta, P., and Branica, M. (1980a). "Kinetics and Mechanism of Trace Metal Chelation in Seawater," *Sci. Total Environ.,* in press.

Raspor, B., Nürnberg, H. W., Valenta, P., and Branica, M. (1980b). "Voltammetric Studies on the Stability of the Zn(II)-Chelates with NTA and EDTA and the Kinetics of their Formation in Lake Ontario Water," *Water Research,* in press.

Reardon, E. J. and Langmuir, D. (1976). "Activity Coefficients of $MgCO_3^0$ and $CaSO_4^0$ Ion Pairs as a Function of Ionic Strength," *Geochim. Cosmochim. Acta* **40,** 549–554.

Riley, J. P. and Taylor, D. (1968). "Chelating Resins for the Concentration of Trace Elements from Sea Water and Their Analytical Use in Conjunction with Atomic Absorption Spectrophotometry," *Anal. Chim. Acta* **40,** 479–485.

Riley, J. P. and Taylor, D. (1972). "Concentrations of Cadmium, Copper, Iron Manganese, Molybdenum, Nickel, Vanadium and Zinc in Part of the Tropical Northeast Atlantic Ocean," *Deep-Sea Res. Oceanogr. Abstr.* **19**, 307–317.

Schnitzer, M. and Hansen, E. H. (1970). *Soil. Sci.* **109**, 333.

Schnitzer, M. and Khan, S. U. (1972). "Reactions of Humic Substances with Metal Ions and Hydrous Oxides." In *Humic Substances in the Environment*. Marcel Dekker, New York, pp. 203–251.

Sillén, L. G. (1961). "The Physical Chemistry of Sea Water." In M. Sears, Ed., *Oceanography*. Am. Assoc. Adv. Sci. Publ. **67**, Washington, D.C., pp. 549–582.

Sipos, L., Magjer, T., and Branica, M. (1974). "An Universal Voltammetric Cell," *Croat. Chem. Acta,* **46**, 35–37.

Sipos, L., Kozar, S., Kontusic, I., and Branica, M. (1978). "Subtractive Anodic Stripping Voltammetry with Rotating Mercury Coated Glassy Carbon Electrode," *J. Electroanal. Chem.* **87**, 347–352.

Sipos, L., Valenta, P., Nürnberg, H. W., and Branica, M. (1979a). "Labile Lead and Cadmium Complexes in Seawater," *Mar. Chem.* in press.

Sipos, L., Raspor, B., Nürnberg, H. W., and Pytkowicz, R. M. (1979b). "Interaction of Metal Complexes with Coulombic Ion pairs in High Salinity Aqueous Media," *Mar. Chem.* in press.

Smith, J. D. and Redmond, J. D. (1971). "Anodic Stripping Voltammetry Applied to Trace Metals in Sea Water," *J. Electroanal. Chem.* **33**, 169–175.

Stiff, M. J. (1971). "The Chemical States of Copper in Polluted Fresh Water and a Scheme of Analysis to Differentiate Them," *Water Res.* **5**, 585–599.

Stoeppler, M. (1978). International Workshop on Monitoring Environmental Materials and Specimen Banking, Berlin, in press.

Strohal, P. and Huljev, D. (1971). "Investigation of Mercury Pollutant Reaction with Humic Acid by Means of Radio Tracers." In *Nuclear Techniques in Environmental Pollution* International Atomic Energy Agency, Vienna.

Stumm, W. and Morgan, J. J. (1970). *Aquatic Chemistry*. Wiley-Interscience, New York.

Sverdrup, H. U., Johnson, M. W., and Fleming, R. H. (1942). *The Oceans: Their Physics, Chemistry and General Biology*. Prentice-Hall, New York.

Thom, N. S. (1971). "Nitrilotriacetic Acid: A Literature Survey," *Water Res.* **5**, 391–399.

Thompson, J. E. and Duthie, J. R. (1968). "The Biodegradability and Treatability of NTA," *J. Water Pollut. Control Fed.* **40**, 306–319.

Valenta, P., Mart, L., and Rützel, H. (1977a). "New Potentialities in Ultratrace Analysis with Differential Pulse Anodic Stripping Voltammetry," *J. Electroanal. Chem.* **82**, 327–343.

Valenta, P., Mart, L., Nürnberg, H. W., and Steoppler, M. (1977b). "Simultaneous Voltammetric Analysis of Traces of Toxic Metals in Sea Water, Inland Waters, Drinking and Consume Waters," *Vom Wasser* **48**, 89–110.

Valenta, P., Rützel, H., Krumpen, P., Salgert, K. H., and Klahre, P. (1978). "Contributions to Automated Trace Analysis. IV: Device for the Automated Simultaneous Voltammetric On-Line Determination of Toxic Trace Metals in Drinking Water," *Fresenius Z. Anal. Chem.* **292,** 120–125.

Vanderzee, C. E. and Dawson, H. J. (1953). "The Stability Constants of Cadmium Chloride Complexes: Variation with Temperature and Ionic Strength," *J. Am. Chem. Soc.* **75,** 5659–5663.

Watson, M. W., Wood, R. H., and Millero, F. J. (1975). "Activity of Trace Metals in Artificial Water at 25°C." In *Marine Chemistry: Coastal Environment.* CACS Symp. Ser., Vol. 18, pp. 112–118.

Windom, H. and Smith, R. (1972). "Distribution of Cadmium, Cobalt, Nickel, and Zinc in southeastern United States Continental Shelf Waters," *Deep-Sea Res.* **19,** 727–730.

Wong, P. T. S., Liu, D., and McGirr, D. J. (1973). "Mechanism of NTA Degradation by a Bacterial Mutant," *Water Res.* **7,** 1367–1374.

Zirino, A. and Yamamoto, S. (1972). "A pH Dependent Model for the Chemical Speciation of Copper, Zinc, Cadmium, and Lead in Sea Water," *Limnol. Oceanogr.* **17,** 661–671.

Zitko, V. and Carson, W. V. (1972). "Release of Heavy Metals from Sediments by NTA," *Chemosphere* **1,** 113–118.

Zur, C. and Ariel, M. (1977). "The Determination of Cadmium in the Presence of Humic Acid by Anodic Stripping Voltammetry," *Anal. Chim. Acta* **88,** 245–251.

Zvonarić, T., Žutić, V., and Branica, M. (1973). "Determination of Surfactant Activity of Sea Water Samples by Polarography," *Thalassia Jugosl.* **9,** 65–73.

7

CADMIUM INTERACTIONS WITH NATURALLY OCCURRING ORGANIC LIGANDS

John P. Giesy, Jr.

Savannah River Ecology Laboratory, University of Georgia, Aiken, South Carolina

1. INTRODUCTION

The physical and chemical state of cadmium is dependent on inorganic water quality and must be considered when assessing cadmium toxicity and availability to aquatic biota. For example, water hardness has an antagnoistic effect on cadmium toxicity to zooplankton (McKee and Wolf, 1963). Similarly, Pickering and Henderson (1966) found increases in Cd 96-hr LC_{50} values with increasing water hardness for all fish tested, and Kinkade and Erdman (1975) reported that organisms accumulated cadmium more slowly from harder water. These results indicate that divalent cationic cadmium is the most toxic form, as is the

Table 1. Binding Capacities of Nominal Molecular Weight Organic Fractions from Three Surface Waters in the Southeastern United States

Station and Nominal Molecular Weight Fraction	Cadmium (ng-atoms/l)	Cadmium (ng-atoms/mg OC)	Cadmium Binding Capacity (μg-atoms/l)	Cadmium Binding Capacity (mg-atoms/mg OC)	Saturation (%)
Upper Three Runs					
I > 300,000	1.8	1.9	0.15	0.16	1.1
300,000 > II > 10,000	0.18	9.0	0.0042	0.21	4.2
10,000 > III > 500	1.1	1.0	0.06	0.056	1.8
IV < 500	1.8	0.48	1.9	1.14	0.10
Total	4.9[a]	—	2.1[a]	0.57[a]	—
Skinface Pond					
I > 300,000	6.3	2.9	0.28	0.13	2.3
300,000 > II > 10,000	0.4	2.2	0.012	0.067	0.33
10,000 > III > 500	6.5	1.4	0.32	0.07	2.0
IV < 500	1.2	0.13	4.6	0.56	0.25
Total	14.0[a]	—	5.2[a]	0.34[a]	—
Fire Pond					
I > 300,000	2.8	2.7	0.11	1.9	2.6
300,000 > II > 10,000	0.4	2.0	0.0074	0.037	5.4
10,000 > III > 500	0.01	0.014	0.098	0.14	0.09
IV < 500	1.3	0.56	1.4	1.4	0.1
Total	4.5[a]	—	1.6[a]	0.54[a]	—

[a] Particulates $\geq$ 0.15 μm removed.

case for copper (Brown et al., 1974; Pagenkopf et al., 1974). Inorganic solubility chemistry would predict that most of the cadmium introduced into the soft-acid waters of the southeastern United States, northeastern United States, and Scandinavia would exist as the free divalent cation (Cd^{2+}) or as aquated ions ($CdOH^+$, $B = 1.5 \times 10^4$; CdO_2^{2-}, $B = 5.8 \times 10^8$) (Weber and Posselt, 1974).

Although extracellular organics from plants and animals may affect cadmium speciation in natural waters, this discussion will deal primarily with the colored, refractory polyphenolic compounds known variously as humics, fulvics, and tannins. These compounds are thought to be the result of chemical polymer-ization and microbial decomposition and synthesis from components of plants such as lignin (Flaig, 1964). Information available in the literature regarding interactions between cadmium and naturally occurring organic ligands is am-biguous. Weber and Posselt (1974) reported that cadmium can form stable complexes with naturally occurring organics, whereas Hem (1972) stated that the amount of cadmium occurring in organic complexes is generally small and that this element is less tightly bound than zinc by most organic ligands. Pittwell (1974) reported that cadmium would be complexed by organic carbon under all pH conditions encountered in normal natural waters, and Levi-Minzi et al. (1976) found cadmium sorption in soils to be correlated with soil organic matter content.

2. NATURE OF CADMIUM IN NATURAL WATERS

McGlynn (1974) observed that between 40 and 60% of the total cadmium dis-solved in natural surface waters existed in the free divalent form, while Mantoura et al. (1978) found less than 11% of the cadmium to be complexed by humic materials. Allen et al. (1978) concluded that little of the cadmium in Lake Michigan exists in the free ionic form and that approximately 75% of a 50 μg Cd/l addition to Lake Michigan water would be bound. Gardiner and Stiff (1975) found that between 10 and 25% of the cadmium complexing capacity of the river waters studied could be attributed to humic materials. Riffaldi and Levi-Minizi (1975) reported that approximately 50% of the total cadmium was bound to humics in coordination complexes, while 50% was in the readily ex-changeable form. Ultrafiltration (Table 1) and ion exchange (Table 2) infor-mation indicates that more than 80% of the soluble cadmium in some waters of the southeastern United States is tightly bound to organics. Bahan et al. (1978) found that cadmium showed relatively little tendency to associate with the fulvic acid fractions of sewage sludge extract. Cadmium, a relatively weak Lewis acid, was more readily complexed by Cl^- than fulvic acid.

The relative amount of cadmium bound in natural waters examined by ion

Table 2.   Percent Cadmium Removed from Water Passing through UM-05 Ultrafilters by Anionic and Cationic Ion Exchange Resins

| Sample | Cadmium Removed by Ion Exchange Resins (%) | |
	Cationic Dowex 50W)	Anionic Dowex 1-8X)
Upper Three Runs, S. C.	17	80
Skinface Pond, S. C.	15	92
Fire Pond, S. C.	50	61
Shaker Pond, Me.	a	a
Great East Lake, Me.	48	51
Monsum Lake, Me.	54	46
Deering Pond, Me.	>99	<1
Salmon Falls River, Me.	51	48
Beaver Dam Pond, Me.	>99	<1
Sunken Pond, Me.	56	55
Estes Lake, Me.	a	a
Saco River, Me.	69	21
Little Pond, Me.	91	9
Bog Pond, Me.	65	37
Sebaga Lake, Me.	b	b

[a] Below detection limit.

[b] Contamination or interference.

exchange techniques ranges from less than 17% to greater than 90% (Table 2). Although not all of this can be attributed to organic binding, inorganic solubility product chemistry predicts that in Upper Three Runs Creek, South Carolina, essentially all of the cadmium would exist as the free Cd^{2+} ion. In Upper Three Runs, a coastal plain stream of high organic content, 80% of the total cadmium was removed onto anionic exchange resin. Approximately 63% of the total cadmium was in association with portions greater than a nominal diameter of 0.9 nm (as measured by ultrafiltration).

Whitfield (1975) stated that values reported for the degree of cadmium complexation by organics in seawater range from 0 to 80%. Duinker and Kramer (1977) found no cadmium-organic complexes in North Sea water, using anodic stripping voltametry. Conversley, Batley and Florence (1976) reported that 75% of the cadmium present in seawater was bound as labile organic complexes removed by Chelex-100 ion exchange resin. This indicates that chloro complexes are relatively unimportant and organic complexes important in the speciation

Table 3.   Effects of Competing Ions on Cadmium Binding Capacity in Three Southeastern Surface Waters

Competing ions tested are listed from left to right in decreasing order of effect on cadmium binding capacity. Percent of cadmium binding capacity remaining in presence of competing ions is listed for each competing ion.

Upper Three Runs Creek					
FI	Al	Mg	Mn	Ni	Ca
	0	30	38	45	94
FII	Ca				
	+100				
FIII	Al	Mg	Mn	Ni	Ca
	0	0	0	0	89
FIV	Al	Mg	Mn	Ni	Ca
	0	8	8	9	89
Skinface Pond					
FI	Al	Mg	Mn	Ni	Ca
	13	33	44	67	100
FII	Ca				
	100				
FIII	Al	Mg	Mn	Ni	Ca
	3	29	48	48	100
FIV	Al	Mg	Mn	Ni	Ca
	0	8	27	28	91
Fire Pond					
FI	Al	Mg	Mn	Ni	Ca
	12	16	18	34	100
FII	Ca				
	100				
FIII	Al	Mg	Mn	Ni	Ca
	0	1	32	33	97
FIV	Al	Mg	Mn	Ni	Ca
	0	8	15	22	91

4.   FACTORS AFFECTING CADMIUM INTERACTION WITH ORGANIC LIGANDS

The relative affinity of cadmium for organics in the presence of competing cations has been determined by titrating organics with cadmium in the presence of 37, 12, 41, 18, and 17 mg-atom/l of Al, Ca, Mg, Mn, and Ni, respectively. All of the metals tested were able to compete successfully with cadmium for some binding sites (Table 3). However, calcium did not compete very actively with cadmium. Aluminum was able to outcompete cadmium for binding sites with

Table 4.   Conditional Stability Constants and Number of Binding Sites for Cadmium-Humic Complexes.

Aldrich humic acids (7.14 mg/l) constant pH = 5.5. Estimate ± asymptotic 95% confidence interval. Non-linear least squares regression $F_{4,65}$ = 722.

K_1[a]	K_2	n_1	n_2
l/g·atom		μg·atom/mg HA	
$5.0 \pm 0.003 \times 10^5$	$2.6 \pm 0.002 \times 10^4$	$2.5 \pm 0.002 \times 10^{-1}$	3.6 ± 0.002

[a] Units of ml/μg·ATOM Cd, assume density of water = 1 to cancel units.

all the organics studied and often resulted in no organic binding of cadmium. In general, Mg, Mn, and Ni reduced cadmium binding by a similar amount, suggesting they were competing for the same binding sites. Organic-specific competition was exhibited. For example, competition for cadmium binding sites was similar for Mg, Mn, and Ni for any particular organic, even though the amount of competition varied between organics.

Stumm and Bilinski (1973) suggested that metal-organic complexes are unimportant in most aquatic systems because of competition for binding sites by calcium, which is generally present in high concentrations. O'Shea and Mancy (1978) have reported that calcium inhibits labile complexation between cadmium and humic acids. At pH 3.5, Cu, Fe, Ni, Pb, and Cd all form more stable complexes with fulvic acids than does calcium (Stevenson and Ardakani, 1972). The results presented here demonstrate that calcium does not effectively compete with cadmium for many of the potential binding sites associated with naturally occurring organics. Calcium, which is present at high concentrations in most surface waters, had a total concentration of only 13 μg-atom/l in Skinface Pond and only 3.6 μg-atom/l in Fire Pond.

Calcium was associated with each of the fractions isolated from all three waters studied here, as well as Okefenokee Swamp water (Giesy and Briese, 1977), indicating that it is bound to organics in these fractions. The amount of calcium associated with each fraction varied among collection sites. Koljonen and Carlson (1975) reported that calcium in Finish humic lakes is associated with humus and causes coagulation and precipitation of these organics, while Beck et al. (1974) reported that calcium was not complexed or was only weakly complexed by organics in the Satilla River, Georgia. More than 99% of the calcium present in FIV of Upper Three Runs Creek could be removed by cationic exchange resins, while less than 1% was removed by anionic exchange resin. This is similar to results obtained by Benes et al. (1976) for Norwegian surface waters and indicates that the calcium in this fraction existed as Ca^{2+} or was loosely bound in labile complexes. Approximately 80% of the dissolved calcium in Upper

Table 5.   Conditional Stability Constants for Cadmium Organic Complexes

Sample	K	pH	Reference
Peat	3.7×10^4	8.0	Mantoura et al. (1978)
Lake	3.7×10^4	8.0	Mantoura et al. (1978)
	5.0×10^4	8.0	Mantoura et al. (1978)
Loch	8.9×10^4	8.0	Mantoura et al. (1978)
	7.4×10^4	8.0	Mantoura et al. (1978)
Seawater	4.9×10^4	8.0	Mantoura et al. (1978)
Soil	1.2×10^5	8.0	Mantoura et al. (1978)
Peat	1.2×10^5	4, 5, 6	Stevenson (1977)
Leonardite humic acid	1.2×10^6	4, 5, 6	Stevenson (1977)
Soil humic acid	1.1×10^6	4, 5, 6	Stevenson (1977)
Humic acid	1.1×10^5	6.8	Guy and Chakrabarti (1976)
Tannic acid	1.8×10^4	6.8	Guy and Chakrabarti (1976)
Lake Michigan			
Nearshore	1.2×10^8	—	Allen et al. (pers. comm.)
Offshore	8.2×10^7	—	

Three Runs water was bound, while more than 50% of the total calcium was retained by ultrafilters of greater than 0.0009-mm nominal pore diameter in all cases. Calcium binding sites were all filled, but the humics still exhibited binding capacity for cadmium. Calcium may be displaced from organics by cadmium or may be bound by different meachnisms, thus not competing for the same sites.

Both relative electronegativity and ionic radius are important parameters in determining the stability of metal-organic complexes (Jameson, 1976). The low affinity of naturally occurring organics for Ca^{2+} can be explained by its large ionic diameter and low relative electronegativity. Aluminum, which was a highly successful competitor of cadmium, has the smallest ionic radium and highest relative electronegativity of the ions studied. Aluminum also has a "rare-gas" octet structure due to its completely filled d shell and forms stable complexes with oxygen-donor ligands (Jameson, 1976). As a trivalent ion, aluminum may also have a greater affinity for organics.

Stability constants reported for cadmium-organic complexes in surface waters range over several orders of magnitude (Tables 4 and 5) and are generally lower than those for copper and lead. True stability constants should be valid for all cadmium concentrations encountered because the ratio of complexed to uncomplexed cadmium is independent of the total concentration of the element, provided that the ligand is in excess. Since humic acids are a mixture and different fractions may have different complexing properties, low concentrations

Table 6.   Number of Cadmium Atoms Bound (N_i) Expressed as Total, Carboxylic, and Phenolic Acidity

Fraction	Total Acidity		Carboxylic Acidity		Phenolic Acidity	
	$n_1{}^a$	$n_2{}^a$	n_1	n_2	n_1	n_2
Par Pond humic acids						
Total	0.00065	0.0020	0.0017	0.0052	0.0010	0.0032
FI > 300,000	0.0055	0.065	0.0094	0.108	0.013	0.0156
FII < 300,000	0.013	0.025	0.026	0.051	0.025	0.049
Aldrich humic acids						
Total	0.0045	0.13	0.032	0.95	0.0052	0.156
FI > 300,000	0.002	0.18	0.011	0.93	0.003	0.23
FII < 300,000	0.0036	0.136	0.023	0.88	0.0042	0.160

[a] Values of n reported as mg-atoms Cd/mequiv humic acid.

of metals may be more highly complexed at a given humic acid concentration than at higher metal concentrations when less active sites are used (Gardiner, 1974; Jenne and Luoma, 1975). Trace metals in natural waters are often not in a state of equilibrium with the organic and inorganic ligands present, thereby greatly complicating equilibrium treatments of metal ions and complexes (Martell, 1975). Gardiner (1974) concluded that the humic concentration was largely unimportant in the range studied (6.6 to 86 mg/l). On the contrary, MacCarthy and Smith (1978) suggested that conditional stability constants for metal-humic interactions are ligand dependent. It is suggested that studies of cadmium-organic interactions be made at several ligand concentrations in the range expected in the environment of interest. Gardiner (1974) found that the fraction of cadmium complexed by humics was only slightly dependent on pH. Hydrogen ion effects are not surprising since we have assumed a competitive interaction between H^+ and M^{2+} (equations 4 and 5).

Binding capacity studies have indicated that the amount of cadmium bound by organics is often small. Guy and Chakrabarti (1976) reported n_i values of 3 and 11 nmol Cd/g organic for the tannic and humic acids listed in Table 5.

When n_i is reported as gram-atoms per gram equivalent weight of total, carboxylic, and phenolic acidity, it can be seen that cadmium does not occupy many of the potential binding sites (Table 6) as compared to lead and copper, which occupy essentially all of the acidic functional groups, when possible binding sites are saturated. Theoretically, if 1:1, 2:1, or mixed complexes are formed, two equivalents should be required to bind Cd^{2+}. Thus, if all of the acidic functional groups were used, a maximum of 0.5 g-atom Cd/g-equiv humics could

Table 7.   Pairwise Correlations for Surface Waters in the
Northeastern United States ($n = 12$)

Parameter	$[Cd]^a$	$[Cd\ BC]^b$
pH	−.55	.68
	***	***
OC^b	.43	−.49
	***	*
Cu BC	.46	−.55
	*	**
Pb BC	.00	.00
	NS	NS
Cu	.57	−.41
	**	NS
Ca	−.47	.72
	*	***
Cd BC	−.78	
	***	

^{a} * $p < .10$, ** $p < .05$, *** $p < .01$, NS = not significant.

be bound, as is the case for lead. If an n_i value is much greater than 0.5, such as that for n_2 calculated for Aldrich humics on a carboxylic acidity basis, that type of functional group model alone is not sufficient to explain the metal binding observed. When a very small number relative to 0.5 is observed, as with the total acidity model for Par Pond humics (Table 5), very few of the model functional groups are actually possible binding sites for cadmium.

Heavy transition elements to the right of the periodic table generally show smaller coordination numbers than do their lighter counterparts, which have greater apparent ionic radii, presumably because their directional valence demands are more stringent. The ionic radius of cadmium (0.92) is large enough that this element does not readily substitute in crystalline structures of clay but is bound to the soil humus fraction directly (Anderson, 1977). Thus, the large difference between lead and cadmium in the number of potential binding sites probably cannot be explained by lead binding to clays that are associated with the humics. Cadmium, however, does have a lower relative electronegativity (1.46) than either copper (1.75) or lead (1.55).

We have discussed cadmium binding by organics in the laboratory and in the unique surface waters of the southeastern United States, but we need to address the broader question of whether organics are important in determining the speciation of cadmium in other surface waters. To address this question binding capacities have been measured and correlated with chemical parameters of the

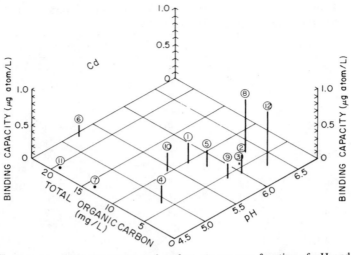

Figure 4. Cadmium binding capacities of surface waters as a function of pH and organic carbon.

surface waters of interest in the northeastern United States (Giesy et al., 1978) and in northwestern Europe (Giesy and Briese, 1978b).

Cadmium binding capacity (Cd BC) in surface waters from the northeastern United States was found to be inversely related to organic carbon (OC) and positively correlated with pH (Table 7). This result, due in part to the fact that pH and OC are inversely correlated in these waters, indicates that Cd BC was controlled by inorganic compounds over wide ranges of pH and OC (Figure 4). Thus, inorganic species are the most important in determining cadmium speciation, and OC is not at all important for the waters studied in the northeastern United States. Organic carbon was also unimportant in explaining a significant amount of the variability in cadmium binding capacity of northern European waters (Giesy and Briese, 1978b). A linear multiple regression model, using CO_3^{2-} and SO_4^{2-} as independent variables, resulted in the prediction equation

$$Cd\ BC = -5.7 \times 10^{-1} + 1.2 \times 10^5 [CO_3^{2-}] + 1.08 \times 10^1 [SO_4^{2-}]$$

with a coefficient of determination (R^2) of .97. Gardiner (1974) reports that for naturally occurring organic ligands to be effective at complexing cadmium at a concentration of 10^{-8} M it would have to have a stability constant of approximately 10^8. The relative importance of naturally occurring organics in regard to cadmium speciation in surface waters is regionally specific, and no completely accurate generalization can be made about all waters.

Gardiner (1974) suggested that, because a substantial proportion of the total dissolved cadmium is expected to be present as the free ion, availability and toxicity in biota can be assessed with reasonable accuracy from a knowledge of total cadmium concentration without reference to the degree of complexation, unless the toxicity of cadmium complexes present is much greater than that of free Cd^{2+}. Giesy et al. (1977) studied the effects of naturally occurring organics on the toxicity of cadmium to the cladoceran *Simocephalus serrulatus* and found that all nominal molecular size organics bound cadmium measurably at the concentrations in which they were observed in the environment. All nominal size organics reduced cadmium toxicity except those less than 0.9 nm (500 M.W., when compared to globular proteins), which increased cadmium toxicity relative to controls. This synergism may be due to cadmium-organic complexes that are more available to these relatively small molecular weight organics by reason of juxtapositioning and uptake facilitation.

ACKNOWLEDGMENT

The preparation of this chapter was supported by Contract EY-76-C-09-0819 between the University of Georgia and the U.S. Department of Energy.

REFERENCES

Anderson, A. (1977). "The Distribution of Heavy Metals in Soils and Soil Material as Influenced by Ionic Radius," *Swed. J. Agric. Res.* **7,** 79–83.

Bahan, J., Ball, N. B., and Sposito, G. (1978). "Gel Filtration Studies of Trace Metal-Fulvic Acid Solutions Extracted from Sewage Sludges," *J. Environ. Qual.* **7,** 181–189.

Batley, G. E. and Florence, T. M. (1976). "Determination of the Chemical Forms of Dissolved Cadmium, Lead and Copper in Seawater," *Mar. Chem.* **4,** 347–363.

Beck, K. C., Reuter, J. H., and Perdue, E. M. (1974). "Organic and Inorganic Geo-chemistry of Some Coastal Plain Rivers of the Southeastern United States," *Geochim. Cosmochim. Acta* **38,** 341–364.

Benes, P., Gjessing, E. J., and Steinnes, E. (1976). "Interactions between Humus and Trace Elements in Freshwater," *Water Res.* **10,** 711–716.

Brown, V. M., Shaw, T. L., and Shurben, D. G. (1974). "Aspects of Water Quality and the Toxicity of Copper to Rainbow Trout," *Water Res.* **8,** 797–803.

Buffle, J., Greter, F. L., and Haerdi, W. (1977). "Measurement of Complexation Properties of Humic and Fulvic Acids in Natural Waters with Lead and Copper Ion-Selective Electrodes," *Anal. Chem.* **49,** 216–222.

Chaberek, S. and Martell, A. E. (1959). *Organic Sequestering Agents.* Wiley, New York, 616 pp.

Crosser, M. L. and Allen, H. E. (1977). "Determination of Complexation Capacity of Soluble Ligands by Ion Exchange Equilibrium," *Soil Sci.* **12**, 176–181.

Duinker, J. C. and Kramer, C. J. M. (1977). "An Experimental Study on the Speciation of Dissolved Zinc, Cadmium, Lead and Copper in River Rhine and North Sea water, by Differential Pulsed Anodic Stripping Voltametry," *Mar. Chem.* **5**, 207–228.

Flaig, W. (1964). "Effects of Microorganisms in the Transformation of Lignin to Humic Substances," *Geochim. Cosmochim.* **28**, 1523–1535.

Gardiner, J. (1974). "The Chemistry of Cadmium in Natural Water I: A Study of Cadmium Complex Formation Using the Cadmium Specific-Ion Electrode," *Water Res.* **8**, 23–30.

Gardiner, J. and Stiff, M. J. (1975). "The Determination of Cadmium, Lead, Copper and Zinc in Ground Water, Estuarine Water, Sewage and Sewage Effluent by Anodic Stripping Voltametry," *Water Res.* **9**, 517–523.

Giesy, J. P. and Briese, L. A. (1977). "Metals Associated with Organic Carbon Extracted from Okefenokee Swamp Water," *Chem. Geol.* **20**, 109–120.

Giesy, J. P. and Briese, L. A. (1978a). "Trace Metal Transport by Particulates and Organic Carbon in Two South Carolina Streams," *Vern. Int. Verein. Limnol.* **20**, 401–417.

Giesy, J. P. and Briese, L. A. (1978b). "Metal Binding Capacity of Northern European Surface Waters," *Organic Geochem.* (in press).

Giesy, J. P., Briese, L. A., and Leversee, G. J. (1978). "Metal Binding Capacity of Selected Maine Surface Waters," *Environ. Geol.,* **2**, 257–268.

Giesy, J. P., Leversee, G. J. and D. R. Williams (1977). "Effects of Naturally Occurring Aquatic Organic Fractions on Cadmium Toxicity to *Simocephalus Serrulatus* (Daphnidae) and *Gambusia Affinis* (Poeciliidae)," *Water Res.* **12**, 1013–1020.

Guy, R. D. and Chakrabarti, C. L. (1976). "Studies of Metal-Organic Interactions in Model Systems Pertaining to Natural Waters," *Can. J. Chem.* **16**, 2600–2611.

Hem, J. D. (1972). "Chemistry and Occurrence of Cadmium and Zinc in Surface Water and Ground Water," *Water Resour. Res.* **8**, 661–679.

Jameson, R. F. (1976). "Selectivity in Metal Complex Formation." In D. R. Williams, Ed. *An Introduction to Bio-inorganic Chemistry.* Charles C Thomas, Springfield, Ill., pp. 29–50.

Jenne, E. A. and Luoma, S. N. (1977). "Forms of Trace Elements in Soils, Sediments and Associated Waters: An Overview of their Determination and Biological Availability." In H. Drucker and R. E. Wilding Eds. *Symposium on Biological Implications of Metals in the Environment.* Energy Research and Development Administration, pp. 213–230.

Kinkade, M. and Erdman, H. E. (1975). "The Influence of Hardness Components (Ca^{2+} and Mg^{2+}) in Water and the Uptake and Concentration of Cadmium in a Simulated Freshwater System," *Environ. Res.* **10**, 308–313.

Koljonen, T. and Carlson, L. (1975). *Behavior of the Major Elements and Minerals in Sediments of Four Humic Lakes in Southeastern Finland.* Societas Geographica Fenniae, Helsinki, 47 pp.

Levi-Minzi, R., Soldatini, G. F., and Riffaldi, R. (1976). "Cadmium Absorption by Soils," *J. Soil Sci.* **27**, 10–15.

McGlynn, J. A. (1974). *Instrumental Methods for the Determination of Trace Metals in Water.* Australian Water Resources Council, Tech. Pap. 8, pp. 53–78.

McKee, J. E. and Wolf, H. W. (1963). *Water Quality Criteria.* California State Resources Central Board, Publ. 3-A, 548 pp.

MacCarthy, P. and G. C. (1978). "Stability Surface Concept. A Quantitation Model for Complexation and Multi-Ligand Mixtures." In E. A. Jenne, Ed. *Chemical Modeling in Aqueous Systems Speciation Sorption Solubility and Kinetics.* ACS Symposium Series 93, pp. 201–224.

Mantoura, R. F. C. and Riley, J. P. (1975). "The Use of Gel Filtration in the Study of Metal Binding by Humic Acids and Related Compounds," *Anal. Chim. Acta* **78**, 193–200.

Mantoura, R. F. C., Dickson, A., and Riley, J. P. (1978). "The Complexation of Metals with Humic Materials in Natural Waters," *Estuarine Coastal Mar. Sci.* **6**, 387–408.

Martell, A. E. (1975). "The Influence of Natural and Synthetic Ligands on the Transport and Function of Metal Ions in the Environment," *Pure Appl. Chem.* **44**, 81–113.

Martell, A. E. and Calvin, A. (1952). *Chemistry of the Trace Metal Chelate Compounds.* Prentice-Hall, New York, 613 pp.

O'Shea, T. A. and Mancy, K. H. (1978). "The Effect of pH and Hardness Metal Ions on the Competitive Interaction between Trace Metal Ions and Inorganic and Organic Complexing Agents Found in Natural Waters," *Water Res.* **12**, 703–711.

Pagenkopf, G. K., Russo, R. C., and Thurston, R. V. (1974). "Effect of Complexation on Toxicity of Copper to Fishes," *J. Fish. Res. Board Can.* **31**, 462–465.

Petruzzelli, G. Giudi, G., and Lubrano, L. (1977). "Cadmium Occurrence in Soil Organic Matter and Its Availability to Wheat Seedlings," *Water, Air, Soil Pollut.* **8**, 393–399.

Pickering, Q. H. and Henderson, C. (1966). "The Acute Toxicity of Some Heavy Metals to Different Species of Warm Water Fishes," *Air Water Pollut. Int. J.* **10**, 453–463.

Pittwell, L. R. (1974). "Metals Coordinated by Ligands Normally Found in Natural Waters," *J. Hydrol.* **21**, 301–304.

Riffaldi, R. and Levi-Minzi, R. (1975). "Adsorption and Desorption of Cd on Humic Acid Fraction of Soils," *Water, Air, Soil Pollut.* **5**, 179–184.

Scatchard, G. (1949). "The Attractions of Proteins for Small Molecules and Ions," *Ann. N.Y. Acad. Sci.* **51**, 660–672.

Schnitzer, M. and Khan, S. U. (1972). *Humic Substances in the Environment.* Marcel Dekker, New York, pp. 37–43.

Stevenson, F. J. (1977). "Nature of Divalent Transition Metal Complexes of Humic Acids as Revealed by a Modified Potentiometric Titration Method," *Soil Sci.* **123**, 10–17.

Stevenson, F. J. and Ardakani, M. S. (1972). "Organic Matter Reactions Involving

Micronutrients in Soil," In J. J. Mortvedt, P. M. Giordano, and W. L. Lindsay, Eds., *Micronutrients in Agriculture*. Soil Science Society of America, Madison, Wis., 666 p.

Stumm, W. and Bilinski, H. (1973). "Trace Metals in Natural Waters: Difficulties of Interpretation Arising from Our Ignorance of Their Speciation." In *Advances in Water Pollution Research*. Pergamon Press, New York, pp. 39–49.

Sunda, W. G. and Lewis, J. M. (1976). *Determination of the Binding of Copper and Cadmium in a Coastal River-Estuarine System Using Ion-Selective Electrodes.* Annual Report, Atlantic Estuarine Fisheries Center, Beaufort, N. C., pp. 53–77.

Wahlgren, M. A., Edgington, D. N., Rowlings, F. F., and Rawls, J. L. (1972). "Trace Element Determinations on Lake Michigan Tributary Water Samples Using Spark Sources Mass Spectrometry." In *Proceedings of the 15th Conference on Great Lake Research*. International Association for Great Lakes Research, pp. 298–305.

Weber, W. J. and Poselt, H. S. (1974). "Equilibrium Models and Precipitation Reactions for Cadmium(II)." In A. J. Rubin, Ed., *Aqueous Environment Chemistry of Metals*. Ann Arbor Science Publishers, Ann Arbor, Mich., 360 pp.

Whitfield, M. (1975). "The Electroanalytical Chemistry of Seawater." In J. P. Riley and G. Skirrow, Eds., *Chemical Oceanography,* 2nd ed., Academic Press, New York, Chapter 20.

Zunino, H. and Martin, J. P. (1977). "Metal-Binding Organic Macromolecules in Soil. 2: Characterization of the Maximum Binding Ability of the Macromolecules," *Soil Sci.* **123**, 188–202.

8

CHEMICAL MOBILITY OF CADMIUM IN SEDIMENT-WATER SYSTEMS

Rashid A. Khalid

Laboratory for Wetland Soils and Sediments, Center for Wetland Resources, Louisiana State University, Baton Rouge, Louisiana

1.   INTRODUCTION

Cadmium is a potentially hazardous pollutant in the environment, and chronic human exposure to low concentrations of this element in the atmosphere, water, or food may cause serious illness and possibly death (Yamagata and Shigematsu, 1970; Friberg et al., 1971, 1973). The production and the industrial utilization of cadmium are continuing to increase throughout the industrialized world with a concurrent increase in the cadmium residues in the environment (Nriagu, this volume, Chapters 1 and 2; Page and Bingham, 1973). As a result there is an urgent need to understand the sources of emissions and their direct or indirect interaction with soils and sediments, water, plants, and animals so that criteria may be developed for assessing potential environmental hazards.

The majority of the recommended standards for drinking water, freshwater, and marine water aquatic life are based on total cadmium concentrations (U.S. Environmental Protection Agency, 1973a, 1976, 1977; Canadian Department of National Health and Welfare, 1969).   However, the biological availability of cadmium depends on the chemical species present both in surface waters and in sediment solids, not on the total metal ion concentrations in the system (Allen et al., 1975; Forstner, 1977). It is therefore imperative that the chemical forms of cadmium be known to permit an accurate prediction of both its chemical mobility and biouptake.

The objectives of this chapter are to review the rapidly expanding literature on (a) the chemical forms of cadmium in fresh and marine waters and sediments, (b) the chemical mobilization of cadmium in sediment-water systems as a function of changes in the controlling physicochemical parameters: pH, oxidation-reduction (redox) potential, salinity, and sediment-seawater mixing zones, and (c) the potential adverse environmental impacts of cadmium release from contaminated sediments during dredging and dredged material disposal in the fresh and marine environments. Based on the available information on the interactions between cadmium and the physicochemical environment during dredging and dredged material disposal, the selection of disposal alternatives to minimize adverse impacts of potentially bioavailable cadmium will also be discussed in this chapter.

2. CHEMICAL FORMS OF CADMIUM IN WATER

Biological criteria, especially acute toxicity to aquatic organisms, are generally based on the total concentration of metal in the experimental medium. Since the effects of a particular concentration will change as the chemical form of the metal changes, it is important also to consider the chemical distribution of trace metals in different environments. The toxicity of copper to several species of algae (Anderson and Morel, 1978) and to a microcrustacean, *Daphnia magna* (Andrew et al., 1977), was shown to be directly related to free cupric ion rather than total copper concentration. In these experiments the effect of varying concentrations of chelating agents, organic and inorganic, on copper toxicity was investigated. An increase in the complexation of copper decreased the toxicity of the metal by decreasing the concentration of free cupric ion. Sunda and co-workers (1978) reported reduced toxicity of cadmium to grass shrimp with increasing salinity and increasing concentration of the chelator nitrilotriacetic acid (NTA). The protective effect of high salinity or NTA was attributed to the complexation of cadmium. Cossa (1976) related the reduced uptake of cadmium by algae to the presence of ethylenediaminetetraacetic acid, (EDTA). The physicochemical parameters—pH, salinity, ionic strength, redox potential, type and concentration of competing metals, and presence of adsorptive surfaces and the total concentration of the metal in question—can affect the concentration of the bioavailable fraction and thus influence the response of an organism to that metal (Bartlett et al., 1974; Phillips, 1977; Sarfield and Mancy, 1977). It is therefore essential to know the chemical speciation of the metal in solution in order to predict the toxicity effect.

Cadmium in the freshwater and marine environment exists in either the dissolved or the particulate form. Dissolved cadmium is the fraction that passes through a membrane filter of some nominal size (usually 0.45 μm), on which particulates are retained. However, it is difficult to clearly distinguish between dissolved, colloidal, and nonsuspended particulate forms (Sibley and Morgan, 1975; Phillips, 1977). Particulate forms include cadmium adsorbed on solid surfaces, incorporated into organisms, or chemically bound to organic matter, as well as solid precipitates (Gardner, 1974a, 1974b; Khalid et al., 1978a; Brannon et al., 1976). The discussion in this section is limited to various aspects of chemical speciation of the dissolved cadmium fraction.

Dissolved cadmium includes both free cationic Cd^{2+} and that complexed with various ligands. The complexed fraction includes both inorganic and organic ligands. The relative magnitude of free ionic and of complexed cadmium is influenced by several processes which include hydrolysis, precipitation, complexation, ligand or ion exchange reactions, adsorption-desorption phenomena, coprecipitation, redox reactions, and biological accumulation or transformation (Leckie and James, 1974; Weber and Posselt, 1974; James and Leckie, 1977; Phillips, 1977).

The methods proposed for the measurement of cadmium species in fresh water, estuarine water, and seawater have been extensively reviewed in various papers (Kester, 1975; Florence and Batley, 1977; Shuman and Fogleman, 1978). These studies indicate that two basic approaches to the quantification of the cadmium species involved are (a) the use of stability constant data, together with known concentrations of various ions and ligands, to compute equilibrium concentrations of various ions and ligands in order to calculate the concentrations of cadmium species (thermodynamic equilibrium models); and (b) the use of physical and chemical separation and reaction techniques, coupled with highly sensitive analytical procedures, to provide direct measurement of cadmium species (chemical separation techniques).

2.1. Thermodynamic Equilibrium Models

Several equilibrium models based on thermodynamic data have been used to predict the chemical mobility of inorganic cadmium species in freshwater, seawater, and freshwater-seawater mixing zones. Hem (1972) presented pH-redox potential diagrams to describe the stability fields of solids and predominant dissolved cadmium species in an inorganic carbonate-sulfide system simulating freshwater conditions (Figure 1). An examination of the data plotted in Figure 1 indicates that under oxidized conditions and at pH values below 7.0 most of the cadmium was present in the free ionic form. As the pH approached 8.0 and above, the concentration of cadmium was governed by solid phase $CdCO_3$. Dissolved cadmium species in this system included Cd^{2+} and $Cd(OH)_3^-$. In the system just described, cadmium solubility would drop below the 10 $\mu g/l$ public drinking water limit (U.S. Environmental Protection Agency, 1973a) only at high pH values (between 8.9 and 10.7) or in reduced systems with complete O_2 depletion. Hem compared the calculated equilibration concentrations and observed concentrations of cadmium in river waters of the United States and concluded that the concentrations of this element that commonly occur in river waters were substantially below the equilibrium solubilities of the carbonate or the hydroxide form of cadmium.

Weber and Posselt (1974) suggested that in aqueous systems containing both hydroxide and carbonate the solid phase controlling dissolved cadmium would

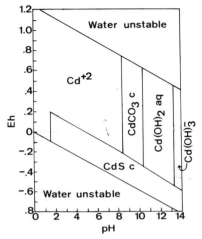

Figure 1. Fields of stability of solids and predominant dissolved cadmium species in system Cd + CO$_2$ + S + H$_2$O at 25°C and 1 atm in relation to *Eh* and pH. Dissolved cadmium activity, 10$^{-7.05}$ mol/l; dissolved CO$_2$ and S species, 10^{-3} mol/l (Hem, 1972.)

depend on the ratio of the concentrations of these ions and of their respective solubility products. The corresponding solubility product expressions are

$$K_{sp} \, Cd(OH)_2 = [Cd^{2+}][OH^-]^2 = 2.2 \times 10^{-14}$$

$$K_{sp} \, CdCO_3 = [Cd^{2+}][CO_3^{2-}] = 5.2 \times 10^{-12}$$

Thus if the ratio of relative concentrations

$$R = \frac{[OH^-]}{[CO_3^{2-}]} < 0.00423 \tag{1}$$

the system will be controlled by CdCO$_3$. These authors concluded from their thermodynamic model that cadmium complexation with chloride and ammonia was insignificant in fresh waters because of the low levels of these substances but would be of great importance in wastewaters. Also, interaction with organic ligands, not considered in this model, will form complexes with dissolved cadmium.

Sibley and Morgan (1975) also used an equilibrium model to calculate cadmium speciation in various freshwater-seawater mixtures. The theoretical computations included the effect of hydrolysis, protonation of ligands, complexation, precipitation of solids oxidation-reduction, and adsorption reactions. The computations showed that solid CdCO$_3$ was an important species in freshwater, while cadmium in the seawater was dissolved and was present as a chloride complex. This transition from fresh water to seawater is clearly illustrated in

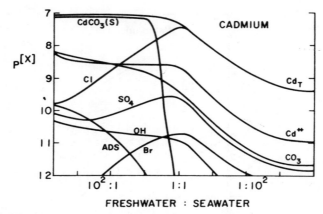

Figure 2.   Speciation of cadmium in various freshwater-seawater mixtures. (Sibley and Morgan, 1975.)

Figure 2. It should be noted that the maximum dissolved concentration of cadmium occurred a seawater/freshwater ratio of 1:1, the point where solid $CdCO_3$ dissolved.

It is evident from these data that a considerable change in cadmium speciation occurs as fresh water flows into the sea. The most significant causes for these speciation changes were suggested to be (*a*) different ionic strengths, (*b*) lower concentrations of adsorbing surfaces in seawater, (*c*) different concentrations of cadmium, or (*d*) different concentrations of major cations and ligands. The concentrations of ligands and major cations were considered to be the controlling mechanisms in seawater. The effects of pH and redox potential were not considered in these calculations, and these two variables were maintained constant.

The aforementioned studies have shown that equilibrium models can be used to calculate cadmium speciation in quite complicated systems if the system can be adequately defined. These equilibrium models are useful for understanding some aspects of speciation in natural environments. They allow comparison of species in different environments, such as the freshwater-seawater interface, and prediction of the effects of different chemical parameters and interactions between different processes on speciation. However, the value of these thermodynamic models may be limited by lack of stability constants for some important complexes or solid phases (Sibley and Morgan, 1975). For example, no data are available on metal ion interactions with many of the natural organic ligands present in the waters, and the concentrations of these ligands are not known. Most equilibrium models assume isothermal, isobaric conditions, whereas natural waters cover a wide range of temperature and pressure. The errors introduced by these assumptions are unknown. Florence and Batley (1977) concluded in a literature review that, because of several limitations and our imperfect knowledge of these systems, even the most sophisticated computer model of trace metal speciation in natural waters bear little, if any, relation to reality.

2.2. Chemical Separation Techniques

The analytical methods employed to identify various metal ion species in the aquatic environment can be classified into two groups. The first group involves potentiometric measurements with ion selective electrode (Gardner, 1974a) and anodic stripping voltametry (ASV) (Chau and Lum-Shue-Chan, 1974; Florence and Batley, 1976; Florence, 1977), which are selective for the free metal ion. The second group involves separation of metal ion species by size, ultrafiltration, dianalysis, and gel permeation chromatography (Guy and Chakrabarti, 1976), and analysis by atomic absorption spectrophotometry or specific ion electrode. The nature of the problems encountered in the use of these techniques, their limitations, and their advantages are discussed in several review articles (Gardner, 1975; Guy and Chakrabarti, 1976; Florence and Batley, 1976; Shuman and Fogleman, 1978). Experimental determination of the forms of cadmium in natural waters is described in detail by Raspor (this volume, Chapter 6).

3. CHEMICAL FORMS OF CADMIUM IN SEDIMENTS

Published literature on cadmium concentrations in surface waters and sediments of the United States and elsewhere (Kopp and Kroner, 1970; Kubota et al., 1974; Iskander and Keeney, 1974; Bryan, 1976; De Groot et al., 1976) indicates that the effect of cadmium contamination from industrial and other sources is reflected in greater levels of this metal in the solid components of sediment-water systems. The sediments act as a sink for cadmium, and the magnitude of this scavenging action of sediments depends on the physical, chemical, and biological properties of the sediments. To appreciate the complexity of cadmium mobility in the sediment-water systems, one must understand the various forms of the element present in the sediment solids, the various chemical processes affecting the availability of sediment-bound cadmium, and the influence of physicochemical parameters on the relative distribution of cadmium in various chemical fractions.

On the basis of mathematical models and experimental data, Krauskopf (1956) concluded that possible sinks for cadmium in marine waters included (*a*) precipitation by sulfides, (*b*) adsorption by various adsorbents present in the sediments, and/or (*c*) organic concentration by living organisms. Presley

Table 1.   Trace Element Sinks and Their Respective Uptake and Release Processes[a]

Sinks	Processes	Reaction Parameter[b]
Oxides (hydrous, amorphic), manganese and iron, aluminum and silicon	Surface exchange	K_{eq}
	Diffusion exchange	K_{eq}, R_{ex}
	Coprecipitation	P, R_{ppt}
Organic substances	Exchange	K_{eq}
	Complexation	K_{eq}
	Chelation	K_{eq}
Biota	"Passive" uptake	R_{gr}
	Exchange, complexation, chelation	K_{eq}
	"Active" uptake	K_{eq}
Carbonates, phosphates, sulfides, basic sulfate and chloride salts	Precipitation	R_{ppt}
	Coprecipitation	P
	Surface (isomorphic) exchange	K_{eq}

[a] (Jenne and Luoma, (1977).
[b] K_{eq} = mass action equilibrium exchange constant; R_{ex} = rate of exchange; P = partitioning coefficient; R_{ppt} = rate of precipitation; R_{gr} = rate of growth.

et al. (1972) attributed the various degrees of cadmium enrichment in the interstitial waters of a reducing fjord to organic complexation and equilibration with unidentified solid mineral phases. Khalid et al. (1977) reviewed the existing literature on the distribution of cadmium in the various sediment components and concluded that mineral colloids, organic matter, carbonates, and sulfides may be the controlling mechanisms that modify the exact distribution and speciation of cadmium in aquatic systems.

Jenne and Luoma (1977) reviewed the state of knowledge on the physicochemical partitioning of trace elements, including cadmium, into various forms and reported that the most general particulate sinks for trace elements were iron and manganese oxides and organic substances. A list of various sediment sinks is presented in Table 1. It was suggested that the release of trace elements from various sediment sinks was controlled by several processes, such as exchange reactions, complexation, and precipitation (Table 1). These authors concluded from the published experimental evidence that equilibrium concentrations of trace metals in overlying waters and their biological availability were regulated via sorption-desorption and dissolution-precipitation reactions.

Cadmium associated with sediment solids may be present in several chemical forms, depending on the sediment composition and physicochemical properties of the sediments, and can be classified into the following groups.

3.1. Exchangeable Phase

Cadmium may be adsorbed by electrostatic attraction to negatively charged ion exchange sites on mineral colloids (clays), organic particulates, and hydrous oxides (Krauskopf, 1956; Jackson, 1958; John, 1971; Andelman, 1973; Haghiri, 1974; Khalid et al., 1978a). This phase is in equilibrium with the interstitial water phase and can be readily adsorbed or desorbed as a consequence of changes in the physicochemical parameters or concentration gradient (De Groot et al., 1976; Chen et al., 1976; Gambrell et al., 1976). The high concentration of cadmium present in this phase in sediments may have an adverse impact on water quality during dredging and disposal operations because of mixing with overlying water (Gambrell et al., 1978).

3.2. Carbonate Phase

Carbonate precipitation may be an important sink for cadmium in sediments high in bicarbonate and alkaline in reaction (Emerson, 1976; Hoeppel et al., 1978). Under slightly reduced to oxidized conditions solid $CdCO_3$ is a major control mechanism for cadmium solubility and is a potentially bioavailable fraction (Weber and Posselt, 1974; Chen et al., 1976).

3.3. Reducible Phase

This phase consists of cadmium adsorbed or coprecipitated with oxides, hydroxides, and hydrous oxides of Fe, Mn, and possibly Al present as coatings on clay minerals or as discrete particles (Jackson, 1958; Gadde and Laitinen, 1974; Lee, 1975; Kinniburgh et al., 1977; Gambrell et al., 1977a; Jenne and Luoma, 1977). Considerable quantities of iron and manganese are present in the interstitial H_2O in reduced sediments. When oxygenated water is mixed with reduced sediments containing ferrous iron, as occurs during dredging and upland disposal, rapid oxidation of the ferrous iron to the amorphous ferric oxyhydroxide form occurs. Coprecipitation and adsorption of heavy metals such as cadmium with the hydrated oxides may remove these metal ions from solution. Freshly precipitated colloidal hydrous oxides are thought to be much more effective in scavenging high concentrations of trace metals because of greater reactive surface area than are aged crystalline materials (Lee, 1975). However, several reports (Khalid et al., 1978a; Brannon et al., 1976; Gambrell et al., 1977a) indicate that cadmium is a weak competitor for adsorption on hydrous metal oxides in the environment and suggest that adsorption by colloidal hydrous oxides may not be a major control mechanism for cadmium in sediments, as suggested by Jenne (1968), Gadde and Laitinen (1974), and Lee (1975).

3.4. Organic Phase

This phase contains complexes that vary in stability from immediately mobile, easily decomposable, and moderately decomposable to resistant to decomposition (Jackson, 1958). Cadmium complexed with the organic fraction may be divided into chelated and organic bound. Chelated cadmium is the fraction that is loosely attached to immediately mobile and easily decomposable organic material and is a good indicator of the easily bioavailable cadmium form (Gambrell et al., 1977a, 1977b). Organic-bound cadmium, on the other hand, is the fraction incorporated into the insoluble organic material and can be solubilized only after intense oxidation of the organic matter (Brannon et al., 1976; Gambrell et al., 1977a). Cadmium complexation with fulvic and humic acid fractions is reported to be an important regulatory process affecting cadmium bioavailability (Stevenson and Ardakani, 1972; Gardiner, 1974b; Giesy et al., 1977).

3.5. Sulfide Phase

This phase represents highly insoluble and stable complexes of cadmium sulfides and exists in reduced sediments (Krauskopf, 1956). Oxidation of reduced sediments results in the transformation of sulfide-bound cadmium into more mobile and potentially available carbonate, exchangeable, and organic-complexed fractions that are in equilibrium with the soluble cadmium fraction (Holmes et al., 1974; Gardiner, 1974b; Gambrell et al., 1977a).

3.6. Mineral Crystalline Lattices Phase

This phase consists of cadmium bound within the crystalline lattices of mineral particles and is essentially unavailable in the sedimentary environment. In many literature reports this phase is termed the lithogenic or residual fraction. A significant part of sediment-bound cadmium is generally present in this unavailable fraction (Brannon et al., 1976). In polluted sediments, however, the fraction of cadmium present in the clay lattices and silicate minerals may be rather small (Serne and Mercer, 1975; Gupta and Chen, 1975; Hoeppel et al., 1978).

3.7. Chemical Fractionation Studies

The various chemical forms with which cadmium may be associated in sediment solids are summarized in the preceding subsections. The quantitative determination of cadmium in the various geochemical forms described above is essential for any studies of the impact of cadmium contamination on sediment-water systems. With regard to water quality and biouptake, it is important to determine whether cadmium is (a) in solution or loosely adsorbed to solids, where it is readily available; (b) complexed with insoluble organic or inorganic substances, where chemical or biological changes are required before it is released; or (c) present in the crystal lattices of deposited or suspended mineral particulates, where it is essentially unavailable. In the last two decades several chemical fractionation procedures have been developed to identify trace metals associated with different components of the sediment-water systems. Notable among the earlier studies are the methods developed by Goldberg and Arrhenius (1958), Chester and Hughes (1967), Presley et al. (1972), and Nissenbaum (1972). These studies were mainly concerned with the sedimentary geochemistry of pelagic sediments. The classification of various sediment components into lithogenic, hydrogenic, biogenic, and related categories was related primarily to their origins. However, these studies do not account for the effects of oxidation-reduction reactions, which may cause phase differentiation of elements within the sediments. Some of the limitations encountered in studying the geochemical partitioning of trace metals are documented in several reports (Serne and Mercer, 1975; Brannon et al., 1976; Khalid et al., 1977).

Engler et al. (1977) developed a selective extraction scheme to separate metals bound in various sediment phases on the basis of their chemical and biological transformations in aerobic and anaerobic sediments. A schematic representation of the sequential chemical fractionation procedure is given in Figure 3. This extraction procedure takes into account the functional properties of various sediment components, rather than the traditional sedimentary geochemistry, and precludes the possibility of any atmospheric oxidation of sediments during the extraction of interstitial water and the exchangeable phases most susceptible to physicochemical changes (Figure 3). The authors suggested that proposed extraction be done on relatively undisturbed sediments, especially avoiding drying, which can change sediment properties and alter metal ion distribution.

Serne and Mercer (1975) and Brannon et al. (1976) applied the selective extraction procedure developed by Engler et al. (1977) to determine various

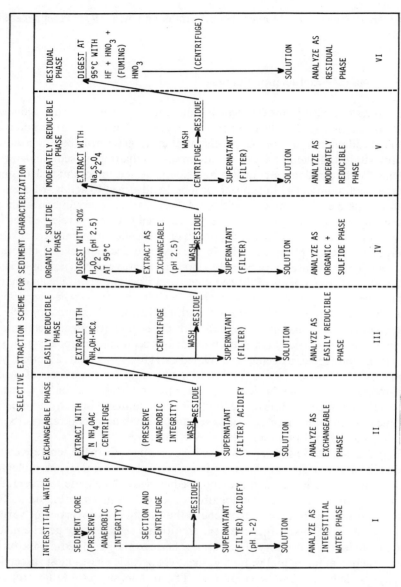

Figure 3. Selective extraction scheme for heavy metals in sediments. (Engler et al., 1977.)

forms of cadmium present in dredged sediments collected from various geographical locations. The cadmium present in each form, expressed as a percentage of the total cadmium in the dredged sediments, is shown in Table 2. The data from San Francisco Bay sediments indicate that 90 to 94% of total cadmium was complexed with insoluble organic matter or precipitated as sulfides. The remainder of the cadmium was associated with clay lattices (residual fraction) or oxides (Serne and Mercer, 1975). A small fraction of cadmium was also present in the exchangeable fraction. The distribution of cadmium in the Ashtabula and Bridgeport Harbor sediments was similar to that in the San Francisco Bay sediments in that most of the metal was associated with the organic + sulfide fraction (Brannon et al., 1976). In the Mobile Bay sediments, however, 90% of total cadmium was associated with the clay lattices in an unavailable form.

Chen et al. (1976) modified the extraction procedure of Engler et al. (1977) to account for cadmium present as $CdCO_3$ solids in the sediments. Analysis of Los Angeles Harbor dredged sediments of varying textural composition indicated that between 86 and 97% of the total cadmium contents in the sediments was derived from external sources, in contrast with that of mineral crystalline structure (the residual portion), soluble only by destruction of crystal lattices. Between 12 and 44% of the nonresidual cadmium was bound with the organics and sulfides. The cadmium content in the organic + sulfide phase increased as the sulfide content of the sediments increased. The remaining nonresidual cadmium was associated with carbonate, manganese oxide, and easily reducible iron oxide. The fraction of cadmium associated with oxides and hydrous oxides of iron, as indicated by a moderately reducible fraction, was essentially zero.

These studies strongly suggest that the most important sink for cadmium in reduced dredged sediments is the nonresidual fraction, which constitutes the reservoir for potential cadmium release into the water column and availability to aquatic organisms. The significant cadmium forms that constitute the nonresidual fraction include organic, sulfide, carbonate, and manganese oxide. Similar results were also reported by Gong et al. (1977), who concluded that in unconsolidated stream sediments cadmium was transported primarily bound to organics and secondarily with manganese oxides. These potentially bioavailable forms of cadmium can be transformed into readily bioavailable water-soluble and exchangeable forms as a result of altered physicochemical conditions, such as would occur during dredging and especially upland disposal.

Another significant finding of these fractionation studies is that cadmium is not associated with oxides and hydrous oxides of iron, as indicated by the lack of cadmium release in the moderately reducible fraction.

Although the extraction procedures developed by Engler et al. (1977) and modified by Chen et al. (1976) are good indicators of cadmium phases that may participate in chemical and biological transformations in dredged sediments, they suffer from an inherent limitation with regard to the overlapping in the

Table 2. **Percentage of Cadmium Extracted in Various Chemical Fractions from Dredged Sediments Collected from Different Dredging Sites in the United States**

Chemical Fraction	San Francisco Bay[a]			Mobile Bay[b]	Ashtabula Harbor[b]	Bridgeport Harbor[b]
	Mare Island	Turning Basin	Inner Harbor			
Interstitial H_2O	0.00	1.60	0.00	0.00	0.00	0.00
Exchangeable	1.80	0.00	0.00	0.00	0.00	0.00
Easily reducible	3.60	0.00	0.00	0.40	5.40	1.90
Organic + sulfide	92.70	90.30	93.60	9.80	82.20	98.10
Moderately reducible	0.00	0.00	0.70	0.00	0.00	0.00
Residual	1.80	8.10	5.70	90.00	12.40	0.00
Total cadmium ($\mu g/g$)	0.55	0.62	1.40	5.02	1.20	8.00

[a] From Serne and Mercer (1975).
[b] From Brannon et al. (1976).

extraction of certain chemical forms. Also certain chemical extracts may not be selective in sediments with different characteristics. For example, extraction with hydroxylamine hydrochloride solution may dissolve large quantities of iron oxides in addition to manganese oxides, and metal ions analyzed in this fraction may not reflect association with manganese compounds only (Khalid et al., 1978b). Another problem in the interpretation of selective fractionation results is that metal ions associated with organics cannot be identified from those precipitated with sulfides because of lack of selective extraction procedures for separating these two sinks for cadmium. Since trace metal distribution in any extraction scheme is operationally defined by the method of extraction rather than by any fundamental property of the metal complexes (McLaren and Crawford, 1973), the data obtained from such studies should be carefully interpreted.

4. EFFECT OF PHYSICOCHEMICAL PARAMETERS ON CADMIUM MOBILITY

It is evident from the preceding discussions that cadmium is present in sediment-water systems in various forms which may differ greatly in their availability to aquatic and benthic organisms. Dissolved cadmium present in overlying and interstitial waters is considered most readily bioavailable. The readily bioavailable fraction includes cadmium adsorbed on cation exchange complexes, which is in equilibrium with the aqueous phase. At the other end of the availability spectrum is cadmium bound within the crystalline lattices of clay minerals. This fraction is unreactive and may slowly mobilize as a result of mineral weathering over geologic time. Between the easily bioavailable and very unreactive forms of cadmium in sediments are a number of chemical forms that are potentially bioavailable. These include inorganic solid phases, $CdCO_3$, $Cd(OH)_2$, CdS, chelated and insoluble organic-bound cadmium, cadmium precipitated or coprecipitated with hydrous oxides of manganese and possibly iron, and several other inorganic constituents of minor importance. Cadmium present in these potentially bioavailable chemical forms can be mobilized to more readily bioavailable fractions as a result of changes in the physicochemical properties of sediments.

Redox potential, pH, and salinity are probably the most important physicochemical parameters controlling cadmium speciation in sediment-water systems and hence availability to biota (Hem, 1972; Hahne and Kroontje, 1973; De Groot et al., 1976; Leland et al., 1978). Reseach studies related to the role of these parameters in affecting the chemical mobility of cadmium in sediment-water systems are reviewed in the following subsections.

4.1   Effects of Redox Potential and pH

The redox potential and pH of a system are interrelated through a definite thermodynamic relationship such that a change in pH is accompanied by a shift in redox potential. This relationship is well documented in the literature (Baas Becking et al., 1960; Krauskopf, 1967; Stumm and Morgan, 1970; Ponnamperuma, 1972) and will not be discussed here.

Surface waters are generally oxidized with a range in redox potential of from 300 to 600 mV. The redox potential in sediments generally ranges from 100 to −400 mV, and the sediments are moderately to strongly reduced. The pH of seawater generally ranges from 7.8 to 8.3 and is relatively uniform, compared to that of fresh water, which may fluctuate from 5.0 to 8.5. The pH of surface sediments may range from 6.1 to 9.0, and a slightly lower pH is encountered in the deeper sediments. The changes in the redox potential and pH of various sediment-water systems and the control mechanisms for these variables are discussed in a review by Khalid et al. (1977).

The redox potential of a sediment-water system is a controlling factor in regulating the chemical form of cadmium, whereas pH influences the stability of the various forms. This relationship is clearly illustrated in Figure 1, where the form as well as the stability of cadmium species is modified by changes in both redox potential and pH. The prediction of redox potential-pH effects on cadmium distribution in an inorganic system where the nature and the concentration of components are known, such as is described in Figure 1 (Hem, 1972), however, may represent an oversimplification of the system. In the natural sediment-water system the chemical forms of cadmium are a net result of interactions between several inorganic and organic components that are in a dynamic state of equilibrium. The thermodynamic data on these components are not available, and thus the effects of redox potential and pH changes on cadmium speciation cannot be accurately determined on the basis of mathematical models (Morel et al., 1973; Florence and Batley, 1977). Selective chemical fractionation schemes have therefore been employed to determine the effects of changes in pH or redox potential, such as would occur during dredging and dredged material disposal, on the bioavailability of sediment-bound cadmium.

Serne and Mercer (1975) studied the effect of oxidized and reduced sediment conditions on the release of cadmium from San Francisco Bay dredged sediment suspensions. Analysis of dissolved cadmium after incubation of sediment suspensions under low O_2 (Eh = <100 mV) and high O_2 (Eh = >350 mV) conditions indicated that significantly higher concentrations of cadmium were released under oxidizing than under reducing conditions. The cadmium content released in the elutriate increased with agitation time. It was suggested that this kinetic effect was due to slow oxidation of sulfide or organically bound cadmium in the reduced sediments over a period before steady state equilibrium conditions

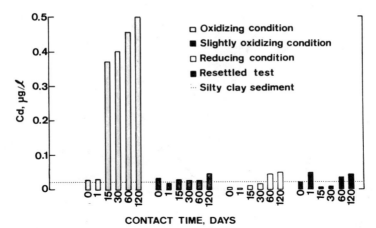

Figure 4. The effect of various oxidation-reduction conditions on the transport of cadmium in sediment-seawater interfaces in a silty clay sediment. (Chen et al., 1976.)

were reached. No significant changes in sediment pH that would alter cadmium speciation occurred in this experiment.

Chen et al. (1976) studied the long-term release of sediment-bound cadmium to the overlying seawater under oxidizing, slightly oxidizing, and reducing conditions in controlled lysimeter experiments. The data plotted in Figure 4 indicate that significant cadmium release occurred only under oxidizing conditions. The cadmium concentrations in the overlying water were increased about 15 times over the original seawater background levels of 0.03 μg/l to about 0.5 μg/l after 4 months of contact. Under slightly oxidizing and reducing conditions no significant change in cadmium concentration occurred.

The geochemical fractionation of sediment solids at the end of the long-term incubation experiment discussed above (Chen et al., 1976) indicated an increase in the water-soluble, exchangeable, and carbonate fractions and a decrease in the easily reducible, organic + sulfide, and residual fractions. On the basis of the $CdCO_3$-$Cd(OH)_2$ relationships of Weber and Posselt (1974) described in equation 1, Chen et al. (1976) concluded that solid $CdCO_3$ was the dominant cadmium species because of the abundance of carbonates in the interstitial water.

Gambrell et al. (1977a) conducted a controlled laboratory study to determine the effects of redox potential and pH on cadmium transformations in estuarine and freshwater sediments. Mississippi River (freshwater) and Barataria Bay (estuarine) sediment suspensions, tagged with [109]Cd, were incubated for 2 weeks under various combinations of pH (5.0, 6.5, and 8.0) and redox potential levels (−150, +50, +250, and +500 mV). A chemical fractionation procedure de-

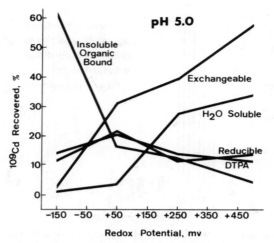

Figure 5.   The effect of redox potential on the distribution of [109]Cd among selected chemical forms in Mississippi River sediment suspensions incubated at pH 5.0. (Gambrell et al., 1977a.)

veloped by the authors was used to determine the distribution of added [109]Cd in the readily bioavailable, potentially bioavailable, and unavailable chemical forms. The data plotted in Figures 5 to 7 for Mississippi River sediments indicate that both pH and redox potential strongly influence the distribution of [109]Cd in various chemical forms. It is evident from these results that the cadmium-organic complex is more stable under reduced conditions at all levels of pH studied. As a result of oxidation, this organic-bound cadmium was mobilized to more bioavailable water-soluble, exchangeable, or chelated fractions. It should be pointed out that total sulfides were not detectable under reduced conditions at any pH level in the Mississippi River sediment suspensions, and thus sulfide precipitation was probably not a controlling factor for cadmium in this study. A general decrease in the water-soluble and exchangeable cadmium fractions with increasing pH was indicative of a shift from easily available to potentially available forms of cadmium. Similar results were obtained in Barataria Bay sediments (Gambrell et al., 1977a). There was no significant change in the [109]Cd associated with hydrous oxides (reducible fraction) as a result of redox potential and pH changes.

In a related study simulating high levels of cadmium contamination, Khalid et al. (1978a) reported that the cadmium present in the easily bioavailable water-soluble and exchangeable fractions decreased sharply with an increase in sediment suspension pH (Figure 8). This was accompanied by a gradual increase in the organically complexed (DTPA-extractable + insoluble organic-bound) cadmium. The fraction of cadmium present in the reducible fraction

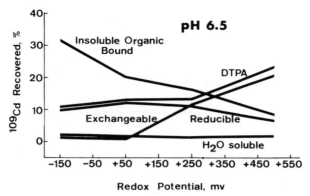

Figure 6. The effect of redox potential on the distribution of [109]Cd among selected chemical forms in Mississippi River sediment suspensions incubated at pH 6.5. (Gambrell et al., 1977a.)

was relatively small and was unaffected by pH changes. On the other hand, hydrous oxides of iron as indicated by the reducible fraction were the most important source of reactive zinc and would release significant amounts of easily bioavailable zinc under acidic sediment conditions, as compared to alkaline conditions (Figure 8). Kinniburgh et al. (1977) also concluded from their studies that cadmium, in comparison to zinc, was a weak competitor for adsorption on hydrous metal oxides.

These studies strongly suggest that adsorption by colloidal hydrous oxides of iron may not be a major control mechanism for cadmium, as alleged in the literature (Jenne, 1968; Gadde and Laitinen, 1974; Lee, 1975). Under the conditions of these experiments, adsorption of cadmium on the exchange sites and complexation with insoluble organic material constitute the major pools of reactive cadmium influencing its availability in sediment-water systems in the absence of sulfides. The important role of clay minerals and organics in cadmium fixation is well documented in the literature (John, 1971; Stevenson and Ardakani, 1972; Haghiri, 1974; Gardiner, 1974b).

Adsorption of cadmium by freshwater and estuarine sediments is favored with increased pH (Gambrell et al., 1977a). At low levels of cadmium addition, more cadmium was adsorbed under reduced than under oxidized conditions; this effect was more pronounced at pH 5.0 and 6.5 and especially in Barataria Bay estuarine sediments. Barataria Bay sediments contained substantial quantities of total sulfides at pH 5.0 and 6.5, which may have resulted in the precipitation of added cadmium as CdS. Very little of the adsorbed cadmium was extracted in the exchangeable fraction in the estuarine sediments incubated under very reduced to moderately reduced conditions. The adsorption of cadmium on hydrous iron oxides, although low in magnitude, is also reported to increase at a pH value of

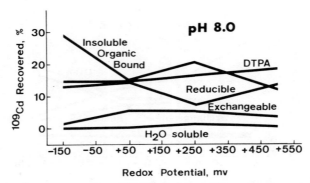

Figure 7. The effect of redox potential on the distribution of ^{109}Cd among selected chemical forms in Mississippi River sediment suspensions incubated at pH 8.0. (Gambrell et al., 1977a.)

7.0 (Gadde and Laitinen, 1974). Kinniburgh et al. (1977) reported that adsorption of cadmium by fresh aluminum gels increased from 4.5 to 50% when the pH of the medium was raised from 6.0 to 7.0.

Frost and Griffin (1977) studied the effect of pH on the adsorption of cadmium from landfill leachates by kaolinite and montmorillonite clay minerals. The cadmium removal by montmorillonite at pH values of 4 to 6 was linear, suggesting that an adsorption mechanism was taking place. The essentially complete removal of added cadmium by montmorillonite at pH 6.5 and 7.0 indicated the possibility of cadmium precipitation. Calculations based on solubility product indicated that $Cd(OH)_2$ was too soluble to account for cadmium precipitation. Instead, it was suggested that the formation of solid $CdCO_3$ at higher pH values and cadmium concentrations was responsible for the removal of cadmium by montmorillonite. Montmorillonite sorbed approximately 5 times more cadmium from solution than did kaolinite because of greater cation exchange capacity.

4.2. Effects of Salinity and Sediment-Seawater Mixing Zone

The effect of salinity as a major control mechanism in increasing cadmium mobility is related to the formation of chloride, sulfate, and carbonate complexes of greater solubility than hydroxide, sulfide, and organic complexes. Hahne and Kroontje (1973) concluded from stability constant computations that, at seawater chloride concentration of 20,000 ppm and pH of less than 8.5, the chloride complexes override the hydroxy complexes, the main species being $CdCl_2^0$ and $CdCl_3^-$. They concluded from their study that chloride complexes were more

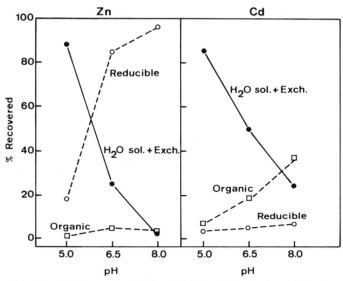

Figure 8.   The effect of pH on the distribution of added cadmium (260 μg/g) in selected chemical fractions in Mississippi River sediment suspensions. (Khalid et al., 1978a.)

mobile and persistent complexing agents than the biodegradable organic complexes and played an important role in the mobilization of cadmium.

Lee et al. (1975) attributed the effect of salinity and salinity changes to sorption-desorption, ion exchange, alteration of clay crystal structure, flocculation of organic matter, and clay particles. Serne and Mercer (1975) reported that under O_2-rich conditions San Francisco Bay sediment suspensions released significantly greater amounts of cadmium as the water salinity increased from 1 to 29%. They suggested that the formation of soluble inorganic complexes with the increased chlorides, carbonates, and sulfates was responsible for cadmium release. Also, the release of cadmium bound to ion exchange sites due to greater competition from high concentrations of Ca, Mn, and Na for exchange sites may be responsible for increased cadmium release with increasing salinity. The effect of reducing conditions on cadmium release was evident only at 1% salinity. This low-salinity treatment was a mixture of estuarine and distilled water which reflected altered physicochemical conditions that may have resulted in the release of sediment-bound cadmium. De Groot (1973) and De Groot and Allersma (1975) reported that over 90% of the cadmium bound to Rhine River sediments was mobilized upon entering the North Sea. They attributed this release of the element to chloride and organic complexation.

Windom (1976) studied the effect of salinity changes in seven southeastern estuaries on the concentration of dissolved cadmium and reported a lack of any

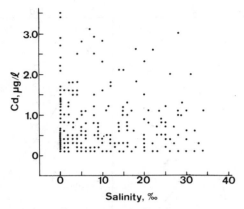

Figure 9.   Relationship between cadmium concentration and salinity in estaurine waters. (Windom, 1976.)

consistent relationship (Figure 9). However, examination of the data plotted in Figure 9 indicates generally greater cadmium concentration between 0 and 5% salinity levels.

The mixing of polluted freshwater sediments with oxygenated seawater, as would occur during river flow or open water dredged material disposal, may result in a new sediment-seawater equilibrium and affect the mobilization of sediment-bound cadmium. Helz et al. (1975) investigated the fate of trace metals associated with wastewater effluent discharged into the Chesapeake Bay estuary. An initial decrease in cadmium concentration within 2 to 3 km of the fall was followed by release of cadmium in the seaward direction, suggesting remobilization of the metal. A decrease in the Cd/Zn ratio in the sediments in the seaward direction indicated that cadmium was decreasing at a faster rate, thus confirming the remobilization theory. Remobilization of cadmium from the sediments was attributed to cation exchange, decomposition of metal-rich organic material in the sediment, or the solubilization of cadmium in the form of aqueous complexes. Windom (1976) developed a model to assess the pollutant transfer from river to estuaries and then to ocean. The net release of metal ions to the ocean was mediated by a series of processes occurring at the river-estuarine interface and in the estuaries. Windom concluded, based on evaluation of the proposed transport model, that estuaries do not serve as sinks for cadmium and mercury and that these two metals were transported to the coastal waters. From the rate of input and the metal concentration in coastal waters, Windom calculated the residence times of cadmium and mercury in the coastal waters to be 3 and 17 weeks, respectively.

Rohatgi and Chen (1975) studied the release of cadmium from waste effluents on mixing with seawater at effluent/seawater ratios of 1:5 to 1:200. Up to 95% of cadmium was released from the suspended solids to the oxygenated seawater. The cadmium desorption was observed to occur in two stages: an initial rapid release, followed by a slower, long-term release. These authors attributed this of cadmium release to (a) the oxidation of organic matter or CdS; (b) desorption from suspended solids, which depended mostly on the dilution ratio and pH of seawater; and (c) the formation of metallic chloride as well as organometallic complexes. Lu and Chen (1977) suggested that the release of cadmium from the polluted reduced sediments upon mixing with the seawater increased as the environment became oxidizing. The slow oxidation of reduced sediments resulted in a change of controlling solids from very insoluble sulfides to higher solubility carbonates, which contributed to greater cadmium release with time.

Research studies conducted on the effect of mixing contaminated Rhine River sediments with the estuarine sediments indicated a decrease in the trace metal concentration in the seaward direction (De Groot, 1973; Muller and Forstner, 1975; Salomons and Mook, 1977). However, the interpretation of this decrease has been the subject of much controversy. De Groot (1973) and Salomons and De Groot (1977) have proposed models wherein mobilization of cadmium from the deposited estuarine sediments is caused by formation of complexes between the metal ion sorbed to the sediments and organic ligands released by the intensive decomposition of organic matter. Muller and Forstner (1975) have proposed that the mixing of uncontaminated marine sediments with contaminated fluvial sediments may equally well explain this phenomenon. More recently, Duinker and Nolting (1977) presented evidence for the occurrence of mixing and of precipitation processes in the Rhine Estuary. They suggested that, if the mixing ratio of marine to fluvial sediments and the relevant trace metal concentrations are known, it is possible to calculate the trace metal concentrations in the estuarine deposits. Calculated values exceeding the observed ones would point to mobilization processes. Lower calculated values, on the other hand, would indicate the occurrence of precipitation or adsorption processes. Using natural tracer (Mg) and primary stable isotopes (^{18}O, ^{13}C) to distinguish between marine and fluvial sediments, Salomons and Mook (1977) showed that a decrease in cadmium and other trace metals in the seaward direction in the estuaries of the rivers Ems and Rhine was caused by the mixing of marine (low metal concentration) and fluvial (high metal concentration) sediments.

5. CADMIUM MOBILITY DURING DREDGING AND DREDGED SEDIMENT DISPOSAL

Waste discharges from industrial and municipal sources have tremendously increased the loading of especially hazardous pollutants such as cadmium, mercury, and pesticides in many rivers, harbors, and lakes in the industrial world. There is a growing concern that dredging and disposal of these contaminated sediments to maintain channel navigation may adversely affect water quality and introduce toxic materials into the food chain (Gambrell et al., 1978). The disposal of dredged materials contaminated with high concentrations of cadmium, such as those from Rotterdam Harbor, Netherlands (Royal Adriaan Volker Group, 1977); Bridgeport Harbor, Connecticut; the Hudson River, New York; and the Houston Ship Channel, Texas (Brannon et al., 1978), in the aquatic or on upland disposal sites holds great potential for the release of cadmium and its subsequent uptake by benthic and aquatic organisms, as well as terrestrial plant species inhabiting the disposal site. It is essential that the behavior of cadmium under these disposal conditions be fully investigated in order to select environmentally acceptable disposal alternatives that will minimize further contamination of the environment and reduce the accumulation of sublethal chronic levels of cadmium by the consumer organisms. Research studies on the chemical mobility of sediment-bound cadmium under subaqueous and upland disposal conditions are summarized in the following subsections, and the potential for bioaccumulation in evaluated.

5.1. Aquatic Disposal

The U.S. Army Corps of Engineers and the U.S. Environmental Protection Agency jointly developed a procedure called the elutriate test to predict the release of chemical contaminants from subaqueous disposal and thus to establish criteria for the disposal of contaminated materials (U.S. Environmental Protection Agency, 1973b, 1973c). The elutriate test is a part of the overall ecological evaluation guidelines established to regulate the open water discharge of dredged or fill material containing chemical contaminants that may adversely affect the water column and benthic organisms. This test is a simplified simulation of the dredging and disposal process wherein predetermined amounts of dredging site water and sediment are mixed together to approximate a dredged material slurry. The supernatant, after approximate mixing and settling, is analyzed for dissolved contaminants. This elutriate concentration, when modified to reflect the dilution or dispersion characteristics at the proposed discharge site, can be compared with the water quality criteria to establish a short-term environmental impact. Detailed discussions of the rationale, factors affecting the elutriate concentrations, validity of the proposed test, standardization procedures, and interpretation of test results are given elsewhere (Keeley and Engler, 1974; Lee et al., 1975; Cheam et al., 1976; Environmental Effects Laboratory, 1976). Elutriate tests have been extensively employed in the Dredged Material Research Program of the U.S. Army Corps of Engineers to determine the potential release of cadmium from dredged sediments from various locations in the United States (Lee et al., 1975; Brannon et al., 1978; Baumgartner et al., 1978; Wyeth and Sweeney, 1978).

Lee et al. (1975) conducted elutriate tests on dredged sediments from both freshwater and estuarine locations in the United States containing total cadmium concentrations of 1.1 to 15.4 mg/kg. The elutriate analyses showed that cadmium remained consistently at ambient water levels under anoxic as well as aerated sediment-water mixing conditions. Baumgartner et al. (1978) conducted elutriate tests on Duwamish River dredged material to be disposed of in Elliot Bay, Puget Sound, Washington, and reported that cadmium concentrations in the dredged material exceeded the 1.5 pollution criterion in several cases. The cadmium concentrations in the elutriate ranged from less than 0.2 to 2.7 μg/l and were comparable to those in the disposal site samples. Elutriate test results reported by Wyeth and Sweeney (1978) indicate Ashtabula River dredged sediments exceeded the pollution criteria and had higher concentrations of soluble cadmium than did the disposal site in Lake Erie. However, if a tenfold dilution of the elutriate upon open lake deposition is assumed, cadmium concentrations would not increase significantly. Brannon et al. (1978) reported a small net release of cadmium in elutriate tests conducted on contaminated and uncontaminated dredged sediments collected from various geographical locations in the United States.

It is evident from these studies that no or insufficient release of cadmium, as determined by elutriate tests, will not adversely affect water quality at the dredging site. The monitoring of actual open water field disposal in the studies cited above indicated that slightly elevated cadmium concentrations in the surface waters at the disposal site would persist for only a short time, and then decrease to predisposal levels within minutes to a few hours. The very short contact time with the surface waters during the settling and mixing of contaminated sediments with the disposal site water usually results in no short-term chemical impact on aquatic organisms.

The long-term release of sediment-bound cadmium from the dredged material deposited on the ocean floor has also been investigated in a few controlled laboratory and disposal site studies. In a simulated long-term study of open water disposal Blom et al. (1976) observed a significant increase in the overlying and interstitial water cadmium concentrations over a 4-month period when freshwater dredged material was disposed of in seawater media. Oxidized conditions were maintained in this experiment to simulate actual overlying water conditions. It was suggested that the higher dissolved O_2, ionic strength, and chloride concentrations of the seawater environment were responsible for cadmium release to the interstitial and overlying water. In a long-term quiescent leaching study of 32 dredged sediments with disposal site water maintained under oxidized conditions, Brannon et al. (1978) reported only a slight release of sediment-bound cadmium after 4 months of equilibration. The concentration of cadmium in the 8-month sampling, however, dropped to less than the site water value for the element, resulting in a net mass release of essentially zero for the study period.

The results of these long-term studies suggest that the disposal of cadmium-contaminated sediments under subaqueous conditions would not cause significant long-term water quality problems. However, cadmium associated with sediment solids settled on the ocean floor inhabited by economically and ecologically important benthic organisms may be accumulated in high concentrations and recycled in the food chain. The implications of this cadmium bioaccumulation from deposited sediments will be discussed in Section 6.

5.2. Upland Disposal

A rapid shift from open water disposal to confined upland disposal has been occurring in recent years because of the potential adverse environmental impact of open water disposal on the aquatic and benthic organisms (Boyd et al., 1972). However, extensive confined disposal area effluent and leachate monitoring studies have been conducted only sporadically, and specific studies have usually been limited to a single site. May (1974) reported that confined land disposal of sediments containing high total cadmium (6 $\mu g/g$) did not degrade water quality in the effluent receiving waters. The levels of soluble cadmium in the return waters from the confined dikes ranged from 10 to 20 $\mu g/l$ and were not significantly different from the cadmium levels in the stream. May proposed that the lack of significant changes in water quality following dredging and dredged material disposal may be due to the following possibilities:

1. Although chemicals in the mud are influenced by redox conditions, the processes are either too slow or too quickly reversible to allow sediment constituents to become dissolved when dredged material is deposited in confined areas or in open water.
2. Trace metals are so strongly adsorbed to inorganic or organic matter that reducing conditions have little effect on their release.
3. The materials are largely in refractory forms, such as lignins, sulfide compounds, organic complexes, and hydroxides, or in the elemental state and so are not readily soluble under the conditions created by dredging.

May suggested that these possibilities should be examined under rigidly controlled laboratory conditions.

Ritchie and Speakman (1973) studied the effect of the settling time of dredged sediments on the quality of effluents from confined disposal facilities. Analysis of influent slurries and effluent filtrate showed that cadmium was reduced from an initial toxic level concentration of 0.5 mg/l to less than 0.03 mg/l in 1 hr. Mudroch and Cheam (1974) monitored changes in the pore water chemistry of dredged spoil disposal on Pilot Island, Mitchell Bay, Lake St. Clair, for a period of 8 months after disposal in the test bay had ceased. The results showed that the cadmium concentrations in the effluent pore water and the Mitchell Bay lake water were almost identical and suggested no cadmium release during confinement.

As part of the U.S. Army Corps of Engineers' Dredged Material Research Program, laboratory and field investigations were commissioned to determine the pollution potential of toxic materials resulting from the land confinement of dredged material. The results of these studies indicate that the removal efficiency of cadmium in effluents from upland diked disposal sites from both freshwater and brackish water sediments was directly related to total solids removal (Hoeppel et al., 1978; Lu et al., 1978; Mang et al., 1978).

Hoeppel et al. (1978) studied changes in cadmium mobility at nine dredged material land containment areas, located at upland, lowland, and island sites, during hydraulic dredging operations in river, lake, and estuarine environments. Analysis of the total cadmium in the influent and effluent samples suggested cadmium removal efficiency of about 96% at the disposal sites for freshwater and estuarine dredged materials. The total cadmium in influents from various disposal sites ranged from 0.002 to 7.17 mg/l, while effluents contained 0.055 to 0.37 mg/l. Soluble phase cadmium showed no significant overall change in concentration in dredged material slurries during confined disposal area retention, averaging 0.004 and 0.003 mg/l in influents and effluents, respectively. There was a slight increase in dissolved O_2 concentrations in the effluents, compared with those in the influent slurries, with mean values of 5.3 and 3.8 mg/l, respectively. Moderately reduced to oxidized conditions measured in the disposal sites may significantly affect the bioavailability of sediment-bound cadmium. The geochemical phase partitioning of influent and effluent solids shows that cadmium concentrations increased significantly in the exchangeable, easily reducible, and carbonate phases as a result of land confinement (Table 3). This increase in readily available cadmium forms was a direct result of a decrease in the organic plus sulfide phase, where most of the cadmium in the influent slurries was associated.

This chemical fractionation study indicates that the retention of dredged material under oxidizing conditions will shift cadmium from less available organic- and sulfide-bound to more readily bioavailable forms, including exchangeable, carbonate, and easily reducible fractions. The exchangeable fraction is considered to be readily available to aquatic organisms and plant species, while carbonate and easily reducible complexes will also be readily available as a result of pH and redox potential changes. The effects of redox potential and pH changes

Table 3. Geochemical Phase Partitioning of Cadmium in Influent and Effluent Solids from Four Confined Land Disposal Areas[a]

Geochemical Phase	Cadmium ($\mu g/g$)[b,c]		Percent of Total Cadmium
	Range	Mean	
1. Exchangeable[d]			
Influent	0.002–0.07	0.03	21.0
Effluent	0.000–1.30	0.29	18.0
2. Carbonate[e]			
Influent	0.008–0.11	0.07	21.4
Effluent	0.007–5.50	1.70	56.7
3. Easily reducible[f]			
Influent	0.000–0.15	0.07	9.2
Effluent	0.000–2.10	0.44	11.8
4. Remaining phases			
Influent	0.000–3.70	0.86	49.3
Effluent	0.000–2.40	0.92	13.5

[a] Hoeppel et al. (1978).
[b] All values are expressed on oven dry basis.
[c] Based on five influent and effluent samples.
[d] Ammonium acetate extractable.
[e] 1 M acetic acid extractable.
[f] 0.1 M hydroxylamine hydrochloride in 0.01 M nitric acid extractable.

on cadmium mobility have been extensively studied by Khalid et al. (1978a) and Gambrell et al. (1977a) and were discussed in Section 4.1.

Lu and associates (1978) studied the chemical mobility of cadmium and other toxic materials during confined land disposal of dredged materials at two disposal site locations; Pinto Island, Mobile, Alabama, and Grassy Island, Detroit, Michigan. Cadmium was determined in the 0.45-μ and 0.05-μ filtrates and in the total samples, both in the influent slurries and the dewatering effluents. The concentration of cadmium in the 0.05-μ filtrates in the influent slurries was 4- to 40-fold higher than in the background waters, whereas the total cadmium concentration increased 37 to 340 times over the background levels. It was suggested that the significant release of soluble cadmium during dredging operations was due to the following:

1. Diffusion from the interstitial water.
2. Oxidation of reduced metallic sulfide solids, which are generally highly insoluble, to more soluble oxidized solids.
3. Formation of soluble metal complexes due to an increase of metal ligands in the soluble phase (such as the high levels of chloride, dissolved organic carbon, and nitrogen compounds in the influent samples.

4. Ion exchange.
5. Oxidation and decomposition of organic compounds.
6. Desorption from clay minerals or other solid species.

The concentration of total cadmium was directly related to greater solids in the influents.

The removal efficiencies of the disposal sites were 18 to 100% for total cadmium and 26 to 81% for the soluble fraction (0.05-μ filtrate) in the Pinto Island and Grassy Island sites, respectively. The lower cadmium removal efficiency at Pinto Island was essentially due to a greater fraction of finer particles remaining in suspension at the disposal site.

The geochemical partitioning of influent and effluent solids at the Pinto Island disposal site indicated a significant increase in the exchangeable cadmium fraction. The increase of cadmium in the carbonate fraction, although noticeable, was not statistically significant, indicating that the $CdCO_3$ solid formed as a result of sulfide oxidation under oxidizing conditions either was unstable or had a slow rate of formation. This may have resulted in a large cadmium increase in the exchangeable fraction. On the basis of the ion-ratio method Lu (1976), it was suggested that $CdCO_3$ was a controlling solid under oxidized conditions in the disposal site.

Based on the thermodynamic equilibrium among controlling solids and the easily released fraction of cadmium, as suggested by Jackson (1958), Lu (1976), Chen et al. (1976), Hoeppel et al. (1978), and Lu et al. (1978), the following relationship can be established:

$$\text{Water-soluble Cd} \rightleftharpoons \text{exchangeable Cd} \rightleftharpoons \text{controlling solid}$$

However, the ability of such a model to predict the soluble cadmium concentration may be limited because of the exclusion of soluble cadmium complexed with organic ligands, for which thermodynamic data are not available.

Mang et al. (1978) conducted a laboratory lysimeter study to determine the potential adverse impacts on groundwater of leachates generated from dredged material in land containment areas and to assess the influence of physicochemical parameters such as pH and redox potential on the mobility of cadmium and other toxic metals. The results of this study indicate that leachate quality may be governed by both the dredged material and the underlying soil. The redox potential of the overlying dredged material leachate (interstitial water) was generally oxidizing to slightly oxidizing. The analytical data showed cadmium levels in the dredged material leachate to be variable but decreasing to low levels at the end of 4 months of leaching. Mang et al. suggested that the soluble cadmium level in the interstitial water of various types of dredged material during leaching was controlled primarily by three mechanisms: solubility, adsorption, and complexation. Using the ion ratio method of Weber and Posselt (1974) and Lu (1976), they calculated that the free cadmium ion concentration was controlled by solid phase $CdCO_3$ rather than $Cd(OH)_2$ or CdS. Interstitial water cadmium was present as free Cd ions as well as cadmium chloride complexes in freshwater sediment material. However, in saline water dredged material cadmium chloride and cadmium-organic complexes were responsible for the high soluble cadmium levels.

The results of the leaching studies also showed that cadmium released from the dredged material was attenuated by the underlying soils (Mang et al., 1978). It was suggested that the major control mechanisms for cadmium retention by the soils were sulfide precipitation under reduced soil conditions and adsorption by clay minerals, sesquioxides, and humic material.

The results of simulated laboratory and actual field studies of upland disposal of cadmium-contaminated dredged sediments suggest that the movement of cadmium between dredged material and disposal site soils is influenced by several unknown factors. Thermodynamic and kinetic influences make transport phenomena difficult to explain and predict. Interaction between some of the major known mechanisms, such as solubilization, solid solution, sorption, complexation, bioreaction, and dilution, may exert some influence on the mobilization and subsequent migration of sediment bound cadmium into adjacent surface waters and groundwater underlying upland disposal sites (Lee and Plumb, 1974; Lu, 1976; Lu and Chen, 1977; Huang et al., 1977). Changes in the physicochemical parameters, especially redox potential, pH, and salinity, may significantly influence the direction of chemical transformation and consequently the levels of bioavailable cadmium (Zirino and Yamamoto, 1972; Hahne and Kroontje, 1973; Khalid et al., 1978a; Gambrell et al., 1977b).

6. POTENTIAL IMPACTS OF DREDGED MATERIAL DISPOSAL

The disposal of cadmium-contaminated dredged sediments in open water and at confined land disposal sites has a potential for both short-term and long-term adverse impacts. The short-term effects include cadmium release in the oxygenated overlying water at the subaqueous disposal sites and the discharge of oxidized effluent waters containing soluble and easily bioavailable cadmium from the upland disposal sites into the adjacent receiving waters. This released cadmium may be toxic to the aquatic organisms under both disposal conditions and exceed water quality standards. The long-term potential effects of bioavailable cadmium in overlying waters and sediment-bound cadmium in the dredged materials settled at the ocean floor will be excessive uptake of the metal by the aquatic and benthic organisms and, hence, recycling in the food chain. The potential long-term impacts of cadmium at upland disposal sites include its uptake by the plant communities established at the disposal sites, groundwater leaching of cadmium solubilized under oxidized, acid conditions, and the transport, due to erosion, of suspended material rich in cadmium to the adjacent surface waters. For example, noncalcareous dredged sediments high in sulfide and organic material become very acidic when subjected to oxidized conditions under upland disposal. This results in the transformation of cadmium from very insoluble sulfide and organic complexes to more mobile carbonate, exchangeable, and soluble fractions. This effect of oxidized acidic sediment conditions on cadmium mobility was documented earlier in the report by Gambrell et al. (1977a). These environmental impacts of the cadmium present in dredged sediments are more fully discussed in several reports published under the Dredged Material Research Program of the U.S. Army Corps of Engineers (Royal Adriaan Volker Group, 1977; Brannon, 1978; Chen et al., 1978; Burks and Engler, 1978; Gambrell et al., 1978).

The bioaccumulation of cadmium by aquatic and benthic organisms and plant uptake under various environmental conditions are described in several chapters in this book and hence will not be discussed in detail here. However, it is appropriate to document very briefly the evidence of cadmium accumulation and recycling by the consumer organisms. For example, Ayling (1974) and Wier and Walter (1976) reported the accumulation of high concentrations of cadmium, which may cause health hazards, from sediments by oysters and snails. The concentration of cadmium in oysters raised in cadmium-contaminated waters of Derwent Estuary, Tasmania, exceeded the Australian Government limit of 2.0 μg/g for food consumption and was as high as 19.8 μg/g (Woodruff et al., 1974). The effects of various physicochemical factors on the accumulation of cadmium by benthic organisms from dredged sediments have been discussed in a recent report (Neff et al., 1978).

Upland disposal sites are generally characterized by oxidized soil conditions, which have a profound effect on cadmium availability to terrestrial and marsh plants. Ito and Iimura (1975) and Reddy and Patrick (1977) reported significantly greater cadmium uptake by rice plants under oxidized than under reduced conditions. The effect of oxidized soil conditions on increased cadmium uptake was accentuated under acidic soil conditions. In controlled laboratory and greenhouse experiments, Gambrell et al. (1977b) observed a substantial increase in the cadmium contents of marsh plants under oxidized soil conditions. B. L. Folsom, Jr., and C. R. Lee (unpublished data; U. S. Army Corps of Engineers Waterways Experiment Station, CE, Vicksburg, Mississippi) demonstrated a significantly higher cadmium uptake by *Cyperus esculantus* from dredged sediments under upland than under flooded conditions. De Groot (1977) studied the uptake of cadmium by several agricultural crops from highly polluted Rotterdam Harbor sediments dumped at upland disposal sites. The crops grown on dredged sediments showed a significantly higher content of cadmium than the crops grown on natural river clays. An exponential increase in cadmium uptake was observed with increasing concentration of the element, especially in leafy vegetable and rootlike crops such as lettuse, radishes, potatoes, and carrots. On the basis of literature reports and laboratory investigations, an upper limit of 5 μg Cd/g in the dredged material was recommended for agricultural crops. Elevated levels of cadmium in polluted agricultural soils have also been reflected in high levels of the metal in pasture grass and in liver and kidney tissues of cattle and swine (Munshower, 1977).

Several criteria have been proposed to assess the potential impact of cadmium under subaqueous disposal conditions. The standard elutriate test originally developed by the U.S. Environmental Protection Agency and the U.S. Army Corps of Engineers (U.S. Environmental Protection Agency, 1973b, 1973c) and modified later (Lee et al., 1975; Environmental Effects Laboratory, 1976) has been used to determine the pollution potential during dredging and open water disposal of sediments. This test, however, predicts only the initial short-term chemical impacts of toxin release and is not suitable for assessing long-term effects. Bioassays of the elutriate have also been proposed to assess the impact of chemical release on metal uptake by the aquatic test organisms (Lee et al., 1975; Shuba et al., 1977). A committee of the U.S. Environmental Protection Agency and the U.S. Army Corps of Engineers (Engler and Wilkes, 1977) has recently prepared a manual of guidelines for a comprehensive ecological evaluation of ocean dumping with emphasis on potential impacts on marine organisms and human beings. In addition to the chemical analyses, bioassays on the liquid phase, suspended particles, and solid phase are required as part of the evaluation process.

Water quality criteria for drinking water supplies and freshwater and marine water aquatic life have been established by several agencies (U.S. Public Health Service, 1962; U.S. Environmental Protection Agency, 1973a, 1976, 1977). These water quality standards are applicable to surface waters and are based on total metal concentration. The validity of these criteria in their application to regulate subaqueous dredged material disposal has been criticized on the grounds that total metal concentration does not represent the easily bioavailable fraction taken up by aquatic organisms (Lee et al., 1975; Brannon et al., 1976).

Most of the aforementioned water quality criteria and test procedures are applicable to open water disposal. Criteria to assess the long-term release and leaching of cadmium and other toxic materials from upland disposal sites are essentially nonexistent. Mang and co-workers (1978) have proposed three evaluatory indices, based on the established water quality criteria, to ascertain relative mobilities, possible toxicities, and the potential degree of hazard caused by the leaching of various constituents from the dredged material at an upland disposal site.

6.1. Mobility Index

The mobility index (MI) is defined as the ratio of the soluble concentration of cadmium in the experimental dredged material disposal site soil leachate to that in the dredged material interstitial water:

$$MI = \frac{\text{soluble Cd conc. in disposal site leachate}}{\text{Cd conc. in dredged material interstitial water}}$$

The higher the MI value, the higher is the amount of cadmium released from the disposal site. The mobility index measures the attenuative capacity of the disposal site soil, which may be influenced by chemical characteristics of the soil, hydraulic properties, redox potential, pH, salinity, organic matter, and several other parameters affecting chemical transformations.

6.2. Evaluatory Index

The evaluatory index (EI) relates the concentration of cadmium in the disposal site leachate to the U.S. EPA drinking water standard and is defined by the following equation:

$$EI = \frac{\text{soluble Cd conc. in disposal site leachate}}{\text{Cd conc. in EPA drinking water standard}}$$

Since dilution is not considered, EI values give a "worst case" concept of the potential impact of cadmium, based on the criteria developed for drinking water.

6.3. Impact Index

The impact index (II) is the product of the mobility index and the evaluatory index for cadmium within the experimental period and is expressed as

$$II = (MI)(EI)$$

The impact index was developed to aid in evaluating the potential impact of dredged material leachates on the environment.

The value of each the three indices under the experimental conditions was less than 1.0, indicating that cadmium would not be a potential problem in the dredged material leachates at the disposal site (Mang et al., 1978). However, it must be pointed out that the interfacing soils in this laboratory-controlled experiment remained reduced during the length of the experiment and cadmium was precipitated as an insoluble CdS compound or as a complex with insoluble organic matter. In a typical upland disposal site with established vegetation, the disposal site sediments would slowly oxidize as a result of effluent discharge and O_2 diffusion. This might result in the transformation of insoluble CdS or organic-bound cadmium into more soluble, exchangeable, and carbonate compounds and hence to greater cadmium leaching out of the disposal site (Hoeppel et al., 1978).

These criteria can be used to qualitatively or semiquantitatively assess the initial impact of cadmium released during subaqueous or upland disposal. But criteria to predict the long-term effects of sediment-adsorbed cadmium on the benthic organisms in subaqueous disposal, or of uptake by plants, contamination of groundwater through leaching, and slow release of easily bioavailable cadmium to the adjacent surface waters under upland disposal are lacking. Therefore considerable attention should be focused on providing a reasonable estimate of cadmium release and subsequent biomagnification through the food web with time.

7. DISPOSAL ALTERNATIVES

The findings presented above strongly suggest that the disposal alternatives for dredged sediments contaminated with high concentrations of cadmium should minimize interaction with ecologically and economically important terrestrial, aquatic, and benthic populations. Gambrell et al. (1978) have prepared a manual of guidelines to aid in the selection of environmentally acceptable disposal alternatives that will minimize the adverse impact of toxic chemicals, including cadmium, under various disposal conditions. They recommend confinement of cadmium-contaminated dredged sediments in deep, low-energy waters where the metal will be less subject to wide dispersal, once deposited. Desirable sites include deep ocean and low-energy subaqueous depressions that would minimize cadmium uptake because of less abundant benthic populations. If freshwater lakes such as the Great Lakes are the only disposal alternatives, cadmium-contaminated dredged material should be contained in relatively low-energy, deep waters and should be covered with clean and cohesive material to minimize dispersal. A report on the chemical impacts of dredging and dredged material disposal in European waters also suggests careful study of the impact of heavy metals, especially cadmium, on lake water quality (Royal Adriaan Volker Group, 1977). Contaminated materials dumped in the lakes were covered with clean sand, silt, or plastic to reduce dispersion.

Based on the studies of dredging and open water disposal of San Francisco Harbor sediments into San Francisco Bay, The U.S. Army Corps of Engineers

(1977) suggests that unconfined wide dispersal of contaminated sediments in the high-energy estuarine zone would dilute toxic materials to very low levels because of the very low concentrations of toxic substances in the ocean water. Gambrell et al. (1978) contended that unconfined dispersal of cadmium-rich dredged materials in high-energy estuarine waters would result in greater interaction with O_2-rich overlying water over a large area and convert highly insoluble, solid-bound cadmium to more mobile forms. This readily bioavailable cadmium may be accumulated by aquatic organisms and enter the food chain under altered physicochemical conditions. Discussing the disposal of toxic metal wastes in estuaries, Davey (1976) warned against the disposal of contaminants in productive estuaries. Because heavy metals, especially cadmium and mercury, can be concentrated by geological, chemical, and especially biological processes in the sea, uptake by aquatic and benthic organisms may result in biomagnification in the food chain. Therefore the disposal of contaminated sediments in ecologically productive high-energy nearshore, estuarine, and inlet zones should be avoided.

Rhoads et al. (1978) suggest that dredged materials can be dumped into protected estuaries and embayments such as Long Island Sound in water depths greater than 20 m without much dispersal due to storm-generated currents. Capping contaminated materials with cleaner material was recommended to minimize bioptake of contaminants. However, no direct reference to cadmium was made in this article.

Confined land disposal of contaminated dredged materials has the greatest potential for releasing large amounts of cadmium in the environment. Cadmium may be transported to adjacent surface waters through effluent discharges containing high concentrations of more mobile cadmium forms associated with the suspended fine particles (Hoeppel et al., 1978; Gambrell et al., 1978). Also, because of slow oxidation of dredged material at the disposal site, cadmium may leach down the profile, contaminating groundwater. The eventual colonization of land disposal sites by plant species, animals, and birds may result in the accumulation of large quantities of cadmium and cause chronic health problems in the consumer organisms. Any land disposal method, therefore, must alleviate or minimize these potential adverse effects. Gambrell et al. (1978) suggest that an effective confined upland disposal site should be designed to allow enough residence time for the suspended particles to settle down since essentially all of the cadmium in the dredged slurry is associated in the particulates. For coarse-textured dredged materials, Mang et al. (1978) recommend the use of liners on the bottom and the dikes of the confined facility to alleviate the possibility of leaching and eventual contamination of groundwater. Since plant colonization at the disposal site is inevitable, the contaminated material should be covered with a thick layer of clean soil or dredged material to avoid excessive cadmium uptake. The agricultural use of dredged materials containing high concentrations of cadmium should be avoided because of excessive cadmium recycling in the food chain with potential harmful effects (Royal Adriaan Volker Group, 1977; Gambrell et al., 1978).

Carefully planned management practices to reduce cadmium release to the surrounding environment are essential for the successful operation of ecologically acceptable confined land disposal sites on a long-term basis. The nature of possible problems encountered subsequent to disposal operations and suggested solutions to minimize the impacts are more fully discussed in a report by Gambrell et al. (1978).

8. SUMMARY

The biological availability of cadmium depends on the chemical species present in both surface waters and sediment solids, not on the total metal ion concentration in the system. This study was undertaken, therefore, to review the existing knowledge on the chemical forms of cadmium present in sediment-water systems, and the effect of physicochemical parameters—pH, redox potential, salinity, and sediment-seawater mixing zone—on the chemical mobility of the cadmium in the system. The potential adverse environmental impact of cadmium release from contaminated dredged sediments during dredging and dredged material disposal was also discussed.

Studies using thermodynamic stability constants and the concentration of inorganic ions and ligands in the system indicate that dissolved cadmium in fresh water is generally present as the free Cd^{2+} ion. Cadmium carbonate and CdS were suggested as the solubility controlling solids under oxidized and reduced conditions, respectively. Stability constant computations also suggested that cadmium was strongly associated with chloride complexes in seawater. The effect of pH on cadmium speciation in seawater in the pH range of 7.0 to 9.0, generally encountered in the marine environment, was negligible because of the dominance of chloride ions. Thermodynamic equilibrium models based on inorganic systems were also used to calculate the chemical speciation of cadmium in freshwater-seawater mixtures. Increasing chloride concentration was accompanied by greater chloride complexing, and the main cadmium species predicted for the estuarine environment were $CdCl_2^0$, $CdCl_4^{2-}$, and $CdCl_3^-$. However, these thermodynamic models have limited application in cadmium speciation studies because of lack of stability constant data for natural organic ligands present in natural waters.

Chemical analysis of dissolved cadmium species in fresh waters by cadmium-specific ion electrode and anodic stripping voltametry confirm the results of thermodynamic stability constant computations in that cadmium was present as the free Cd^{2+} ion. The proportion of cadmium complexed with organic ligands under most natural conditions was small. However, an increase in the concentration of dissolved organic carbon, such as would occur during disposal of municipal wastewater effluents, resulted in higher concentration of organic-complexed cadmium and a decrease in free Cd^{2+} ion. Chemical analysis of various cadmium species in seawater with anodic stripping voltametry, by a recently developed analytical scheme using combinations of ultraviolet irradiation and separation by chelating agents, has shown that a larger fraction of dissolved cadmium was present as the labile organic species, while the cadmium bound as nonlabile organic and inorganic complexes was relatively small. These studies indicate that cadmium species in estuarine waters varied with water properties and were not dominated by chloride complexes, as indicated by stability constant computations. Practical applications of various methods employed to describe cadmium species in fresh and estuarine waters, and their advantages and limitations, are discussed in this chapter.

Cadmium associated with sediment solids may be present in several chemical forms, depending on the sediment composition and physicochemical properties. The predominant cadmium forms can be classified into exchangeable, carbonate, reducible, organic, sulfide, and mineral crystalline lattices phases. Chemical fractionation studies indicated that the most important sinks for cadmium in the reduced sediments were organics, sulfides carbonates, and manganese oxides. The fraction of cadmium associated with oxides and hydrous oxides of iron was small.

Redox potential, pH, and salinity are probably the most important physicochemical parameters controlling cadmium speciation in the sediment-water system and availability to biota. The results of research studies show that oxidation of reduced sediments results in the transformation of cadmium associated with insoluble sulfides and organic matter to more readily and potentially available exchangeable, carbonate, and organic-chelated fractions. A shift in sediment pH from alkaline to acidic conditions was also accompanied by an increase in the water-soluble, exchangeable, and chelated fractions and a decrease in cadmium adsorption. Thus the development of oxidized acidic conditions in contaminated sediments, such as may occur at upland disposal sites, may increase the potential for cadmium bioavailability. An increase in salinity may result in cadmium mobilization because of greater complexation of cadmium with chloride, and greater competition for exchange sites with high concentrations of Ca, Mg, and Na in saline waters. However, the effect of salinity on increased cadmium desorption may be offset by dilution of cadmium-contaminated freshwater sediments with estuarine waters of very low background cadmium levels. Studies on the effects of salinity and the sediment-seawater mixing zone on cadmium mobilization present conflicting results, and more research is needed for a better understanding of these interactions.

The research studies sponsored by the Dredged Material Research Program of the U.S. Army Corps of Engineers and others show that disposal of cadmium-contaminated sediments under subaqueous conditions does not cause significant short- or long-term water quality problems because of no or very little cadmium release. The upland disposal of cadmium-contaminated dredged materials may, however, result in the oxidation of reduced sediments and in a shift from unavailable sulfide- and organic-bound cadmium to more readily bioavailable forms, including exchangeable, carbonate, and easily reducible fractions. The potential short- and long-term adverse impacts of upland disposal may include cadmium uptake by plant communities established at the disposal site, groundwater leaching of cadmium solubilized under acid, oxidized conditions, and transport, due to erosion, of suspended material rich in cadmium to adjacent surface waters. The various criteria proposed to assess the potential impacts of cadmium under subaqueous and upland conditions are also discussed in this chapter.

The literature reports strongly suggest that disposal methods for dredged sediments contaminated with high concentrations of cadmium should be selected to minimize interaction with ecologically and economically important terrestrial, aquatic, and benthic populations. The suitability of various disposal alternatives to minimize the adverse environmental impact of cadmium is evaluated in this chapter.

ACKNOWLEDGMENTS

This study was supported by the Dredged Material Research Program of the Environmental Laboratory, U.S. Army Corps of Engineers Waterways Experiment Station, Vicksburg, Mississippi, under Contract DACW39-77-C-0054. The author wishes to express his appreciation to Drs. R. P. Gambrell, W. H. Patrick, Jr., and C. N. Reddy of the Center for Wetland Resources, Louisiana State University, for their critical review of the manuscript and constructive comments.

REFERENCES

Allen, H. E., Matson, W. R., and Mancy, K. H. (1970). "Trace Metal Characterization in Aquatic Environments by Anodic Stripping Voltametry," *J. Water Pollut. Control Fed.* **42**, 573–581.

Allen, H. E., Crosser, M. L., and Brisbin, T. D. (1975). "Metal Speciation in Aquatic Environments." In R. W. Andrew, P. V. Hodson, and D. E. Konosewich, Eds., *Toxicity to Biota of Metal Forms in Natural Waters*. Great Lakes Research Advisory Board, Minnesota, pp. 33–57.

Andelman, J. B. (1973). "Incidence, Variability, and Controlling Factors for Trace Elements in Natural Fresh Waters." In P. C. Singer, Ed., *Trace Metals and Metal-Organic Interactions in Natural Waters.* Ann Arbor Science Publishers, Ann Arbor, Mich., pp. 57–88.

Anderson, D. M. and Morel, F. F. M. (1978). "Copper Sensitivity of *Gonyaulax tamarensis,*" *Limnol. Oceanogr.* **23,** 283–295.

Andrew, R. W., Biesinger, K. E., and Glass, G. E. (1977). "Effect of Inorganic Complexing on the Toxicity of Copper to *Daphnia magna,*" *Water Res.* **11,** 309–315.

Ayling, G. M. (1974). "Uptake of Cadmium, Zinc, Copper, Lead, and Chromium in the Pacific Oysters, *Crassostrea gigas,* Grown in the Tamar River, Tasmania," *Water Resour. Res.* **8,** 729–738.

Baas Becking, L. G. M., Kaplan, I. R., and Moore, D. (1960). "Limits of the Natural Environment in Terms of pH and Oxidation-Reduction Potential," *J. Geol.* **68,** 243–284.

Bartlett, L., Rabe, F. W., and Funk, W. F. (1974). "Effects of Copper, Zinc, and Cadmium on *Salanastrum capricornutum,*" *Water Res.* **8,** 179–185.

Batley, G. E. and Florence, T. M. (1976). "Determination of the Chemical Forms of Dissolved Cadmium, Lead, and Copper in Seawater," *Mar. Chem.* **4,** 347–363.

Batley, G. E. and Garnder, D. (1978). "A Study of Copper, Lead, and Cadmium Speciation in Some Estuarine and Coastal Marine Waters," *Estuarine Coastal Mar. Sci.* **7,** 59–70.

Baumgartner, D. J., Shults, D. W., and Carkin, J. B. (1978). Aquatic *Disposal Field Investigation, Duwamish Waterways Disposal Site, Puget Sound, Washington. Appendix D: Chemical and Physical Analyses of Water and Sediment in Relation to Disposal of Dredged Material in Elliot Bay.* U.S. Army Engineer Waterways Experiment Station, Contract Rep. D-77-24, CE, Vicksburg, Miss.

Blom, B. E., Jenkins, T. F., Leggett, D. C., and Murrmann, R. P. (1976). *Effect of Sediment Organic Matter on Migration of Various Chemical Constituents during Disposal of Dredged Material.* U.S. Army Engineer Waterways Experiment Station, Contract Rep. D-76-7, CE, Vicksburg, Miss.

Boyd, M. B., Saucier, R. T., Keeley, J. W., Montgomery, R. L., Brown, R. D., Mathis, D. B., and Guice, C. L. (1972). *Disposal of Dredge Spoil; Problem Identification and Assessment and Research Program Development.* U.S. Army Engineer Waterways Experiment Station, Tech. Rep. H-72-8, CE, Vicksburg, Miss.

Brannon, J. M. (1978). *Evaluation of Dredged Material Pollution Potential.* U.S. Army Engineer Waterways Experiment Station, Tech. Rep. DS-78-6, CE, Vicksburg, Miss.

Brannon, J. M., Engler, R. M., Rose, J. R., Hunt, P. G., and Smith, I. (1976). *Selective Analytical Partitioning of Sediments to Evaluate Potential Mobility of Chemical Constituents during Dredging and Disposal Operations.* U.S. Army Engineer Waterways Experiment Station, Tech. Rep. D-76-7, CE, Vicksburg, Miss.

Brannon, J. M., Smith, I., and Plumb, R. H., Jr. (1978). *The Long-Term Release of Contaminants from Dredged Material.* U.S. Army Engineer Waterways Experiment Station, Tech. Rep. D-78-49, CE, Vicksburg, Miss.

Breger, I. A. (1968). "What You Don't Know Can Hurt You; Organic Colloids and Natural Waters," In D. W. Hood, Ed., *Proceedings of a Symposium on Organic Matter in Natural Waters*. University of Alaska, Fairbanks, Alaska, pp. 563–574.

Bryan, G. W. (1976). "Heavy Metal Contamination in the Sea." In R. Johnston, Ed., *Marine Pollution*. Academic Press, London, pp. 185–302.

Burks, S., and Engler, R. M. (1978). *Water Quality Impacts of Aquatic Dredged Material Disposal (Laboratory Investigations)*. U.S. Army Engineer Waterways Experiment Station, Tech. Rep. DS-78-4, CE, Vicksburg, Miss.

Canadian Department of National Health and Welfare (1969). *Canadian Drinking Water Standards and Objectives*. Department of National Health and Welfare, Ottawa, Ont.

Chau, Y. K. and Lum-Shue-Chan, K. (1974). "Determination of Labile and Strongly Bound Metals in Lake Water," *Water Res.* **8,** 383–388.

Cheam, V., Mudroch, A., Sly, P. G., and Lum-Shue-Chan, K. (1976). "Examination of the Elutriate Test, a Screening Procedure for Dredging Regulatory Criteria," *J. Great Lakes Res.* **2,** 272–282.

Chen, K. Y., Gupta, S. K., Sycip, A. Z., Lu, J. C. S., Knezevic, M., and Choi, Won-Wook (1976). *Research Study on the Effect of Dispersion, Settling, and Resedimentation on Migration of Chemical Constituents during Open-Water Disposal of Dredged Materials*. U.S. Army Engineer Waterways Experiment Station, Contract Rep. D-76-1, CE, Vicksburg, Miss.

Chen, K. Y., Eichenberger, B., and Mang, J. L. (1978). *Confined Disposal Area Effluent and Leachate Control (Laboratory and Field Investigations)*. U.S. Army Engineer Waterways Experiment Station, Tech. Rep. DS-78-7, CE, Vicksburg, Miss.

Chester, R. and Hughes, M. J. (1967). "A Chemical Technique for the Separation of Ferro-manganese Minerals, Carbonate Minerals, and Adsorbed Trace Elements from Pelagic Sediments," *Chem. Geol.* **2,** 249–262.

Cossa, D. (1976). "Sorption of Cadmium by a Population of the Diatom *Phaeodactylum tricornutum* in Culture," *Mar. Biol.* **34,** 163–167.

Cumme, G. A. (1973). "Calculation of Chemical Equilibrium Concentrations of Complexing Ligands and Metals: A Flexible Computer Program Taking into Account Uncertainty in Formation Constants," *Talanta* **20,** 1009–1016.

Davey, E. W. (1976). "Trace Metals in the Oceans: Problem or Not?" In *Water Quality Criteria Research of the U.S. Environmental Protection Agency*. U.S. Environmental Protection Agency, EPA-600/3-76-079, Corvallis, Ore., pp. 13–22.

De Groot, A. J. (1973). "Occurrence and Behavior of Heavy Metals in River Deltas with Special Reference to the Rivers Rhine and Ems." In R. E. Goldberg, Ed., *North Sea Science*. M.I.T. Press, Cambridge, Mass., pp. 308–325.

De Groot, A. J. (1977). *Assessment of Certain European Dredging Practices and Dredged Material Containment and Reclamation Methods*. Appendix E: *Heavy Metals in the Dutch Delta, an Integrated Program for Research*. Final Report, Contact DAJA 37-75-C-0382, U.S. Army Engineer Waterways Experiment Station, CE, Vicksburg, Miss.

De Groot, A. J. and Allersma, E. (1975). "Field Observations on the Transport of Heavy Metals in Sediments," *Prog. Water Technol.* **7,** 85–95.

De Groot, A. J., Salomons, W., and Allersma, E. (1976). "Processes Affecting Heavy Metals in Estuarine Sediments." In J. D. Burton and P. S. Liss, Eds., *Estuarine Chemistry.* Academic Press, London, pp. 131–157.

Duinker, J. C. and Kramer, J. C. M. (1977). "An Experimental Study on the Speciation of Dissolved Zinc, Cadmium, Lead, and Copper in River Rhine and North Sea Water, by Differential Pulsed Anodic Stripping Voltametry," *Mar. Chem.* **5,** 207–228.

Duinker, J. C. and Nolting, R. F. (1977). "Dissolved and Particulate Trace Metals in the Rhine Estuary, and the Southern Bight," *Mar. Pollut. Bull.* **8,** 65–69.

Emerson, S. (1976). "Early Diagenesis of Anaerobic Lake Sediments: Chemical Equilibria in Interstitial Waters," *Geochim. Cosmochim. Acta* **40,** 925–934.

Engler, R. M. and Wilkes, F. G. (1977). *Ecological Evaluation of Proposed Discharge of Dredged Material into Ocean Waters.* U.S. Army Engineer Waterways Experiment Station, CE, Vicksburg, Miss.

Engler, R. M., Brannon, J. M., and Rose, J. (1977). "A Practical Selective Extraction Procedure for Sediment Characterization." In T. F. Yen, Ed., *Chemistry of Marine Sediment.* Ann Arbor Science Publishers, Ann Arbor, Mich., pp. 163–171.

Environmental Effects Laboratory (1976). *Ecological Evaluation of Proposed Discharge of Dredged or Fill Material into Navigable Waters.* U.S. Army Engineer Waterways Experiment Station, Misc. Pap. D-76-16, CE, Vicksburg, Miss.

Florence, T. M. (1977). "Trace Metal Species in Fresh Waters," *Water Res.* **11,** 681–687.

Florence, T. M. and Batley, G. E. (1976). "Trace Metals Species in Seawater. I: Removal of Trace Metals from Seawater by a Chelating Resin," *Talanta* **23,** 179–186.

Florence, T. M. and Batley, G. E. (1977). "Determination of the Chemical Forms of Trace Metals in Natural Waters, with Special Reference to Copper, Lead, Cadmium, and Zinc," *Talanta* **24,** 151–158.

Forstner, U. (1977). "Forms and Sediment Associations of Nutrients (C, N, and P), Pesticides and Metals: Trace Metals." In H. Shear and A. E. P. Watson, Eds., *Proceedings of Workshop on the Fluvial Transport of Sediment-Associated Nutrients and Contaminants.* Research Advisory Board, International Joint Commission, Kitchener, Ont., Canada, pp. 219–232.

Friberg, L., Piscator, M., and Nordberg, G. (1971). *Cadmium in the Environment.* CRC Press, Cleveland, Ohio, pp. 1–166.

Friberg, L., Piscator, M., Nordberg, G., and Kjellstrom, T. (1973). *Cadmium in the Environment,* II. U.S. Environmental Protection Agency, EPA-R2-73-190, Research Triangle Park, North Carolina, pp. 1–147.

Frost, R. R. and Griffin, R. A. (1977). "Effect of pH on Adsorption of Copper, Zinc, and Cadmium from Landfill Leachates by Clay Minerals," *J. Environ. Sci. Health* **A12,** 139–156.

Gadde, R. R. and Laitinen, H. A. (1974). "Studies of Heavy Metal Adsorption by Hydrous Iron and Manganese Oxides," *Anal. Chem.* **46,** 2022–2026.

Gambrell, R. P., Khalid, R. A., and Patrick, W. H., Jr. (1976). "Physicochemical Parameters That Regulate Mobilization and Immobilization of Toxic Heavy Metals." In P. A. Krenkel, J. Harrison, and J. C. Burdick, Eds., *Dredging and Its Environmental Effects*. American Society of Civil Engineers, New York, pp. 418–434.

Gambrell, R. P., Khalid, R. A., Verloo, M. G., and Patrick, W. H., Jr. (1977a). *Transformations of Heavy Metals and Plant Nutrients in Dredged Sediments as Affected by Oxidation-Reduction Potential and pH. Vol. II: Materials and Methods, Results and Discussion*. U.S. Army Engineer Waterways Experiment Station, Contract Rep. D-77-4, CE, Vicksburg, Miss.

Gambrell, R. P., Collard, V. R., Reddy, C. N., and Patrick, W. H., Jr. (1977b). *Trace and Toxic Metal Uptake by Marsh Plants as Affected by Eh, pH, and Salinity*. U.S. Army Engineer Waterways Experiment Station, Tech. Rep. D-77-40, CE, Vicksburg, Miss.

Gambrell, R. P., Khalid, R. A., and Patrick, W. H., Jr. (1978). *Disposal Alternatives for Contaminated Dredged Material as a Management Tool to Minimize Adverse Environmental Effects*. U.S. Army Engineer Waterways Experiment Sation, Tech. Rep. DS-78-8, CE, Vicksburg, Miss.

Gardiner, J. (1974a). "The Chemistry of Cadmium in Natural Water. I: A Study of Cadmium Complex Formation Using the Cadmium Specific Ion Electrode," *Water Res.* **8**, 23–30.

Gardiner, J. (1974b). "The Chemistry of Cadmium in Natural Water. II: The Adsorption of Cadmium on River Muds and Naturally Occurring Solids," *Water Resour. Res.* **8**, 157–164.

Gardiner, J. (1975). "The Complexation of Trace Metals in Natural Waters." In T. C. Hutchinson, Ed., *International Conference on Heavy Metals in the Environment*. University of Toronto, Canada, pp. 303–318.

Gardiner, J. and Stiff, M. J. (1975). "The Determination of Cadmium, Lead, Copper, and Zinc in Groundwater, Estuarine Water, Sewage and Sewage Effluent by Anodic Stripping Voltammetry," *Water Res.* **9**, 517–523.

Giesy, J. P., Leversee, G. L., and Williams, D. R. (1977). "Effects of Naturally Occurring Aquatic Organic Fractions on Cadmium Toxicity to *Simochephalus serrulatus* (Daphnidae) and *Gambusia affinis* (Polciludae)," *Water Res.* **11**, 1013–1020.

Goldberg, E. D. and Arrhenius, G. B. S. (1958). "Chemistry of Pacific Pelagic Sediments," *Geochim. Cosmochim. Acta* **13**, 153–212.

Gong, H., Rose, A. W., and Suhr, N. H. (1977). "The Geochemistry of Cadmium in Some Sedimentary Rocks," *Geochim. Cosmochim. Acta* **41**, 1692–1696.

Gupta, S. K. and Chen, K. Y. (1975). "Partitioning of Trace Metals in Selective Chemical Fractions of Nearshore Sediments," *Environ. Lett.* **10**, 129–158.

Guy, R. D. and Chakrabarti, C. L. (1975). "Analytical Techniques for Speciation of Trace Metals." In T. C. Hutchinson, Ed., *International Conference on Heavy Metals in the Environment*. University of Toronto, Canada, pp. 275–293.

Guy, R. D. and Chakrabarti, C. L. (1976). "Analytical Techniques for Speciation of Heavy Metal Ions in Aquatic Environment," *Chem. Can.* **28**, 26–29.

Haghiri, F. (1974). "Plant Uptake of Cadmium as Influenced by Cation Exchange Capacity, Organic Matter, Zinc, and Soil Temperature," *J. Environ. Qual.* **3,** 180–183.

Hahne, H. C. H. and Kroontje, W. (1973). "Significance of pH and Chloride Concentration on Behavior of Heavy Metal Pollutants: Mercury(II), Cadmium(II), Zinc(II), and Lead(II)," *J. Environ. Qual.* **2,** 444–450.

Helz, G. R., Huggett, R. J., and Hill, J. M. (1975). "Behavior of Mn, Fe, Cu, Zn, Cd, and Pb Discharged from a Wastewater Treatment Plant into an Estuarine Environment," *Water Res.* **9,** 631–636.

Hem. J. D. (1972). "Chemistry and Occurrence of Cadmium and Zinc in Surface Water and Groundwater," *Water Resour. Res.* **8,** 661–679.

Hoeppel, R. E., Meyers, T. E., and Engler, R. M. (1978). *Physical and Chemical Characterization of Dredged Material Influents and Effluents in Confined Land Disposal Areas.* U.S. Army Engineer Waterways Experiment Station, Tech. Rep. D-78-24, CE, Vicksburg, Miss.

Holmes, C. W., Slade, E. A., and McLerran, C. J. (1974). "Migration and Redistribution of Zinc and Cadmium in Marine Estuarine System," *Environ. Sci. Technol.* **8,** 255–259.

Huang, C. P., Elliot, H. A., and Ashmead, R. M. (1977). "Interfacial Reaction and the Fate of Heavy Metals in Soil-Water Systems," *J. Water Pollut. Control Fed.* **40,** 745–756.

Iskander, I. K. and Keeney, D. R. (1974). "Concentration of Heavy Metals in Sediment Cores from Selected Wisconsin Lakes," *Environ. Sci. Technol.* **8,** 165–170.

Ito, H. and Iimura, K. (1975). "Adsorption of Cadmium by Rice Plants in Response to Change of Oxidation-Reduction Conditions of Soils," *J. Sci. Soil Manure Jap.* **46,** 82–88.

Jackson, M. L. (1958). *Soil Chemical Analysis.* Prentice-Hall, Englewood Cliffs, N.J., pp. 183–204.

James, A. D. and Leckie, J. O. (1977). "The Effect of Complexing Ligands on Trace Metal Adsorption at the Sediment/Water." In W. E. Krumbein, Ed., *Environmental Biogeochemistry and Geomicrobiology,* Vol. 3: *Methods, Metals and Assessment.* Ann Arbor Science Publishers, Ann Arbor, Mich., pp. 1009–1024.

Jenne, E. A. (1968). "Controls of Mn, Fe, Co, Ni, Cu, and Zn Concentrations in Soils and Water: The Significant Role of Hydrous Mn and Fe Oxides." In R. A. Baker, Ed., *Trace Inorganics in Water.* Advances in Chemistry Series 73 American Chemical Society, Washington, D.C., pp. 337–388.

Jenne, E. A. and Luoma, S. N. (1977). "Forms of Trace Elements in Soils, Sediments and Associated Waters: An Overview of Their Determination and Biological Availability." In H. Drucker and R. E. Wildung, Eds., *Biological Implications of Metals in the Environment.* ERDA Symposium Series 42, Energy Research and Development Administration, Oak Ridge, Tenn., pp. 110–143.

John, M. K. (1971). "Influence of Soil Characteristics on Adsorption and Desorption of Cadmium," *Environ. Lett.* **2,** 173–179.

Keeley, J. W. and Engler, R. M. (1974). *Discussion of Regulatory Criteria for Ocean Disposal of Dredged Materials: Elutriate Test Rationale and Implementation Guidelines.* U.S. Army Engineer Waterways Experiment Station, Misc. Pap. D-74-14, CE, Vicksburg, Miss.

Kester, D. R. (1975). "Chemical Speciation in Seawater—Group Report." In E. D. Goldberg, Ed., *The Nature of Seawater.* Report of the Dahlem Workshop, Dahlem Konferenzen, Berlin, pp. 17–41.

Khalid, R. A., Gambrell, R. P., Verloo, M. G., and Patrick, W. H., Jr. (1977). *Transformations of Heavy Metals and Plant Nutrients in Dredged Sediments as Affected by Oxidation-Reduction Potential and pH,* Vol. I: *Literature Review.* U.S. Army Engineer Waterways Experiment Station, Contract Rep. D-77-4, CE, Vicksburg, Miss.

Khalid, R. A., Gambrell, R. P., and Patrick, W. H., Jr. (1978a). "Chemical Transformations of Cadmium and Zinc in Mississippi River Sediments as Influenced by pH and Redox Potential." In D. C. Adriano and I. L. Brisbin, Eds., *Environmental Chemistry and Cycling Processes.* Proceedings of 2nd Mineral Cycling Symposium, Energy Research Development Administration, Oak Ridge, Tenn., pp. 417–433.

Khalid, R. A., Patrick, W. H., Jr., and Gambrell, R. P. (1978b). "Effect of Dissolved Oxygen on Chemical Transformations of Heavy Metals, Phosphorus and Nitrogen in an Estuarine Sediment," *Estuarine Coastal Mar. Sci.* **6,** 21–35.

Kinniburgh, D. G., Sridhar, K., and Jackson, M. L. (1977). "Specific Adsorption of Zinc and Cadmium by Iron and Aluminum Hydrous Oxides," In H. Drucker and R. E. Wildung, Eds., *Biological Implications of Metals in the Environment.* ERDA Symposium Series 42, Energy Research and Development Administration, Oak Ridge, Tenn., pp. 231–239.

Kopp, J. F. and Kroner, R. C. (1970). *Trace Metals in the Waters of the United States: A Five Year Summary of Trace Metals in Rivers and Lakes of the United States.* U.S. Department of Interior, Federal Water Pollution Control Administration, Division of Pollution Surveillance, Cincinnati, Ohio.

Krauskopf, K. B. (1956). "Factors Controlling the Concentration of Thirteen Rare Metals in Sea Water," *Geochim. Cosmochim. Acta* **9,** 1–32.

Krauskopf, K. B. (1967). *Introduction to Geochemistry.* McGraw-Hill, New York, pp. 236–280.

Kubota, J., Mills, E. L., and Oglesby, R. T. (1974). "Lead, Cd, Zn, Cu, and Co in Streams and Lake Waters of Cayuga Lake Basin, New York," *Environ. Sci. Technol.* **8,** 243–248.

Leckie, J. O. and James, R. O. (1974). "Control Mechanisms for Trace Metals in Natural Waters." In A. J. Rubin, Ed., *Aqueous Environmental Chemistry of Metals.* Ann Arbor Science Publishers, Ann Arbor, Mich., pp. 1–76.

Lee, G. F. (1975). "Role of Hydrous Metal Oxides in the Transport of Heavy Metals in the Environment." In P. A. Krenkel, Ed., *Heavy Metals in the Aquatic Environment.* Pergamon Press, Oxford, England, pp. 137–154.

Lee, G. F. and Plumb, R. H. (1974). *Literature Review on Research Study for the De-*

velopment of Dredged Material Disposal Criteria. U.S. Army Engineer Waterways Experiment Station, Contract Rep. D-74-1, CE, Vicksburg, Miss.

Lee, G. F., Piwoni, M. D., Lopez, J. M., Mariani, G. M., Richardson, J. S., Homer, D. H., and Saleh, F. (1975). *Research Study for the Development of Dredged Material Disposal Criteria.* U.S. Army Engineer Waterways Experiment Station, Contract Rep. D-75-4, CE, Vicksburg, Miss.

Lee, C. R., Smart, R. M., Sturgis, T. C., Gordon, R. N., Sr., and Landin, M. C. (1978). *Evaluation of Heavy Metal Uptake by Marsh Plants in Relation to Chemical Extraction Procedures for Productive Purposes.* U.S. Army Engineer Waterways Experiment Station, Tech. Rep. D-77-6, CE, Vicksburg, Miss.

Leland, H. V., Luoma, S. N., Elder, J. F., and Wilkes, D. T. (1978). "Heavy Metals and Related Trace Elements," *J. Water Pollut. Control Fed.* **50,** 1469–1512.

Long, D. T. and Angino, E. E. (1977). "Chemical Speciation of Cd, Cu, Pb, and Zn in Mixed Freshwater, Seawater and Brine Solutions," *Geochim. Cosmochim. Acta* **41,** 1183–1191.

Lu, J. C. S. (1976). "Studies on the Long-Term Migration and Transformation of Trace Metals in the Polluted Marine Sediment-Water System." Ph.D. Dissertation, University of Southern California, Los Angeles, Calif., pp. 1–310.

Lu, J. C. S. and Chen, K. Y. (1977). "Migration of Trace Metals in Interfaces of Seawater and Polluted Surficial Sediments," *Environ. Sci. Technol.* **11,** 174–181.

Lu, J. C. S., Eichenberger, B., and Chen, K. Y. (1978). *Characterization of Confined Disposal Area Influent and Effluent Particulate and Petroleum Fractions.* U.S. Army Engineer Waterways Experiment Station, Tech. Rep. D-78-16, CE, Vicksburg, Miss.

Maljovic, D. and Branica, M. (1971). "Polarography of Seawater. II: Complex Formation of Cadmium with EDTA," *Limnol. Oceanogr.* **16,** 779–785.

Mang, J. L., Lu, J. C. S., Lofy, R. J., and Stearns, R. P. (1978). *A Study of Leachate from Dredged Material in Upland Areas and/or in Productive Uses.* U.S. Army Engineer Waterways Experiment Station, Contract Rep. D-78-20, CE, Vicksburg, Miss.

Mantoura, R. F. C., Dickson, A., and Riley, J. P. (1978). "The Complexation of Metals with Humic Materials in Natural Waters," *Estuarine Coastal Mar. Sci.* **6,** 387–408.

May, E. B. (1974). "Effects of Water Quality When Dredging a Polluted Harbor Using Confined Spoil Disposal," *Ala. Mar. Res. Bull.* **10,** 1–8.

McLaren, R. G. and Crawford, D. V. (1973). "Studies on Soil Copper. I: The Fractionation of Copper in Soils," *J. Soil Sci.* **24,** 172–181.

Morel, F., McDuff, R. E., and Morgan, J. J. (1973). "Interactions and Chemostasis in Aquatic Chemical Systems: Role of pH, pE, Solubility, and Complexation." In P. C. Singer, Ed., *Trace Metals and Metal-Organic Interactions in Natural Waters.* Ann Arbor Science Publishers, Ann Arbor, Mich., pp. 157–200.

Mudroch, A. and Cheam, V. (1974). "Chemical Changes in Pore Water from Dredge

Spoil Disposed on the Pilot Island Mitchell Bay, Lake St. Clair, Ontario." In *House Report.* Canada Center for Inland Waters, Burlington, Ont.

Muller, G. and Forstner, V. (1975). "Heavy Metals in Sediments of the Rhine and Elbe Estuaries: Mobilization or Mixing Effect," *Environ. Geol.* **1**, 33–39.

Munshower, F. F. (1977). "Cadmium Accumulation in Plants and Animals of Polluted and Non-Polluted Grasslands," *J. Environ. Qual.* **6**, 411–417.

Neff, J. W., Foster, R. S., and Slowey, J. F. (1978). *Availability of Sediment-Adsorbed Heavy Metals to Benthos with Particular Emphasis on Deposit-Feeding in Fauna.* U.S. Army Engineers Waterways Experiment Station, Tech. Rep. D-78-42, CE, Vicksburg, Miss.

Nissenbaum, A. (1972). "Distribution of Several Metals in Chemical Fractions of Sediment Core from the Sea of Okhotsk," *Israel J. Earth-Sci.* **21**, 143–154.

Page, A. L. and Bingham, F. T. (1973). "Cadmium Residues in the Environment," Residue Rev. **48**, 1–44.

Phillips, D. J. H. (1977). "The Use of Biological Indicator Organisms to Monitor Trace Metal Pollution in Marine and Estuarine Environment—A Review," *Environ. Pollut.* **13**, 281–317.

Poldoski, J. E. and Glass, G. E. (1975). "Considerations of Trace Element Chemistry for Streams in the Minnesota-Ontario Border Area." In T. C. Hutchinson, Ed., *International Conference on Heavy Metals in the Environment,* University of Toronto, Canada, pp. 901–921.

Ponnamperuma, F. M. (1972). "The Chemistry of Submerged Soils," *Adv. Agron.* **24**, 29–88.

Presley, B. J., Kolodry, Y., Nissenbaum, A., and Kaplan, I. R. (1972). "Early Diagenesis in a Reducing Fjord, Saanich Inlet, British Columbia. II: Trace Element Distribution in Interstitial Water and Sediment," *Geochim. Cosmochim. Acta* **36**, 1073–1090.

Reddy, C. N. and Patrick, W. H., Jr. (1977). "Effect of Redox Potential and pH on the Uptake of Cadmium and Lead by Rice," *J. Environ. Qual.* **6**, 259–262.

Rhoads, D. C., McCall, P. L., and Yingst, J. Y. (1978). "Disturbance and Production on the Estuarine Seafloor," *Am. Sci.* **66**, 577–586.

Ritchie, G. A. and Speakman, J. N. (1973). "Effects of Settling Time on Quality of Supernatant from Upland Dredge Disposal Facilities." In *Proceedings of the 16th Conference on Great Lakes Research,* pp. 321–328.

Rohatgi, N. and Chen, K. Y. (1975). "Transport of Trace Metals by Suspended Particulates on Mixing with Seawater," *J. Water Pollut. Control. Fed.* **47**, 2298–2316.

Royal Adriaan Volker Group (1977). *Assessment of Certain European Dredging Practices and Dredged Material Containment and Reclamation Method.* Final Report, Contract DAJA 37-75-0382, U.S. Army Engineer Waterways Experiment Station, CE, Vicksburg, Mississippi.

Salomons, W. and De Groot, A. J. (1977). *Pollution History of Trace Metals in Sediments, as Affected by the Rhine River.* Delft Hydraulics Laboratory, Pub. 84, Delft, The Netherlands.

Salomons, W. and Mook, W. G. (1977). "Trace Metal Concentrations in Estuarine Sediments: Mobilization, Mixing or Precipitation," *Netherlands J. Sea Res.* **11**, 199–209.

Sarfield, L. J. and Mancy, K. H. (1977). "The Properties of Cadmium Complexes and Their Effect on Toxicity to a Biological System." In H. Drucker and R. E. Wildung, Eds., *Biological Implications of Metals in the Environment.* ERDA Symposium Series 42, Energy Research and Development Administration, Oak Ridge, Tenn., pp. 335–345.

Serne, R. J. and Mercer, B. W. (1975) *Dredge Disposal Study, San Francisco Bay and Estuary.* Appendix F: *Crystalline Matrix Study.* U.S. Army Engineer District, San Francisco, Calif.

Sharp, J. H. (1973). "Size Classes of Organic Carbon in Seawater," *Limnol. Oceanogr.* **18**, 441–447.

Shuman, M. S. and Fogleman, W. W. (1978). "Nature and Analysis of Chemical Species—Inorganics," *J. Water Pollut. Control Fed.* **50**, 1000–1021.

Sibley, T. H. and Morgan, J. J. (1975). "Equilibrium Speciation of Trace Metals in Freshwater:seawater Mixtures." In T. C. Hutchinson, Ed., *International Conference on Heavy Metals in the Environment.* University of Toronto, Canada, pp. 319–340.

Siegel, A. (1971). "Metal Organic Interactions in the Marine Environment." In S. D. Faust and J. V. Hunter, Eds., *Organic Compounds in Aquatic Environments.* Marcel Dekker, New York, pp. 265–275.

Stevenson, F. J. and Ardakani, M. S. (1972). "Organic Matter Reactions Involving Micronutrients in Soils." In R. C. Dinauer, Ed., *Micronutrients in Agriculture.* Soil Science Society of America, Madison, Wis., pp. 79–114.

Stumm, W. and Morgan, J. J. (1970). *Aquatic Chemistry.* Wiley-Interscience, New York, pp. 300–382.

Sunda, W. G., Engel, D. W., and Thuotte, R. M. (1978). "Effect of Chemical Speciation on Toxicity of Cadmium to Grass Shrimp, *Palaemonetes pugio:* Importance of Free Cadmium Ion," *Environ. Sci. Technol.* **12**, 409–413.

U.S. Army Corps of Engineers (1977). *Dredge Disposal Study, San Francisco Bay and Estuary.* Final Report, U.S. Army Engineer District, San Francisco, Calif.

U.S. Environmental Protection Agency (1973a). *Proposed Criteria for Water Quality,* 2 vols. Washington, D.C.

U.S. Environmental Protection Agency (1973b). *Ocean Dumping Criteria.* Fed. Reg. 38, 12872–12877.

U.S. Environmental Protection Agency (1973c). *Ocean Dumping Final Criteria.* Fed. Reg. 38, 28610–28621.

U.S. Environmental Protection Agency (1976). *Quality Criteria for Water.* EPA 440/9-76-023, Washington, D.C.

U.S. Environmental Protection Agency (1977). *National Secondary Drinking Water Regulations.* Fed. Reg. 42, 17143–17147.

U.S. Public Health Service (1962). *Drinking Water Standards.* U.S. PHS Publ. 956. Washington, D.C.

Weber, W. J. and Posselt, H. S. (1974). "Equilibrium Models and Precipitation Reactions for Cadmium(II)." In A. J. Rubin, Ed., *Aqueous-Environmental Chemistry of Metals.* Ann Arbor Science Publishers, Ann Arbor, Mich., pp. 255–290.

Wier, C. F. and Walter, W. M. (1976). "Toxicity of Cadmium in the Freshwater Snail, *Physa gyrina* Say," *J. Environ. Qual.* **5,** 359–362.

Windom, H. L. (1976). *Geochemical Interactions of Heavy Metals in Southeastern Salt Marsh Environment.* U.S. Environmental Protection Agency, EPA-600/3-76-023, Corvallis, Ore.

Woodruff, B., Horwood, D., Lee, B., Lehance, R., and Lumbers, J. (1974). "Toxic Metals in Tasmanian Rivers," *CSIRO Environ. Res.* **1,** 3–10.

Wyeth, R. K. and Sweeney, R. A. (1978). *Aquatic Disposal Field Investigations, Ashtabula River Disposal Site, Ohio.* Appendix C: *Investigation of Water Quality and Sediment Parameters.* U.S. Army Engineer Waterways Experiment Station, Tech. Rep. D-77-42, CE, Vicksburg, Mississippi.

Yamagata, N. and Shigematsu, I. (1970). "Cadmium Pollution in Perspective. *Inst. Publ. Health Tokyo Bull.* **19,** 1–27.

Zirino, A. and Yamamoto, S. (1972). "A pH-Dependent Model for the Chemical Speciation of Copper, Zinc, Cadmium and Lead in Seawater," *Limnol. Oceanogr.* **17,** 661–671.

9

CADMIUM IN POLLUTED SEDIMENTS

Ulrich Förstner

Institut für Sedimentforschung der Universität, Heidelberg, Federal Republic of Germany

1.   INTRODUCTION

Water samples, suspended particulates, bottom sediments, and indicator organisms such as algae and mussels are some of the various media analyzed to assess, monitor, and control pollution effects in aquatic systems. Although the concentration of a pollutant in the aqueous phase usually is more relevant with regard to its possible toxic effects than are the contents of the respective solid phases, the assessment of water quality data and the identification of pollution sources from water analysis are often very time consuming and sometimes inconclusive. This is mainly due to fluctuations, within short time intervals, of dissolved constituents. Such vacillation is attributable to a large number of variables, such as daily and seasonal variations in water flow, surreptitious local discharges of effluents, changing pH and redox conditions, detergent levels in the water, salinity, and temperature.

It has been established from hundreds of investigations, however, that *sediment analyses* can be particularly useful in detecting pollution sources (and in the selection of critical sites for routine water sampling) for contaminants, which, upon being discharged to surface waters, do not remain solubilized, since they are rapidly adsorbed by particulate matter, and may thus escape detection by water analyses. During periods of reduced rates of flow, suspended material settles to the bed of the river, lake, or sea, becoming partially incorporated into the bottom sediment. By virtue of their composition, sediments "express the state of a water body" (Züllig, 1956). With lateral distributions (quality profiles), local sources of pollution can be determined and evaluated. Vertical sediment profiles (cores) also are useful because they often uniquely preserve the historical sequence of pollution intensities, and at the same time enable a reasonable es-

Table 1. **Average Stream Sediment and Water Concentrations of Selected Trace Elements in Mineralized and Unmineralized Areas as a Basis for the Evaluation of Tentative Threshold Values in Stream Sediments (Last Column)**[a]

Element	Tributary A (mineralized)		Tributary B (unmineralized)		HDL Value (μg/l)	Tentative Threshold Sediment Value (ppm)
	Sediment (ppm)	Water (μg/l)	Sediment (ppm)	Water (μg/l)		
Zinc	2750	2000	320	12	5000	2000
Copper	3245	515	89	8	50	1000
Lead	291	4	89	1.5	50	500
Cadmium	3	5	2	1	10	10

[a] After Aston and Thornton (1977).

timation to be made of the *background* level and the input variations of a pollutant over an extended period of time; this approach is particularly useful if the rate of sedimentation is known (Syers et al., 1973; Förstner, 1976).

2. SEDIMENT-WATER INTERRELATIONS

Although sediment analyses do not furnish quantitative data on the absolute degree of pollution, they can play a key role in ascertaining relative factors of enrichment whereby sources of pollution in the aquatic environment may be traced and monitored. Investigations performed by Aston and Thornton (1977) on stream sediments of both mineralized and unmineralized areas in Great Britain have shown that sediment analyses may also provide a qualitative approximation of the metal concentrations of associated waters (Table 1). The salient factors are the degree of "contrast" and the assignment of "threshhold" values; "contrast" is defined as the highest anomalous concentration divided by the average background concentration. It was found that high contrast values are readily identified for contaminated areas; hence the most useful contrast value relates to the "highest desirable level" (HDL) for the particular trace element in water. Table 1 indicates the concentration levels in sediments at which it is likely that associated waters may occasionally exceed the HDL value (Aston and Thornton, 1977). However, because of considerable variations in the physical and chemical environments of the catchment areas as well as of the aquatic system proper, it is difficult to establish universally acceptable contrast and threshhold levels. In the case of *cadmium,* the tentative threshhold is reached at approximately 5 to 10 ppm Cd, depending on the drinking water standard,

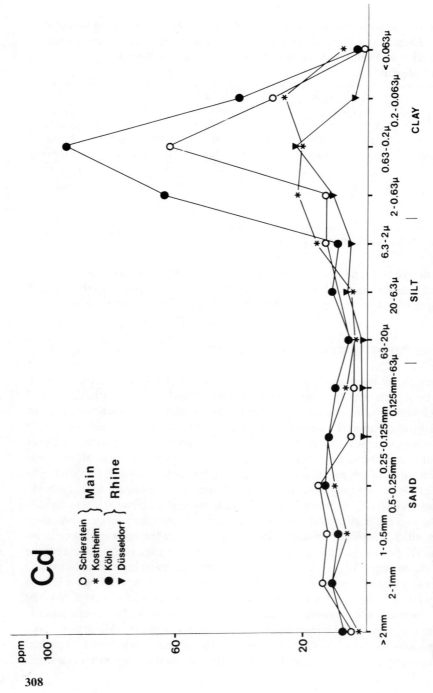

Figure 1. Grain size dependencies of cadmium concentrations in sediment samples from the Main and Rhine Rivers, Federal Republic of Germany. Samples collected by members of the working group "Metals," German Research Society, 1976. (Förstner, 1977a.)

Table 2.   Reduction of Grain Size Effects[a]

Grain size distribution (μm)	
Extrapolation (<16)	De Groot et al. (1971)
Separation (<2)	Banat et al. (1972a,b)
(<63)	Allan et al. (1972)
(<200)	Aston et al. (1974)
Acid-soluble fraction	
0.1 M hydrochloric acid	Piper (1971)
0.3 M hydrochloric acid	Malo (1977)
HCl-HNO$_3$ (1–1) digestion	Anderson (1974)
Mineral separation	
Quartz correction	Thomas (1972)
Comparison with conservative elements	
Ratio of X to aluminum	Bruland et al. (1974)
	Kemp et al. (1976)

[a] From Förstner (1977b).

which is now commonly set at either 5 or 10 μg/l in different countries (Eriksson, 1977; Hattingh, 1977).

3.   GRAIN SIZE EFFECTS INFLUENCING METAL CONCENTRATIONS

Two effects must be considered, however, when sediment analyses are used for the identification of sources of pollution, as well as for the evaluation of the historical record of pollution intensities. Both of these effects are particularly important in fluviatile environments, but may also be relevant in some nearshore marine and lacustrine deposits. First, under conditions of high water discharge, erosion of the bottom sediments can take place, generally leading to a lower degree of local contamination. Second (a detail that is often overlooked), it is imperative to base metal analyses of sediments on a standardized procedure with regard to particle size, since there is a marked decrease in the content of metals as sediments particle size increases. An example is given in Figure 1, which indicates the concentrations of cadmium in different grain size fractions of sediments from the highly polluted Rhine and Main rivers in the Federal Republic of Germany. Maximum concentrations of cadmium occur in the fractions of 0.2 to 0.6 μm; from there the metal contents decrease toward both finer and coarser grain diameters (relatively high levels of cadmium in the sand fractions are probably due to the input of coarse waste particles).

It is clearly demonstrated that, without a correction for grain size influences, metal analyses from bulk sediment samples may be of little significance for monitoring purposes, particularly samples from river deposits undergoing strong granulometric changes. In practice, several methods (examples are given in Table 2) are used to reduce grain size effects, that is, to include most of the substances active in metal enrichment, such as hydrated oxides, sulfides, and amorphous and organic material, but to exclude coarser grained sediment fractions, which are largely chemically inert. In Sections 7 to 9, where a collection of typical examples of sediment studies on cadmium pollution from different aquatic environments is presented (Tables 5 to 7), the specific methods to reduce grain size effects will be indicated.

4.   FACTORS INFLUENCING THE DISTRIBUTION OF CADMIUM IN NATURAL SEDIMENTS

In addition to characteristic influences of grain size, the interpretation of sedimentological data as functions of the aquatic medium must take into account a number of geological, hydrological, biological, and mineralogical factors. According to their effects on metal distribution, these factors can be subdivided into the following categories: (*a*) allochthonous influences, which reflect mainly the lithology of a particular watershed; (*b*) synsedimentary mechanisms (e.g., processes of precipitation or sorption) within the area of deposition; and (*c*) post depositional transformations and their diagenetic effects. These processes and mechanisms were presented in detail in an earlier chapter (Nriagu, this volume, Chapter 1) and will be discussed only briefly here in respect to the *geochemical background* of polluted sediments.

4.1.   Lithogenic Influences

According to analytical data from Marowsky and Wedepohl (1971), there is not much difference in the overall abundance of cadmium in mafic and granitic rocks (Table 3). However, an accumulation of cadmium takes place in the sedimentary environment, in addition to the amounts contributed by rock weathering. If the shale + clay concentration is taken to be representative of the average abundance of cadmium in sedimentary rocks, a 2.4-fold increase is measured as compared to magmatic rocks of the upper continental crust. Marowsky and Wedepohl have explained that this effect is due to an accumulation occurring through the degassing of the earth. A further slight increase in cadmium concentration observed in pelagic clay is partly explained by the high absorptive capacity of sedimentary iron and manganese compounds. In comparison with magmatic rocks and shales, concentrations of cadmium in

Table 3. Cadmium Concentrations in Magmatic Rocks and Sediments (Examples)

Rock/Sediment	Cadmium (ppm)	Remarks	Reference
Granitic rocks	Avg. 0.075–0.100	Gamma spectroscopy	Marowsky and Wedepohl (1971)
Basaltic rocks	0.130	Gamma spectroscopy	Marowsky and Wedepohl (1971)
Ultramafic rocks	0.026	Gamma spectroscopy	Marowsky and Wedepohl (1971)
Shale, clay	0.030	Gamma spectroscopy	Marowsky and Wedepohl (1971)
Pelagic clay	0.405	Gamma spectroscopy	Marowsky and Wedepohl (1971)
Sandstones	0.020	Literature compilation	Horn and Adams (1966)
Limestones	0.035	Literature compilation	Turekian and Wedepohl (1961)
Limestones	0.048	Literature compilation	Horn and Adams (1966)
Limestones	0.090	Literature compilation	Marowsky and Wedepohl (1971)
Diatomaceous ooze	0.39	$n = 5$, $HF/HNO_3/HClO_4$	Mullin and Riley (1956)
Globigerina ooze	0.42	$n = 3$, optical spectrographic analysis	Mullin and Riley (1956)
Radiolarian ooze	0.45	$n = 4$, optical spectrographic analysis	Mullin and Riley (1956)
Red clay	0.56		Mullin and Riley (1956)
Green mud	0.27		Mullin and Riley (1956)
Calcareous ooze	0.57		Mullin and Riley (1956)
Organic mud	0.39		Mullin and Riley (1956)

sandstones and limestones are diluted (Table 4). Because of crystal chemical properties, the cadmium concentrations in limestones are expected to be controlled by their clay fraction (Marowsky and Wedepohl, 1971). This explains the relatively large differences in the average values for cadmium in limestones shown in Table 4.

4.2. Synsedimentary Mechanisms

Whereas the precipitation of carbonate minerals generally seems to effect a dilution of cadmium concentrations (the opposite effect is noticed in sediments heavily polluted with cadmium, where there is a relatively strong affinity for carbonate associations; see below), the coprecipitation and sorption onto hydrous iron or manganese oxides are typical mechanisms of cadmium enrichment in both freshwater and seawater sedimentary environments. Mullin and Riley (1956) found 8.40 and 5.06 ppm Cd in manganese nodules from the North and South Pacific, respectively; Damiani et al. (1977) reported cadmium concentrations of 0.8 to 6.4 ppm (average, 3.0 ppm) in iron and manganese concretions from different parts of Quinte Bay, Lake Ontario. Significant (>99%) correlation between manganese and cadmium was observed from the statistical evaluation of sediment data from 74 lakes in south and western Australia, South America, eastern Africa and South Africa, central and eastern Europe, and western Asia (Förstner, 1977c). Altogether, the different chemical associations do not seem to lead to significant differences in the cadmium content in natural sediment, as is shown from data in Table 4 on siliceous, iron-rich, calcareous, and organic-rich marine sediments (Mullin and Riley, 1956). Generally significant enrichment has been observed only from marine phosphorites (×50 compared to average shale: Altschuler, 1978).

A high positive correlation between the content of organic material and heavy metal concentrations has often been observed in aquatic sediment. This, however, may not always be interpreted as being due to direct bonding of heavy metals to organic substances. In highly polluted areas, for instance, contamination by heavy metals, such as cadmium, simply coincides with the accumulation of organic substances, both being derived from urban, industrial, or agricultural sewage effluent. Extraction of metals from fulvic and humic acids in sediments from "undeveloped" Lake Malawi indicates that only Zn, Cu, and V are significantly accumulated in organic material; Cd and Ni accumulation is largely negligible (Fe, Mn, Cr, Pb and Co are more or less diluted by organic substances). Jonasson (1976) found a significant enrichment of cadmium in organic material of deep-water sediments from Perch Lake, Lanark County, Ontario. He suggests that gel-like coagulating colloids—as precipitates of aging dissolved organic acids, spores, pollen, and very finely dispersed vegetation debris—may

Table 4. Cadmium Concentrations in Sediments from Relatively Less Polluted Aquatic Environments (Examples)

Sediment Sources	Cadmium (ppm)	Remarks	Reference
Scarcely influenced sea sediment in Japan	Avg. 0.45		Murakami et al. (1975)
Fishing ground in the Inland Sea of Japan	Avg. 0.5		Momoyama et al. (1975)
Principal river mouths in Japan	0.51	$n = 91$	Shibahara et al. (1975)
Tor Bay, England	0.37 (0.2–0.7)	$n = 44$, HNO_3/HCl	Taylor (1974)
Avon Estuary, England	0.22	HNO_3	Bryan and Hummerstone (1973)
Fal Estuary, England	0.32	HNO_3	Bryan and Hummerstone (1973)
Unpolluted lakes in South America, Asia, Africa, and Australia	Avg. 0.354 (0.04–0.84)	$n = 66$, $<2 \ \mu m$, HCl/HNO_3	Förstner (1978)
Jubilee Creek, Ill.	0.14 ± 0.07 (0.08–0.23)	Total sediment, $HNO_3/HClO_4$	Enk and Mathis (1977)
Redstone River, Northwest Territories, Canada	0.25	Suspended sediments, HNO_3/HF-bomb	Wagemann et al. (1977)
Mississippi River, stream mile 239,222,175			
Sands	0.143	$n = 7$, $1 \ N$ HCl	Hartung (1974)
Silts	0.300	$n = 28$, extraction	Hartung (1974)
Clays	0.488	$n = 4$, AAS	Hartung (1974)
"Undeveloped" tributaries around the Great Lakes	0.7	Total sediment, $n = 25$, $HNO_3/HClO_4$	Fitchko and Hutchinson (1975)

act as a trap for certain migrating ions (e.g., Cd). From partitioning studies on sediments of the River Blyth, Cooper and Harris (1974) determined the characteristic accumulation of cadmium in organic phases; in a highly polluted sample containing 14% organic substances, 6% of the total cadmium found was fixed in organic material. Polluted sediments from the lower Rhine in Germany revealed a more than 50-fold enrichment of cadmium in humic substances as compared with the content of 0.1 N NaOH-extractable organic material (see below).

4.3. Diagenetic Effects

The distribution of metals such as Mn, Fe, and Co in recent sediments is often controlled to a large extent by redox conditions. Manganese, in particular, reveals a great diagenetic mobility and "can be supplied to the surface sediments from above and from below" (Delfino and Lee, 1968). Enrichment of manganese in the near-surface sediments commonly results from syndiagenetic dissolution of manganese compounds within the lower reducing part of the sedimentary column. Upward migration is due to sediment compaction and subsequent precipitation at the oxidizing sediment-water interface. The behavior of other metals during redox changes and syndiagenetic processes is still open to controversy. There are indications that Cu, Zn, and Cd are partially released from reducing sediments into surface waters at low oxygen contents (see below). These processes, however, seem to be controlled more by bacterial activity and organocomplex formations than by ion migration, since the calculated stability constants for the sulfide compounds are drastically exceeded by the concentrations of Cu, Cd, Zn, Pb, and other metals in the pore water (Brooks et al., 1968; Cline and Upchurch, 1973; Nissenbaum and Swaine, 1975). Studies of Elderfield and Hepworth (1975) on the fluxes of metals in anoxic sediments of the Conway Estuary revealed that ~2% (Zn) to 22% (Ni) of the metal enrichment on the surface of sediments results from diffusion processes, as compared with deeper sections. Because of the crystal chemical similarity of cadmium and zinc, the diagenetic contributions to surface enrichment of cadmium can also be expected to be relatively small. Core sediment studies in Lake Malawi did not reveal a characteristic increase in cadmium concentrations even at sites where distinct diagenetic accumulation of Fe, Mn, and Co occurred (Förstner, unpublished data).

The geochemistry of cadmium in natural sedimentary cycles can be summarized as follows (Gong et al., 1977): During the weathering, transportation, and deposition phases, selection processes operate in such a way as to enrich certain materials (e.g., dark shales, soils, sediments containing organic substances and/or sulfides) in cadmium while depleting it in others (e.g., red shales, sandstones, limestones) relative to the original source rocks.

5. NATURAL BACKGROUND CONCENTRATIONS OF CADMIUM IN RECENT SEDIMENT

Several procedures have been proposed to evaluate the geochemical background concentrations of trace metals in natural sediments, which could serve as a basis of comparison for anthropogenic effects (Förstner and Müller, 1974a): (a) global standard value (e.g., from average shale composition); (b) fossil sediments in respect to the environmental characteristics of rivers, lakes, and oceans; (c) recent deposits from unpolluted regions; (d) bottom sections from short, dated sediment cores; and (e) stormwater sediments.

5.1. Average Shale Composition

Because of the large areal distribution and fine-grained texture, argillaceous sediments seem to be particularly suitable as a global reference basis. From a computer synthesis of analytical data published since Clarke (1924), Horn and Adams (1966) found an average value for cadmium in shales of 0.18 ppm; Wakita and Schmitt (1970) computed 0.74 ppm; and Marowsky and Wedepohl (1971) determined an average of 0.22 ppm Cd from 36 shale samples of Paleozoic age from Germany, Spain, Switzerland, and France.

5.2. Fossil Sediments from Different Environments

Cadmium analyses performed by Aston et al. (1972) on deep-sea sediments from the North Atlantic indicate an average composition of 0.225 ppm Cd. Fossil river sediments from a bore-hole in the lower reaches of the Rhine River contained 0.21 to 0.63 ppm Cd with an average (12 samples) of 0.37 ppm (Förstner and Müller, 1974a). Lake sediments from the Tertiary Ries Crater in southern Germany showed an average cadmium concentration of 0.24 ppm (25 samples; $<2\mu$m fraction: Förstner, 1977d).

5.3. Recent Deposits from "Undeveloped" Aquatic Systems

Examples of unpolluted or scarcely influenced recent sediments from coastal marine, lacustrine, and fluviatile environments are given in Table 4. From the

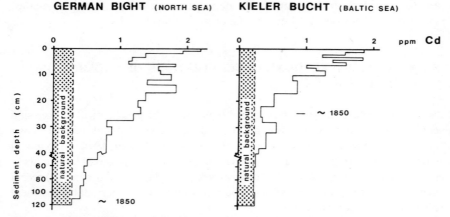

Figure 2. Chronological development of cadmium concentrations in the North Sea and Baltic Sea as derived from analyses of sediment cores from the German Bight (Förstner and Reineck, 1974) and Kieler Bucht (Erlenkeuser et al., 1974). 1850 (left)—German Bight from ^{210}Pb data (Dominik et al., 1978); 1850 (right)—Kieler Bucht, estimated according to coal residue and ^{14}C data (Suess and Erlenkeuser, 1975).

data on Mississippi River sediments (Hartung, 1974), it can clearly be seen that grain size exerts characteristic influences on cadmium concentrations.

5.4.   Short Sedimentary Cores

An approximately 1-m-long core profile from recent deposits usually covers a time period of at least 200 years; the pollution history can be traced by virtue of the metal contents in the individual sediment layers. Two examples are given in Figure 2 from marginal seas of northern Germany; one from the German Bight, North Sea (Förstner and Reineck, 1974), the second from the Kieler Bucht, Baltic Sea (Erlenkeuser et al., 1974). In both examples the natural background in the deeper section of the cores for cadmium is approximately 0.25 to 0.30 ppm, that is, similar to the average cadmium content of shale; from that point the cadmium-concentrations increase about 7 to 8 times to reach their present levels in the upper layers of the core profiles.

5.5.   Stormwater Suspended Matter

High water and increased flow in rivers mobilize bottom sediments deposited during periods of low water flow. If some of the solid bedrock is eroded, as is often

MIDDLE RHINE (KOBLENZ)

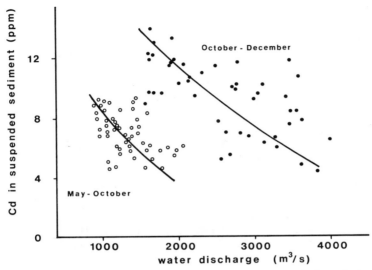

Figure 3. Cadmium concentrations in suspended sediment versus water discharge in the middle Rhine (Schleichert, 1975). Higher levels of cadmium in the autumn/winter period result from partial erosion of polluted sediment particles from the Main and Neckar Rivers.

the case during stormwater events, the suspended sediments tend to exhibit relatively low concentrations of metal pollutants. On the basis of this effect, Hellmann and Griffatong (1972) tried to determine the natural trace element concentrations of solid matter in the lower section of the Rhine River. In practice, however, results from such investigations are affected by various influences that prevent the analysis of real background data. By way of example, the general development of the cadmium concentrations in the polluted Rhine River system at Koblenz is shown in Figure 3, where analyses were made on suspended matter for different water discharge rates (Schleichert, 1975). The values for the spring/summer period are marked by open circles; those for the autumn/winter period, by dots. For both categories a clear dependency on river flow can be seen. However, even the moderate cadmium concentrations in the winter period exceed the maximum value for the spring/summer, despite the much higher water discharge. None of the cadmium values, from either the summer or the winter period, fell below approximately 4 ppm, which is more than 10 times greater than the tentative natural background content of cadmium in the suspended

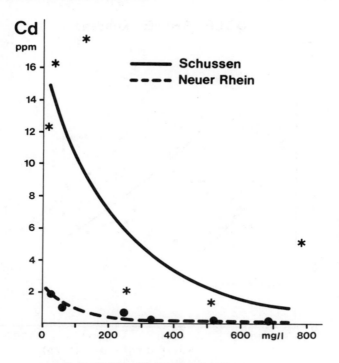

Figure 4. Cadmium concentrations versus concentrations of suspended sediment from two tributaries of Lake Constance. At higher sediment discharges, heavy metal concentrations approximate the background levels. (Wagner, 1976, and unpublished data by Dr. G. Wagner, Institut für Seenforschung, Langenargen.)

solids of the Rhine River system. Such a development is due to the suspended sediments, rich in cadmium, being held back in periods of poor flow, for example, mainly during the summer, in the lock-regulated Neckar and Main rivers (see below). In autumn/winter the tendency is reversed; the material is carried by high water flows into the Rhine in increased quantities. The metal concentrations in suspended solids also depend on the load of suspended matter, which usually increases at higher water discharge. An example has been given by Wagner (1976) for both heavier and lesser contaminated rivers feeding into Lake Constance (Figure 4). As the concentration of suspended matter increases from approximately 20 to more than 700 mg/l, the concentration of cadmium in the solids decreases from 15 to 1 ppm in the Schussen River (polluted from both domestic and industrial sources) and from 2 to approximately 0.2 ppm in the Rhine. The latter value can be taken as representative for the natural background. It is clear, however, that for the Schussen River, even at extreme loads of suspended sediments, the natural background will not be reached.

6. CHEMICAL EXTRACTION AND ANALYSIS OF CADMIUM IN POLLUTED SEDIMENTS

Numerous investigations using sediment analysis in pollution control have shown that it is not necessary to obtain full digestion of all sediment components, including metals bound into the internal structures of silicates and other detrital minerals, since the pollution effects usually occur at the surface of the sediment particles and in the authochthonous precipitates. Metals associated with the degradable organics and with surface coatings of mineral particles may be brought into solution, for example, by the intrusion of water of significantly different quality, and particularly by changes in the oxygen regime such as occur in stagnant waters (Malo, 1977).

Various methods have been developed for the extraction of the more readily available portions of trace metals in polluted sediments: Jones (1973) has suggested digestion with concentrated HNO_3; Anderson (1974) has proposed digestion with $HCl-HNO_3$ (1:1); Malo (1977) prefers a 0.3 M HCl extraction procedure for routine use in monitoring the readily acid-soluble minor elements in aquatic sediments. The aqua regia digestion ($HCl/HNO_3 = 3:1$) is very commonly used in the extraction of metals from polluted sediments. There is a major advantage to the aqua regia extraction for the analysis of the more volatile elements such as cadmium, since the samples are not boiled as in the case of hydrofluoric acid in combination with nitric, perchloric, or sulfuric acid (silicate analysis; e.g., from Bennett and Hawley, 1965). A compilation of laboratory methods that have been proposed to measure cadmium in soil, sewage sludge, and usually sediments, with special emphasis on procedures for the assessment of cadmium available to plants, has been given by Symeonides and McRay (1977). The authors have shown that a 1-hr shaking with 1 N ammonium nitrate solution at a soil/solution ratio of 1:10 (wt/vol), whereby a certain amount of metal is extracted, is the most sensitive indicator procedure for cadmium uptake by plants.

Most of the cadmium determinations in sediments, collected and described in the following sections, have been performed by atomic absorption spectroscopy (AAS), either by conventional AAS or with graphite furnace equipment. Both procedures give satisfying precision for the evaluation of distinct pollution effects.

7. CADMIUM IN POLLUTED COASTAL MARINE AND ESTUARINE SEDIMENTS (TABLE 5)

It was shown in the early 1970s that the distribution of waste materials in the sea can be determined from the amount of trace metals in the bottom sediments. Klein and Goldberg (1970) and Applequist et al. (1972) utilized the concen-

Table 5a. Cadmium in Polluted Sediments from Marine Coasts (Examples)

Sediment Source	Cadmium (ppm)	Remarks	Reference
Israel, Europe			
Coast of Israel	0.7	<200 μm	Roth and Hornung (1977)
	(0.3–2.2)		
Côte d'Azur	<0.2–0.9	<63 μm	Vernet et al. (1977)
Coast of Portugal and Spain	0.9–4.1	HNO_3	Stenner and Nickless (1975)
Firth of Clyde			
Estuary	0.4–1.5	<204 μm	Steele et al. (1973)
Clyde background	0.4		MacKay et al. (1972)
Deposit area	6.4	$HNO_3/HClO_4$	Halcrow et al. (1973)
	(3–7)		
Garroch head	0.6–8.5		Steele et al. (1973)
Solway Firth			
Urr water	0.9	$n = 19$	Taylor 1976)
	(0.8–1.3)		
Salton Parton beaches	<1.0	HNO_3/HCl	Perkins et al. (1973)
	(ND-2.4)	<208 μm	
Allonby Beekfoot	2.9	$HNO_3/HClO_4$	Perkins et al. (1973)
	(2.4–4.0)		
Mersey Estuary	4.2	$H_2SO_4/HClO_4$	Leatherland and Burton (1974)

Location	Value	Method	Reference
Cardigan Bay	1.1 (0.2–3.4)	Conc. HNO_3	Jones (1973)
Swansea Bay, Wales	11–25	XRF, "mud"	Bloxam et al. (1972)
Bristol Channel	9.7	Suspended sediments, HNO_3	Abdullah and Royle (1974)
Severn Estuary	3.7 (1.6–4.3)	$HF/HClO_4$	Butterworth et al. (1972)
Southwest England estuaries			
Restronguet Creek	3.1	HNO_3	Bryan and Hummerstone (1973)
Gannel Estuary	4.2	HNO_3	Bryan and Hummerstone (1973)
Plym Estuary	9.3	HNO_3	Bryan and Hummerstone (1973)
Poole Harbor, Dorset	<1–12	<200 μm	Boyden (1975)
The Solent, Southampton	0.55–0.69	$H_2SO_4/HClO_4$	Leatherland and Burton (1974)
Baltic Sea, Sweden	1.1 (<0.01–8.1)	HCl	Olausson et al. (1977)
Lübecker Bucht	1.7	$HNO_3/HClO_4$	Brügmann (1977)
Norwegian fjords	0.3–1.5	Hot HNO_3	Stenner and Nickless (1974)
Sörfjord	16–850	XRF	Skei et al. (1972)
Japan, Australia, Tasmania			
Tokyo Bay	3.1–40.4		Goto (1973) Ishibashi et al. (1970)
Port Phillip Bay (bottom)	0.04–1.6		Talbot et al. (1976)
Victoria, Australia (top)	0.15–9.9		Talbot et al. (1976)
Derwent Estuary, Tasmania	0.8–862 (max. 1400)	Conc. HNO_3	Bloom and Ayling (1977)

Table 5b. Cadmium in Polluted Sediment from U.S. Marine Coast (Examples)

Sediment Source	Cadmium (ppm)	Remarks	Reference
U.S. West Coast			
Santa Barbara Basin	1–2	$HF/HCl/HNO_3$	Bruland et al. (1974)
San Francisco Bay			
Bottom	0.93	HCl/H_2SO_4	Moyer and Budinger (1974)
	(0.14–3.91)		
Top	1.22	<2 mm	Moyer and Budinger (1974)
	(0.06–4.69)		
Santa Monica Canyon	Avg. 21.3 (max. 65)		Schafer and Bascom (1976)
	Background 0.22		
South Coast and			
U.S. East Coast			
Gulf of Mexico			
Mississippi Delta	0.2	$n = 72, 16\,N$	Trefry and Presley (1976)
	(0.02–0.44)	$HNO_3 + 6N$ HCl	
San Antonio Bay	0.3	$n = 51,$	Trefry and Presley (1976)
	(0.02–0.7)	$16\,N\,HNO_3$	

Location	Value	Method	Reference
Harbor Island	(0.4–1.1)	16 N HNO$_3$	Lytle et al. (1973)
Corpus Christi Bay	(0.05–1.95)	$n = 44$,	Holmes et al. (1974)
Galveston Bay	<0.6	HNO$_3$-leach	Hann and Slowey (1972)
Houston ship channel	(<0.2–4.9)	$n = 24$,	
	<2.9	HNO$_3$-leach	Hann and Slowey (1972)
Corpus Christi Harbor	(<0.1–10.7)	16 N HNO$_3$	Holmes et al. (1974)
	2.0–130		Segar and Pellenbarg (1975)
Florida/Bermuda	0.07–0.8		Helz (1976), Villa and Johnson (1974)
Chesapeake Bay	0.3–6.5		Greig and McGrath (1977)
Raritan Bay	<1.0–15.0	50% HNO$_3$	Greig et al. (1977)
Long Island Sound	1.0–4.2	50% HNO$_3$	Brannon et al. (1976)
Bridgeport, Conn.	17.6	HF/HNO$_3$	
Rhode Island			
Narragansett Bay	0.7		Goldberg et al. (1977)
Quonset Point	0.47	Conc. HNO$_3$	Eisler et al. (1977)
Massachusetts Bay	(0.06–3.92)		Issac and Delaney (1973)
	2.0		
Boston Harbor	7.5		Issac and Delaney (1973)

trations of mercury in sediment to estimate the rate of accumulation of sewage sludge and to trace the sources of pollution near marine coasts and in harbor areas. Of all the minor elements studied by Gross et al. (1971) in sediments of New York Bight, lead proved to be the most useful indicator of waste materials. Rutherford and Church (1975) proposed the additional use of silver and zinc to trace sewage sludge dispersal in coastal waters.

The list of indicator metals for sewage material can be enlarged by adding cadmium, since this element is characteristically enriched in both domestic and industrial effluents. Sewage sludge samples from Swedish treatment plants indicate median concentrations of 7 ppm Cd (range, 2 to 172 ppm; Berggren and Odén, 1972); from sewage plants in Michigan, median values of 12 ppm Cd (2 to 1100) were reported (Blakesley, 1973); and examples from England and Wales (Berrow and Webber, 1972) contained 60 up to 1500 ppm Cd. At an average concentration of 10 ppm, the cadmium in sewage materials normally surpasses the standard shale composition by factors of 30 to 40 (Förstner and Stiefel, 1978).

Upon inspecting the cadmium data for marine sediments (Table 5) it becomes obvious that an enrichment of between 1 and 10 ppm Cd is commonly due to the *proximity of domestic and industrial sewage outfalls.* This is true for the coastal waters of *Israel,* as well as for several examples of Great Britain (Solway Firth, Mersey Estuary, Bristol Channel, including Severn Estuary, Poole Harbor, and the Solent) and from the United States (Chesapeake Bay, Raritan Bay, Long Island Sound, Quonset Point [R.I.], and Massachusetts Bay). The correspondence between sewage disposal and increased cadmium concentration is especially clear for the Firth of Clyde (Great Britain) and the coast of southern California. The combined annual mass emission rate of southern California's five largest municipal effluents has been calculated at approximately 50 tons Cd (Schafer, 1976), which is on the same order of magnitude as the annual cadmium discharge in sediments of the lower Mississippi (85 tons: Hartung, 1974).

Major sources of cadmium enrichment in coastal marine sediments are the river inputs from *mineralized areas* and effluents from *smelting industries.* Through these influences a very high enrichment of cadmium in sediments is sometimes reached. Samples from the mouth of the *Rio Tinto, Spain,* provide an example of moderate enrichment. This river has cut a valley through rocks that contain very rich deposits of copper (and other metals), which have been worked for many centuries (Stenner and Nickless, 1975). The source of elevated levels of cadmium in sediment of *Swansea Bay, Wales,* is suggested by the very high past and present concentration of metal processing plants, which use both Welsh and imported ores, along the River Tawe (Bloxam et al., 1972). Enrichment of cadmium in sediments from estuaries of *southwest England* is explained by river inputs draining the old metalliferous mining areas of Cornwall and Devon (Byran and Hummerstone, 1973). Particularly strong pollution by cadmium (up to 850 ppm) and other metals has been found in the *Sörfjord of western Norway,* a 40-km-long north-south extension of the Hardanger Fjord; in the town of Odda, three industrial plants, including a zinc smelter, are reported to have released their metalbearing wastes, including 10 tons Cd/year, into a relatively shallow area of the fjord (Skei et al., 1972). Discharges of metallurgical effluents into the *Derwent Estuary of Tasmania* began several decades ago with an electrolytic zinc refining plant; dust containing more than 100 ppm Cd (max. 1450 ppm) was found to have been blown over the harbor of Hobart. Cadmium concentrations in sediment near the zinc refining company wharf were as high as 1400 ppm and reached 860 ppm in the adjacent estuary, compared to background concentrations of approx. 1 ppm Cd (Bloom and Ayling, 1977). There are such high concentrations of cadmium in some shellfish that they may be toxic when eaten (Ayling, 1975; Thrower and Eustace, 1973); condemned beds of commercial oysters contained more than 200 ppm Cd dry weight (Ratkowsky et al., 1974), which is on the same order of magnitude as the level in rice of the epidemic itai-itai area of Japan (120 to 350 ppm; Yamagata and Shigematsu, 1970).

Harbor areas, because of the usual nearby industrial intensity, are especially hard hit by anthropogenic pollution effects. Table 5*b* presents some examples from the United States, ranging from moderate to very heavy cadmium enrichment. In cases of Boston Harbor, Massachusetts, Bridgeport, Connecticut, and Corpus Christi Harbor, Texas, the origin of the increased cadmium content has yet to be localized. A detailed investigation of the New Bedford Harbor (Massachusetts) area, one of the most heavily industrially polluted coastal regions in the United States, was carried out by Stoffers et al. (1977). Figure 5 is based partially on unpublished data of the Institute for Sediment Research at the University of Heidelberg, where the surface distribution of cadmium in clay-sized sediments ($<2\mu$m fraction, 92 samples) was found to indicate a gradual seaward decrease from more than 100 ppm Cd (max. 130 ppm) in the estuary of the Acushnet River to less than 0.5 ppm at the edge of Buzzard Bay (Figure 5*a*). Within the core profiles—3 examples were selected in Figure 5*b* from 19 cores—marked increases in the concentrations of cadmium occur at burial depths of 100 cm in core *A*, 50 cm in core *B*, and at about 20 cm in core *C*; the cadmium-rich waste deposit is thickest in the deeper part of the harbor, thinning both toward the edge of the harbor and, exponentially, toward the ocean. Examination of the relationships between copper and other metals in the cores, such as cadmium, shows that these metals covary in a consistent manner. It appears that the harbor is the common source of these metals, persumably from metallurgical industries (Summerhayes et al., 1977). On the assumption that, according to the sediment core data and available U.S. Environmental Protection Agency records, the discharges of cadmium amount to approximately 2% of the

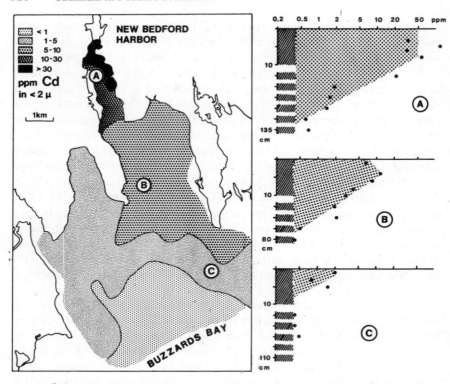

Figure 5. Cadmium distribution in surface sediments and vertical sediment profiles (examples) from New Bedford Harbor, Massachusetts. Note the logarithmic scale of cadmium values in the core sediments. (Unpublished data, Institut für Sedimentforschung, Heidelberg.)

copper emissions, a total input of 30 tons of Cd from anthropogenic sources has occurred during the 80-year history of waste metal accumulation from industrial plants in New Bedford Harbor.

8.   CADMIUM IN POLLUTED SEDIMENTS FROM LAKES AND DAMS (TABLE 6)

In recent years metal investigations on lacustrine sediments have become increasingly of interest, on the one hand, because of the importance of lakes as drinking water reservoirs, and, on the other, because of the application of lake sediment geochemistry as a guide to mineralization: "Both the exploration and environmental geochemist can be looking for the same type of areas, those with high metal concentrations, but obviously from a different motivation" (Allan,

1974). These two aspects overlap in some cases where mineral exploration has been followed by large-scale mining and processing activities. Table 6 presents examples from the *Sudbury Lakes* area, where the atmospheric inputs from smokestacks have sometimes led to a very high enrichment of cadmium (and other metals, above all nickel; Semkin and Kramer, 1976; Allan, 1977). On the other hand, cadmium concentrations in Clearwater Lake, downstream from Viburnum Trend, the world's largest lead mining district, did not significantly surpass normal sedimentary levels (0.3 to 0.5 ppm; Gale et al., 1976). Investigations on sediment cores have proved especially effective in determining and differentiating geochemical and civilizational influences in such areas; in other cases of metal pollution in lakes and dams, the temporal development of the pollution can be very effectively followed on vertical profiles as well. Such investigative results on sediment cores are indicated in Table 6 by a superior letter *c*.

A general increase in cadmium contents is observed in the *Great Lakes of North America;* which is lowest in Lake Superior followed by Lake Huron, Lake Erie, and Lake Ontario in that order, although local deviations may occur. On the one hand, Johnston (1977) measured on 38 core profiles from the Kingston Basin-upper St. Lawrence River very small mean levels of cadmium from 0.7 ± 1.05 ppm, whereas, on the other, high sediment cadmium enrichments occurred in some river mouths (see the next section), and especially in harbor areas. A sample from Ashtabula Harbor on Lake Erie contained 4.1 ppm Cd (Brannon et al., 1976); in the middle section of a sediment profile taken from Cleveland Harbor, more than 10 ppm Cd was measured (Walters et al., 1974). These accumulations of cadmium (together with chromium and nickel) are most probably due to the numerous (over 100) *electroplating facilities* in the Cleveland area, from which pollution inputs have occurred, particularly in the years 1948–1955. Effluents from a plating plant are responsible for the strongest enrichment of cadmium in lacustrine sediments yet reported: up to 2640 ppm was found in *Palestine Lake* sediments, Palestine, Indiana (Wentsel and Berry, 1974). High rates of heavy metal increase were also found in sediments of the *Moste hydroelectric power plant* on the upper course of the Sava River (Slovenia, Yugoslavia). Not only domestic sewage from the municipality Jesenice, but also effluents from the Javornik *steelwork and smelter complex,* flow into the Sava (Štern and Förstner, 1976). Compared to the levels found in Sava sediments where the river empties into the reservoir, the actual deposits in the dam are 10 to 20 times higher in cadmium.

Various sources of moderate enrichment (5 to 10 ppm) of cadmium in lake sediments are shown in Table 6. For example, the reservoir at Fort Gibson in Oklahoma and the Wahnbach reservoir in West Germany both lie downstream of *lead and zinc mining* and mineral areas. Drainage from a *heavily traveled highway* and *city streets* is the major contributor of cadmium to the sediments

Table 6. Cadmium in Polluted Sediments from Lakes and Dams (Examples)

Sediment Source	Cadmium (ppm)	Remarks	Reference
North America			
Lake Huron	0.5	HNO_3HClO_4	Brown and Chow (1977)
Lake St. Clair	1.7[a] (1)[b]	n = 50, 0.5 mm, HNO_3/HCl	Thomas et al. (1977)
Lake Michigan (Green Bay)	2.3 (1.9)[b]	<200 mm, $HNO_3/HClO_4$	Pezzetta and Iskandar (1975)
Wisconsin lakes	3.0[a] (1.6)[c]	n = 10, <200 μm, $HNO_3/HClO_4$	Iskandar and Keeney (1974)
Lake Erie	4 (0.5)[c]	n = 74, conc. HCl	Kemp et al. (1976), Nriagu et al. (1979)
Cleveland Harbor	Max. 11 (<0.2)[c]		Walters et al. (1974)
Lake Ontario	5.2[a] (1.4)[c]	Hot conc. HCl	Kemp and Thomas (1976)
Wapato Lake, Wash.	Max. 6	Conc. HNO_3	Wisseman and Cook (1977)
Fort Gibson Reservation, Okla.	Max. 9	$HF/HNO_3/HClO_4$	Pita and Hyne (1975)
Sudbury area lakes	0.5–19.4	<80 μm, n = 65, HNO_3/HCl	Semkin and Kramer (1976)
Palestine Lake, Ind.	3.2–2640	HNO_3, effluents	Wentsel and Berry (1974), Yost (1978)
Africa, Asia, Australia			
Hartbeespoort Dam, South Africa	(0.14–0.54)	<2 μm, HNO_3/HCl	Wittmann and Förstner, (1975)

Palace moat, Tokyo	Max. 5 (1.7)[c]	60 cm core	Goldberg et al. (1976)
Lake Biwa, Japan	1.7[a] (0.5–5.5)	$n = 58$	Tatekawa et al. (1975)
Lake of Tunis	0.1–13.3	HNO_3/H_2SO_4	Harbridge et al. (1976)
Lake Illawarra, Australia	7.8 (0.02)[c] max. 45.0	$HNO_3/HClO_4$	Depers (1978)
Europe			
Uusikaupunki Reservoir, Finland	0.75[a] (0.27)[c]	8 cores, $HNO_3/HClO_4$	Hinneri (1974)
Lake Vänern, Sweden	<1–2.4		Håkanson (1977)
Lakes in Bavaria	1.0[a] (0.54–1.85)	$n = 8$, $<2\ \mu m$	Förstner (1978)
Bodensee, FRG	0.68 (0.21)[c]	HNO_3/HCl	Förstner and Müller (1974)
Near Schussen inflow	Max. 1.57	HNO_3/HCl	Müller (1977)
Wahnbach Reservoir, FRG	2.0–3.4	Dil. HNO_3	Mihm et al. (1976)
Lake Geneva, Switzerland	1.0 (0.25)[c]	$<03\ \mu m$, HNO_3/HCl	Vernet (1976)
	Max. 8.51		Davaud et al. (1978)
Lago Maggiore, Italy	1.3[a] (0.23–3.8)	$n = 10$	Ravera et al. (1973)
Lake Bled, Yugoslavia	Max. 3.2 (0.6)[c]	$<2\ \mu m\ HNO_3$	Molnar et al. (1978)

[a] Average concentration in surface sediments.
[b] Background concentrations from data of less polluted areas.
[c] Background values of deeper parts of sediment cores.

of Wapato Lake near Tacoma, Washington. *Atmospheric fallout,* magnified by the sediment concentrations of polluted soil debris, is probably the major source of cadmium enrichments in deposits, for example, of a moat surrounding the palace of the Emperor in central Tokyo. Among the main sources of such cadmium-pollution are fossil fuel mobilization (Ruch et al., 1973) and cement production (Goldberg, 1976). Other sources may also dominate: Mathis and Kevern (1973) have described the case of Wintergreen Lake in southwest Michigan, where feces of migratory water fowl have lead to a greater increase of cadmium in the lake deposits than can be attributed to atmospheric contributions.

Significant enrichment of cadmium in lake sediments can be expected from *sewage effluents* from communal sources. In sediments of Lake Tunis, a eutrophic marine lagoon in northern Africa, cadmium values clearly tend to be higher near the city. Mixed inputs from atmospheric sources and direct effluents are probably responsible for moderate (two- to tenfold) increases of cadmium concentrations in surface sediments from large freshwater lakes in densely populated areas of northern and central Europe (Vänern, Bodensee, Léman, Maggiore), since there is a similar accumulation of other pollutants such as nitrogen and phosphorus compounds, DDT, PCB, polycyclic aromatic hydrocarbons and radioactive elements (Ravera, 1964; Ravera and Premazzi, 1972; Förstner et al., 1974; Vernet, 1976; Grimmer and Böhnke, 1977; Håkanson, 1977).

An estimation of man's contribution to the heavy metal load of lakes with little industry in their catchment areas has been given by Roberts et al. (1977, cited in Stumm and Baccini, 1978) in studies on Lake of Alpnach in Switzerland. Approximately one-third of the Cd content is "natural", 45% stems from waste waters and about 25% from the atmosphere (precipitation). In highly industrialized areas the atmospheric contribution should become more and more significant (See Nriagu et al., 1979).

9.   CADMIUM CONCENTRATIONS IN POLLUTED RIVER SEDIMENTS (TABLE 7)

Towards the end of the 1960s, the possibility of applying geochemical methods of exploration to water quality assessment was recognized both in North America and in Europe. Turekian and Scott (1967) studied the concentrations of Cr, Ag, Mo, Ni, and Co in suspended solids from 18 rivers in southern and eastern sections of the United States. They were able to determine a distinct accumulation of Ag, Ni, and Co, particularly in the Susquehanna River, which, it was suggested, was the result of industrial contamination.

In Europe the monitoring of river pollution by means of sediments began in the 1960s in the Netherlands by De Groot and co-workers. Their studies centered on the transport of sediments from the Rhine and Ems rivers into the North Sea and on the behavior of trace metals in the mixing zone of fresh water and sea-water (De Groot, 1966; De Groot et al., 1971). At the same time in the Federal Republic of Germany, Holluta et al. (1968), Herrig (1969), and Hellmann and co-workers (Hellmann, 1970, 1972; Hellmann and Griffatong, 1972) were carrying out metal investigations on the Rhine River and its tributaries in order to ascertain to what extent the process of water reclamation from bank filtrates was endangered by the increasing river water pollution, as well as to establish the origin of troublesome mud deposits. In the latter studies, special emphasis was given to the role of zinc as a characteristic element in domestic waste effluents.

Systematic investigations on river sediments carried out by Förstner and colleagues (Banat et al., 1972a, 1972b; Förstner and Müller, 1973a, 1973b, 1974, 1976; Müller and Förstner, 1975, 1976) led to the detection of the most significant centers of heavy metal pollution in the Federal Republic of Germany. Enrichment of cadmium in pelitic fractions ($<2 \mu$m) of the sediments samples studied (Figure 6) points to three major sources of pollution: domestic waste-water effluents (e.g., in the Elbe River), mineralized areas (e.g., in the Diemel and Werre, both tributaries of the Weser River), and industrial effluents. The most striking example of an industrial source was found in the lower Neckar River, a tributary of the Rhine. Cadmium accumulations of more than 300 ppm in sediments and more than 200 μg /l in water and severe fish poisoning (up to 230 μg Cd/kg wet weight: Müller and Förstner, 1973) were registered. The main source was a pigment dye production plant on the Neckar that released an annual total of approximately 10 to 20 tons of Cd in dissolved and particulate form into the river. As a result of improved techniques of wastewater treatment, since February 1973 there has been a decrease in the cadmium contents of the water (Förstner and Müller, 1975), sediment (Förstner, 1976b), and fish (Müller and Prosi, 1977). Because of the numerous lock reservoirs in sections of the river where highly polluted sediments settle, regeneration necessarily proceeds at a very slow rate. There is approximately 50 tons of cadmium in the mud of the middle and lower sections of the Neckar. In the past a portion of this metal reached the nearby fields as dredged Neckar River mud was employed there to improve the soil. (Dredging was done to keep the shipping canal open.) In the fall of 1978 a citizens' initiative raised the alarm; over a total of 45 hectares of land, cadmium concentrations of up to 73.5 ppm were measured. Early in 1979 agricultural planting was stopped, and compensation (about $1000/ha) was offered to the affected growers (*Stern Magazine,* March 22, 1979). The problem is much greater, however, as after a flooding catastrophe in May 1978 extensive farm areas were covered with water from the Neckar, whereby decimeter-thick layers of mud were deposited.

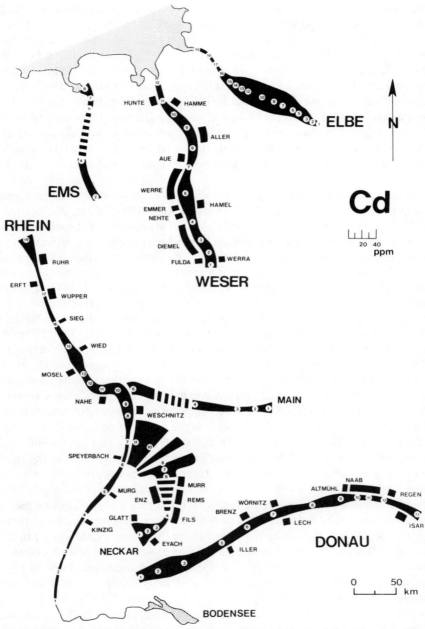

Figure 6. Cadmium concentrations in pelitic sediments (fraction <2 μm) from rivers in western Germany. (Banat et al., 1972.)

Still greater are the difficulties encountered in the Rhine delta in The Netherlands (only partially affected, however, by cadmium pollution from the Neckar), where, for example in Rotterdam Harbor, 12 million m³ of sludge is yearly dredged. Most difficult to dispose of are the 4 million m³ of deposits in Waalhaven, Maashaven, and Rijnhaven, which contain 20 to 30 ppm Cd (dry matter) (*Rotterdam Europoort Delta,* January, 1979, pp. 10–12).

A number of other examples of the source types mentioned above have been compiled in Table 7. As for marine sediments and lacustrine deposits, the mode of grain size separation and metal extraction must be considered. The suspended solids show generally higher contents than do the corresponding bottom sediments. This effect is due chiefly to the smaller grain size of the suspended particles.

9.1. Domestic Effluents

Cadmium enrichments of between 5 and ca. 10 ppm in river sediments (in suspended particles correspondingly higher) are often caused by communal wastewater effluents, sometimes with an industrial component, for instance, from the electroplating industry. The first six North American examples in Table 7 show an increasing influence of urban-industrial activity. In the Milwaukee River, Wisconsin, this effect for the most part is dominant over domestic sources; effluents from brewing, tanning, incineration, the chemical industry, foundries, metal works, manufacturing, and so on are common sources of cadmium (Fitchko and Hutchinson, 1975). Particularly high concentrations of cadmium have been analyzed by Chen et al. (1974) in suspended silts of river runoff in the southern California coastal region. The suspended silts in the dry weather flows (February to April 1973) from the urbanized Los Angeles basin and those of Ballona Creek (830 ppm Cd), the Los Angeles River (860 ppm), and the Dominguez Channel (950 ppm Cd) bear significantly higher amounts of cadmium and other heavy metals than the suspended silts in the flows from the northerly and southerly areas (approx. 100 ppm) draining more natural regions. During stormwater events, the concentrations of cadmium are significantly reduced. Sediment contents in Santa Clara River (natural example) decrease from 32 ppm Cd during dry weather to 3 ppm (0.3 to 6 ppm). In storm flows, suspended sediments in Balboa Creek contain 15 ppm Cd (2 to 50 ppm), in the Los Angeles River 30 ppm (5 to 200 ppm).

9.2 Mining Activities

The enrichment of cadmium in mineralized areas varies greatly (Table 7). Whereas, for example, the suspended particles in the effluents of the Witwaters

Table 7. Cadmium in Polluted River Sediments (Examples)

Sediment Source	Cadmium (ppm)	Remarks	Reference
North America			
Susquehanna River, Harrisburg, Pa.	1.68	<2 mm, 0.3 M HCl	Malo (1977))
Grand River, Mich.	3.5	Total sediment, $HNO_3/HClO_4$	Fitchko and Hutchinson (1975)
Grand Calumet River, Ind.	9.7	Total sediment, HNO_3	Hess and Evans (1972), Romano (1976)
Murderkill River, Del.	3.1–7.9	<63 μm, dil. HCl	Bopp et al. (1973)
Illinois River	0.8–8.7	Total sediment, $n = 73$	Mathis and Cummings (1973)
	2.0	$HNO_3/HClO_4$	
	(0.2–12.1)		
Rideau River, Ont.	0.3–15	0.5 N HCl	Agemian & Chau (1977)
Lake Cayuga tributaries	15.6	Suspended sediments, $HClO_4$	Kubota et al. (1974)
Saginaw River, Mich.	28		Hesse and Evans (1972)
Coeur d'Alene River	Max. 80		Maxfield et al. (1974)
Milwaukee River, Wis.	16.6	$HNO_3/HClO_4$	Fitchko and Hutchinson (1975)
	Max. 149	Total sediment	
Tennessee River	Max. 227	Suspended sediments, HNO_3 /HF/$HClO_4$	Perhac (1972)
Los Angeles River, Calif.	860	Suspended sediments, HNO_3	Chen et al. (1974)
Hudson River estuary, N.Y.	2.3	Suspended sediments,	Vaccaro et al. (1972)
Foundry Cove, N.Y.	Max. 50,000	Suspended sediments,	Kneip et al. (1974) Kneip (1978, Bower et al. (1978)
South Africa, Australia, Japan			
Gold mine drainage, South Africa	0.21 (0.05–1.0)	<2 μm, HCl/HNO_3	Wittmann and Förstner (1976a)
Jukskei River, South Africa	0.25–4.9	<2 μm, HCl/HNO_3	Wittmann and Förstner (1976b) Australian Government Technical Commission (1974)
Molonglo River, Australia	0.8–3.3	Hot HNO_3	
Tamar River, Tasmania	3.6 (<0.1–6.0)		Ayling (1974)

Location	Value	Method / Notes	Reference
South Esk River, Tasmania	Max. 153	Acid digestion	Tyler and Buckney (1973)
Tama River, Tokyo	0.7–9.8	$n = 26$	Suzuki et al. (1975)
Jintsu River, Toyama Pref.	3.27 (0.16–5.0)		Goto (1973)
Takahara River (near Kamioka mine)	121 (4.1–238)		Kiba et al. (1975)
Rivers around Himeji City (west of Osaka)	0.56–10.4 Max. 129	$n = 30$, conc. HNO_3	Azumi and Yoneda (1975)
Rivers in the Hitachi area, northeast Tokyo	Max. 368	Acid digestion	Asami (1974)
Israel, Europe			
Gadura River (Bay of Haifa, Israel)	Max. 123	<0.17 mm, 1 M $NH_2OH\cdot HCl$, acetic acid	Kronfeld and Navrot (1975)
Lake Geneva tributaries, Switzerland	1.4 (0.09–12.4)	<63 μm, HNO_3/HCl	Vernet (1976)
Upper Rhone, Switzerland	0.1–73	<63 μm, HNO_3/HCl	Viel et al. (1978)
Elbe, FRG	2.9–19.4	25% <2 μm	Ribordy (1978)
Sajo River, Hungary	Max. 20	Total sediment, HNO_3/HCl	Lichtfuss & Brümmer (1977)
Blies, Saar, FRG	0.5–24.0	<63 μm, HNO_3/HCl	Literáthy and László (1977)
Bavarian rivers, FRG	<0.05–29.2	HNO_3/H_2SO_4	Becker (1976) Bayerische Landesanstalt für Wasserforschung (1977)
Main River, FRG	17–151	Suspended sediments	Schleichert and Hellmann (1977)
Ginsheimer Altrhein, FRG	2–95	<2 μm, HNO_3/HCl	Laskowski et al. (1975)
River Conway, Great Britain, mineralized areas	21 (3–95)	<200 μm, $HNO_3/HClO_4$	Thornton et al. (1975)
River Tawe, Wales	Max. 355	<2 mm, $HNO_3/HClO_4$	Vivian and Massie (1977)
Sava basin, Yugoslavia, Voglajna River	Max. 66	<2 μm, HNO_3/HCl	Štern and Förstner (1976)
Stola River, Poland	Max. 116		Pasternak (1974)
River Maas, Belgium	Max. 230		Bouquieaux (1974)
River Vesdre, Belgium (near Liège)	Max. 430		Boudquieaux (1974)

and gold mines show practically no increase of cadmium and the sediments of the polluted Molongo River (Australia) and Tamar River (Tasmania) are only moderately enriched, there are other areas where very threatening increases in cadmium contents occur, chiefly as a result of past and present mining activities. Here may be mentioned regions of sulfidic lead-zinc mineralization, where with oxidation acidic waters arise, in which heavy metals such as cadmium may occur in high solution concentrations. The area around Coeur d'Alene River in Idaho, the Tennessee River near Knoxville in Tennessee, Wales, southwest England, and especially the Takahara River-Jintsu River region of Japan in which the catastrophic itai-itai disease occurred are examples of the lead-zinc mining effect. The discharge of effluents from tin and tungsten mines has caused severe pollution of the South Esk River in northeastern Tasmania.

9.2. Industrial Effluents

Characteristic point-source emissions of cadmium from industrial plants have been recorded in sediments of the Hudson River and in a river of the Hitachi area, northeast of Tokyo. Maximum concentrations of 3000 to 50,000 ppm Cd occur near the effluent of a nickel-cadmium battery plant in Foundry Cove near Cold Springs, New York (Kneip et al., 1974). Up to 368 ppm Cd was measured in the sediments of the irrigation ditch (up to 45,500 ppm at the wastewater outlet and settling tank) of a Braun-tube factory. Strong contamination by cadmium was detected in sediments from European rivers draining highly industrialized areas, for example, in the Sajo River of northern Hungary, in the tributaries and reservoirs in the vicinity of Miasteczko Slaskie, the zinc and lead works at the upper reaches of Mala Panew River (Poland), in the Voglajna River near Celje in Slovenia (Yugoslavia), in the Ginsheimer Altrhein near Mainz (Federal Republic of Germany) and in the Maas and Vesdre rivers at Liège (Belgium). There is particularly strong cadmium pollution in the catchment area of River Tawe in the lower Swansea Valley, Wales, where coal has been mined since the fourteenth century and provides the basis for smelting and other industries (Vivian and Massie, 1977). In Israel a case of cadmium contamination in the Qishon River was reported by Kronfeld and Navrot (1974), demonstrating the various aspects to be taken into consideration in the final evaluation of pollution effects. Although metal values in the sediments were especially high, reaching more than 120 ppm Cd near the tributary Nahar Gadura, only small amounts of heavy metals were found in the aqueous phase, except in the vicinity of an oil refinery, where acidic effluents occur. The otherwise high pH values encountered in the Nahar Gadura effectively remove most of the metals from solution, thus diminishing the toxicity of the metal discharges. The authors state therefore that "ironically a sudden stoppage of the pollution of effluents with

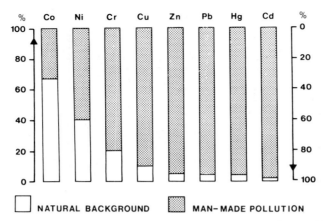

Figure 7.   Natural and anthropogenic contributions to the concentrations of selected heavy metals in polluted sediments from the lower Rhine.

high pH before other corrective measures are taken, appears to present a greater biological danger to the Qishon River and Haifa Bay than a continuance of the pumping of these wastes into the river."

10. CADMIUM ENRICHMENT IN SEDIMENTS—COMPARISON WITH OTHER METALS

After examples have been compiled of cadmium enrichment in polluted sediments, the question arises as to the extent of the increase in cadmium in respect to the rise in other toxic metals. If the "index of relative pollution potential" is taken as the ratio of metal consumption to average metal content in a specific sphere (e.g., the pedosphere), it is evident that Cd, along with Hg, Pb, Cu and Zn, is one of the most prevalent of the metal contaminants in the environment (Förstner and Müller, 1973a). Another approach to the problem is a comparison of metal data from actual and fossil sediments (e.g., from sediment cores). In Table 8 selected examples of sediment samples from lacustrine and coastal marine environments are presented. It is evident that cadmium, next to mercury, shows generally the greatest factor of enrichment. Suess and Erlenkeuser (1975) have pointed out that the order of enrichment of anthropogenically influenced sediments, Cd > Pb > Zn > Cu (not considering Hg), corresponds to the accumulation of metals in fossil fuel residues. The latter is also reflected in the metal enrichment of airborne particulates.

An example of anthropogenic metal enrichment in fluviatile sediments is shown in Figure 7, which compares the metal concentrations in actual Rhine

Table 8. Sediment Enrichment Factors (SEFs) of Heavy Metals in Selected Examples from Lacustrine and Marine Environments

Source	Element									Reference[a]
	Cd	Co	Cu	Cr	Fe	Hg	Ni	Pb	Zn	
Lake Vänern	2.6	NA[b]	1.2	NA	NA	(38)	0.8	2.3	3.7	(1)
Lake Constance	2.9	0.5	1.1	3.0	1.0	2.0	0.9	2.7	3.1	(2)
Lake Geneva	4.6	1.5	2.7	1.3	NA	15	1.0	2.3	3.6	(3)
Lake Erie	7.3	1.6	3.7	2.9	1.5	8.3	2.1	4.7	3.6	(4)
Kieler Bight (Baltic Sea)	7.5	1.4	2.0	NA	0.9	NA	1.6	4.1	2.7	(5)
German Bight (North Sea)	7.5	2.2	1.8	1.5	1.4	9.4	1.3	8.2	4.0	(6)
Mean SEF	5.3	1.4	2.1	2.2	1.2	8.7	1.3	4.1	3.5	

[a] (1) Håkanson (1977), mean values from eight surficial samples (0—1 cm) versus natural background ($\bar{x}$ from 68 preindustrial samples + 1 standard deviation); Cd background is estimated; Hg probably represents local enrichment. (2) Förstner and Müller (1974), sediment core from central part (250 m water depth). (3) Vernet (1976), mean SEF from 13 coastal stations. (4) Walters et al. (1974), sediment core from central basin (No. 13-2); Pb value from Kemp et al. (1976), mean of SEF from two cores of central basin. (5) Erlenkeuser et al. (1974), core A-GC from 28 m water depth. (6) Försnter and Reineck (1974), core 1/315 from 22 m water depth.

[b] NA = not available.

338

RHINE (NETHERLANDS)

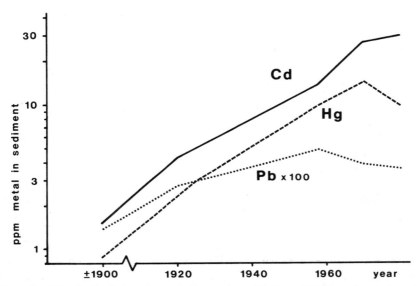

Figure 8.   Historical evolution of metal pollutants (e.g., Cd, Hg, and Pb) in sediments from the Rhine River in the Netherlands since 1922. (From Salomons and DeGroot, 1978.)

sediments with those found in deposits from the Cologne area (26-m core). It has been found that more than 90% of the concentration levels for Cu, Zn, Pb, Hg, and Cd in the sediments of the lower Rhine originate from human sources. Only 2% of the cadmium in the fine-grained sediments is supplied from natural sources; 98% results from anthropogenic influences, in particular, from the highly cadmium-contaminated tributaries Neckar and Main (see above).

This type of pollution is apparently still on the increase. By analyzing sediments collected since 1922, Salomons and De Groot (1978) were able to trace the development of metal pollution in the lower Rhine in the Netherlands over a period of more than 50 years (Figure 8). The samples taken from the floodplain at the beginning of this century were already anthropogenically influenced with respect to Pb, Cd, and Mg, as can be shown by comparison with sediment data from polders reclaimed in the fifteenth and eighteenth centuries. Between 1920 and 1958 all trace element concentrations studied in sediment from the Rhine River increased. Whereas the concentrations of mercury and lead decreased between 1958 and 1975, cadmium continued to increase up to a level that is now 100 times greater than that in sediments of Rhine polders reclaimed in 1788.

The consequences of such a development in the case of cadmium is especially damaging for the Netherlands. For one thing, it is becoming increasingly difficult to dispose of cadmium-contaminated sediments from harbors such as Rotterdam in near-cost areas (Oslo Convention). For another, research being conducted by, among others, A. J. De Groot in association with the Adriaan Volker Dredging Company (1976; see also Lee et al., 1976) has shown that there is an especially distinct relationship between the cadmium content of crops grown on contaminated sediment from the Rhine River and the concentration of the substrate. With a cadmium level of 15 ppm in dredged material, the corresponding values of the element in lettuce increases up to 5 to 6 ppm (see also Chapter 15 by Jastrow and Koeppe, this volume).

11. CHEMICAL ASSOCIATIONS OF CADMIUM IN POLLUTED SEDIMENTS

The availability of trace metals for metabolic processes is closely related to their chemical species both in solution and in particulate matter. The types of chemical associations between metals and sediment constituents have therefore become of particular interest in connection with problems arising from the disposal, that is land applications, of contaminated dredged material. Chemical extraction procedures from soil studies and sedimentary geochemistry—for example, how to distinguish between detrital and authigenic phases in limestones (Hirst and Nicholls, 1958), in shales (Gad and Le Riche, 1966) and in pelagic manganese or iron-concretions (Chester and Hughes, 1967)—have been developed to a level whereby it has become possible to estimate the toxicity potential of a sediment. At present, the most advanced techniques include the successive extraction and determination of the metal contents in interstitial water and in ion-exchangeable, easily reducible, moderately reducible, organic, and residual sediment fractions (Table 9).

In this respect, "oxidizable," "reducible," and "residual" phases were first differentiated in polluted sediments by Bruland et al. (1974). Comparison of data from the Soledad Basin off California (less polluted) with those from Whites Point, located near a sewer outfall, shows a definite increase in the oxidizable and, in particular, the reducible fraction in anthropogenic contamination by copper and zinc. This development concurs with the results of studies by Gupta and Chen (1975) on sediment from Los Angeles Harbor, by Patchineelam (1975) on deposits from the lower Rhine River, by Brannon and colleagues (1976) on contaminated harbor sediments from Lake Erie, Mobile Bay, Alaska, and Bridgeport, Connecticut, and by Patchhineelam and Förstner (1977) on a sediment core from the German Bight. The examples of cadmium associations in these sediments given in Table 10 suggests a general decrease in the residual bonding form, that is, the predominantly inertly fixed cadmium content, as the

Table 9. Selective Chemical Extraction of Trace Metal Associations in Sediments

Form of Chemical Association	Extraction Procedures	Reference
Detrital/authigenic phases	Treatment with complexing agents (EDTA) or dilute acids (0.1 M HCl or HNO_3)	Goldberg and Arrhenius (1958), Piper (1971), Malo (1977)
		Chester and Hughes (1967)
Reducible phases (sorption, carbonate, hydrous Fe or Mn oxide)	1 M $NH_2OH \cdot HCl$ + acetic acid "acid-reducing agent"	
Exchangeable cations	$BaCl_2$, $MgCl_2$, NH_4OAc	Jackson (1958)
Carbonate phases	CO_2 treatment, exchanger-column	Patchineelam (1975) Deurer et al. (1978)
		Chao (1972)
Easily reducible phases (carbonate, Mn oxides)	0.1 M hydroxylaminhydrochloride + 0.01 M HNO_3	
	Na dithionite/citrate,	Holmgren (1967)
Moderately reducible phases (hydrous Fe oxides)	0.04 M $NH_2OH \cdot HCl$ + 25% Ac	Gupta and Chen (1975)
Oxidizable phases (sulfides, organic substances)	30% H_2O_2 (95°), extraction with 1 N ammonium acetate	Engler et al. (1974)
Humic substances	0.5 N NaOH, 0.1 N NaOH	Rashid (1971)
Lipids, asphalt	Benzene/dichloromethane/methanol (1:1:1)	Cooper and Harris (1974)
Organic residues	Na hypochlorite, citrate-dithionite-extraction	Gibbs (1973)
Detrital silicates	Digestion with $HF/HClO_4$	Riley (1958)

Table 10. Chemical Associations of Cadmium (%) in Polluted Sediments.

Extracted Phase	German Bight (North Sea)[a]	Los Angeles Harbor[b]	Ashtabula (Lake Erie)[c]	Bridgeport, Conn.[c]	Lower Rhine[d]
	Total Cadmium Concentration (ppm)				
	1.2	2.2	4.1	15.2	28.4
Cation exchange	30	1	1	1	4
Organics, sulfides	10	38	76	96	47
Carbonate	21	5			43
Easily reducible		9	2	4	
Moderately reducible	11	34	1	1	6
Residual	28	14	22	1	1

[a] Patchineelam and Förstner (1972).
[b] Gupta and Chen (1975).
[c] Brannon et al. (1976).
[d] Patchineelam (1975).

342

anthropogenic metal enrichment increases. There is a characteristic affinity of cadmium for organic substances and sulfides in polluted sediment at lower carbonate concentrations. At higher carbonate contents the associations with carbonate minerals—either as discrete cadmium carbonate or as coprecipitates with calcite—seem to provide the major mechanism of immobilization of elevated cadmium concentrations in the effluents (De Groot and Salomons, 1977; Förstner and Patchineelam, 1978; Salomons and Mook, 1978).

Experiments using sand filter columns, fed with both polluted river water and metal dosages (Förstner et al., in prep.) indicate that the major portion of dissolved cadmium is eliminated within the upper 10-15 cm of the sediment material in the form of carbonate precipitates. Partitioning data from the reducible fractions suggest that this is mainly due to the formation of discrete carbonate phases.

12. RELEASE OF CADMIUM FROM POLLUTED SEDIMENTS

Trace metals temporarily immobilized in the bottom sediments and suspended matter of aquatic systems may be released as a result of physicochemical changes, such as (a) increased salinity, (b) lowering of pH, (c) increased input of organic chelators, (d) microbial activity, and (e) an alteration in the redox conditions (Förstner and Patchineelam, 1976; Khalid, this volume, Chapter 8).

12.1. River-Sea Interface

Increased salinity in a water body leads to competition between dissolved cations and absorbed trace metal ions and can result in partial replacement of the latter. Such effects should be expected particularly in the estuarine environment. However, apart from the highly complicated situation there (Burton and Liss, 1976), it was suggested by Müller and Förstner (1975) Duinker and Nolting (1977), and Salomons and Mook (1977) that the decrease in heavy metal concentration at the river-sea interface—particle-associated cadmium in the Elbe Estuary fell from 38 to less than 3 ppm, in the Rhine Estuary from 28 to less than 1 ppm (De Groot et al., 1973)—is caused by mixing processes of polluted river particulates with relatively "clean" marine sediments, rather than by solubilization and/or desorption. Experiments performed by Van der Weijden et al. (1977) with artificial seawater indicate desorption of trace metals from particulate matter, presumably by inorganic complex formation (highest for cadmium). Similar developments have been described by Rohatgi and Chen (1975) from experiments on the desorption of metals from sludge material diluted in seawater; 93% of the original cadmium content of sewage particles was released after 4 weeks' treatment.

12.2.   Acidic Waters

A lowering of pH leads to the dissolution of carbonate and hydroxide minerals and (as a result of hydrogen ion competition) to an increased desorption of metal cations. Long-term changes in pH conditions have been observed in waters poor in bicarbonate ions, which are affected by atmospheric SO_2 emissions. Significant increases in the cadmium concentrations were reported for waters of the Sudbury, Canada, area (Beamish, 1976). Cadmium enrichment in acidic mine effluents usually is relatively low (in South African gold mines 10^2 times higher than in unpolluted river water, in comparison with metals such as Fe, Mg, Ni, Co, Zn, and Cu, for which a 10^3- to 10^4-fold increase occurs; Förstner and Wittmann, 1976, 1979).

12.3.   Organic Complexing

Interaction with organic substances plays an increasingly important role in the transport of heavy metals in both surface water and groundwater, since the amount of organic complexing material increases further because of secondary sewage treatment effluents. An even more serious impact on heavy metal remobilization from polluted sediments may result (a matter of considerable controversy during the last few years) from the growing use of synthetic complexing agents (e.g., nitrilotriacetic acid, NTA) in detergents to replace polyphosphates. Experiments performed by Banat et al. (1974) with polluted river sediments indicate a high percentage of cadmium mobilization. The results of a test made by Chau and Shiomi (1972) on various NTA-metal complexes in natural waters of Lake Ontario show a very delayed degradation of cadmium chelates. Thus a potential danger seems to arise for drinking water obtained from bank filtration or artificial recharge processes (Schöttler, 1975).

12.4.   Redox Changes

A change in redox conditions is usually caused by an increased input of nutrients. Oxygen deficiency in sediments leads to an initial dissolution of manganese oxides, followed by that of hydrous iron oxides. Since these metals are readily

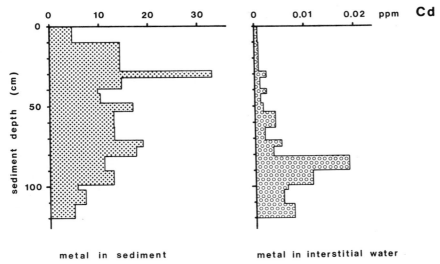

Figure 9. Cadmium concentrations of sediments and interstitial solutions in a vertical profile from the Besigheim/Neckar lock reservoir. The data suggest the remobilization and release of cadmium (and of other metals) from the sediments into the surface waters. (Reinhard and Förstner, 1976.)

soluble in their divalent states, any coprecipitates with metallic coatings become partially remobilized. On the other hand, the presence of sulfide under reduced conditions will precipitate toxic metals, which will be released, by conversion of sulfide to sulfate, under oxidizing conditions (Lu and Chen, 1976).

Isotope studies performed by Gambrell et al. (1976) with Mississippi River sediment material indicate that exchangeable [109]Cd levels were strongly pH-redox potential dependent. A much greater proportion of the incubated cadmium isotope was recovered in the readily bioavailable forms than was obtained for any other potentially toxic heavy metal studied. Incubation experiments by Schindler and Albert (1977) with ascorbic acid and Na_2S indicate that cadmium probably associates with the organic fraction of the sediment during reduction treatment.

Mobilization of cadmium has been studied on core profiles from lock reservoirs of the highly polluted Neckar River (see above); the example shown in Figure 9 stems from the Besigheim reservoir, a few hundred meters upstream from the former cadmium-rich effluent of a pigment plant. The highest cadmium concentrations both in the sediment (100-fold increase) and in the interstitial solutions (200-fold increase compared to unpolluted river waters) occur in the deeper parts of the core. In the near-surface sediment layers the metal contents of the pore solutions do not differ greatly from the metal concentration in the

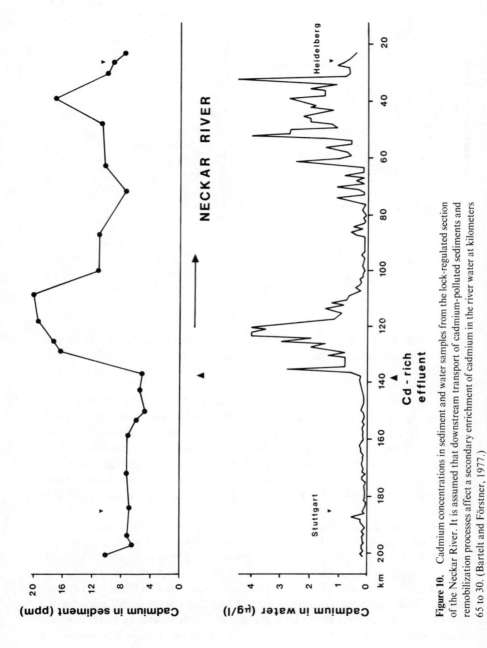

Figure 10. Cadmium concentrations in sediment and water samples from the lock-regulated section of the Neckar River. It is assumed that downstream transport of cadmium-polluted sediments and remobilization processes affect a secondary enrichment of cadmium in the river water at kilometers 65 to 30. (Bartelt and Förstner, 1977.)

surface water. The development of such a concentration gradient is explained by an exchange of cadmium at the sediment-water interface; the metal is continuously supplied from the sediment by dissolution and complexation, thus causing its depletion in the surface layers relative to the deeper sediments.

The characteristic effects of cadmium remobilization can be seen from a hydrologic long-section of the lock-regulated Neckar River (Figure 10). The former waste inputs of cadmium-contained materials near kilometer 60 are still partially present in the sediment, even though the extreme enrichment of more than 300 ppm Cd has been reduced to approximately 20 ppm. Another area of increased cadmium concentrations in sediments is in the lower section of the river. Since no cadmium-containing effluent is present in this region, it can be assumed that the increase is due to the partial discharge of cadmium-rich sediments downstream from the former point source near Besigheim. In both river sections, significant enrichment of cadmium in the soluble phase occurs, quite probably because of its release from the highly polluted sediments.

Possible exchange mechanisms from metal-rich interstitial solutions into surface water include diffusion processes, upward migration during compaction, resuspension of surface sediments in stormwater, and bioturbation. The release of elevated concentrations of heavy metals such as cadmium into the surface water may also be affected by dredging activities. From dredge experiments in San Francisco Bay (Pacific Northwest Laboratories, 1975) it was found that larger concentrations of cadmium were released to the water column under oxygen-rich that under oxygen-deficient conditions; under oxidized conditions larger releases of cadmium were measured as salinity increased. The first effect is attributed to the release, upon oxidation, of trace metals bound to sulfide phases; the second observation can be explained by the enhanced formation of soluble inorganic complexes and by the greater cation competition of more saline waters. Investigations performed by Patrick et al. (1977) on the solubility and potential availability of toxic metals in aquatic systems have shown that cadmium is much more active under acid-oxidizing than under non-acid-reducing conditions. (Levinson, 1974, noted that cadmium possesses moderate mobility in oxidizing acidic and neutral to alkaline environments, and low mobility in a reducing environment.) The general observation that the effects of dredging on both water quality and biota are relatively insignificant because of the slowness of the processes, the largely refractory bonding of metals, or a rapid reprecipitation of heavy metals (Windom, 1973; May, 1974; Lee, 1977; Patrick at al., 1977; Sly, 1977) should be reexamined with respect to the behavior of cadmium. Studies by Gilbert et al. (1976) on the influence of the sediment-water interface on the aquatic chemistry of Cr, Ag, and Cd have indicated that with short-term resuspension no significant release of all three metals will occur but with prolonged resuspension of anoxic sediments some cadium may redissolve. Because of the high toxicity of cadmium, therefore, local conditions should be carefully studied, and further experimental work is strongly needed to understand the various aspects of the interactions between water, sediments, and biota in areas polluted with cadmium.

REFERENCES

Abdullah, M. I. and Royle, L. G. (1974). "Cadmium in some British Coastal and Freshwater Environments." In *Problems of the Contamination of Man and His Environment by Mercury and Cadmium.* Proceedings of an International Symposium, Luxembourg, July 3–5, 1973, pp. 69–81.

Adriaan Volker Dredging Company (1976). *Assessment of Certain European Practices and Dredged Material Containment and Reclamation Methods.* First Interim Report, Dredged Material Research Program. U.S. Army Engineer Waterways Experiment Station, Vicksburg, Miss. Cited by C. R. Lee et al. (1976).

Agemian, H. and Chau, A. S. Y. (1977). "A Study of Different Analytical Extraction Methods for Nondetrital Heavy Metals in Aquatic Sediments," *Arch. Environ. Contam. Toxicol.* **6**, 69–82.

Allan, R. J. (1974). *Metal Contents of Lake Sediment Cores from Established Mining Areas: an Interface of Exploration and Environmental Geochemistry.* Geol. Surv. Can; Pap. 74-1/B, pp. 43–49.

Allan, R. J. (1977). "Natural versus Unnatural Heavy Metal Concentrations in Lake Sediments in Canada." In *Proceedings of the International Conference on Heavy Metals in the Environment, Toronto, Oct. 27–31, 1975,* Vol. II/2, pp. 785–808.

Allan, R. J., Cameron, E. M., and Durham, C. C. (1972). *Reconnaissance Geochemistry Using Lake Sediments of a 36,000 Square-Mile Area of the Northwestern Canadian Shield. Geol. Surv. Can.* Pap. 72/50, 70 pp.

Altschuler, Z. S. (1978). "Trace Elements as Discriminants of Origin in Marine Phosphorites." In *Abstracts, 10th International Congress on Sedimentology, Jerusalem, July 9–14, 1978,* pp. 16–17.

Anderson, J. (1974). "A Study of the Digestion of Sediment by the HNO_3-H_2SO_4 and the HNO_3-HCl Procedures," *Atomic Abstr. Newslett.* **13**, 31.

Applequist, M. D., Katz, A., and Turekian, K. K. (1972). "Distribution of Mercury in the Sediments of New Haven (Conn.) Harbor," *Environ. Sci. Technol.* **6**, 1123–1124.

Asami, T. (1974). "Environmental Pollution by Cadmium and Zinc Discharged from a Braun-Tube Factory," *Ibaraki Daigaku Nogakubu Gakujutsu Hokaku (Jap.)* **22**, 19–23.

Aston, S. R. and Thornton, I. (1977). "Regional Geochemical Data in Relation to Seasonal Variations in Water Quality," *Sci. Total Environ.* **7**, 247–260.

Aston, S. R., Chester, R., Griffiths, A., and Riley, J. P. (1972). "Distribution of Cadmium in North Atlantic Deep-Sea Sediments," *Nature* **239**, 393.

Aston, S. R., Thornton, I., Webb, J. S., Purves, J. B., and Milford, B. L. (1974). "Stream

Sediment Composition, an Aid to Water Quality Assessment," *Water, Air, Soil Pollut.* **3,** 321–325.

Australian Gov. Technical Commission (1974). *Mine Waste Pollution of the Molonglo River.* Final Report on Remedial Measures, June, 68 pp.

Ayling, G. M. (1974). "Uptake of Cadmium, Zinc, Copper, Lead, and Chromium in the Pacific Oyster, *Crassostrae Gigas,* Grown in the Tamar River, Tasmania," *Water Res.* **8,** 729–738.

Ayling, G. M. (1975). "The Environmental Influence on the Uptake of Cadmium by Shellfish." In *Environment '75, Sydney (Australia), July 1–4,* Vol. III, pp. 237–246.

Azumi, T. and Yoneda, A. (1975). "Environmental Analysis. I: Content of Heavy Metals in the Sediment of Rivers near Himeji City." *Himeji Kogyo Daigaku Kenkyu (Jap.)* **28A,** 110–114.

Banat, K., Förstner, U., and Müller, G. (1972a). "Schwermetalle in den Sedimenten des Rheins," *Umschau Wiss. Tech.* **72,** 192–193.

Banat, K., Förstner, U., and Müller, G. (1972b). "Schwermetalle in Sedimenten von Donau, Rhein, Ems, Weser und Elbe in Bereich der Bundesrepublik Deutschland," *Naturwissenschaften* **12,** 525–528.

Banat, K., Förstner, U., and Müller, G. (1974). "Experimental Mobilization of Metals from Aquatic Sediments by Nitrilotriacetic Acid," *Chem. Geol.* **14,** 199–207.

Bartelt, R. D. and Förstner, U. (1977). "Schwermetalle im staugeregelten Neckar. Untersuchungen an Sedimenten, Algen und Wasserproben," *Jber. Mitt. Oberrhein. Geol. Ver.* **59,** 247–263.

Bayerische Landesanstalt für Wasserforschung (1977). *Untersuchungen über die Belastung bayerischer Gewässer mit Cadmium (1972–1977),* 78 pp.

Beamish, R. J. (1976). "Acidification of Lakes in Canada by Acid Precipitation and the Resulting Effects on Fishes," *Water, Air, Soil Pollut.* **6,** 501–514.

Becker, R. (1976). "Geochemische Untersuchung der Sedimente des Flusses Blies, Saarland," *Naturwissenschaften* **63,** 144.

Bennett, H. and Hawley, W. G. (1965). *Methods of Silicate Analysis.* Academic Press, London and New York.

Berggren, B. and Odén, S. (1972). *Analyseresultat rorande fungmetallen och klorerade kolväten i rötslam from Svenska reningsverk 1968–1971.* Inst. Markvetenskap Lantbrukshögskolan, Uppsala/Sweden. Cited by A. L. Page (1974).

Berrow, M. L. and Webber, J. (1972). "Trace Elements in Sewage Sludges," *J. Sci. Food Agric.* **23,** 93–100.

Blakesley, P. A. (1973). *Monitoring Considerations for Municipal Wastewater Effluent and Sludge Application to the Land.* U.S. Environmental Protection Agency, U.S. Department of Agriculture, University Workshop, III, Champaign, Urbana. Cited by A. L. Page (1974).

Bloom, H. and Ayling, G. M. (1977). "Heavy Metals in the Derwent Estuary," *Environ. Geol.* **2,** 3–22.

Bloxam, T. W., Aurora, S. N., Leach, L., and Rees, T. R. (1972). "Heavy Metals in Some River and Bay Sediments near Swansea," *Nature (Phys. Sci.)* **239,** 158–159.

Bopp, F. and Biggs, R. B. (1973). "Trace Metal Environments near Shell Banks in Delaware Bay," *Delaware Bay Rep. Ser.* **3**, 23–69.

Bouquieaux, J. (1974). "Mercury and Cadmium in the Environment—First Results of an Enquiry on an European Scale." In *Problems of the Contamination of Man and his Environment by Mercury and Cadmium,* Proceedings of an International Symposium, Luxembourg, July 3–5, 1973, pp. 23–46.

Bower, P. M., Simpson, H. J., Williams, S. C., and Li, Y. H. (1978). "Heavy Metals in the Sediments of Foundry Cove, Cold Spring, New York," *Environ. Sci. Technol.* **12**, 683–687.

Boyden, C. R. (1975). "Distribution of Some Trace Metals in Poole Harbour, Dorset," *Mar. Pollut. Bull.* **6**, 180–187.

Brannon, J. M., Engler, R. M., Rose, J. R., Hunt, P. G., and Smith, I. (1976). "Distribution of Toxic Heavy Metals in Marine and Freshwater Sediments." In *Proceedings of a Special Conference on Dredging and Its Environmental Effects, Mobile, Ala., Jan. 26–28,* 1976, ASCE, pp. 455–495.

Brooks, R. R., Presley, B. J., and Kaplan, I. R. (1968). "Trace Elements in the Interstitial Waters of Marine Sediments," *Geochim. Cosmochim. Acta* **32**, 397–414.

Brown, J. R. and Chow, L. Y. (1977). "Heavy Metal Concentrations in Ontario Fish," *Bull. Environ. Contam. Toxicol.* **17**, 190–195.

Brügmann, L. (1977). "Zur Verteilung einiger Schwermetalle in der Ostsee—eine Übersicht," *Acta Hydrochim. Hydrobiol.* **5**, 3–21.

Bruland, K. W., Bertine, K., Koide, M., and Goldberg, E. D. (1974). "History of Metal Pollution in Southern California Coastal Zone," *Environ. Sci. Technol.* **8**, 425–432.

Bryan, G. W. and Hummerstone, L. G. (1973). "Adaptation of the Polychaete *Nereis Diversicolor* to Estuarine Sediments Containing High Concentrations of Zinc and Cadmium," *J. Mar. Biol. Assoc. U.K.* **53**, 839–857.

Burton, J. D. and Liss, P. S. (1976). *Estuarine Chemistry.* Academic Press, London.

Butterworth, J., Lester, P., and Nickless, G. (1972). "Distribution of Heavy Metals in the Severn Estuary," *Mar. Pollut. Bull.* **3**, 72–74.

Chao, L. L. (1972). "Selective Dissolution of Manganese Oxides from Soils and Sediments with Acidified Hydroxylamine Hydrochloride," *Soil Sci. Soc. Am. Proc.* **36**, 764–768.

Chau, Y. K. and Shiomi, M. T. (1972). "Complexing Properties of Nitrilotriacetic Acid in the Lake Environment," *Water, Air, Soil Pollut.* **1**, 149–164.

Chen, K. Y., Young, T. K. J., Jan, T. K., and Rohatgi, N. (1974). "Trace Metals in Waste Water Effluents," *J. Water Pollut. Control Fed.* **46**, 2663–2675.

Chester, R. and Hughes, M. J. (1967). "A Chemical Technique for the Separation of Ferro-Manganese Minerals, Carbonate Minerals and Adsorbed Trace Elements from Pelagic Sediments," *Chem. Geol.* **2**, 249–262.

Clarke, F. (1924). *The Data of Geochemistry.* U.S. Geol. Surv. Bull. *770,* 841 pp.

Cline, J. T. and Upchurch, S. B. (1973). "Mode of Heavy Metal Migration in the Upper

Strata of Lake Sediments." In *Proceedings of the 16th Conference on Great Lakes Research,* pp. 349–356.

Cooper, B. S. and Harris, R. C. (1974). "Heavy Metals in Organic Phases of River and Estuarine Sediment," *Mar. Pollut. Bull.* **5,** 24–26.

Damiani, V., Ferrario, A., Gavelli, G., and Thomas, R. L. (1977). "Trace Metal Composition and Fractionation of Mn, Fe, S, P, Ba and Si in the Bay of Quinte Freshwater Ferro-Manganese Concretions, Lake Ontario." In H. L. Golterman, Ed. *Proceedings of the International Symposium on Interactions between Sediments and Freshwater, Amsterdam, Sept. 6–10, 1976.* Junk B. V. Publ., The Hague and Pudoc, Wageningen, pp. 83–93.

Davaud, E., Viel, M., and Vernet, J.-P. (1978). "Contamination des Sédiments Côtiers par les Métaux Lourds." In J.-P. Vernet and G. Scolari, Eds. *Etude de la Pollution des Sédiments du Léman et du Bassin du Rhône.* Rapports Commission Internationale pour la Protection des Eaux du Léman Contre la Pollution. Campagne 1977. 16 pp.

De Groot, A. J. (1966). "Mobility of Trace Elements in Delta." In *Transactions of the International Society of Soil Science Conference, Aberdeen,* Vol. II/IV, pp. 267–297.

De Groot, A. J., De Goeij, J. J. M., and Zengers, C. (1971). "Contents and Behaviour of Mercury as Compared with Other Heavy Metals in Sediments from the Rivers Rhine and Ems," *Geol. Mijnbouw* **50,** 393–398.

De Groot, A. J., Allersma, E., and Van Driel, W. (1973). "Zware Metalen in Fluviatile en Marine Ecosystemen." In *Symp. Waterloopkunde in Dienst van Industrie en Milieu,* Publ. 110 N, 27 pp.

De Groot, A. J. and Salomons, W. (1977). "Influence of Civil Engineering Projects on Water Quality in Deltaic Regions." In *Proceedings of the Symposium on Effects of Urbanization and Industrialization on the Hydrological Regime and on Water Quality, Amsterdam, October 1977.* IAHS Publ. 123, pp. 351–357.

Delfino, J. J. and Lee, G. F. (1968). "Chemistry of Manganese in Lake Mendota," *Environ. Sci. Technol.* **2,** 1094–1100.

Depers, A. M. (1978). "Man's Influence on the Trace Element Concentrations of the Sediments in Lake Illawarra, New South Wales, Australia." *Abstracts, 10th International Congress on Sedimentology, Jerusalem, July 9–14, 1978,* pp. 171–172.

Deurer, R., Förstner, U., and Schmoll, G. (1978). "Selective Chemical Extraction of Carbonate-Associated Metals from Recent Lacustrine Sediments," *Geochim. Cosmochim. Acta* **42,** 425–427.

Dominik, J., Förstner, U., Mangini, A., and Reineck, H.-E. (1978). "Pb-210 and Cs-137 Chronology of Heavy Metal Pollution in a Sediment Core from the German Bight (North Sea)," *Senckenberg. Marit.* **10,** 213–227.

Duinker, J. C. and Nolting, R. F. (1977). "Dissolved and Particulate Trace Metals in the Rhine Estuary and the Southern Bight," *Mar. Pollut. Bull.* **8,** 65–69.

Elderfield, H. and Hepworth, A. (1975). "Diagenesis, Metals and Pollution in Estuaries," *Mar. Pollut. Bull.* **6,** 85–87.

Engler, R. M., Brannon, J. M., Rose, J., and Bigham, G. (1974). "A Practical Selective Extraction Procedure for Sediment Characterization." 168th Meeting, American Chemical Society, Atlantic City, N.J.

Enk, M. D. and Mathis, B. J. (1977). "Distribution of Cadmium and Lead in a Stream Ecosystem," *Hydrobiologia* **52**, 153–158.

Eriksson, E. (1977) "Water Chemistry and Water Quality," *Ambio* **6**, 27–30.

Erlenkeuser, H., Suess, E., and Willkomm, H. (1974). "Industrialization Affects Heavy Metal and Carbon Isotope Concentration in Recent Baltic Sea Sediments," *Geochim. Cosmochim. Acta,* **38**, 823–842.

Fitchko, J. and Hutchinson, T. C. (1975). "A Comparative Study of Heavy Metal Concentrations in River Mouth Sediments around the Great Lakes," *J. Great Lakes Res.* **1**, 46–78.

Förstner, U. (1976a). "Lake Sediments as Indicators of Heavy-Metal Pollution," *Naturwissenschaften* **63**, 465–470.

Förstner, U. (1976b). "Schwermetallverunreinigungen in Gewässern und ihre Erfassung durch Sedimentuntersuchungen." In *Polizei Technik Verkehr 1976*, pp. 25–32.

Förstner, U. (1977a). "Forms and Sediment Associations of Trace Metals." In H. Shear and A. E. P. Watson, Eds., *The Fluvial Transport of Sediment-Associated Nutrients and Contaminants.* Proceedings of Workshop IJC/PLUARG, Kitchener, Ont., Oct. 20–22, 1976, pp. 219–233.

Förstner, U. (1977b). "Metal Concentrations in Freshwater Sediments—Natural Background and Cultural Effects." In H. L. Golterman, Ed., *Proceedings of the International Symposium on Interactions between Sediments and Fresh Water, Amsterdam, Sept. 6–10, 1976.* Junk B.V. Publ., The Hague and Pudoc, Wageningen, pp. 94–103.

Förstner, U. (1977c). "Metal Concentrations in Recent Lacustrine Sediments," *Arch. Hydrobiol.* **80**, 172–191.

Förstner, U. (1977d). "Geochemische Untersuchungen an den Sedimenten des Ries-Sees (Forschungsbohrung Nördlingen 1973)," *Geol. Bavarica* **75**, 37–48.

Förstner, U. (1978). *Metallanreicherungen in rezenten See-Sedimenten—geochemischer Background und zivilisatorische Einflüsse. Mitt. Nationalkomitee Bundesrepublik Deutschland für das Internationale Hydrologische Programm der Unesco,* Heft 2. Koblenz, 66 pp.

Förstner, U. and Müller, G. (1973a). "Heavy Metal Accumulation in River Sediments: A Response to Environmental Pollution," *Geoforum* **14**, 53–62.

Förstner, U. and Müller, G. (1973b). "Anorganische Schadstoffe im Neckar," *Ruperto Carola* **51**, 67–71.

Förstner, U. and Müller, G. (1974a). *Schwermetalle in Flüssen und Seen als Ausdruck der Umweltverschmutzung.* Springer, Berlin, 225 pp.

Förstner, U. and Müller, G. (1974b). "Schwermetallanreicherungen in datierten Sedimentkernen aus dem Bodensee und aus dem Tegernsee," *Tschermaks Min. Petr. Mitt.* **21**, 145–163.

Förstner, U. and Müller, G. (1975). "Hydrochemische Beziehungen zwischen Flusswasser und Uferfiltrat," *Gas-Wasserfach* **116**, 74–80.

Förstner, U. and Müller, G. (1976). "Heavy Metal Pollution Monitoring by River Sediments," *Forschr. Mineral.* **53,** 271–288.

Förstner, U. and Patchineelam, S. R. (1976). "Bindung und Mobilisation von Schwermetallen in fluviatilen Sedimenten," *Chem-Ztg.* **100,** 49–57.

Förstner, U. and Patchineelam, S. R. (1978). "Chemical Associations of Heavy Metals in Polluted Sediments from the Lower Rhine River." In *Abstracts, 10th International Congress on Sedimentology, Jerusalem, July 9–14, 1978,* pp. 217–218.

Förstner, U. and Reineck, H.-E. (1974). "Die Anreicherung von Spurenelementen in den rezenten Sedimenten eines Profilkerns aus der Deutschen Bucht," *Senckenbergiana Marit.* **6,** 175–184.

Förstner, U. and Stiefel, R. (1978). "Umweltprobleme durch Metallanreicherungen in kommunalen Abwässern," *Chem.-Ztg.* **102,** 161–168.

Förstner, U. and Wittmann, G. T. W. (1976). "Metal Accumulations in Acidic Waters from Gold Mines in South Africa," *Geoforum* **7,** 42–49.

Förstner, U. and Wittmann, G. (1979). *Metal Pollution in the Aquatic Environment.* Springer-Verlag, Berlin, 486 pp.

Förstner, U., Müller, G., and Wagner, G. (1974). "Schwermetalle in Sedimenten des Bodensees—natürliche und zivilisatorische Anteile," *Naturwissenschaften* **61,** 270.

Gad, M. A. and Le Riche, H. H. (1966). "A Method for Separating the Detrital and Non-detrital Fractions of Trace Elements in Reduced Sediments," *Geochim. Cosmochim. Acta* **30,** 841–846.

Gale, N. L., Bolter, E., and Wixson, B. G. (1976). "Investigations on Clearwater Lake as a Potential Sink for Heavy Metals from Lead Mining in Southeast Missouri." In D. D. Hemphill, Ed., *Trace Substances in Environmental Health,* Vol. X. University of Missouri, Columbia, pp. 187–196.

Gambrell, R. P., Khalid, R. A., and Patrick, W. H., Jr. (1976). "Physicochemical Parameters That Regulate Mobilization and Immobilization of Toxic Heavy Metals." In *Proceedings of the Special Conference on Dredging and Its Environmental Effects, Mobile, Ala., Jan. 26–28, 1976,* ASCE, pp. 418–434.

Gibbs, R. J. (1973). "Mechanisms of Trace Metal Transport in Rivers," *Science* **180,** 71–73.

Gilbert, T. R., Clay, A. M., and Leighty, D. A. (1976). *Influence of the Sediment/Water Interface on the Aquatic Chemistry of Heavy Metals.* Environmental Chemistry Research Division, Tyndall Air Force Base, Rep. AFCEC-TR-6-22, Florida, 89 pp.

Goldberg, E. D. (1976). *The Health of the Oceans.* UNESCO Press, Paris, 172 pp.

Goldberg, E. D. and Arrhenius, G. O. S. (1958). "Chemistry of Pacific Pelagic Sediments," *Geochim. Cosmochim. Acta* **13,** 153–212.

Goldberg, E. D., Hodge, V., Koide, M. and Griffin, J. J. (1976). "Metal Pollution in Tokyo as Recorded in Sediments of the Palace Moat," *Geochem. J.* **10,** 165–174.

Goldberg, E. D., Gamble, E., Griffin, J. J. and Koide, M. (1977). "Pollution History of Narragansett Bay as Recorded in Its Sediments," *Estuarine Coastal Mar. Sci.* **5,** 549–561.

Gong, H., Rose, A. W., and Suhr, N. H. (1977). "The Geochemistry of Cadmium in Some Sedimentary Rocks," *Geochim. Cosmochim. Acta* **41**, 1687–1692.

Goto, M. (1973). "Inorganic Chemicals in the Environment—with Special Reference to the Pollution Problems in Japan," *Environ. Qual. Safety* **2**, 72–77.

Greig, R. A. and McGrath, R. A. (1977). "Trace Metals in Sediments of Raritan Bay," *Mar. Pollut. Bull.* **8**, 188–192.

Greig, R. A., Reid, R. N., and Wenzloff, D. R. (1977). "Trade Metal Concentrations in Sediments from Long Island Sound," *Mar. Pollut. Bull.* **8**, 183–188.

Grimmer, G. and Böhnke, H. (1977). "Schadstoffuntersuchungen an datierten Sedimentkernen aus dem Bodensee. I: Profile der polycyclischen Kohlenwasserstoffe," *Z. Naturforsch.* **32c**, 703–707.

Gross, M. G., Black, J. A., Kalin, R. J., Schramel, J. R., and Smith, R. N. (1971). *Survey of Marine Waste Deposits, New York Metropolitan Region.* State University of New York, Marine Science Research Center, Tech. Rep. 8, 72 pp.

Gupta, S. K. and Chen, K. Y. (1975). "Partitioning of Trace Metals in Selective Chemical Fractions of Nearshore Sediments," *Environ. Lett.* **10**, 129–158.

Håkanson, L. (1977). "Sediments as Indicators of Contamination—Investigations in the Four Largest Swedish Lakes," *Naturvårdsverkets Limnologiska Undersökning*, **92** (SNV PM 839), 159 pp.

Halcrow, W., MacKay, D. W., and Thornton, I. (1973). "The Distribution of Trace Metals and Fauna in the Firth of Clyde in Relation to the Disposal of Sewage Sludge," *J. Mar. Biol. Assoc. U.K.* **53**, 721–739.

Hann, R. W., Jr., and Slowey, J. F. (1972). *Sediment Analysis—Galveston Bay.* Texas A & M University, Environmental Engineering Division, Tech. Rep. 24, 57 pp.

Harbridge, W., Pilkey, O. H., Whaling, P., and Swetland, P. (1976). "Sedimentation in the Lake of Tunis: a Lagoon Strongly Influenced by Man," *Environ. Geol.* **1**, 215–225.

Hartung, R. (1974). "Heavy Metals in the Lower Mississippi." In *Proceedings of the International Conference on Persistent Chemicals in Aquatic Ecosystems, Ottawa, May 1–3, 1974*, pp. I-93–98.

Hattingh, W. H. J. (1977). "Reclaimed Water: a Health Hazard?" *Water S.A. (S. Afr.)* **3**, 104–112.

Hawkes, H. E. and Webb, J. S. (1962). *Geochemistry in Mineral Exploration.* Harper & Row, New York.

Hellmann, H. (1970). "Die Charakterisierung von Sedimenten auf Grund ihres Gehaltes an Spurenelementen," *Dtsch. Gewässerkd. Mitt.* **14**, 160–164.

Hellmann, H. (1972). "Herkunft der Sinkstoffablagerungen in Gewässern. 2: Überlegungen und Ergebnisse aus der Sicht der Abwassertechnik." *Dtsch. Gewässerkd. Mitt.* **16**, 137–141.

Hellmann, H. and Griffatong, A. (1972). "Herkunft der Sinkstoffablagerungen in Gewässern. 1: Chemische Untersuchungen der Schwermetalle," *Dtsch. Gewässerkd. Mitt.* **16**, 14–18.

Helz, G. R. (1976). "Trace Element Inventory for the Northern Chesapeake Bay with Emphasis on the Influence of Man," *Geochim. Cosmochim. Acta* **40**, 573–580.

Herrig, H. (1969). "Untersuchungen an Flusswasser-Inhaltstoffen," *Gas- Wasserfach* **110,** 1385–1391.

Hesse, J. L. and Evans, E. D. (1972). *Heavy Metals in Surface Water, Sediments and Fish in Michigan.* Michigan Water Resources Commission, Department of Natural Resources, 58 pp.

Hinneri, S. (1974). "Enrichment of Elements, Especially Heavy Metals, in Recent Sediments of the Freshwater Reservoir of Uusikaupunki, SW Coastland of Finland," *Turun Yliopiston Julkaisuja Ann. Univ. Turkuensis,* Ser. A (Biol. Geograph. Geolog.), **56,** 30–44.

Hirst, D. M. and Nicholls, G. D. (1958). "Techniques in Sedimentary Geochemistry. 1: Separation of the Detrital and Non-detrital Fractions of Limestone," *J. Sediment. Petrol.* **28,** 461–468.

Holluta, J., Bauer, L., and Kölle, W. (1968). "Über die Einwirkung steigender Flusswasserverschmutzung auf die Wasserqualität und die Kapazität der Uferfiltrate," *Gas- Wasserfach* **109,** 1406–1409.

Holmes, C. W., Slade, E. A., and McLerran, C. J. (1974). "Migration and Redistribution of Zinc and Cadmium in Marine Estuarine System," *Environ. Sci. Technol.* **8,** 255–259.

Holmgren, G. S. (1967). "A Rapid Citrate-Dithionite Extractable Iron Procedure," *Soil Sci. Soc. Am. Proc.* **31,** 210–211.

Horn, M. K. and Adams, J. A. S. (1966). "Computer-Derived Geochemical Balances and Element Abundances," *Geochim. Cosmochim. Acta* **30,** 279–297.

Isaac, R. A. and Delaney, J. (1973). *Toxic Element Survey.* Massachusetts Water Resources Commission, Division of Water Pollution Control, Prog. Rep. 1, Publ. 6108.

Ishibashi, M., Ueda, S., and Yamamoto, Y. (1970). "The Chemical Composition and the Cadmium, Chromium and Vanadium Contents of Shallow-Water Deposits in Tokyo Bay," *J. Oceanogr. Soc. Jap.* **26,** 189–194.

Iskandar, K. and Keeney, D. R. (1974). "Concentration of Heavy Metals in Sediment Cores from Selected Wisconsin Lakes," *Environ. Sci. Technol.* **8,** 165–170.

Jackson, M. L. (1958). *Soil Chemical Analysis.* Prentice-Hall, Englewood Cliffs, N.J., 498 pp.

Johnston, L. M. (1977). "Geochemistry of Selected Sediment Cores from the Kingston Basin—Upper St. Lawrence River Area." In *Abstracts, Conference on Great Lakes Research,* 1977.

Jonasson, I. R. (1976). *Detailed Hydrogeochemistry of Two Small Lakes in the Greenville Geological Province.* Geol. Surv. Can. Pap. 76-13, 37 pp.

Jones, A. S. G. (1973). "The Concentration of Copper, Lead, Zinc and Cadmium in Shallow Marine Sediments, Cardigan Bay, Wales," *Mar. Geol.* **14,** M1-M9.

Kemp, A. L. W. and Thomas, R. L. (1976). "Impact of Man's Activities on the Chemical Composition in the Sediments of Lakes Ontario, Erie and Huron," *Water, Air, Soil Pollut.* **5,** 469–490.

Kemp, A. L. W., Thomas, R. L., Dell, C. I., and Jaquet, J. M. (1976). "Cultural Impact on the Geochemistry of Sediments in Lake Erie," *J. Fish. Res. Board Can.* **33,** 440–462.

Kiba, T., Terada, K., Honjo, T., Matsumoto, R., and Ameno, K. (1975). "Generic Relationships among Samples of River Sediments with the Aid of Concentration Correlation Matrix," *Bunseki Kagaku* **24,** 18–25.

Klein, D. H. and Goldberg, E. D. (1970). "Mercury in the Marine Environment," *Environ. Sci. Technol.* **4,** 765–768.

Kneip, T. J. (1978). "Effects of Cadmium in an Aquatic Environment." In *Proceedings of the First International Cadmium Conference,* San Francisco, Jan. 31-Feb. 2, 1977, pp. 120–124.

Kneip, T. J., Re, G., and Hernandez, T. (1974). "Cadmium in an Aquatic Ecosystem: Transport and Distribution." In D. D. Hemphill, Ed., *Trace Substances in Environmental Health,* Vol. VIII. University of Missouri, Columbia, pp. 173–177.

Kronfeld, J. and Navrot, J. (1974). "Aspects of Trace Metal Contamination in the Coastal Rivers of Israel," *Water, Air, Soil Pollut.* **4,** 127–134.

Kubota, J., Mills, E. L., and Oglesby, R. T. (1974). "Lead, Cadmium, Zinc, Copper and Cobalt in Streams and Lake Waters of Cayuga Basin, New York," *Environ. Sci. Technol.* **8,** 243–248.

Laskowski, N., Kost, T., Pommerenke, D., Schäfer, A., and Tobschall, H. J. (1975). "Heavy-Metal and Organic-Carbon Content of Recent Sediments near Mainz," *Naturwissenschaften* **62,** 136.

Leatherland, T. M. and Burton, J. D. (1974). "The Occurrence of Trace Metals in Coastal Organisms with Particular Reference to the Solent Region," *J. Mar. Biol. Assoc. U.K.* **54,** 457–468.

Lee, C. R., Engler, R. M., and Mahloch, J. L. (1976). *Land Application of Waste Materials from Dredging, Construction and Demolition Processes.* U.S. Army Engineers Waterways Experiment Station, Misc. Pap. D-76-5, Vicksburg, Miss., 42 pp.

Lee, G. F. (1977). "Summary of Studies on the Release of Contaminants from Dredged Sediments and Openwater Disposal." In H. L. Golterman, Ed., *Interactions between Sediment and Fresh Water.* Junk B.V. Publ., The Hague, and Pudoc, Wageningen, pp. 444–446.

Levinson, A. A. (1974). *Introduction to Exploration Geochemistry.* Applied Publ.

Lichtfuss, R. and Brümmer, G. (1977). "Schwermetallbelastung von Elbe-Sedimenten," *Naturwissenschaften* **64,** 122–125.

Literáthy, R. and László, F. (1977). "Uptake and Release of Heavy Metals in the Bottom Silt of Recipients." In H. L. Golterman, Ed., *Interactions between Sediments and Fresh Water.* Junk B.V. Publ., The Hague, and Pudoc, Wageningen, pp. 403–409.

Lu, C. S. J. and Chen, K. Y. (1977). "Migration of Trace Metals in Interfaces of Seawater and Polluted Surficial Sediments," *Environ. Sci. Technol.* **11,** 174–182.

Lytle, T. F., Lytle, J. S., and Parker, P. L. (1973). In *Gulf Res. Rep.* **4,** 214. Cited by I. Roth and H. Hornung (1977).

MacKay, D. W., Halcrow, W., and Thornton, I. (1972). "Sludge Dumping in the Firth of Clyde," *Mar. Pollut. Bull.* **3,** 7–11.

Malo, B. A. (1977). "Partial Extraction of Metals from Aquatic Sediments," *Environ. Sci. Technol.* **11**, 277–282.

Marowsky, G. and Wedepohl, K. H. (1971). "General Trends in the Behaviour of Cd, Hg, Tl and Bi in Some Major Rock Forming Processes," *Geochim. Cosmochim. Acta* **35**, 1255–1267.

Mason, B. (1966). *Principles of Geochemistry,* 3rd ed. Wiley, New York.

Mathis, B. J. and Cummings, T. F. (1973). "Selected Metals in Sediments, Water and Biota in the Illinois River," *J. Water Pollut. Control Fed.* **45**, 1573–1583.

Mathis, B. J. and Kevern, N. R. (1973). *Distribution of Mercury, Cadmium, Lead and Thallium in an Eutrophic Lake.* Michigan State University, Institute of Water Research, Tech. Rep. 34, 23 pp.

Maxfield, D., Rodriguez, J. M., Buettner, M., Davis, J., Forbes, L., Kovacs, R., Russel, W., Schultz, L., Smith, R., Stanton, J., and Wai, C. M. (1974). "Heavy Metal Pollution in the Sediments of the Coeur d'Alene River Delta," *Environ. Pollut.* **7**, 1–6.

May, E. B. (1974). "Effects on Water Quality When Dredging a Polluted Harbor Using Confined Spoil Disposal," *Ala. Mar. Resour. Bull.* **10**, 1–8.

Mihm, U., Botzenhart, K., and Noeske, G. (1976). "Die Bestimmung einiger Metalle und des Phosphats im Sediment der Vorsperre zur Wahnbachtalsperre," *Dtsch. Gewässerkd. Mitt.* **20**, 47–51.

Molnar, F. M., Rothe, P., Förstner, I., Štern, J., Ogorelec, B., Šervelj, A., and Culiberg, M. (1978). "Lakes Bled and Bohinj-Origin, Composition, and Pollution of Recent Sediments," *Geologija (Ljubljana)* **21**, 93–164.

Momoyama, K., Kobayashi, T., Yoshitsugu, K., and Takayama, S. (1975). "Sediment of the Fishing Grounds of Yamaguchi Prefecture in the Inland Sea of Japan," *Yamaguchi-ken Naikai Suisan Shinkenjo Hokoku (Jap.)* **5**, 79 pp.

Moyer, B. R. and Budinger, T. F. (1974). "Cadmium Levels in the Shoreline Sediments of San Francisco Bay." In D. D. Hemphill, Ed., *Trace Substances in Environmental Health,* Vol. VIII. University of Missouri, Columbia, pp. 127–135.

Müller, G. (1977). "Schadstoffuntersuchungen an datierten Sedimentkernen aus dem Bodensee. II: Historische Entwicklung von Schwermetallen-Beziehung zur Entwicklung polycyclischer aromatischer Kohlenwasserstoffe," *Z. Naturforsch.* **32c**, 913–919.

Müller, G. and Förstner, U. (1973). "Cadmium-Anreicherung in Neckar-Fischen," *Naturwissenschaften* **60**, 258–259.

Müller, G. and Förstner, U. (1975). "Heavy Metals in Sediments of the Rhine and Elbe Estuaries: Mobilization or Mixing Effect?" *Environ. Geol.* **1**, 33–39.

Müller, G. and Förstner, U. (1976). "Schwermetalle in den Sedimenten der Elbe bei Stade: Veränderungen seit 1973," *Naturwissenschaften* **63**, 242–243.

Müller, G. and Prosi, F. (1977). "Cadmium in Fischen des mittleren und unteren Neckars: Veränderungen seit 1973," *Naturwissenschaften* **64**, 530.

Mullin, J. B. and Riley, J. P. (1956). "The Occurrence of Cadmium in Seawater and in Marine Organisms and Sediments," *J. Mar. Res.* **15**, 103–122.

Murakami, T., Kida, A., Nakai, M., and Matsunaga, S. (1975). "Heavy Metals in Sea Sediments," *Eisei Kaguku* (*Jap.*) **21**, 275–281.

Nissenbaum, A. and Swaine, D. J. (1976). "Organic Matter-Metal Interactions in Recent Sediments: the Role of Humic Substances," *Geochim. Cosmochim. Acta* **40**, 809–816.

Nriagu, J. O., Kemp, A. L. W., Wong, H. K. T., and Harper, N. (1979). "Sedimentary Record of Heavy Metal Pollution in Lake Erie," *Geochim. Cosmochim. Acta* **43**, 247–258.

Olausson, E., Gustafsson, O., Mellin, T., and Svensson, R. (1977). *Current Level of Heavy Metal Pollution and Eutrophication in the Baltic Proper. Medd. Marinegeol. Labor. Göteborg 9*, 28 pp.

Pacific Northwest Laboratories (1975). *Characterization of Pollutant Availability for San Francisco Bay Dredge Sediments Crystalline Matrix Study*. Report submitted to U.S. Army Engineering District, San Francisco, Calif. Cited by Patrick et al. (1977).

Page, A. L. (1974). *Fate and Effects of Trace Elements in Sewage Sludge When Applied to Agricultural Lands: A Literature Review Study*. U.S. Environmental Protection Agency, Office of Research and Development, National Environmental Research Center, Cincinnati, Ohio, 98 pp.

Pasternak, K. (1974). "The Accumulation of Heavy Metals in the Bottom Sediments of the River Biala Przemsza as an Indicator of Their Spreading by Water Courses from the Centre of the Zinc and Lead Mining Smelting Industries," *Acta Hydrobiol.* **16**, 51–63.

Patchineelam, S. R. (1975). "Untersuchungen über die Hauptbindungsarten und die Mobilisierbarkeit von Schwermetallen in fluviatilen Sedimenten." Dissertation, University of Heidelberg, Germany, 136 pp.

Patchineelam, S. R. and Förstner, U. (1977). "Bindungsformen von Schwermetallen in marinen Sedimenten," *Senckenberg. Marit.* **9**, 75–104.

Patrick, W. H., Jr., Gambrell, R. P., and Khalid, R. A. (1977). "Physicochemical Factors Regulating Solubility and Bioavailability of Toxic Heavy Metals in Contaminated Dredged Sediment," *J. Environ. Sci. Health* **A12**, 475–492.

Perhac, R. M. (1972). "Distribution of Cd, Co, Cu, Fe, Mn, Ni, Pb, and Zn in Dissolved and Particulate Solids from Two Streams in Tennessee," *J. Hydrol.* **15**, 177–186.

Perkins, E. J., Gilchrist, J. R. S., Abbott, O. J., and Halcrow, W. (1973). "Trace Metals in Solway Firth Sediments," *Mar. Pollut. Bull.* **4**, 59–61.

Pezzetta, J. M. and Iskandar, I. K. (1975). "Sediment Characteristics in the Vicinity of the Pulliam Power Plant, Green Bay, Wisconsin," *Environ. Geol.* **1**, 155–165.

Piper, D. Z. (1971). "The Distribution of Co, Cr, Cu, Fe, Mn, Ni and Zn in Framvaren, a Norwegian Anoxic Fjord," *Geochim. Cosmochim. Acta* **35**, 531–550.

Pita, F. W. and Hyne, N. J. (1975). "The Depositional Environment of Zinc, Lead and Cadmium in Reservoir Sediments," *Water Res.* **9**, 701–706.

Premazzi, G. and Ravera, O. (1977). "Chemical Characteristics of Lake Lugano Sediments." In H. L. Golterman, Ed., *Interactions between Sediments and Fresh Water*. Junk B.V. Publ., The Hague, and Pudoc, Wageningen, pp. 121–124.

Rashid, M. A. (1971). "Role of Humic Acids of Marine Origin and Their Different Molecular Weight Fractions in Complexing Di- and Trivalent Metals," *Soil Sci.* **111,** 298–305.

Ratkowsky, D. A., Thrower, S. J., Eustace, I. J., and Olley, J. (1974). "A Numerical Study of the Concentration of Some Heavy Metals in Tasmanian Oysters," *J. Fish. Res. Board Can.* **31,** 1165–1171.

Ravera, O. (1964). "The Radioactivity of *Viviparus ater, christ,* and *jan.,* Freshwater Molluscs, in Relation to That of the Sediment," *Bull. FEPE,* **10,** 61–65.

Ravera, O. and Premazzi, G. (1971). "A Method to Study the History of any Persistent Pollution in a Lake by the Concentration of Cs-137 from Fallout." *Proceedings of the International Symposium on Radioecology Applied to the Protection of Man and His Environment, Rome,* pp. 703–719.

Ravera, O., Gommes, R., and Muntau, H. (1973). "Cadmium Distribution in Aquatic Environment and Its Effects on Aquatic Organisms." In *Problems of the Contamination of Man and His Environment. Proceedings of an International Symposium,* Luxembourg, July 3–5, 1973, pp. 317–331.

Reinhard, D. and Förstner, U. (1976). "Metallanreicherungen in Sedimentkernen aus Stauhaltungen des mittleren Neckars," *N. Jb. Geol. Paläontol. Mh.* **1976,** 301–320.

Ribordy, ·E. (1978). *Le Rhône Amont et Ses Affluents.* Cited in Davaud et al. (1978).

Rohatgi, N. K. and Chen, K. Y. (1976). "Fate of Metals in Wastewater Discharge to Ocean." *J. Environ. Eng. Div. ASCE* **102,** 675–685.

Romano, R. R. (1976). "Fluvial Transport of Selected Heavy Metals in the Grand Calumet River System." In D. D. Hemphill, Ed., *Trace Substances in Environmental Health,* Vol. X. University of Missouri, Columbia, pp. 207–216.

Roth, I. and Hornung, H. (1977). "Heavy Metal Concentrations in Water, Sediments and Fish from Mediterranean Coastal Area, Israel," *Environ. Sci. Technol.* **11,** 265–269.

Ruch, R. R., Gluskoter, H. J., and Shimp, N. F. (1973). *Occurrence and Distribution of Potentially Volatile Trace Elements in Coal.* Environ. Geol. Notes, [Ill. Geol. Survey] 61, 43 pp.

Rutherford, F. and Church, T. (1975). "Use of Silver and Zinc to Trace Sewage Sludge Dispersal in Coastal Waters." In T. M. Church, Ed., *Marine Chemistry in the Coastal Environment,* American Chemical Society Symposium Series 18, 440–452.

Salomons, W. and De Groot, A. J. (1978). "Pollution History of Trace Metals in Sediments, as Affected by the Rhine River." In W. E. Krumbein, Ed., *Environmental Biogeochemistry,* Vol. 1. Ann Arbor Science Publishers, Ann Arbor, Mich., pp. 149–162.

Salomons, W. and Mook, W. G. (1977). "Trace Metal Concentrations in Estuarine Sediments: Mobilization, Mixing or Precipitation," *Neth. J. Sea Res.* **11,** 119–129.

Salomons, W. and Mook, W. G. (1978). "Processes Affecting Trace Metals in Lake

IJssel." In *Abstracts, 10th International Congress on Sedimentology, Jerusalem, July 9–14, 1978*, pp. 569–570.

Schafer, H. A. (1976). *Characteristics of Municipal Wastewater Discharges, 1976.* Annual Report, Southern California Coastal Water Research Project, pp. 57–60.

Schafer, H. A. and Bascom, W. (1976). *Sludge in Santa Monica Bay.* Annual Report, Southern California Coastal Water Research Project, pp. 77–82.

Schindler, J. E. and Albert, J. J. (1977). *Behavior of Mercury, Chromium and Cadmium in Aquatic Systems.* U.S. Environmental Protection Agency, Rep. 600/3-77-023, 70 pp.

Schleichert, U. (1975). "Schwermetallgehalte der Schwebstoffe des Rheins bei Koblenz im Jahresablauf—eine gewässerkundliche Interpretation," *Dtsch. Gewässerkd. Mitt.* **19**, 150–157.

Schleichert, U. and Hellmann, H. (1973). *Auftreten und Herkunft von Zink in Gewässern.* Literature Report, 1972/1973. Bundesanstalt für Gewässerkunde (F.R.G.), Koblenz, 32 pp.

Schleichert, U. and Hellmann, H. (1977). *Bilanzierung und Herkunft Toxischer Schwermetalle für das Rhein-Einzugsgebiet.* Interim Report. Bundesanstalt für Gewässerkunde, Koblenz, 64 pp.

Schöttler, U. (1975). "Das Verhalten von Schwermetallen bei der Langsamsandfiltration," *Z. Dtsch. Geol. Ges.* **126**, 373–384.

Segar, D. A. and Pellenbarg, R. E. (1973). "Trace Metals in Carbonate and Organic Rich Sediments," *Mar. Pollut. Bull.* **4**, 138–142.

Semkin, R. G. and Kramer, J. R. (1976). "Sediment Geochemistry of Sudbury Area Lakes," *Can. Mineral.* **14**, 73–90.

Shibahara, M., Yamazaki, R., Nishida, K., Suzuki, J., Suzuki, S., Nishida, H., and Tada, F. (1975). "Heavy Metals in Rivers of the Toyama Prefecture," *Eisei Kaguku (Jap.)* **21**, 173–182.

Skei, J. M., Price, N. B., and Calvert, S. E. (1972). "The Distribution of Heavy Metals in Sediments of Sörfjord, West Norway," *Water, Air, Soil Pollut.* **1**, 452–461.

Sly, P. G. (1977). "Some Influence of Dredging in the Great Lakes." In H. L. Golterman, Ed., *Interactions between Sediments and Fresh Water.* Junk B.V. Publ., The Hague, and Pudoc, Wageningen, pp. 435–443.

Steele, J. H., McIntyre, A. D., Johnston, R., Baxter, I. G., Topping, G., and Dooley, H. D. (1973). "Pollution Studies in the Clyde Sea Area," *Mar. Pollut. Bull.* **4**, 153–157.

Stenner, R. D. and Nickless, G. (1974). "Distribution of Some Heavy Metals in Organisms in Hardangerfjord and Skjerstadfjord, Norway," *Water, Air, Soil Pollut.* **3**, 279–291.

Stenner, R. D. and Nickless, G. (1975). "Heavy Metals in Organisms of the Atlantic Coast of South-west Spain and Portugal," *Mar. Pollut. Bull.* **6**, 89–92.

Štern, J. and Förstner, U. (1976). "Heavy Metals Distribution in the Sediment of the Sava Basin in Slovenia," *Geologija (Ljubljana)* **19**, 259–274.

Stoffers, P., Summerhayes, C., Förstner, U., and Patchineelam, S. R. (1977). "Copper and Other Heavy Metal Contamination in Sediments from the New Bedford Harbor, Mass.: A Preliminary Note," *Environ. Sci. Technol.* **11**, 819–821.

Stumm, W. and Baccini, P. (1978). "Man-made Chemical Perturbation of Lakes." In A. Lerman, Ed., *Lakes—Chemistry, Geology, Physics.* Springer-Verlag, New York, pp. 91–126.

Suess, E. (1978). "How Can We Distinguish between Natural and Anthropogenic Materials in Sediments, and Can We Predict the Effects of Man's Additions? A Review of Past Attempts." Workshop on the Biogeochemistry of Estuarine Sediments, Melreux/Belgium, Nov. 29–Dec. 3, 1976, pp. 224–237.

Suess, E. and Erlenkeuser, H. (1975). "History of Metal Pollution and Carbon Input in Baltic Sea Sediments," *Meyniana (Kiel)* **27**, 63–75.

Summerhayes, C. P., Ellis, J. P., Stoffers, P., Briggs, S. R., and Fitzgerald, M. G. (1977). *Fine-Grained Sediment and Industrial Waste Distribution and Dispersal in New Bedford Harbor and Western Buzzard Bay, Mass.* Woods Hole Oceanographic Institution, Rep. 76-115, 110 pp.

Suzuki, M., Yamada, T., Miyazaki, T., and Kawazoe, K. (1975). "Accumulation of Cadmium in the Sediment of Tama River," *Seisan Kankyu (Jap.)* **27**, 108–112.

Syers, J. K., Iskandar, I. K., and Keeney, D. R. (1973). "Distribution and Background Levels of Mercury in Sediment Cores from Selected Wisconsin Lakes," *Water, Air, Soil Pollut.* **2**, 105–118.

Symeonides, C. and McRay, S. G. (1977). "The Assessment of Plant-available Cadmium in Soils," *J. Environ. Qual.* **6**, 120–123.

Talbot, V., Magee, R. J., and Hussain, M. (1976). "Distribution of Heavy Metals in Port Phillip Bay," *Mar. Pollut. Bull.* **7**, 53–55.

Tatekawa, M., Nakamura, M., and Nakano, S. (1975). "The Pollution of Lake Biwa in the Light of the Distribution of Heavy Metals." In *Proceedings of the International Congress on the Human Environment, Kyoto, Japan,* pp. 402–407.

Taylor, D. (1974). "The Natural Distribution of Trace Metals in Sediments from a Coastal Environment, Tor Bay, England," *Estuarine Coastal Mar. Sci.* **2**, 417–424.

Taylor, D. (1976). "Distribution of Heavy Metals in the Sediment of an Unpolluted Estuarine Environment," *Sci. Total Environ.* **6**, 259–264.

Thomas, R. L. (1972). "The Distribution of Mercury in the Sediment of Lake Ontario," *Can. J. Earth Sci.* **9**, 636–651.

Thomas, R. L., Jaquet, J.-M., and Mudroch, A. (1977). "Sedimentation Processes and Associated Changes in Surface Sediment Trace Metal Concentrations in Lake St. Clair, 1970–1974." In *Proceedings of the International Conference on Heavy Metals in the Environment, Toronto, Oct. 27–31, 1975,* Vol. II/2, pp. 691–708.

Thornton, I., Watling, H., and Darracott, A. (1975). "Geochemical Studies in Several Rivers and Estuaries for Oyster Rearing," *Sci. Total Environ.* **4**, 325–345.

Thrower, S. J. and Eustace, I. J. (1973). "Heavy Metals Accumulation in Oysters Grown in Tasmanian Waters," *Food Technol. Aust.* **25**, 546–553.

Trefry, J. H. and Presley, B. J. (1976). "Heavy Metals in Sediment from San Antonio Bay and the Northwest Gulf of Mexico," *Environ. Geol.* **1**, 283–294.

Turekian, K. K. and Scott, M. R. (1967). "Concentrations of Cr, Ag, Mo, Ni, Co and Mn in Suspended Material in Streams," *Environ. Sci. Technol.* **1**, 940–942.

Turekian, K. K. and Wedepohl, K. H. (1961). "Distribution of the Elements in Some Major Units of the Earth's Crust," *Bull. Geol. Soc. Am.* **72**, 175–192.

Tyler, P. A. and Buckney, R. T. (1973). "Pollution of a Tasmanian River by Mine Effluents," *Int. Rev. Ges. Hydrobiol.* **58**, 873–883.

Vaccaro, F. R., Grice, G. D., Rowe, G. T., and Wiebe, P. H. (1972). "Acid-Iron Waste Disposal and the Summer Distribution of Standing Crops in the New York Bight," *Water Res.* **6**, 251–256.

Van der Weiden, C. H., Arnoldus, M. J. H. L., and Meurs, C. J. (1977). "Desorption of Metals from Suspended Material in the Rhine Estuary," *Neth. J. Sea Res.* **11**, 130–145.

Vernet, J.-P. (1976). *Etude de la Pollution des Sediments du Léman et du Bassin du Rhône: Rapport sur les Études et Recherches Entreprises dans le Bassin Lémanique.* Commission Internationale pour la Protection des Eaux du Lac Léman contre la Pollution, 1976, pp. 247–321.

Vernet, J.-P., Rapin, F., Faverger, P. Y., and Fernex, F. (1977). "Contamination des Sédiments Marins (Côte d'Azur) par les Métaux Lourds (Hg et Cd)," *Rev. Int. Oceanogr. Med.* **47**, 91–95.

Viel, M., Davaud, E., and Vernet, J.-P. (1978). *Contamination par les Métaux des Sédiments de Quelques Affluents du Léman et des Principales Rivières du Canton de Genève.* Commission Internationale pour la Protection des Eaux du Léman contre le Pollution. 13 pp.

Villa, O., Jr. and Johnson, P. G. (1974). *Distribution of Metals in Baltimore Harbor Sediments.* U.S. Environmental Protection Agency, Annapolis Field Office, Tech. Rep. EPA-903/9-74-012, 71 pp.

Vivian, C. M. G. and Massie, K. S. (1977). "Trace Metals in Waters and Sediments of the River Tawe, South Wales, in Relation to Local Sources," *Environ. Pollut.* **14**, 47–61.

Wagemann, R., Brunskill, G. J., and Graham, B. W. (1977). "Composition and Reactivity of Some River Sediments from the Mackenzie Valley, N.W.T., Canada," *Environ. Geol.* **1**, 349–358.

Wagner, G. (1976). "Die Untersuchung von Sinkstoffen aus Bodenseezuflüssen," *Schweiz. Z. Hydrol.* **38**, 191–205.

Wakita, H. and Schmitt, R. A. (1970). "Cadmium." In K. H. Wedepohl, Ed., *Handbook of Geochemistry.* Springer, New York, Chapter 48.

Walters, L. J., Jr., Wolery, T. J., and Myser, R. D. (1974). "Occurrence of As, Cd, Co, Cr, Cu, Fe, Hg, Ni, Sb and Zn in Lake Erie sediments." In *Proceedings of the 17th Conference on Great Lakes Research,* pp. 219–234.

Wentsel, R. S. and Berry, J. W. (1974). "Cadmium and Lead Levels in Palestine Lake, Palestine, Indiana," *Proc. Indiana Acad. Sci.* **84**, 481–490.

Windom, H. L. (1973). *Investigations of Changes in Heavy Metals Concentrations Resulting from Maintenance Dredging of Mobile Bay Ship Channel, Mobile Bay, Alabama.* Report for U.S. Army Corps of Engineers, Mobile District, 46 pp. Cited in Patrick et al. (1977).

Wisseman, R. W. and Cook, S. F., Jr. (1977). "Heavy Metal Accumulation in the Sediments of a Washington Lake," *Bull. Environ. Contam. Toxicol.* **18,** 77–82.

Wittmann, G. T. W. and Förstner, U. (1975). "Metal Enrichment of Sediments in Inland Waters—The Hartbeespoort Dam," *Water S.A. (S. Afr.)* **1,** 76–82.

Wittmann, G. T. W. and Förstner, U. (1976a). "Heavy Metal Enrichment in Mine Drainage. II. The Witwatersrand Goldfields," *S. Afr. J. Sci.* **72,** 365–370.

Wittmann, G. T. W. and Förstner, U. (1976b). "Metal Enrichment in Sediments in Inland Waters—the Jukskei and Hennops River Drainage System," *Water S.A. (S. Afr.)* **2,** 67–72.

Wittmann, G. T. W. and Förstner, U. (1977). "Heavy Metal Enrichment in Mine Drainage. IV: The Orange Free State Goldfield," *S. Afr. J. Sci.* **73,** 374–378.

Yamagata, N. and Shigematsu, I. (1970). "Cadmium Pollution in Perspective," *Bull. Inst. Public Health, Tokyo* **19,** 1–27.

Yost, K. J. (1978). "Some Aspects of the Environment Flow of Cadmium in the United States." In *Proceedings of the First International Cadmium Conference,* San Francisco, Jan. 31-Feb. 2, 1977, pp. 147–164.

Züllig, H. (1956). "Sedimente als Ausdruck des Zustandes eines Gewässers," *Schweiz. Z. Hydrol.* **18,** 7–143.

10

CADMIUM RETENTION BY CLAYS AND OTHER SOIL OR SEDIMENT COMPONENTS

William F. Pickering

Department of Chemistry, The University of Newcastle, New South Wales, Australia

I.   INTRODUCTION

Recognition of the potential health hazards associated with cadmium entering the food chain has focused attention on the levels of this element in plants and soils.

Although uncontaminated soils contain very little cadmium (e.g., 0.1 ppm), impurities in agricultural chemicals can lead to some buildup, and in industrial areas fallout and/or contact with wastewaters can produce an accumulation level of up to 100 ppm.

Losses due to leaching by natural waters tend to be minimal because both metallic cadmium and its divalent ion have low solubility in solutions containing phosphate or carbonate. On the other hand, plants are capable of taking up the toxic material from the surface soil, and it has been noted that the ratio of cadmium in the plant to cadmium in the soil tends to fall in the range of 1 to 10 (Hodgson, 1970).

The proportion taken up appears to depend greatly on the adsorption capacity of the soil. This soil property involves contributions from components such as clays, organic matter, and hydrous metal oxides, and to predict the availability of cadmium one must possess knowledge of the types and strengths of metal ion interactions with individual soil components.

Soils differ in respect to the proportions of individual adsorbents present; thus, while the emphasis in this chapter is on the adsorption of cadmium by clays and other soil or sediment components, it is equally important to consider the combined effect as observed using selected soil samples. Accordingly, Section 6 summarizes the trends observed in a number of "cadmium sorption by soil" studies.

Some subdivision within the broad classification of soil components is also necessary because the structure and composition of materials can significantly influence the adsorption behavior of the solid. For example, there are many different types of clay mineral, the three best known being possibly kaolinite, illite, and montmorillonite. Even "pure" samples of these clays can contain over 10% of other types of mineral species (derived from the initial parent rock). In addition, different degrees of isomorphous substitution alter the exchange capacity, and a change in the nature of chemical counterions can influence the affinity of the surface for different species. Other factors that can influence the total sorptive capacity are the surface area, the solution phase pH, and the chemical form of the sorptive species. Soil clays tend to be a mixture of several basic types of clay, with individual characteristics masked by the presence of organic material.

The organic components of soil vary in nature from comparatively pure cellulosic material to the complex polymers known collectively as humic matter. On the average, over 40% of the organic matter in sediments can be classified as humic acids, and a major proportion of heavy metals (e.g., copper and zinc) tends to be associated with this component (Jackson, 1975; Nissenbaum and Swaine, 1976).

The properties of humic materials vary with source. There are differences in polymer chain length, nature of associated functional groups, and solubility. The lower molecular fractions, which are water "soluble" and are not precipitated on reducing the pH to unity, are known as fulvic acid. The larger humic acid molecules can exist either as colloidal particles or sorbed on the surface of other components (e.g., clay suspensions). The humic acids derived from soil have a structure similar to that of benzene carboxylic acids and phenolic acids and are held together by hydrogen bonding. Marine humic acids appear to contain fewer, carboxyl and phenolic groups, but possess more aliphatic and heterocyclic structures and more sulfur and nitrogen.

The cadmium contents of a series of humic substances derived from marine environments have been found to be too low for detection (Nissenbaum and Swaine, 1976). On the other hand, correlation between the cadmium content and the organic content of soils has been noted by a number of workers. In fact, it has been proposed (Petruzzelli et al., 1977) that the linkage of cadmium to alkali-soluble humic substances is so strong that the element is not removed by seedlings until the total amount present exceeds the toxic level. The bound cadmium appears to be associated with the fulvic acid fraction, which contains most of the amino acid compounds. This suggests that interaction between cadmium species and proteic substances, frequently reported in animal tissue studies, is active in soil systems too.

The colloid fractions of soils and sediments can be contaminated with partial layers of hydrous oxides, such as those of Fe, Al, and Mn. The oxides appear to exert a chemical activity effect that is out of proportion to the amount present, and it has been proposed (Jenne, 1968) that these materials play a significant role in controlling aqueous metal ion levels.

Since, ultimately, equilibrium has to be established between sorbed species and the adjacent "soil solution" or "interstitial water," the availability of cadmium to plants and organisms depends on all the factors that can influence the phase distribution process, for example, the pH, chemical nature of the metal ion species, stability of the metal complexes, binding power of the functional groups, ionic strength of the solutions, and presence of competing ions. From the discussions in subsequent sections, it will become apparent that many of these aspects are still only partially understood.

2. THE SORPTION-DESORPTION OF CADMIUM BY CLAYS

2.1. Clay Characteristics

Clays are major components of many soils, the suspended matter in streams, and sediments, and as they are capable of adsorbing heavy metal ions, it has been proposed that a high proportion of the metal loading of aquatic systems may be associated with the clay fraction.

Clay minerals are difficult to define adequately in simple terms. The basic

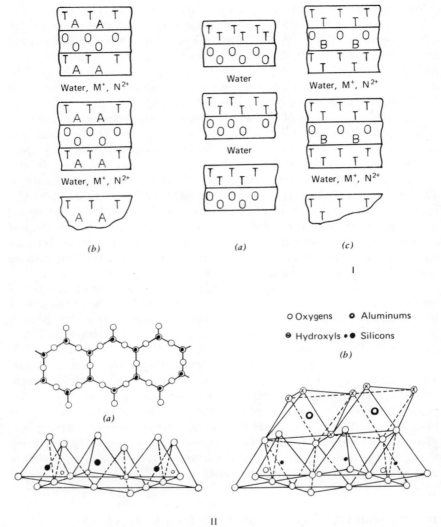

Figure 1. I. Schematic representation of the three principal types of clay mineral: (*a*) kaolinite, (*b*) illite, and (*c*) montmorillonite. Symbols used: O = octahedral aluminate layers; T = tetrahedral silicate layers. Isomorphous substitution indicated by letters A (tetrahedral sites) and B (octahedral sites). Balancing cations represented by M^+, N^{2+}. II. Representation of structure of (*a*) a tetrahedral silicate layer, and (*b*) kaolin.

structural component is a continuous sheet of silica tetrahedra, formed through adjacent units sharing three oxygen atoms. To establish electrical neutrality within the lattice, the residual oxygen of each SiO_4^{2-} unit needs to be linked to an external cation, and in clay minerals this is achieved through condensation with a sheet hydroxide structure, such as the ones formed by the hydroxides of divalent (e.g., magnesium) or trivalent (e.g., aluminum) elements.

Clay minerals are thus composed of small, platelike particles, ranging in diameter from a few hundredths of a micron to several microns. In a two-layer clay, such as kaolin, the platelets are composed of one sheet of silica and one sheet of alumina. The basal faces of these clays are thus half siloxane and half hydroxylated alumina. Three-layer clays, such as illites and montmorillonites, consist of a sheet of octahedral alumina or magnesia sandwiched between the two sheets of tetrahedral silica.

The basic structures of these clay types are shown schematically in Figure 1.

Electrokinetic studies of clay solutions have shown that, at pH values greater than 2 or 3, the particles carry a net negative charge (normally compensated for by the presence of positive counterions). This charge is believed to originate from lattice imperfections, exposed structural hydroxyl groups, broken bonds at edges of particles, and isomorphic substitution within the lattice. With kaolinites the first three sources predominate; with illites and montmorillonites most of the charge or exchange capacity is attributable to isomorphous substitution.

In montmorillonite or bentonite type minerals, extensive isomorphic ion replacement occurs mainly in the central octahedral layer (e.g., replacement of Al^{3+} by Mg^{2+} or Fe^{2+}). This creates a significant charge deficiency, but since the source is located in the central layer, the counterions present in the surface film are not strongly bound and are readily exchanged, for example,

$$\text{Clay, } Ca^{2+} + Cd^{2+} \rightleftharpoons \text{clay, } Cd^{2+} + Ca^{2+}$$

In minerals like illite, isomorphic substitution occurs mainly in the outside tetrahedral layers (e.g., silicon replaced by aluminum), and the attractive forces between the negatively charged surfaces and the balancing cations are much stronger. The structure resists expansion in aqueous suspensions, and this leads to a low rate of exchange and exclusion of some large cations.

There is strong evidence that at pH < 7 the edges of clay particles are positively charged. These positive charges are believed to arise from bond-breaking reactions of the type

$$-Al-O-Al- \rightarrow (-Al-O^- + -Al^+) \xrightarrow{H_2O, H^+} 2Al-OH_2^+$$

The presence on clay particles of oppositely charged faces and edges leads to considerable edge-to-face flocculation in aqueous suspensions.

Partly as a result of competing interparticle forces, ion exchange in clay minerals is nonstoichiometric, and the cation exchange capacity depends somewhat on the chemical composition and type of clay. An extensive study of a wide range of clays has been made by the American Petroleum Institute

(1951), and examination of the data shows that the cation exchange capacities of montmorillonites are on the order of 85 to 160 mequiv/100 g clay, whereas the exchange capacities of most kaolinites fall in the 3 to 10 mequiv/100 g clay range. Illite values tend to fall between these two sets of limits.

The equilibrium position, or exchange thermodynamics, in clay-solute interactions depends on many factors, including the ionic size and degree of hydration of both the initial counterion and the displacing ion, the solution composition (e.g., pH, ionic strength, anions), the exchanging ion concentration, and any pretreatment of the clay sample (e.g., destruction of organic impurities or predrying).

The relationship between uptake per unit mass (x/m) and solution concentration (c) can usually be described in terms of a standard type of isotherm such as the Langmuir or Freundlich, that is,

$$\frac{x}{m} = \frac{kk'c}{1 + kc} \qquad \text{or} \qquad \frac{x}{m} = kc^{1/n}$$

where k, k', and n are constants.

Cation exchange rates depend on the surface area of the solid and the speed of diffusion of ions into (or out of) the interplanar spaces, and tend to follow the order kaolinite > montmorillonite ≫ illite. In some systems the degree of reversibility decreases with time of contact, and this has been attributed to slow incorporation of sorbed cations into the lattice structure. Such ions are said to be specifically, or irreversibly, sorbed.

The tendency of a cation to exchange onto a negative surface appears to increase with (a) increasing charge on the counterion, (b) decreasing ease of cation hydration, and (c) decreasing hydrated radius and increasing polarizability. Observed variations from the affinity order predicted from these generalities have been ascribed to factors such as specific interactions, the formation of hydroxy species (e.g., MOH^+), or steric hindrance.

The weak positive charge that can be present on the edge of clay particles is liable to attract anions, whereas the negative faces repel such species. As a result, a variation in charge distribution (e.g., due to a pH change) can lead to observation of both positive and negative adsorption of anions.

In the presence of acid species, protonation of some of the hydroxy groups associated with surface aluminum atoms can occur, for example,

$$\text{Clay—Al—OH} + \text{HCl} \rightarrow \text{clay—AlOH}_2^+ \ldots \text{Cl}^-$$

In such systems, interchange of chloride, nitrate, or bromide ion appears to take place readily, and the process has been described as nonspecific anion sorption.

Other anions interact with the clay surface and are not so readily exchanged, for example,

$$\text{Clay—Al—OH} + H_2PO_4^- \xrightarrow{H^+} \text{clay—Al—}H_2PO_4 + H_2O$$

The preference for anion exchange on clays appears to be

$$CNS^- < I^- < NO_3^- < Br^- \approx CH_3COO^- < Cl^- < H_2PO_4^- < OH^- < F^-$$

It has been noted, for example, that acetate ions sorbed on montmorillonite (H^+ counterions) are not displaced by nitrate or chloride ions but are desorbed in the presence of hydroxyl, fluoride, or dihydrogen phosphate ions (Bingham et al., 1965).

The adsorption of anions by clays can have some effect on metal ion uptake, partly through causing deflocculation and partly by creating ligand bridges.

A detailed review of the surface and colloid chemistry of clays has been written by Swartzen-Allen and Matijevic (1974), and summaries of clay ion exchange behavior appear in a number of monographs (e.g., Amphlett, 1964; Grimshaw and Harland, 1975). The brief introduction provided in the preceding pages, however, should serve to emphasise the wide range of variables that can influence the distribution of a metal ion (e.g., Cd^{2+}) between a clay and an adjacent aqueous phase.

2.2. Cadmium Ion Sorption by Clay

Before discussing the effect of solution variables on the uptake of cadmium by clays, it is appropriate to consider the relative affinity of this ion for clay surfaces. Few studies have been made of this aspect, and to date no consistent pattern has emerged.

If one assumes that ion exchange is the controlling process, affinities can be compared in terms of selectivity coefficient (i.e., K) values. [$K = (M_1/M_2)_{aq} \cdot (\overline{M}_2/\overline{M}_1)_{clay}$, M_1 and M_2 being the equilibrium concentrations in solution, and $\overline{M}_1$ and $\overline{M}_2$ the amounts of cation sorbed by the clay.]

Investigations based on the addition of clays (saturated with Pb or Cd or Ca ions) to solutions containing different cations have yielded K values which indicate that Cd ions compete more or less on an even basis with Ca ions, but are less favored than Pb ions (by a factor of 3) (Bittel and Miller, 1974). The average selectivity coefficients for the ion distributions examined, using montmorillonite, illite, and kaolinite suspensions, were reported to be 1.0, 1.0, and 0.9 for Cd-Ca exchange; 0.6, 0.6, and 0.3 for Pb-Cd exchange; and 0.6, 0.4, and 0.3 for Pb-Ca exchange. The coefficient values were imprecise (e.g., 1.0 ± 0.4), but as the distribution appeared to vary in a random fashion with solution concentration ratios, the use of mean values for comparison purposes was considered to be justified. Part of the variability may be ascribed to the fact that solution pH values were not controlled, and it was assumed that changes within the range 4.8 to 6.5 would have little effect.

When clays preconverted to the Na^+ form were added to binary combinations of divalent cations, the selectivity coefficient approach proved to be valid only in the case of montmorillonite suspensions. These yielded an affinity order sequence of Ca > Pb > Cu > Mg > Cd > Zn, and K values for the above combinations of 5.5 (Cd/Ca), 0.4 (Pb/Cd), and 2.0 (Pb/Ca).

With illite and kaolinite suspensions the affinity order appeared to be influenced by relative concentrations, and quoted sequences such as Pb > Cu > Zn > Ca > Cd > Mg (illite) and Pb > Ca > Cu > Mg > Zn > Cd (kaolinite) refer to the specific situation where equal amounts of both competing divalent ions are present in solution at equilibrium, and the solution pH is about 5. As discussed later, the uptake of each metal ion is sensitive to pH, and the process involved may be more adequately described as competitive adsorption than as ion exchange. For example, better data correlation was obtained using log $\overline{M}_1$ versus log $(M_1/M_2)_{aq}$ plots than with the selectivity coefficient approach (Farrah and Pickering, 1977a).

A radioisotope dilution technique has been used by Stuanes (1976) to prepare adsorption isotherms for the uptake of Cd^{2+}, Zn^{2+}, Mn^{2+}, and Hg^{2+} from binary mixtures of the chloride salts of these elements. On clay minerals (i.e., kaolinite, montmorillonite, and vermiculite) the order of decreasing selectivity was found to be Zn > Mn > Cd > Hg. With the minerals albite and labradorite the positions of cadmium and manganese were reversed. Very little mercury was adsorbed by any surface, and it was suggested that metal ion selectivity was affected by the stability of chloro and hydroxy complexes.

The divergent conclusions drawn from the various investigations clearly indicate a need for further detailed studies. It has been suggested that the best parameter for comparing affinities, or relative bonding strength, is the minimum concentration required for specific adsorption (MCSA). This parameter is considered to be directly related to the constant term (k), which may be evaluated from Langmuir adsorption isotherm plots. Using this approach, Wakatsuki et al. (1975) constructed the following sequence for the adsorption of cations on a kaolinite soil clay at pH 6: Cr^{3+} > Fe^{3+} > Al^{3+} > Ca^{3+} > Cu^{2+} > Pb^{2+} > Mn^{2+} > Ni^{2+} > Co^{2+} > Zn^{2+} > Sr^{2+} > Mg^{2+} > NH_4^+, K^+. Cadmium ion was not included in this investigation, but it can be implied from other studies that its position would lie close to that of zinc.

The need to specify pH when quoting relative affinities or metal ion uptakes becomes more obvious when one considers the effect of pH on clay-solute equilibria, as summarized in Figure 2. From this diagram it can be seen that in acid media the amount of Cd ion sorbed increases with increasing pH. At around pH 7 loss of metal ion from solution increases markedly, and by pH 8 virtually all of the initial Cd^{2+} has been taken up, either through sorption or precipitation

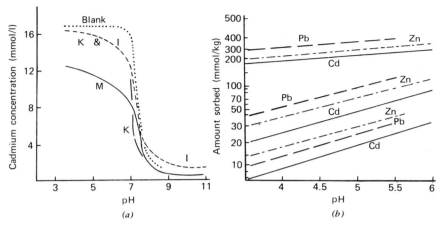

Figure 2. The effect of pH and clay type on the adsorption of Cd ions by clay. (*a*) Aqueous heavy metal ion levels, showing precipitation region. (*b*) Plots of log amount sorbed versus pH, with comparison plots for other heavy metal ions. Lines for copper fall close to those for lead. Initial cadmium concentration, $1.7 \times 10^{-4} M$; temperature, 25°C; clay concentration, 0.025% (w/v) for illite and montmorillonite, 0.05% for kaolin. (Based on Farrah and Pickering, 1977a.)

as a sparingly soluble cadmium hydroxy species (Farrah and Pickering, 1977b).

The activity, or concentration of hydroxyl ions, must be higher at the clay-solution interface than in the bulk solution because precipitation occurs at a solution pH value which is several units lower than that predicted from published solubility product data (e.g., Sillen and Martell, 1964). This surface effect is not restricted to clays or Cd ions, as shown in Table 1.

A high percentage of natural waters are slightly alkaline; hence precipitation of the metal ion could play an extremely important role in controlling aqueous cadmium levels.

The cadmium adsorption values in acid media (Figure 2*a*), when plotted as a log function, vary almost linearly with pH (as shown in Figure 2*b*) (Farrah and Pickering, 1977a). It can be noted that the slopes of the logarithmic plots, using illite and kaolinite suspensions, are similar, and resemble those derived from studies of other heavy metal ions. The slope value (≈ 0.25) suggests that metal ion uptake has a fractional power dependence on hydrogen or hydroxyl ion concentration. For example, it may be associated with the adsorption of hydroxyl ions (possibly with subsequent proton abstraction from surface sites or from water molecules coordinated to sorbed metal ion). The slope observed with montmorillonite is smaller (~ 0.07). This seems to imply that edge sites (which can be initially positive and attractive to hydroxyl ions) play an important role in the overall process. (On montmorillonite the contribution of edge sites

Table 1. pH Values for Precipitation of Metal Hydroxy Species in the Presence of Suspended Solids

Nature of Suspension	pH^a				Reference
	Cd^{2+}	Zn^{2+}	Cu^{2+}	Pb^{2+}	
Kaolin	7.2	6.8	6.1	5.9	Farrah and Pickering (1977)
Illite	7.3	6.6	5.8	5.9	
Montmorillonite	7.5	6.6	5.9	6.1	
Cellulose	6.9	6.8	6.4	6.0	Farrah and Pickering (1978b)
Hydrous Al oxide	6.6	5.6	4.8	5.2	Kinneburgh et al. (1976)
Hydrous Fe oxide	5.8	5.4	4.4	3.1	Kinneburgh et al. (1976)
Goethite, FeOOH	7.7	—	5.2	5.6	Forbes et al. (1976)
	6.8				James et al. (1975)
TiO_2	6.0				James et al. (1975)
Calculated from pK_{so} data	8.8–9.8	7.4–8.5	6.1–6.9	6.1–9.1	Sillen and Martell (1964)

^{a} pH for 50% precipitation (± 0.2), using an initial metal ion concentration of $\sim 10^{-4}$ M.

to total capacity should be much smaller than with the other two types of clays.)

Figure 2b also provides an indication of the relative exchange capacities of the three clay types, because the concentration of Cd ion used (1.7×10^{-4} M) was shown by adsorption isotherm studies to be adequate for saturation of all sites. It should be noted that this concentration (~ 18 ppm) is much greater than is encountered in most natural water systems. With lower concentrations the amount adsorbed is much less (as expected from systems conforming to normal isotherm behavior), and it is to be hoped that in the near future a new series of experiments will be undertaken to ascertain the relationship between uptake and pH when clay surfaces are only partially saturated.

From solubility product calculations one can suggest that the pH for onset of precipitation will be about one pH unit higher for solutions having cadmium concentrations of <1 ppm; and should the free Cd ion concentration be reduced further through the formation of complexes, precipitation may cease to be a significant component of the total adsorption process.

This hypothesis is partially supported by the results obtained in a study of the effect of pH and the presence of a selected range of ligands on the uptake of cadmium by clay suspensions (Figure 3). However, each system has to be considered individually, since there can be complicating factors. For example, it was observed that conversion of the cadmium content into the anionic ethylenediaminetetraacetate complex (Cd/EDTA = 1:1) resulted in no uptake of cadmium by the clay (i.e., no precipitation or sorption) over the pH range 3 to 11. The addition of nitrilotriacetate to the test solutions (Cd/NTA = 5:6) also

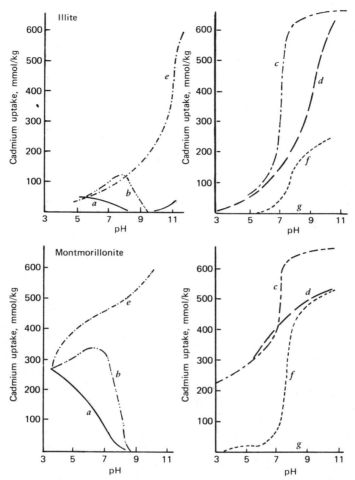

Figure 3. Diagram showing the effect of pH on the uptake of cadmium species by illite and montmorillonite in the presence of complexing agents. Initial Cd^{2+} concentration, 1.7×10^{-4} M; clay concentrations (w/v), 0.025%. *a,* Nitrilotriacetate (5:6); *b,* cyanide ion (1:6); *c,* tartaric acid (1:5); *d,* glycine (1:6); *e,* α,α'-bipyridyl (1:6); *f,* ethylenediamine (1:6); *g,* EDTA (1:1) follows baseline. Values in parentheses indicate Cd:ligand ratio. The effect of complex formation in the presence of kaolin resembles that shown for illite, except that scale values have to be approximately halved. (From Farrah and Pickering, 1977b.)

converts the metal ion into an anionic complex species, but in this case the stability is such that at low pH some of the dissociated metal ions are sorbed, and at high pH there is partial precipitation (Figure 3, curve *a*). The effect of pH on complex stability is even more clearly shown in the case of the cyanide system (Figure 3, curve *b*). In acid media the presence of this ligand has little effect on total metal ion uptake. At pH > 7 the effective stability increases, precipitation is masked, and cadmium uptake becomes virtually zero at pH > 9.

Contrary to expectations, an excess of tartaric acid in solution (curve c) has no observable effect on either the adsorption or the precipitation process.

An excess of the simplest amino acid, glycine, causes precipitation to be displaced to a region of higher pH. It should be noted that the predominant cadmium-glycine complex is positively charged, and the formation of this species can be seen (curve d) to lead to an extension (into alkaline media) of the region in which adsorption capacity increases with pH. A similar effect was noted with the cationic complex formed with α,α'-bipyridyl (curve e). However, in the case of ethylenediamine (Figure 3, curve f), uptake of the cationic complex ion was minimal because of the clays preferentially sorbing protonated ligand species. The basicity of the ligand is such that species such as EnH_2^{2+} and EnH^+ persist up to pH 9.

From studies on cadmium complexes (and other metal complex investigations) it has been concluded that clays do not sorb anionic metal complexes to any significant extent, and the uptake of cationic species can be reduced significantly through competition from protonated ligands. It is also apparent that the masking action of ligands (in respect to precipitation) can be pH sensitive. In solution equilibria calculations, based on published data (e.g., Sillen and Martell, 1964, 1971), both acid dissociation values and·stability constants are regularly utilized, but such calculations have limited value in the presence of a suspended solid, because appropriate data for solute-solid interactions have yet to be tabulated. Table 1 clearly shows that published solubility product values have little significance in the presence of clays or hydrous oxides, and the results summarized in Figure 3 demonstrate that it is difficult to predict relative affinity values, or the impact of sorption on the final equilibrium position. For example, preferential sorption of protonated ligand species can minimize the amount of metal ion adsorbed at the interface. At the same time the removal of the ligand from solution will prevent it from acting as a masking agent, and significant precipitation may occur in slightly alkaline media. This duality between competitive sorption and precipitation can even be observed in acid media. At around pH 4 the presence of an equivalent amount of phosphate or sulfate tends to slightly increase the uptake of cadmium by clays; a hundredfold excess of these acid anions reduces the metal uptake to almost zero (Farrah and Pickering, 1977a).

Phosphates and sulfates are present in most natural waters. Chlorides can also be present in appreciable amounts, especially in estuarine areas, and it has been proposed that the formation of anionic chloro complexes should reduce cadmium uptake by suspended matter. In view of the limited stability of these complexes, when compared with the stability of the other species used in the tests, it would seem that chloro complex formation should have only marginal influence on metal uptake. Displacement effects attributable to enhanced concentrations of diverse cations are probably more important in saltwater conditions.

Small amounts of chelating agents such as EDTA and NTA do find their way into waters from domestic and industrial washing operations, and the presence of chemicals such as these should greatly enhance the mobility of cadmium and could be responsible for high aqueous media levels. This potential effect has been recognized, but few answers have been obtained to the type of questions that automatically follow, for example:

1. Are stable anionic complexes as toxic as hydrated ions?
2. Are complexes readily assimilated by plants and biota?
3. What conditions will promote decomposition or deposition of the stable complex ions?

Much also remains to be learned about the effect of fulvic acids. Analogy suggests that cadmium could react with these low molecular weight polymers to form soluble uncharged complexes, in which case the metal ion could reside mainly in the aqueous phase. On the other hand, fulvic acid is known to adsorb on, and react with, clays, and thus the organic material may equally well act as a bridging unit and enhance the removal of cadmium from waters.

2.3. Extraction of Cadmium Sorbed on Clay

The desorption process is possibly the most relevant aspect of solid-metal ion interactions if one is concerned primarily with the availability of metal ions to plants and biota.

Important factors in this process are competition between cations for adsorption sites, dissolution of metal hydroxy precipitates through proton addition, and formation of stable soluble complexes. Displacement by cations is an equilibrium process which is controlled by the concentration of displacing ion and the relative affinities of the two species for the surface adsorption sites. Total displacement cannot always be assumed; for example, Cd ions are adsorbed from acid solutions containing an excess of protons. Similarly, the efficiency of the complex-forming process can be very pH dependent; dissolution of precipitated material can be slow.

Considerations such as these have not always been recognized, and over a period of years it has become accepted practice to judge availability in terms of a comparison of the amount extracted by some selected reagent versus the total metal ion content of the adsorbent (e.g., soil). Any metal ion not displaced by the nominated reagent (e.g., $0.1\ M$ NH$_4$Ac or $0.1\ M$ HCl) has been classified

Clay Type			Kaolin		Illite		Montmorillonite	
Extracting Solution	pH	Type	S	P	S	P	S	P
0.1 M HNO$_3$	1.4	a						
0.001 M HNO$_3$	3.1	a						
0.1 M HAc	2.9	a(ℓ)						
0.1 M HOx	3.3	aℓp						
0.01 M NH$_4$NO$_3$	5.0	p						
0.1 M NH$_4$Ac	6.9	p(ℓ)						
0.1 M NH$_4$Ox	7.9	pℓ						
0.01 M En	7.6	pℓ						
0.01 M Nacit	7.8	ℓ(p)						
0.001 M EDTA	7.2	ℓ(p)						
0.5 M NaCl	6.5	c						
0.1 M NaNO$_3$	5.6	c						
0.001 M NaNO$_3$	5.8	c						
0.001 M CaCl$_2$	5.7	c						
0.05 M CaCl$_2$	6.0	c						
0.05 M Ca(Ac)$_2$	7.4	c(ℓ)						
0.1 M NH$_3$	10.8	ℓ						

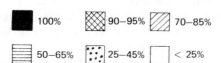

100%   90–95%   70–85%   50–65%   25–45%   < 25%

Figure 4. Diagram indicating the percentage of sorbed cadmium extracted from clay samples by a series of chemical solutions. S = sorbed at pH 5; P = sorbed at pH 7 (i.e., includes precipitated metal ion). Letter symbols: a = acids, p = protonated species present: ℓ = ligand species present; c = competing Na$^+$ or Ca^{2+} ions. (Based on data in Farrah and Pickering, 1978a.)

as specifically sorbed, or fixed, and the selection of a particular reagent has often been justified by showing that there is a distinct correlation between the amount extracted and the amount taken up by plants grown in the soil.

A recent study of the desorbing effects of a range of electrolyte solutions raises doubts about the validity of some of the assumptions previously made. Samples of kaolinite, illite, and montmorillonite were exposed to 10^{-4} M metal ion solutions (including cadmium) adjusted to a pH of about 5 or 7. After equilibrium had been achieved, the clay fraction was isolated on a filter and then placed in a volume of extracting solution. After a day of mixing, the phases were separated and the metal ion content of the filtrate was determined by atomic absorption spectroscopy. The results obtained with cadmium-treated clays are summarized in Figure 4 (Farrah and Pickering, 1978a).

Examination of this diagram shows that Cd ion sorbed at pH 5 may be totally recovered on extraction with a reagent such as 0.1 M oxalic acid (pH 3.3), 0.01 M sodium citrate (pH 7.8), or 0.001 M EDTA (pH 7.2). The last two form anionic complexes (known to be minimally sorbed), while oxalic acid appears to act partly through complex formation and partly through an affinity for surface sites. Mineral acid solutions left some metal ion attached to the clay, as predicted from adsorption behavior in acid media. Effective cationic competition (plus a pH too low for hydroxy species formation) allows cadmium to be totally displaced from illite and montmorillonite by 0.01 M NH_4NO_3 (pH 5); 0.5 M NaCl also totally displaces the Cd ion from montmorillonite.

Adsorption of the cadmium onto the clay at pH 7 results in most of the metal ion being present as sparingly soluble hydroxy species, and to achieve high recovery rates the extractants have to be capable of dissolving this material, as well as displacing the sorbed fraction. Figure 4 shows that none of the extractants tested achieved total recovery. All the reagents mentioned above gave high recoveries, however, and the performance of mineral acid extractants was somewhat improved, because of their ability to dissolve the precipitated segment. Had the samples been dried between deposition and extraction, ageing and dehydration would have occurred, leading to a further reduction in the extraction values. Similar types of behavior have been observed in studies using clays loaded with Pb, Cu, or Zn.

The extraction study results may be generalized in a statement such as the following: "The cadmium sorbed on suspended clay particles may be fully or partially released by any change in environment that renders the solution acidic or brings the suspension into contact with complexing ligands or high concentrations of electrolyte."

In natural systems, such as soils, some of the total metal ion content is associated with the organic matter (e.g., humates), and the factors influencing displacement from these substrates have yet to be elucidated. However, if the trends observed in the clay studies are maintained, it is not surprising that soil chemists are still seeking to validate extraction methods.

It has been noted that acid ammonium oxalate extraction gives the best estimates of available copper and lead (Mishra et al., 1973; Misra and Pandey, 1976), and an EDTA-lime treatment has been found to be effective in removing cadmium from polluted paddy soils, together with the Pb, Zn, and Cu contents

(Kobayashi et al., 1974). These observations are in accord with the results obtained in the clay desorption studies. Conversely, it can be suggested that workers relying on mineral acid extraction procedures, or ammonium acetate extraction values, are achieving only partial recoveries (for chemical reasons), and the concept of specifically sorbed material may be solely an artifact of the procedure selected.

It has been shown (Haghiri, 1976) that the release of cadmium from cadmium-treated H-clays (kaolinite and illite) increases with increasing amendments of potassium and calcium salts. The release of metal ion from both clays was greater in the presence of Ca than of K ions; but, contrary to the results presented in Figure 4, cadmium concentrations in the dialysates increased as the percentage of calcium or potassium saturation of the clays in the suspension decreased. In separate experiments it was shown that the concentration of cadmium in soybean shoots also decreased with increasing percent calcium or percent potassium saturation of cadmium-treated silt loam soil.

Such apparent contradictions clearly show that there is room for more studies in this area.

3. SORPTION BEHAVIOR OF ORGANO-CLAY COMPLEXES

There appears to be a relatively high affinity between clay and humic acids. Gel filtration and infrared studies (Tan, 1975, 1976) indicate that the adsorption of low molecular weight humic acids may be followed by acidolytic attack and gradual decomposition of the —Si—O—Si linkages of kaolinite clay structures. Interaction is considered to be restricted to surface sites, as distinct from interlamellar regions. For example, X-ray diffraction analysis has shown that adsorption of humic acid in the interlamellar regions of bentonite (a montmorillonite type clay) is not aided by saturation of the bentonite with different cations. However, it has been established from studies using ^{14}C-labeled humic acid that the affinity of this organic material for montmorillonite varies with the nature of the principal counterion in the order Na < K < Cs, Ba < Ca < Zn < Co < Cu, and La < Al < Fe (Theng and Scharpenseel, 1976). Interaction between fulvic acid and nonexpanding clay minerals such as kaolin, muscovite, and sepiolite is also believed to occur mainly at surface sites. The amount sorbed decreases with increasing pH but increases with increasing concentration of organic matter in the system (Kodama and Schnitzer, 1974).

The uptake of humic acids by clays varies nonlinearly with concentration and may be described by an equation having the same algebraic form as a Freundlich adsorption isotherm, that is, $y = ac^b$, where y is the amount sorbed per unit mass, c is the solution concentration of humic acid, and a, b are constants. For sorption on kaolinite, vermiculite, and muscovite samples, the values of both a and b are less than unity. Uptake values are greatest when highly dispersed montmorillonite clay is used (Orlov and Pivovarova, 1974).

With simulated seawater conditions (i.e., 3.5% NaCl and pH 8) the adsorption capacities of clay materials have been found to follow the order illite > chlorite > kaolinite > marine sediment. The capacity increases with increasing acidity or salinity of the medium (e.g., < 0.4% by weight in fresh water, increasing to >2.5% in seawater). The fact that only about two thirds of the sorbed material is readily removed by washing tends to confirm the proposal that some specific bonding between clay and humate materials occurs (Rashid et al., 1972).

Besides being strongly sorbed by clay particles, humic acids can complex from 3 to 200 g metal ion/kg organic matter (Rashid, 1971, 1974). Thus they have the potential to play a significant role in the control of aqueous metal ion levels.

The effect of an adsorbed layer of humic acid on the distribution of cadmium between an aqueous phase and clay particles has been investigated by at least two groups, with the aid of radioactive tracers.

In one study, synthetic organo-clay complexes were prepared by allowing humic acid (extracted from soil) to adsorb on <2-μm-diameter particles of montmorillonite. The clay samples used included Na^+-, Al^{3+}-, Fe^{3+}-, and Ca^{2+}-saturated material, and with these the presence of a humate covering made little difference to the metal ion uptake. In another series of tests, calcium-saturated montmorillonite was given a coating of aluminum or iron hydroxide before being treated with humic acid. Cadmium ions were strongly bonded to these suspensions, and coating the hydrous oxides with humic acid reduced the metal ion uptake (Levy and Francis, 1976).

It has been shown (see Section 4) that the adsorptive capacity of humic acid is on the same order as that of montmorillonite, a fact that may explain the limited change in uptake observed when the organic material is coated directly onto the clay. In the case of hydrous oxide surfaces, it would appear that either the adsorption sites for cadmium and humic acid are identical, or the prior adsorption of humic fractions covers available cadmium sites.

With natural organo-clay complexes, isolated from a silt loam, the greatest amount of adsorption was observed when using fractions that contained high quantities of organic matter or sesquiosides. Desorption studies using 0.01 M calcium nitrate solutions showed that the cadmium was adsorbed far more tenaciously by the sesquioxides than by organo-clay fractions.

From this investigation one can conclude that the affinity of cadmium for exchange sites (and the amount adsorbed) decreases in the following order: hydrous oxides of Al or Fe > humic acid $\approx$ montmorillonite > soil clay.

Further insight into the relative affinity sequence has evolved from a study of the adsorption of cadmium by the components of river muds (Gardiner, 1974). With each component, much of the labeled cadmium was taken up in a matter

Table 2.   Concentration Factors for Cadmium on a Range of Solids[a]

Solid	Added Cadmium Concentration (μg/l)	Concentration Factor, $K = C_s/C_l$	
		After 2 hr	After 24 hr
Kaolin	7.3	250 ($\pm$ 30)	380 ($\pm$ 50)
Silica	7.3	430 ($\pm$ 40)	1,000 ($\pm$ 100)
Plant material (watercress)	2.1	800	1,000
Humic acid	2.1–4.3	14,000 ($\pm$ 2000)	18,000 ($\pm$ 3000)
River muds	2.1	3000–17,000	6000–40,000

[a] Based on data from Gardiner (1974).

of minutes, indicating either that vacant sites are readily available or that exchange with previously adsorbed cations is rapid. However, since the amount sorbed increased with time, each substrate must also possess sites at which exchange is sterically hindered or is retarded by the nature of the chemical bonding involved (see Table 2).

The percentage of the total cadmium content sorbed by a solid was found to increase more or less linearly with increasing concentration (w/v) of solid (range, 0 to 200 mg/l^{-1}). With a fixed amount of solid, the percentage uptake decreased with increasing total Cd ion concentration. With very low concentrations (μg/l) of metal ion in solution uptake varied almost linearly with the solution phase concentration, and accordingly it was considered appropriate to compare behavior in terms of concentration factor (or distribution coefficient) values (K), where

$$K = \frac{\text{amount sorbed per unit mass}}{\text{equilibrium solution concentration}}$$

The typical values recorded in Table 2 indicate some similarity between the values obtained with humic acids and with river muds, and from this it may be presumed that the humic acid component of mud is principally responsible for its adsorptive properties. Such conclusions are tentative, however, because the magnitude of the concentration factor evaluated in this manner is a function of many variables, for example, particle size of the suspension, chemical composition of the solid phase, total concentration of metal ion present, and time of contact.

The K value can also be altered by changing the chemical form of the cadmium species. For example, the presence in solution of an equivalent amount of EDTA halved the K value; an excess halved the total uptake. Since in clay adsorption studies excess EDTA appears to completely eliminate metal ion uptake, one has to conclude that binding to some sites on natural clay-organo systems is extremely strong.

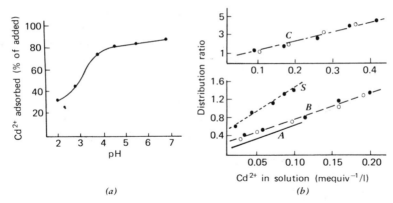

Figure 5. (a) Diagram showing the effect of pH on the uptake of Cd ion by a humic acid sample. (b) Langmuir isotherm plots for Cd^{2+} adsorption on humic acids extracted from A, Rendzina; B, Brown Mediterranean; and C, Podzol, soil samples; 5°C (O) and 25°C (●). Distribution ratio = [Cd in solution (mequiv/l)/adsorbed Cd (mequiv/100 g)] × 10^{-2}. (Based on data from Riffaldi and Levi-Minzi, 1975.) S is the isotherm obtained using a peat substrate soil (Bientina 1). (From Levi-Minzi et al., 1976.)

4. SORPTION OF CADMIUM BY HUMIC ACIDS

Interaction between metal ions and humic acid yields complexes of variable stoichiometry. The organic material is generally polyfunctional, and in the presence of excess metal salt more than one metal ion may become coordinated with each humic acid unit. With excess organic material, M-Org or M-Org$_b$ type complexes tend to be formed, regardless of the polyfunctional charge on the macromolecule (Org) (Zunino et al., 1975).

It has been suggested that the binding of metal ions to humic acids involves chelation of a type similar to that observed with salicylic acid, that is,

with the stability of the metal-humate complex being a function of the nature of the metal ion, the binding energy of the ligand functional groups, and environmental factors such as pH and E_h.

It has also been observed that the lower molecular weight fractions have a greater metal-holding capacity than higher molecular weight fractions.

The uptake of Cd ions by humic acids isolated from three different Italian soils has been shown to be adequately described by Langmuir type adsorption isotherms (as shown in Figure 5). The equation parameters (i.e., bonding constant and maximum adsorption capacity) increased in value with increasing content of oxygen-containing functional groups. For example, functional group concentrations of 55, 60, and 66 mequiv-kg were associated with constant values of 0.2, 0.6, and 0.8 and capacities of around 110, 170, and 190 mequiv Cd/kg, respectively. The metal uptake increased with pH but was independent of temperature (5 and 25°C). The effect of pH (e.g., ~45% of initial concentration sorbed at pH 2, 90% at pH 3.5) indicates that the cation adsorption process is connected with the dissociation of the functional groups on the humic acids (Riffaldi and Levi-Minzi, 1975).

Conversion of the cadmium into the anionic, EDTA complex resulted in zero adsorption of the metal ion.

Nearly all of the cadmium sorbed by the humic acid could be displaced by an 0.25 M copper acetate solution; only half was released by a 1 M ammonium acetate solution. In another study fresh water was found to release about a fifth of the adsorbed cadmium back into the aqueous phase (Gardiner, 1974).

These desorption studies imply that cadmium coordinated to the humic acid functional groups is not as strongly bound as copper, a view supported by the fact that 1 M ammonium acetate removes only 6 to 11% of adsorbed copper. On the other hand, the metal ion is preferred to protons, as shown by the fact that the pH drops to 4.5 on mixing cadmium chloride solution (pH 6) and sodium humate (pH 6).

Since proton displacement is involved in the metal uptake process, one can conclude that the reactivity of cadmium will depend strongly on the acid strength of the functional groups; hence behavior could vary with the source of humic acid. Organic matter possessing readily dissociated acid groupings should adsorb significant amounts of cadmium over a wide pH range, whereas more basic groupings (e.g., —OH) may become reactive only in alkaline media. Association of the humic material with clay suspensions could reduce the total capacity, but the studies made on river muds (cf. Table 2) indicate that the total effect may not be relatively great.

At this point let us consider two extremes in respect to functional group density and diversity.

Peat can be regarded as consolidated humate material, and it has been shown (Bunzl et al., 1976) that two H ions are released for each divalent ion (i.e., Ca^{2+}, Pb^{2+}, Cu^{2+}, Zn^{2+}, or Cd^{2+}) adsorbed. Sorption and desorption of the metal ions occur rapidly, with the rates of uptake decreasing in the order Pb > Cu > Cd > Zn > Ca. Determination of distribution coefficients over the pH range 3.5 to 4.5 leads to a selectivity order of Pb > Cu > Cd $\simeq$ Zn > Ca.

Cellulose possesses a much smaller exchange capacity (e.g., <50 mequiv/kg), but again it has been observed that two protons have to be displaced before metal ions are adsorbed. With cellulosic materials some of the acid functional groups are attached to the matrix (e.g., through oxidation of CH_2OH groups or aldehydic groups at the chain ends), and some functional groups are components of associated materials (e.g., lignins, pectins, hemicellulose). Adsorptive capacity and general behavior can thus vary with the type of cellulosic material involved.

A highly purified cellulose powder has been shown to have acid functional groups of $pKa \sim 4.4$, and a capacity of about 22 mmol M^{2+}/kg. The affinity order for the exchange sites appears to be $H^+ >$ Pb, Cu $>$ Zn, Cd. The uptake of Cd and other metal ions increases with pH, with precipitation of metal hydroxy species in the pH range 6 to 7 (cf. Section 2 on clays, Figure 2). Anionic metal complexes are not sorbed, while cationic complexes are sorbed to a lesser extent than hydrated ions (Farrah and Pickering, 1978b).

Although the adsorptive capacity of cellulose is similar to that of kaolin, metal ion uptake values for clay suspensions are affected to a greater extent by pH changes. Clays appear to be more readily saturated, and there is significant affinity between clays and some soluble ligand species. Despite these differences it can be proposed that sorption of cadmium by cellulosic materials could play a role similar to that of kaolin in controlling aqueous metal ion levels in the environment (cf. Plant material, and Kaolin, Table 2). Humic acids have a larger capacity, but they also bind the metal ion more strongly, and a smaller fraction is released on contact with electrolytes or ligands.

5. THE ROLE OF HYDROUS METAL OXIDES

Hydrous oxides, such as those of iron and manganese, are considered by Jenne (1968) to be capable of playing a dominant role in controlling the concentrations of metal ions (Mn, Fe, Co, Ni, Cu, Zn) in soils and waters. This is due in part to their adsorptive capacity, and in part to the fact that formation or dissolution of such substances can be induced by changes in acidity, oxidation potential, or the presence of complexing agents.

Colloidal hydrous iron oxide surfaces can retain a positive charge up to a pH of about 8, so that some degree of interaction between this material and negatively charged clay surfaces is to be expected. Formation of larger particles through coagulation is apparently minimal, and sorption on smaller particles appears to be favored.

Significant amounts of metal ions are adsorbed on freshly precipitated iron oxide gels even when the extent of cation hydrolysis in solution is $\ll 1\%$, and total adsorption invariably occurs at a pH lower than that required for hydroxide precipitation in particle-free solution (e.g., Cd^{2+} is half precipitated at pH 5.8; cf. Table 1). The selectivity sequence for the adsorption of divalent ions (in the

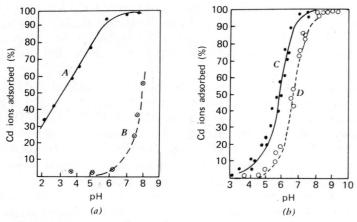

Figure 6. Effect of pH on the adsorption of Cd ions by hydrous metal oxides. (*a*) Percent adsorbed from 100 ml solution on *A*, hydrous manganese oxide (0.436 mmol); and B, hydrous ferric oxide (0.025 mmol). Initial Cd^{2+} concentration, 1 mmol/l. (From Gadde and Laitinen, 1974.) (*b*) Percent adsorbed on *C*, 10 g TiO_2; and *D*, 10 g α-FeOOH from a 2×10^{-4} M cadmium solution. (From James et al, 1975.)

presence of M $NaNO_3$) is reported to be Pb > Cu > Zn > Ni > Cd > Co > Sr > Mg (Kinniburgh et al., 1976).

This sequence is more extensive than that derived in an earlier study, namely, Pb^{2+} > Zn^{2+} > Cd^{2+} > Tl^{+}. The adsorption process releases protons (the ratio of moles H^+ released to moles metal ion sorbed is greater than 1 but less than 2) and appears to be reversible, with respect to both H^+ and other adsorbing ions. The uptake increases with pH, as shown in Figure 6 (Gadde and Laitinen, 1974).

Total retention of heavy metal ions (e.g., Co, Cu, Pb, Zn, Cd) at pH values less than the p^*K_1 predicted has also been observed in studies of adsorption on goethite (FeOOH) (p^*K_1 is the first hydrolysis constant). The amount sorbed increases with pH, and the difference between cations in respect to their adsorption properties is considered to reflect their abilities to form bonds with the oxide surface through hydroxyl bridges. With this adsorbent each mole of metal ion sorbed displaces approximately 2 mol of hydrogen ions from the surface of the solid. The displaced H^+ could equally well originate from the oxide surface or from the primary hydration shell of the adsorbing metal ion, although it appears that there is competition between the metal ions and protons for the surface sites. The intrinsic affinities of the heavy metals for the oxide surface decrease in the order Cu > Pb > Zn > Co > Cd, and it may be noted that this sequence differs from that proposed in the preceding section. The form of the adsorption curves appears to be influenced by surface charge, size of the ions, and adsorption density of the metal ions (Forbes et al., 1976).

In an earlier investigation of specific adsorption reactions on goethite surfaces, it was calculated that approximately one H^+ ion was displaced for every M^{2+} ion adsorbed (Grimme, 1968). The approach used in this study has been criticized on the basis that it cannot be assumed that the equilibrium constant has a single value over the whole pH range examined. In other words, the possibility has been ignored that the distribution coefficient value may be a function of surface charge. In addition, the mass action approach assumes that the adsorption versus pH relationship is the result of a single reaction, whereas one actually measures the additive effect of all possible interfacial reactions (Forbes et al., 1976).

The effect of pH on the uptake of Cd^{2+} by another α-FeOOH sample is shown in Figure 6. It may be observed that the metal ion is removed from solution at a pH value that is approximately three units lower than the value required for precipitation in the absence of a solid phase (cf. Table 1). With increasing pH, concentrations of Cd^{2+} in the solution decrease as the amount of hydrolysis product increases, and the adsorption density (Γ) (i.e., the amount sorbed per unit surface area) increases. From a comparison of the adsorption data and the effect of pH on species distribution, it has been proposed that there may be a relationship of the type

$$\log \Gamma = A + \log \{k_1[Cd^{2+}] + k_2[CdOH^+] + k_3[Cd(OH)_2]\}$$

where A, k_1, k_2, k_3 are constants, and $k_1 \ll k_2 \approx k_3$ (James et al., 1975).

On freshly precipitated hydrous aluminum oxide (point of zero charge, pH 9.4) the pH required to fully precipitate metal ions is slightly higher than that observed with iron gels (e.g., the cadmium half-precipitation value becomes 6.6), but the selectivity sequence resembles that reported for goethite: Cu > Pb > Zn > Ni > Co > Cd > Mg > Sr (Kinniburgh et al., 1976).

Cadmium ions are also specifically sorbed by hydrous manganese oxides (HMO). The amount of metal ion taken up increases almost linearly with pH (until all is sorbed), as shown in Figure 6. To a large extent this pH-dependent adsorption process is reversible. At a fixed pH, plots of the amount sorbed versus concentration have a plateau (as in Langmuir type isotherms), and the adsorption capacity at pH 6 is about 0.2 mol Cd^{2+}/mol HMO (i.e., $\sim$2 mol/kg, or slightly greater than the exchange capacity of montmorillonite suspensions). In view of the reversibility of the adsorption process and the magnitude of the capacity, it is logical to suggest that hydrous manganese oxide suspensions (or coatings) must strongly influence aqueous metal ion levels whenever they are present.

Titanium dioxide (TiO_2) (point zero charge, pH 5.8) has also been used as a substrate in studies of the effect of pH on cadmium uptake. Most of the cadmium is removed from 2×10^{-4} M solutions at pH > 6 (see Figure 6). Plots of calculated distribution coefficient values (i.e., $\Sigma_{\text{adsorbed species}}/[Cd^{2+}]$) against $[H^+]$ have been found to be nonlinear. The curvature can be explained if one postulates surface hydrolysis and simultaneous uptake of the Cd ion and hydrolysis products (cf. the FeOOH equation given above). The experimental adsorption data were used to determine surface hydrolysis constant and surface stability constant values, and from these a distribution diagram for cadmium species on the TiO_2 surface was calculated. This diagram has surface concentrations of Cd^{2+} declining from a maximum at pH $<$ 4 to nearly zero at pH 6; surface $CdOH^+$ has a maximum concentration value at around pH 5.5; and $Cd(OH)_2$ levels increase in importance at pH $>$ 6. If this surface hydrolysis model is valid, proton release per mole of cadmium sorbed should increase from nearly zero (pH 4) to two; with an ion exchange model involving preferential sorption of $CdOH^+$, at least one proton should be released over the full pH range.

The surface hydrolysis constants and stability constants calculated from the TiO_2 study differ in magnitude from those evaluated from FeOOH data (pH_{pzc} = 9.0) (James et al., 1975). This discrepancy further confirms that the nature of the surface is an important variable in these heterogeneous reactions.

The uptake of hydrolyzable metal ions at solid-aqueous solution interfaces may be represented by a number of models, but no one model has proved to be fully satisfactory. This is due to experimental difficulties and to the fact that all models tend to yield algebraic equations of similar form. Complete description of the interfacial system by any model is hindered by our lack of knowledge in respect to the nature of the solid phases being formed (e.g., particle size can affect solubility); the sequence of hydrolysis products (e.g., there can be a transition from monomers to low molecular weight polymers to high molecular weight polymers, to colloidal particles, to crystalline hydroxides, etc.); and the influence of pH, ligands, and metal ions on the zeta potential of the oxide-solution interface.

In the simple "ion exchange model" reaction of free metal ion with surface hydroxyl groups is considered to yield a coordinate metal-oxygen bond with the release of H_2O and H^+. In another model it is proposed that hydrolysis in the solution phase (promoted by pH increases) is followed by specific adsorption of the hydrolysis species. Alternatively, adsorption of the metal ion may promote the loss of a proton from a coordinated water molecule, that is, surface hydrolysis. A fourth, related model invokes the adsorption of the low to high molecular weight polymeric hydrolysis products which are the intermediates and precursors of solid hydroxide formation. These polymeric species could be subject to large van der Waals attractive force effects, and the postulated preferential adsorption of hydrolyzed species from a suite of complexes could arise from a combination of coulombic, solvation, and chemical energy changes, that is,

$$\Delta G_{ads\ n} = \Delta G_{coul\ n} + \Delta G_{solv\ n} + \Delta G_{chem\ n}$$

as adsorbates approach the interface (James and Healy, 1972).

For the adsorption of cadmium on TiO_2 or α-FeOOH, all the models yielded percentage adsorption against pH relationships of similar form. The James-Healy model gave calculated values that were slightly low at low pH, but this may have been due to errors in ΔG_{solv} estimates.

It has been concluded that studies of the adsorption of relatively simple metal aquohydroxy complexes do not distinguish unambiguously between the possible mechanisms in regard to the importance of each. In the future, however, it may prove possible to exploit some of the differences in the mathematical expressions for the different models, and so further improve our understanding of the uptake of hydrolyzable metal ions by solids.

6. SOIL, SEDIMENT, AND PLANT UPTAKE

For the environmentalist the most important questions to be answered are probably:

1. How mobile are cadmium species in soils and waterways?
2. What factors minimize uptake by plants?

Accordingly, for many workers, data derived from real soil and sediment samples (as distinct from isolated components) are of greater interest than the preceding discussions.

The toxic nature of cadmium species and concern about its transmission along food chains has prompted several groups to investigate the sorption and desorption of this metal by soils.

The adsorption behavior of cadmium on soils can usually be described in terms of a Freundlich type isotherm. Over restricted concentration ranges, the Langmuir relationship is also applicable, and this form of adsorption equation has tended to be preferred since it can be manipulated to provide information on both saturation capacities (Q) and bonding energies (expressed as Langmuir constants, k). Some typical values are listed in Table 3.

In one study of over 30 soils the adsorption maxima were found to be similar for all soils and to correlate with the amount of aluminum and zinc extracted by $0.01\ M$ $CaCl_2$. The bonding energy coefficients generally decreased in the following order: organic soils > heavy clay > sandy and silt loam > sandy soil, and the values could be related to the total aluminum + iron content. Neither of the adsorption equation parameters appeared to be significantly related to the cation exchange capacity (CEC), organic content, or clay content (John, 1972). In direct contrast the Q and k values of 10 Tuscany soils (Levi-Minzi et al., 1976) and some alkaline Indian soils (Singh and Sekhon, 1977) have been found to be significantly correlated with soil CEC values and organic contents.

Table 3.   Langmuir Coefficients for the Adsorption of Cadmium by Soils[a]

Soil	pH	Sand (%)	Clay (%)	Organic Carbon (%)	CaCO₃ (%)	CEC (mequiv/ 100/g)	Q (mequiv/ 100/g)	K (l/mequiv)
Gurdaspur	7.5	55.6	10.1	0.61	0.9	6.96	1.5	1.0
Jatwan	8.8	38.1	31.2	1.26	22.2	18.23	4.4	5.3
Langanwal	9.1	55.2	24.6	0.75	13.1	12.18	2.5	1.7
Langrian	8.9	87.3	5.5	0.33	1.05	3.83	1.3	1.1
Casciana	7.7		7.0	7.8	25.7	32.5	18.1	12.6
Spazzavento	7.9		16.1	2.7	0.8	27.5	15.7	3.7
Santo Pietro	8.6		34.7	1.0	5.6	20.0	13.7	3.1
Navacchio	8.5		10.0	3.1	8.0	17.5	9.4	7.6
Bientina I	8.1		24.9	4.5	5.8	31.2	18.9	11.6

[a] Based on data from Levi-Minzi et al. (1976) and Singh and Sekhon (1977).

With the Indian soils the parameter values also correlated with the clay and calcium contents.

Some discrepancies in derived relationships are not surprising when one considers the varying proportions of different adsorbents that can be present in different soils (cf. Table 3) and when allowance is made for the fact that each adsorbent may contribute to the overall effect in a particular way. All of the soils listed in Table 3 were alkaline, and for each group a value for the solubility product of $Cd(OH)_2$, that is, $[Cd^{2+}][OH^-]^2$, has been calculated. The Italian soils yielded concentration products ranging between 10^{-17} and 10^{-18}, values that are consistent with the surface precipitation data recorded in Table 1, but very different from the quoted value for precipitation from solution: $\sim 10^{-14}$. For the Indian soils the product values ranged from 10^{-12} to 10^{-15}.

The Q and k values determined at 5° have been shown to differ from those determined at 25°C; for example, Q may change from 100 to 120 mequiv Cd/kg as k changes from 7 to 11. This suggests that cadmium sorption by soils is at least partly of chemical nature, a view supported by the slow rate at which equilibrium is achieved (e.g., >16 hr at 5°C, >8 hr at 25°C). The Langmuir plots for the two temperatures had the same shape, so one may suggest that temperature does not change the nature of the interactions, but rather controls the number of sites and their affinity for Cd^{2+}.

Only a small proportion (e.g., 20 to 30%) of the cadmium taken up by the alkaline Indian soils was released by treatment with 0.1 M KCl, but 0.05 M copper acetate displaced about three quarters of the sorbed ion.

Pretreatment of soils with ions of different valence can alter the cadmium uptake (Lagerwerff and Brower, 1972). The amount adsorbed from solutions (initially containing 20 to 170 μequiv/l) has been found to be greater in the presence of Ca^{2+} than in the presence of Al^{3+}, and in both cases uptake decreased with increasing concentrations of competing ion (e.g., 5 to 50 mequiv/l). With NaCl-treated soils the adsorption of Cd^{2+} decreased with decreasing salt concentration. In this situation it has been suggested that products derived from alkaline corrosion of the soil matrix competed for sites and so reduced the exchange capacity.

In the presence of the higher valence cations, the distribution of ions fitted a Gappon type equation, that is,

$$\frac{\overline{Cd}}{\overline{Al}} = k_1 \left(\frac{\sqrt{Cd}}{\sqrt[3]{Al}}\right) \qquad \text{or} \qquad \frac{\overline{Cd}}{\overline{Ca}} = k_2 \left(\frac{\sqrt{Cd}}{\sqrt{Ca}}\right)$$

where $\overline{Cd}$, $\overline{Al}$, and $\overline{Ca}$ represent the amounts sorbed, and Cd, Al, and Ca are the equilibrium solution concentrations.

The soils used in this study contained significant amounts of clay, but this component was usually a mixture of several mineral types. For example, one clay fraction was composed of 45% kaolinite and 25% vermiculite with the rest undefined; another fraction consisted of 47% montmorillonite and 22% mica; a third clay segment was 80% illite. It was noted in Section 2 that clay types differ in their sorption characteristics; hence it is possible that correlation of metal uptake with total clay content will be observed only with soils in which a single type of clay mineral predominates.

The influence of different components on the uptake of cadmium by river muds was considered in Section 3. The author of this study (Gardiner, 1974) drew attention to the fact that the results obtained from a given sample are not necessarily characteristic of the particular river sampled, because much depends on the surface area of the particles and the bulk composition of the test sample. For example, samples taken at different times from locations within a few yards of each other in one stream yielded concentration factors that ranged in value from 6000 to 25,000 (using an added cadmium concentration of 2 μg/l).

There is some evidence that sediments effectively remove metal ions from wastewater. For example, it has been shown that, when metal ions are discharged from a wastewater treatment plant into an estuarine environment, the concentrations of metal ions (in μg/l) decrease within 2 to 3 km from the outfall as follows: Cd, 3.5 to 0.5; Pb, 31 to <4; Cu, 53 to 7; Mn, >120 to 90; Zn, 280 to 9; and Fe, >570 to 300. With the possible exception of the Mn and Fe ions, these decreases are much greater than could be ascribed to simple dilution, and the sediment concentrations of Cd, Pb, Cu, and Zn appear to be an order of magnitude higher than are normally found in uncontaminated areas. With cadmium the initial decrease in aqueous Cd ion level is followed by a slight rise as one moves further seaward from the pollution source. This suggests that some remobilization from the sediments occurs with increasing salinity (Helz et al.,

1975). The displacement effect may be due to a combination of the formation of chloro complexes and the presence of higher concentrations of Na ions.

It has been suggested that drops in concentration, such as those noted above, are due to precipitation of insoluble salts (e.g., carbonate, hydroxide, phosphate, or sulfide). However, the concentrations of cadmium in fresh water are usually much lower than the maximum permitted by the solubility product of the carbonate, which is probably the least soluble salt in most natural waters. Sulfide formation requires anaerobic conditions, and these are more likely to occur in sediments buried under successive layers of fresh material.

Under certain conditions, however, the formation of sparingly soluble compounds can affect the mobility of cadmium in soils. For example, it has been shown that the extractability of cadmium present as the chloride or sulfate is higher than that observed in the presence of cadmium sulfide. The degree of extractability increases with decreasing pH and E_h values and is greater from sands than from loams or calcareous soils (Tanaka and Ueta, 1974; Santillan-Medrano and Jurinak, 1975). In systems containing high cadmium levels, the cadmium concentration in the aqueous phase may be regulated by the solubility of solid phases such as $CdCO_3$ and $Cd_3(PO_4)_2$. Thus in one recent study it was predicted that the formation of $CdCO_3$ was controlling Cd^{2+} activity in sandy soils which possessed low organic matter contents and low cation exchange capacities, but a pH > 7 (Street et al., 1977). Under alkaline conditions the metal ion activity has been observed to decrease about one hundredfold for every unit increase in pH.

In another investigation it has been noted that the amount of cadmium released by treatment with 0.01 M HCl is less from soils treated with $(NH_4)_2HPO_4$ than from untreated soil, and the difference in behavior has been attributed to the formation of the sparingly soluble $CdHPO_4$. By contrast, the presence of $(NH_4)_2SO_4$ caused a decrease in the amount of cadmium taken up during adsorption studies, but sulfate-treated soils released more cadmium to acid extractants. Differences in pH and solubility may be responsible for this, and for the observation that the amount of cadmium extracted from soils by ammonium salt solutions decreases in the order $(NH_4)_2SO_4$ > H_2O > $(NH_4)_2HPO_4$ (Nakajima and Kawamura, 1977).

The lower pH required to form sparingly soluble hydroxy compounds in the presence of solids (cf. Table 1) has not been taken into account in these soil studies, an omission that may be unfortunate because the results on clays imply that this can be an important, and sometimes dominant, process.

At pH < 7 there seems to be agreement that cadmium levels are controlled by adsorption equilibria, with the total effect being adequately described by a Freundlich type isotherm.

Although uptake of cadmium by plants is the subject of a later chapter (see J. D. Jastrow and D. E. Koeppe, this volume, Chapter 15), it is desirable to briefly mention this aspect here to show the relationship (if any) to adsorption-desorption processes.

The cadmium content of plants increases with increasing amounts of this metal ion in the soil, but the quantitative relationship between uptake and soil content is determined by soil type and chemical form of the cadmium.

The effect of soil type has been demonstrated in a study of cadmium uptake by oats plants. The soils used were a peaty mixture, a sandy soil, and a clay soil, and the metal content of the plants decreased along this sequence. Lime additions retarded the uptake process (Sortebery, 1976). Further evidence is provided by the results obtained in an investigation of cadmium uptake by wheat plants. Two soil types were used, one classified as suitable for wheat growth (S), the other described as unsuitable (U). When 0.5 ppm Cd (as $^{115}CdSO_4$) was added to the soils, it was found that the shoots took up 0.5% of the applied cadmium from the good soil and 4.5% from the unsuitable soil. This larger uptake by the plants grown in the latter soil was maintained in all derived products, as shown by the comparative U soil values recorded in parentheses; for example, dry straw from soil S contained 0.33 ppm Cd (3 ppm, U soil); chaff, 0.15 (2.9); whole grains, 0.1 (1.1); bran, 0.2 (2.8); and flour, 0.07 (0.55) (Oberlaender and Roth, 1976).

The possible role of chemical form is illustrated by the observation that the amount of cadmium taken up by corn seedlings is greater when inorganic cadmium salts are used as a soil additive, than when an equivalent addition is made in the form of sewage sludge or cadmium-spiked sludge (Street et al., 1977). This group of workers found that the cadmium concentrations in the corn seedlings were highly correlated with the amount of cadmium that could be extracted from the soil with diethylenetriaminepentaacetic acid (DTPA). Other investigators, using different soils or marker elements or types of plants, have found correlations to exist with a range of extractants, and it would appear that at this point in time each system has to be treated individually. As knowledge develops, a useful generalized approach may evolve, and so facilitate the very necessary distinction betwen "total" and "available" levels.

7. IMPLICATIONS AND EXTENSIONS

The important role that may be played by adsorption on clays and other sediment components in controlling cadmium levels in the environment should be quite apparent from the preceding discussion.

It would seem that most investigators recognize the complexity of the total distribution process and the limitations of sectional studies. Interest in this field is comparatively new (e.g., most of the references are to papers published in the last 5 years); hence much of the effort has been associated with evaluation of the problem and elucidation of the possible contributions of individual factors. Future studies involving new systems, extended basic investigations, or refined techniques, possibly with mixed discipline team approaches, should add much to our understanding of the basic processes, the mobility of this metal ion, and the health implications of pollution by cadmium species.

Concurrent studies using other heavy metal ions (e.g., Cu, Pb, Zn, Hg) have revealed many points of similarity, but also areas of distinct differences. For example, mercury is less prone to adsorption because of the stability and solubility of its chloro and hydroxy complexes; copper is more strongly bound to sediment components than the other metal ions; lead seems to be the most readily precipitated species; and cadmium appears to be the species most sensitive to changes in the chemical environment.

In the case of cadmium, a number of areas warrant further investigation. For example, studies should be made of the effect of cationic competition; the influence of ligands introduced by human activities (e.g., sewage components); the uptake by a range of natural sediments; and so on. The presence of silicate ions has been found to enhance zinc uptake by clays; heating and drying processes have been shown to reduce the mobility of copper; lead levels may be altered through precipitation of sulfate or phosphate. The effects of these variables on cadmium behavior have yet to be elucidated. Similarly, there have been few physical method studies, such as X-ray diffraction examination of cadmium-saturated solids, infrared or laser Raman examination of cadmium-surface interactions, or thermal decomposition investigations. A reexamination of the sorption behavior of cadmium on different surfaces could help to identify the most appropriate model for adsorption at oxide-water interfaces.

Continued effort in this field can be readily justified by quoting a few one-sentence extracts from papers listed in the reference section:

Of the metals studied, Cd presents the greatest potential for serious pollution because its input from waste water probably exceeds fluvial input, it appears to be readily remobilized from sediments, and it is known to be toxic to many organisms (Helz et al., 1975).

Adsorption and desorption processes are likely to be major factors in controlling the concentration of cadmium in natural waters and will tend to counteract changes in the concentration of the metal ion (Cd^{2+}) in solution (Gardiner, 1974).

Aquatic organisms may encounter toxic metals in high concentrations adsorbed on particles in suspension or in sediments. Plant roots are known to exude organic substances that act as a complexing agent. These aid in the absorption of trace metal nutrients and at the same time expose the growing plant to toxic metals (Gadde and Laitinen, 1974).

In brief, sorption processes can play a significant role in the biogeochemical distribution of cadmium in the environment.

REFERENCES

American Petroleum Institute (1951). Research Project 49, "Reference Clay Minerals." Columbia University, New York.

Amphlett, C. B. (1964). *Inorganic Ion Exchangers.* Elsevier, Amsterdam, pp. 15–41.

Bingham, F. T., Page, A. L., and J. R. Sims, (1965). "Retention of Acetate by Montmorillonite," *Soil Sci. Soc. Am. Proc.* **29**, 670–672.

Bittel, J. E. and Miller, R. J. (1974). "Pb, Cd and Ca Selectivity Coefficients on a Montmorillonite, Illite and Kaolinite," *J. Environ. Qual.* **3**, 250–253.

Bunzl, K., Schmidt, W., and Sansoni, B. (1976). "Kinetics of Ion Exchange in Soil Organic Matter. IV: Adsorption and Desorption of Pb^{2+}, Cu^{2+}, Cd^{2+}, Zn^{2+} and Ca^{2+} by Peat," *J. Soil Sci.* **27**, 32–41.

Farrah, H. and Pickering, W. F. (1977a). "Influence of Clay-Solute Interactions on Aqueous Heavy Metal Ion Levels," *Water, Air, Soil Pollut.* **8**, 189–197.

Farrah, H. and Pickering, W. F. (1977b). "The Sorption of Lead and Cadmium Species by Clay Minerals," *Aust. J. Chem.* **30**, 1417–1422.

Farrah, H. and Pickering W. F. (1978a). "Extraction of Heavy Metal Ions Sorbed on Clays," *Water, Air, Soil Pollut.,* **9**, 491–498.

Farrah, H. and Pickering, W. F. (1978b). "The Effect of pH and Ligands on the Sorption of Heavy Metal Ions by Cellulose," *Aust. J. Chem.,* **31**, 1501–1509.

Forbes, E. A., Posner, A. M., and Quirk, J. P. (1976) "The Specific Adsorption of Divalent Cd, Co, Cu, Pb and Zn on Goethite," *J. Soil Sci.* **27**, 154–166.

Gadde, R. R. and Laitinen, H. A. (1974). "Studies of Heavy Metal Adsorption by Hydrous Iron and Manganese Oxides," *Anal. Chem.* **46**, 2022–2026.

Gardiner, J. (1974). "The Chemistry of Cadmium in Natural Water II: The Adsorption of Cadmium on River Muds and Naturally Occurring Solids," *Water Res.* **8**, 157–164.

Grimme, H. (1968). "Die Adsorption von Mn, Co, Cu and Zn durch Goethit aus verduennten Loesungen," *Z. Planzem Düng. Bodenk* **121**, 58–65.

Grimshaw, R. W. and Harland, C. E. (1975). *Ion Exchange: Introduction to Theory and Practice.* The Chemical Society, London, pp. 4–19.

Haghiri, F. (1976). "Release of Cd from Clays and Plant Uptake of Cd from Soils as Affected by K and Ca amendments," *J. Environ. Qual.* **5**, 395–397.

Helz, G. R., Huggett, R. J., and Hill, J. M. (1975). "Behaviour of Mn, Fe, Cu, Zn, Cd and Pb Discharged from a Wastewater Treatment Plant into an Estuarine Environment," *Water Res.* **9**, 631–636.

Hodgson, J. F. (1970). In D. D. Hemphill, Ed., *Trace Substances in Environmental Health,* Vol. VIII. University of Missouri Press, Columbia, pp. 45–58.

Jackson, T. A. (1975). "Humic Matter in Natural Water and Sediment," *Soil Sci.* **119**, 56–64.

James, R. O. and Healy, T. W. (1972). "Adsorption of Hydrolyzable Metal Ions at the Oxide-Water Interface. III: A Thermodynamic Model of Adsorption," *J. Colloid Interface Sci.* **40**, 65–81.

James, R. O., Stiglich, P. J., and Healy, T. W. (1975). "Analysis of Models of Adsorption of Metal Ions at Oxide/Water Interfaces," *Disc. Faraday Soc.* **59,** 142–156.

Jenne, E. A. (1968). Chapter 21 in R. F. Gould Ed., *Trace Inorganics in Water.* Advances in Chemistry Series 73, American Chemical Society, Washington, D.C.

John, M. K. (1971). "Influence of Soil Characteristics on Adsorption and Desorption of Cadmium," *Environ. Lett.* **2,** 173–179.

John, M. K. (1972). "Cadmium Adsorption Maxima of Soils as Measured by the Langmuir Isotherm," *Can. J. Soil Sci.* **52,** 343–350.

Kinniburgh, D. G., Jackson, M. L., and Syers, J. K. (1976). "Adsorption of Alkaline Earth, Transition and Heavy Metal Cations by Hydrous Oxide Gels of Iron and Aluminium," *Soil Sci. Soc. Am. Proc.* **40,** 796–799.

Kobayashi, J., Morii, F., and Muramoti, S. (1974). "Removal of Cadmium from Polluted Soil with EDTA," *Trace Subst. Environ. Health* **8,** 179–192

Kodama, H. and Schnitzer, M. (1974). "Adsorption of Fulvic Acid by Non-expanding Clay Minerals," *Tr. Mazhduner. Kongr. Pochvoved* **2,** 51.

Lagerwerff, J. V. and Brower, D. I. (1972). "Exchange Adsorption of Trace Quantities of Cadmium in Soils Treated with Chlorides of Aluminum, Calcium and Sodium," *Soil Sci. Soc. Am. Proc.* **36,** 734–737.

Levi-Minzi, R., Soldatini, G. F. and Riffaldi, R. (1976). "Cadmium Adsorption by Soils," *J. Soil Sci.* **27,** 10–15.

Levy, R. and Francis, C. W. (1976). "Adsorption and Desorption of Cadmium by Synthetic and Natural Organo-Clay Complexes," *Geoderma* **15,** 361–370.

Mishra, P. C., Mishra, M. K., and Misra, S. G. (1973) "Evaluation of Methods for Estimating Available Copper in Soils," *Indian J. Agric. Sci.* **43,** 609–610.

Misra, S. G. and Pandey, G. (1976). "Evaluation of Suitable Extractants for Available Lead in Soils," *Plant Soil* **45,** 693–696.

Nakajima, T. and Kawamura, S. (1977). "Effect of Diammonium Phosphate and Ammonium Sulfate on Cadmium Adsorption by Soils," *Kinki Daigaku Nogakubu Kiyo* (10), 45–51.

Nissenbaum, A. and Swaine, D. J. (1976). "Organic Matter-Metal Interactions in Recent Sediments: the Role of Humic Substances," *Geochim. Cosmochim. Acta.* **40,** 809–816.

Oberlaender, H. E. and Roth, K. (1976). "Radiometric Determination of Cadmium Accumulated in the Endosperm of Wheat after the Addition of Cadmium to the Soil," *Naturwissenschaften* **63,** 483.

Orlov, D. S. and Pivovarova, I. A. (1974). "Selective Adsorption of Humic Acid Fractions by Clay Minerals: Regression Equations for Quantitative Soil Humic Acid Adsorption by Minerals," *Pochvovedenie* **29,** 59–64 and 96–99.

Petruzzelli, G., Guidi, G., and Lubrano, L. (1977). "Cadmium Occurrence in Soil Organic Matter and Its Availability to Wheat Seedlings," *Water, Air, Soil Pollut.* **8,** 393–400.

Rashid, M. A. (1971). "Role of Humic Acids of Marine Origin and Their Different Molecular Weight Fractions in Complexing Di- and Trivalent Metals," *Soil Sci.* **111,** 298–306.

Rashid, M. A., Buckley, D. E., and Robertson, K. R. (1972). "Interactions of Marine Humic Acids with Clay Minerals and a Natural Sediment," *Geoderma* **8**, 11–27.

Rashid, M. A. (1974). "Adsorption of Metals on Sedimentary and Peat Humic Acids," *Chem. Geol.* **13**, 115–123.

Riffaldi, R. and Levi-Minzi, R. (1975). "Adsorption and Desorption of Cadmium on Humic Acid Fraction of Soils," *Water Air, Soil Pollut.* **4**, 179–184.

Santillan-Medrano, J. and Jurinak, J. J. (1975). "The Chemistry of Lead and Cadmium in Soil: Solid Phase Formation," *Soil Sci. Soc. Am. Proc.* **39**, 851–856.

Sillen, L. G. and Martell, A. E. (1964, 1971). *Stability Constants of Metal-Ion Complexes.* The Chemical Society, Spec. Publ. 17 and 25, London.

Singh, B. and Sekhon, G. S. (1977). "Adsorption, Desorption and Solubility Relationships of Lead and Cadmium in Some Alkaline Soils," *J. Soil Sci.* **28**, 271–275.

Sortebery, A. (1976). "Effects of Some Heavy Metals on Soil and Crops," *Nord. Jordbrugsforsk.* **58**, 122–123 (CA. 85:158678).

Street, J. J., Lindsay, W. L., and Sabey, B. R. (1977). "Solubility and Plant Uptake of Cadmium in Soils Amended with Cadmium and Sewage Sludge," *J. Environ. Qual.* **6**, 72–77.

Stuanes, A. (1976). "Adsorption of Mn^{2+}, Zn^{2+}, Cd^{2+} and Hg^{2+} from Binary Solutions by Mineral Materials," *Acta Agric. Scand.* **26**, 243–250.

Swartzen-Allen, S. L. and Matijevic, E. (1974). "Surface and Colloid Chemistry of Clays," *Chem. Rev.* **74**, 385–400.

Tan, K. H. (1975). "The Catalytic Decomposition of Clay Minerals by Reaction with Humic and Fulvic Acids," *Soil Sci.* **120**, 188–194.

Tan, K. H. (1976). "Complex Formation between Humic Acid and Clays as Revealed by Gel Filtration and IR Spectroscopy," *Soil Biol. Biochem.* **8**, 235–239.

Tanaka, A. and Ueta, H. (1974). "Movement of Cadmium in Soil under Various Conditions," *Tottoriken Nogyo Shikenjo Kenkya Hokoku* **14**, 31–46 (CA 85, 187347).

Theng, B. K. G. and Scharpenseel, H. W. (1976). "The Adsorption of Carbon-14-Labeled Humic Acid by Montmorillonite." In S. W. Bailey, Ed., *Proceedings of The International Clay Conference, 1975,* Applied Publishers, Wilmette, Ill., pp. 643–653.

Wakatsuki, T., Furukawa, H., and Kawaguchi, K. (1975). "Specific Adsorption of Cations on Kaolin and Kaolinic Soil Clays," *Soil Sci. Plant Nutr. (Tokyo)* **21**, 351–360.

Zunino, H., Peirano, P., Aguilera, M., and Schalscha, E. B. (1975). "Measurement of Metal Complexing Ability of Polyfunctional Molecules: A Discussion of the Relationship between the Metal Complexing Properties of Extracted Soil Organic Matter and Soil Genesis and Plant Nutrition," *Soil Sci.* **119**, 210–216.

11

BIOGEOCHEMICAL CYCLING OF CADMIUM IN A MARSH ECOSYSTEM

Robert E. Hazen

Theodore J. Kneip

New York University Medical Center, Institute of Environmental Medicine, New York, New York

1. INTRODUCTION

The widespread use of cadmium in a large number of products and industrial processes has resulted in discharges of the element to aquatic ecosystems (Fleisher et al., 1974), with considerable potential for impacts on ecosystems and human health. An example of such an effect is the occurrence of itai-itai ("ouch-ouch") disease in Japan, resulting from chronic exposure of a human population to environmental cadmium contamination (Friberg et al., 1974). Because of past discharges and continuing potential hazards, there is concern that further unforeseen effects may occur in aquatic biota or in human populations. For this reason there is a critical need to define the role of aquatic ecosystems in the accumulation, transfer, effects, and ultimate fate of this element.

These factors have been investigated in an estuarine marsh-cove ecosystem. Sediments at this location were reported to contain as much as 50,000 mg Cd/kg (F. T. Brezenski, unpublished results), offering an unparalleled opportunity for study of the problem.

The site of the contamination was Foundry Cove, which is located about 85 km north of Battery Park (New York City) on the Hudson River. The cove is on the east shore of the river, south of the town of Cold Spring and across the river from the U.S. Military Academy at West Point. Contamination occurred during the 1950s and 1960s from waste discharged by a nickel-cadmium battery factory. Concentrated efforts by the U.S. Environmental Protection Agency in 1971 failed to demonstrate significant effects on the ecosystem; nevertheless, a consent agreement in the U.S. District Court for the Southern District of New York required removal of sediments believed to exceed about 1000 mg Cd/l wet sediment (Gregor, 1973). (This is equivalent to about 1490 mg Cd/kg dry sediment.) Nickel as well as cadmium contamination occurred in the cove, but the effects of nickel are not believed to be significant at the concentrations observed (Fennikoh et al., 1978).

The location of Foundry Cove and the geographical features of the location are indicated in Figures 1 and 2. The most contaminated portion (the east cove) is partially isolated from the west cove and the Hudson River by a railroad roadbed. The connection of the east cove and the river is limited to a 10-m passage under the trestle, and a channel system connects the cove to a wildlife sanctuary east of Constitution Island.

The predominant influence on water circulation between the cove and the river is a tide of 1 to 1.5 m, which exposes a considerable portion of the east cove bottom at low water. Estimates of the cove area and the tidal rise indicates that approximately 125,000 m^3 (125 × 10^6 liters) of water are exchanged between the cove and the river in each tidal cycle. This is probably more than one half of the total volume of water in the cove at high tide. Flow patterns observed in-

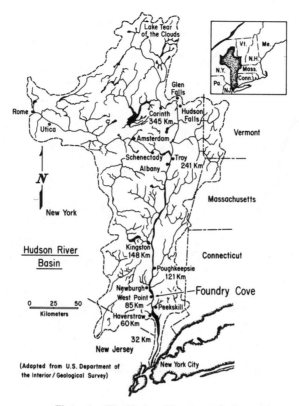

Figure 1.   The Hudson River watershed.

dicate that most of the exchange of water occurs by way of the channel under the trestle.

To conform with the court order the cove was dredged in two stages, the outfall area in 1972 and a portion of the east cove and channel in 1973. We studied the area before, during, and after the removal effort. Our studies were designed to define the following:

1. The extent of cadmium interchange between sediments and water.
2. The effects of the dredging.
3. The mechanisms of transport within the cove.
4. The rate of transfer from the cove to the river.
5. The effects of tide and salinity on the transfer of cadmium.
6. The extent and routes of transfer through the food web.
7. The effects or potential effects on the ecosystem.
8. Possible human exposures and hazards therefrom.

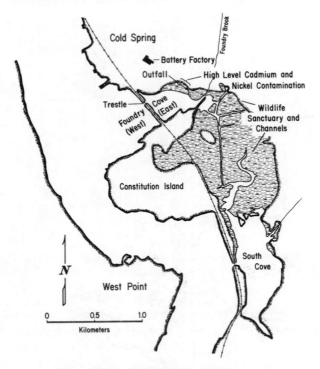

Figure 2.   Hudson River—Foundry Cove region.

Cadmium concentrations were measured in the sediments, water, and biota over a period of several years to obtain the information needed. Populations of benthic organisms and plankton were also studied (Bondietti et al., 1974; Kneip et al., 1974a, 1974b; Hazen and Kneip, 1976; Kneip, 1978; Fennikoh et al., 1978).

2. CADMIUM IN THE SEDIMENTS

Estimates of the total cadmium in Foundry Cove have ranged from 23 tons (Gregor, unpublished data) to 18 tons (Bower et al., 1978). Since the waste discharge was controlled in 1970–1971, the major source of cadmium to the Foundry Cove ecosystem has been the cadmium that was incorporated in the sediments. For these reasons efforts were focused first on achieving an understanding of these deposits. Sampling and analytical methods for sediments, water, and biota have been described in detail elsewhere (Kneip et al., 1974a, 1974b). The cadmium contamination of Foundry Cove sediments is notable because of the extremely high levels within 100 m of the outfall, where con-

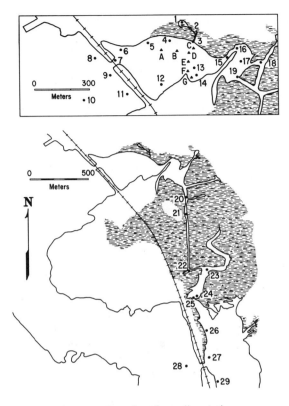

Figure 3.   Location of sampling stations.

centrations above 10,000 mg/kg dry weight are common. This compares to a worldwide background level of cadmium in sediments of about 1 mg/kg dry weight (Moyer and Budinger, 1974). Core samples taken in 1976 show that contamination is largely limited to surface sediment in the east cove (Table 1), and Bower et al. (1978) have reported similar findings for the west cove (stations are shown in Figure 3). The cadmium is present in a minimum of two chemically distinct forms (Bondietti et al., 1974). Near the outfall, where the concentrations are the greatest, cadmium is in a mixed calcium-cadmium carbonate system. Further from the outfall cadmium is associated with the organic fraction of the sediments.

Concentrations (in mg Cd/kg dry sediment) are shown in Table 2 for different stations over several years. The stations are listed in order of their distance from the outfall by the most direct water route. In 1974 and 1976 enough sediment samples were obtained to develop concentration contours of the inner cove. In

Table 1. Cadmium Concentrations in Core Samples (mg/kg dry weight)

	Distance from Outfall (m)						
	175	200	250	295	300	325	390
				Station			
Depth (cm)	C	D	E	B	F	G	A
0	3820	2850	6780	160	2	8.6	3.2
10							
						0.5	0.5
				19			
		32			0.5		
20	64		5				
							0.3
30	62	95					
					2		
					1.5		
40			4				
						2.4	
50							

both years a small area (less than 100 m^2) exceeded 40,000 to 50,000 mg Cd/kg, and an area of about 50,000 m^2 exceeded 1000 mg/kg.

It has been reported previously that the dredging failed to remove a significant amount of the contamination from the cove (Kneip et al., 1974b), although there is an estimated 5.5 metric tons in the dredge spoil (New York State Department of Environmental Conservation, unpublished data). The strong coincidence of cadmium contours in 1974 and 1976 (Hazen and Kneip, 1976), as well as the presence of highly contaminated areas from 1974 to 1978, indicates that cadmium is not being removed by the effects of solution, burial, shifting sediments, or biological accumulation at a rate sufficient to alter the sediment concentrations to a detectable extent. The total amount of cadmium remaining in the cove has been calculated on the assumption that the measured surface concentration (upper 5 cm) drops linearly to background levels at a depth of 20 cm. Core data from both Bower et al. (1978) and New York University (NYU) (Table 1) confirm 20 cm as a good estimate for the penetration of the contamination. The cadmium concentration was calculated for a depth of 10 cm and was taken to represent the average concentration of the entire 20 cm. A calculation was necessary in most cases because surface samples of the upper 5 cm rather than core samples were taken. Approximately 9 metric tons is estimated to be in the east cove. The inclusion of the estimate by Bower et al. (1978) of 2 metric tons in the west cove results in a total of 11 metric tons, which in addition to the 5.5 metric tons in the dredge spoil is within 50% of the original estimate of 23 metric tons.

Table 2.   Cadmium (mg/kg dry weight) in Sediment

Station	Distance from Outfall (m)	1971	1973	1974	1975	1976
1	10	30,500[a]	9,630	42,000[a]	12,000	11,500
2	60	18,500[a]	10,680			
3	175	10,400[a]	25,700	28,300		3,200
4	300		450	1,680	290	1,600
13	310				160	2
14	360					8
15	400		92			
5	420			760	430	3
16	480		150			
12	480		550			
17	580					780
6	600	33	490			
7	610					
19	660		430			
18	690		870			
8	710	26	100			
9	720	420				
20	760		340			
11	860	140				
10	880	180				
21	900		350			5
22	1400					
23	1530					
24	1790[b]					
25	1800		35			
26	2100		14			12
27	2330		77			
28	2510		16			
29	2530					2

[a] Median.
[b] No data available.

3. CADMIUM IN THE WATER OF FOUNDRY COVE

The major source of cadmium in the water, at least since 1971, has been either dissolution or resuspension of the cadmium-containing sediments. Concentrations of cadmium in the water column at various stations over a number of years are shown in Table 3. Both filterable and particulate cadmium were measured. Foundry Cove water is often elevated in cadmium content (this was particularly true in the early years of the study) over a Hudson River background concentration of about 1 μg/ (Durum et al., 1970).

Table 3. Cadmium in Water

Station	Date	n	Filterable Cadmium (μg/l) $\overline{X}_{ebb}$	Particulate Cadmium (μg/l) $\overline{X}_{ebb}$	n	Filterable Cadmium (μg/l) $\overline{X}_{flood}$	Particulate Cadmium (μg/l) $\overline{X}_{flood}$
1	8/15/72					1.0	
	8/22/72		27.8				
	8/24/72[a]		20.9				
	8/30/72[a]		51.5		2	3.1	
	9/5/72	2	26.3			7.8	
			$\overline{X}$ 31.6			$\overline{X}$ 3.9	
	7/16/73					18.1	67.8
	9/27/73					7.0	21.2
	10/25/73		8.5	7.5			
			$\overline{X}$ 8.5	$\overline{X}$ 7.5		$\overline{X}$ 12.6	$\overline{X}$ 44.5
3	8/15/72		<0.1			0.4	
	8/17/72		<0.5			1.0	
	8/22/72	2	9.2				
	8/24/72[a]	2	91.3				
	8/30/72[a]	2	18.2			3.7	
	9/5/72	3	17.0				
			$\overline{X}$ 11.0			$\overline{X}$ 1.7	
	7/28/73	2	24.2	199.2		5.0	12.1
	7/16/73					9.5	16.1
	9/27/73					2.4	3.5
	10/11/73	6	16.0	16.1	4	7.5	10.5
	10/24/73	6	2.4	4.8	4	2.6	4.6
	10/25/73		4.9	3.0		7.5	10.5
			$\overline{X}$ 11.9	$\overline{X}$ 55.8		$\overline{X}$ 5.6	$\overline{X}$ 9.6
	6/24/74	4	6.3	33.0	5	2.4	17.6
7	5/22/74	6	2.3	4.2	4	2.6	1.6
	6/24/74	4	49.2	7.2	5	26.4	1.8
	10/24/74	7	1.7	1.1	5	1.2	1.2
			$\overline{X}$ 17.8	$\overline{X}$ 4.2		$\overline{X}$ 7.1	$\overline{X}$ 1.5
	3/8/76	8	2.0	(3.8)[b]	4	0.8	(1.0)
	9/2/76	6	2.1	(3.9)	6	1.4	(1.4)
7 bottom							
	9/2/76	6	2.3	(4.5)	6	0.8	(1.4)
			$\overline{X}$ 2.1	$\overline{X}$ 4.1		$\overline{X}$ 1.0	1.3
16	10/24/74	6	2.4	10.8	5	12.0	22.0
	9/2/76	3	3.6				

[a] During dredging.
[b] Numbers in parentheses are total metal in unfiltered water.

Table 4.   Range of Water Quality Parameters at Stations 3, 9, 10, 25

Parameter	1973[a]	1974[b]
Temperature (°C)	6–27	4–28
Dissolved O_2 (mg/l)	5–13	6–15
pH	7.0–7.6	7.1–7.4
Alkalinity (mg/l) (Phenolphthalein)	30–60	45–70
Total hardness (mg/l) as ($CaCO_3$)		68–205
Chloride (mg/l)		25–600

[a] Fourteen samples at each station from March to December.

[b] Eight samples from March to October.

Water quality measurements are summarized in Table 4. The observed pH is normally above 7, and alkalinity consistently exceeds 30 mg/l as $CaCO_3$. The normal range in salinity as calculated from the chloride range is from about 0.07 to 1.1 parts per thousand (ppt).

A striking feature of the cadmium concentrations in water (either filterable or particulate) is that they are much less elevated over background (approximately 10 to 50-fold) than are the sediment concentrations (1000 to 50,000-fold). This is in agreement with the observed stability of the cadmium content of the sediments, as well as expectations based on the known water chemistry of cadmium (Hem, 1972).

4. POTENTIAL RELEASES

As most of the east cove water is funneled through the trestle on both ebb and flood tide, it was convenient to measure the cadmium contained in water samples at this location in order to obtain an upper estimate of the net loss of cadmium from the east cove to the river. Cadmium that leaves the east cove under the trestle, however, does not necessarily stay in solution or suspension and proceed to the river. In fact, it can be hypothesized that the contamination of the west cove must be the result of the deposition of cadmium of east cove origin.

The results of a typical tidal cycle sampling are shown in Figure 4. Samples were collected once an hour during a 12-hr tidal cycle. The impact of the cove cadmium on water column cadmium concentration is apparent in the higher values recorded at ebb tide when water was flowing from the cove to the Hudson River.

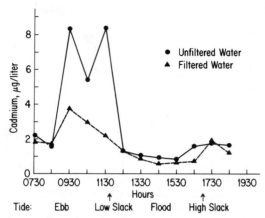

Figure 4.   Cadmium in tidal water samples at the Trestle (Station 7) (Mar. 8, 1976).

From tidal cycle data it is possible to calculate a potential net effect on the cadmium concentration of the Hudson River. Based on averages of three tidal cycles in 1974 and two in 1976, there were differences of 13.0 and 2.8 µg total Cd/l, respectively, at the trestle between ebb and flood tidal samples. When these concentrations are multiplied by an estimated 125,000 m³ of water exchanged per cycle, the result is 1.6 kg (1974) and 0.35 kg (1976) of cadmium moved from the cove to the river (inside the trestle to outside) per tidal cycle.

In drought condition the Hudson River has a flow of about 55 m³/sec and approximately 10 times this amount in normal years. Assuming that all the cadmium moved out of the west cove remains in suspension or solution, and dividing the mass of cadmium (1.6 kg, 1974; 0.35 kg, 1976) by the volume of water in 12 hr (a tidal cycle) of drought flow (2.37 × 10⁹ liters), one obtains a potential increase in Hudson River cadmium concentration of 0.68 µg/l in 1974 and 0.15 µg/l in 1976. These concentrations would be reduced by a factor of 10 in years of normal river flow; therefore it is not likely that discharges from this location account for a major fraction of the 1 µg Cd/l found in the Hudson River in 1970 by Durum et al. (1970).

Because of the estuarine nature of the Foundry Cove region an increase in salinity might occur during a period of severe drought. Experiments conducted with this eventuality in mind demonstrated a two- to threefold increase in the release of cadmium from the sediments in water containing 5 ppt sea salt (ocean water has approximately 35 ppt). Thus continued stability of the contamination depends on the maintenance of existing chemical and physical conditions in the water and sediments. An increase in salinity could have the effect of mobilizing cadmium to some extent.

The worst-case effect on Hudson River water can be calculated by assuming the unlikely event of the release of the entire remaining 11 metric tons of Foundry Cove cadmium during one tidal cycle. (This is not a foreseeable event under any projected set of changing water quality parameters.) In this case, because the process is not a continuous one, the calculation is based on dilution by all the water that passes Foundry Cove in a tidal cycle. This amount of water is much greater than the net freshwater flow, and a rough estimate based on the average depth and width of a 6-km* length of the river is 11×10^7 m^3. The solution of 11 metric tons Cd in this volume of water would result in a concentration of about 100 μg/l. Mixing with the river flow would lead to dilution and possible transfer to suspended and bottom sediments. The rates and routes of transfer and ultimate fate and effects of such a release cannot be predicted because of lack of knowledge of the behavior of cadmium in an estuarine system. Short-term toxic effects would be likely, but long-term predictions cannot be made. The actual gradual increase in salinity that a drought would cause would be more likely to lead to a gradual release and redistribution of cadmium to the river.

The physical situation presented to the Foundry Cove ecosystem is therefore one in which the sediments are extremely contaminated (thousands of times the background level) with a stable distribution pattern, while the water column is relatively well flushed and usually less than 10 times background in recent years.

5. CADMIUM IN THE MARSH BIOTA

The route of cadmium uptake by vascular plants is via solution from the sediments and subsequent uptake by roots. The sources to animals are ingestion of sediments or water, uptake of cadmium in water via gills, and ingestion of food organisms that have themselves already taken up cadmium.

5.1. Plants

Leaf, female inflorescence (seed head), and seed samples were collected from *Typha angustifolia* (cattail) along a transect extending from the outfall across the marsh to the open water of the inner cove (see Figure 5). Sediment samples were collected from the location of each plant sample. Because of the very high sediment cadmium concentration in this area only plant portions above the high tide level were collected. Most of the female inflorescence consists of hairs that provide for windborne seed dispersal. Analyses were performed on both the hairs and the seeds for most of the samples. In some cases the seeds were separated

---

* Six kilometers is the approximate distance of the tidal flow.

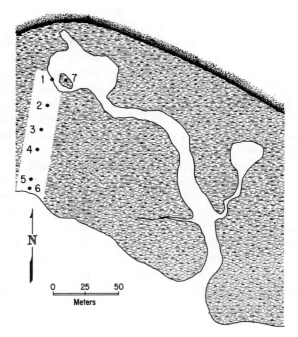

Figure 5.   Special Foundry Cove marsh stations.

from the hairs and analyzed separately. Separation was accomplished by shaking the inflorescence with water. The seeds drop out of the carpels and sink, whereas the hairs float (Krattinger, 1975).

The results are shown in Table 5 for Foundry Cove and other similar control coves. The concentrations in the leaves and female inflorescence generally reflect the concentrations in the soil from which they were growing. The leaf concentrations are for the most part higher than the inflorescence concentrations, which in turn appear higher than the concentrations in seeds. Vanderberg Cove is 56 km north of Foundry Cove, and Iona Bird Sanctuary is about 15 km south. Both are marsh-cove ecosystems.

Cadmium concentrations in other plant species are shown in Table 6. *Peltandra virginica* (arrow arum) and *Pentederia coradata* (pickerel weed) are emergent marsh plants growing in open water. *Myriophyllum* sp. is also found in open water areas, but it is submergent and has finely divided leaves. Stations 1 to 26 are in the Foundry Cove area, and Vanderberg Cove serves as a control. The cadmium concentrations above 30 μg/g dry weight probably result from the influence of the Foundry Cove contamination. There is not as good correlation between the cadmium in these Foundry Cove plants and the distance from the outfall, however, as there is for the sediment cadmium concentrations versus

Table 5.   Cadmium Concentrations (μg/g dry weight) in Marsh Soil and
Typha angustifolia

Station	Soil	Leaf	Female Inflorescence	Seeds Only
Foundry Cove Marsh				
1	5820	48	17	1.7 ± 0.86[a]
2	4340	24	15	
3	1320	26		
4	313	22	1.4	
5	247	1.5	3.5	
6	247	3.8	7.8	
Vanderberg Cove				
(EPA unpublished data, 1971)		14.0		
Iona Bird Sanctuary				0.46 ± 0.28[b]

[a] Seven samples from different plants.
[b] Two samples from different plants.

Table 6.   Cadmium (μg/g dry weight) in Plants, 1973

Distance from Outfall (m)	Station	*Peltandra virginica*	*Pentederia coradata*	*Myriophyllum* sp.
10	1	37–239		
60	2	13		
175	3	62–513		
300	4		5	
400	15		70	
480	12		12	
480	16	13–18		
600	19	8–13		
690	18	13–55		
710	8			213
720	9		2	22
760	20	90–124		
880	10	14	121	26
2100	26			36
Vanderberg Cove[a]		28	15	

[a] Data from the EPA, 1971. Unpublished.

distances. Submerged and emergent aquatic plants are necessarily exposed externally to suspended sediments in the water. Values for these plants may reflect cadmium adsorbed onto surface tissue, but in any case are probably a true indication of the potential dose to an herbivore.

An unusual cattail female inflorescence morphology was observed near the

Figure 6.   Abnormal *Typha angustifolia* female inflorescence.

outfall in the area of the highest cadmium and nickel contamination (nickel concentrations are generally of the same order of magnitude as cadmium concentrations). Figure 6 illustrates an example of the abberation collected from Station 7 in the marsh (see Figure 5 for location). The fact that the spikes are joined at the tip means that the doubling and sometimes tripling probably result from splitting of the central axis. These aberrant spikes have a frequency of about 1% within 50 m of the outfall, but have not been observed anywhere else in this cove or in a control marsh in South Cove. Germination trials (using tap water) were performed with seeds from these deformed spikes, normal spikes from the outfall, and seeds from control spikes from Iona Bird Sanctuary. The seeds were checked for germination at 3 days and 10 days, and growth (shoot length) at 10 days (see Table 7). The germination was greater from the abnormal cattails, while the average shoot length was nearly the same in all three groups.

Table 7.   Germination and Cadmium Content in *Typha angustifolia*

Seed Head Description	Number of Seeds Tested	Germination (%) in 3 Days	Germination (%) in 10 Days	Average Shoot Length in 10 Days (mm)	Cadmium (μg/g dry weight) in Seeds Alone
IONA (control)	1361	12	19	13.2 ± 1.4	0.46 ± 0.28
Foundry Cove					
Normal	2819	15	28	11.1 ± 3.2	1.06 ± 0.28
Abnormal	2248	21	38	13.7 ± 1.0	1.68 ± 0.86

5.2.   Animals

Data on concentrations of cadmium in various Foundry Cove organisms were determined by the Environmental Protection Agency (EPA) in 1971 prior to the removal efforts. Cadmium is known to concentrate in the liver and kidney. These organs generally showed an elevation in Foundry Cove organisms, compared to the organs of organisms from control areas (Table 8).

We collected *Fundulus diaphanus* (banded killifish) from Foundry Cove and at Stony Point, an uncontaminated area 20 km south of the cove, in 1976, 4 years after the dredging. There is a significant elevation in cadmium concentration at the 95% level of confidence in the gills, the 99% level in the gut and liver, and the 90% level for the kidney of *F. diaphanus* seined from Foundry Cove, as compared to Stony Point (Table 8). A predominant member of the cove macroplankton, *Gammarus tigrinus,* has a higher concentration than that of the sunfish, which is the next higher level of the food chain. The carp and white perch kidney values reported by the EPA in 1971 are considerably higher than the concentration in the NYU sunfish kidney analyzed in 1976. The muscle, liver, and gill values, however, are not very different between the NYU and EPA fish. Fish from control areas, such as the *Fundulus diaphanus* from Stony Point and finfish reported by the National Marine Fisheries Service (1975), do not exceed a few micrograms cadmium per gram dry weight in any tissue, whereas the fish caught by Foundry Cove by both NYU and the EPA show organ burdens one to two orders of magnitude higher in cadmium.

The eel analyzed by NYU in 1976 and the one analyzed by the EPA in 1971 differ considerably in the liver value. The 1976 liver was a few hundred times higher in cadmium concentration. One pickerel frog (Table 8) shows elevation in all tissue and an extremely high kidney level. This animal, which was caught in the marsh within 10 m of the outfall, shows a higher organ burden than the second pickerel frog, caught on the opposite shore of the cove, where the nearest

Table 8. Cadmium (μg/g dry weight[a]) in Organisms from Foundry Cove and Control Areas

Location and Organism	Muscle	Liver	Kidney	Gills	Gut and Contents	n
East Foundry Cove						
NYU 1976 data						
Blue Crab (*Callinectes sapidus*)	0.4–32	26		24	25	4
Fundulus diaphanus	0.6	15.6	26.9	6.8	16.6	4
Eel (*Anguilla rostrata*)	3.8	121	103	10.3	27	1
Sunfish	0.6	28.8	33.8	3.1	21.3	7
Gammarus tigrinus	59(b)					
Pickerel frog (*Rana palustris*)[b]	12	267	640		800	1
Pickerel frog (*Rana palustris*)[c]	4.2	13	37		62	1
EPA 1971 data						
Eel	0.76	0.35	83	21		1
Carp	0.37	28	600	7		1
White perch	0.45		450	12		1
Control areas						
NYU 1976 data						
Blue crab[d]	0.06–0.17	1.34–1.91				3
Fundulus diaphanus[e]	1.8	4.2***[h]	3.0*[h]	2.3**[h]	1.6***	9
National Marine Fisheries Service data[f]						
Blue crab	0.8–1.2					9
Finfish	<0.1–<0.4[g]					1838

[a] For wet weight basis divide by 4.
[b] Caught May 14, 1976, over marsh where cadmium concentration ranged between 1000 and 5000 μg/g.
[c] Caught Aug 31, 1976, over marsh where cadmium concentration ranged between 70 and 300 μg/g.
[d] Caught in Hackensack River and New Jersey waters.
[e] Caught at Stoney Point, 20 km south of Foundry Cove on the Hudson River.
[f] Specimens from U.S. coastal waters.
[g] Only 1 of 44 species exceeded 0.1 μg.
[h] * = significant at 90% LC, ** = significant at 95% LC, *** = significant at 99% LC.

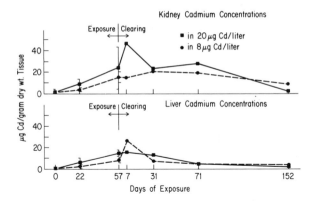

Figure 7.   Laboratory uptake and clearance of cadmium in goldfish.

recorded sediment value was considerably less than the marsh value. There is a high degree of confidence in these numbers as muscle values for the two frogs were similar and served as a procedural control ruling out the possibility of tissue contamination by sediment. Blue claw crabs (*Callinectes sapidus*) fished from Foundry Cove show considerable elevation in both muscle and liver (hepato-pancreas) tissue over controls analyzed by both NYU and the National Marine Fisheries Service.

The finding of cadmium accumulation in the blue claw crab in Foundry Cove is the only evidence to date of a potential human health hazard. The crabs are scavengers which are likely to ingest contaminated sediments during feeding. In contrast to fish, the liver of the crab is considered a delicacy and is consumed by fishermen and their families. The crabs are also cooked whole, a factor that, again in contrast to fish, allows for the transfer of contamination from the organs to edible portions such as thoracic muscle.

5.3. Controlled Exposures

Laboratory exposures to cadmium (for various lengths of time) were carried out with *Carassius auratus* (goldfish) in a static test system at concentrations approximating what might be expected in Foundry Cove water. Two exposure concentrations were used for uptake, 20 and 8 μg Cd/l. The concentrations of cadmium in the liver and kidney are shown in Figure 7. The kidney concentrations were higher in the fish that were exposed to 20 μg Cd/l than those exposed to 8 μg Cd/l, during both the 57-day exposure period and for all but the last data point for the 152-day clearing period (<1 μg Cd/l water).

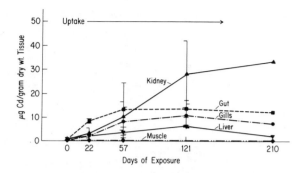

Figure 8.   Long-term cadmium uptake in goldfish.

There is not much difference in the liver concentrations between the 20 μg Cd/l and the 8 μg/l exposure. The more distinct dose-effect relation seen for the kidney as compared to the liver is consistent with the kidney being the final organ of concentration. (The liver may approach steady state faster than the kidney.) The kidney values also continue to rise for some time after the initiation of clearing. This is undoubtedly the result of the transfer of body cadmium stores to the kidney.

Data from a continuous exposure of *C. auratus* to 7 μg Cd/l for 210 days are shown in Figure 8. Data for successive dates represent analyses of several animals. The liver appears to have reached steady state between 22 and 57 days, whereas the kidney reached steady state only after 121 days. The gut and contents and the gills also reach steady state at about 57 days. Muscle tissue did not show an elevation at any time.

5.4.   Route of Exposure

The route of exposure of fish in the field is a matter of significant interest. The sediments, bottom and suspended, are at high cadmium concentrations and may be ingested, while exposure via the gills is to relatively low concentrations of dissolved cadmium. Laboratory experiments were performed by exposing *Carassius auratus* (goldfish) to cadmium either by water, which resulted in uptake via the gills, or by food, which resulted in uptake via the gut. After 3 weeks the resulting gill and gut concentrations reflected the route of exposure (Figure 9). The gills contained a higher percentage of the body burden of cadmium than the gut in the fish exposed via water, while the reverse was true in the food-exposed animals. The concentrations of exposure (240 μg Cd/l water and 24 μg Cd/g food) in this experiment were relatively high. *Fundulus diaphanus* exposed to a low concentration of cadmium in water (10 μg/l) showed little

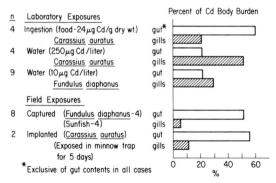

Figure 9. Comparison of Cadmium tissue distribution in laboratory and field-exposed animals.

difference from the fish exposed to 250 μg Cd/l in the percent of the body burden in the gut, but a considerably lower fraction for gills.

Specimens of *Carassius auratus* were placed in a plastic minnow trap in the east cove and recovered 5 days later. Body burden distributions between gills and gut are similar to those of the fish captured in the cove (see Figure 8). Of the experimentally exposed fish, those exposed to cadmium in food show the greatest similarity in gill-gut cadmium distribution to the field-implanted or captured fish. This shows that Foundry Cove fish are receiving cadmium principally by ingestion of contaminated food organisms or of sedimental material that is contaminated with cadmium.

6. BENTHIC ORGANISM POPULATIONS

In 1973 benthic grab samples were collected at 9 stations twice a month from April through October, and once a month in March, November, and December (Waller, unpublished results, this laboratory). Three replicate samples were collected at each station by means of an Ekman dredge with a 15-cm² opening. Each sample was washed through a No. 30 mesh screen and preserved in a 10% formalin solution, to which rose bengal had been added to stain the organisms and make them easier to separate and count. In the laboratory the samples were again washed and sorted for purposes of identification. The dominant organisms were chironomid larvae (Diptera) and oligochaete (segmented) worms.

Stations 3 through 6 were located within east Foundry Cove, while Station 8 was located in the west cove. Station 21 was located between Foundry Cove and South Cove in one of a series of canal-like passages, Stations 25 through 27 were located in South Cove, and Station 28 was located outside South Cove in the shallows of the river. The stations in South Cove were chosen to serve as

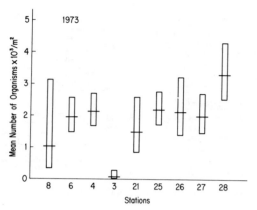

Figure 10.   Numbers of benthic organisms in 1973.

controls for the cadmium-contaminated stations in Foundry Cove. As can be seen in Table 9, Stations 3, 6, 24 all had similar particle size distributions. Station 24 and 27 were particularly good controls for Stations 3 and 6, respectively, in terms of percent organic matter.

The results of a one-way analysis of variance (ANOVA) on square-root-transformed data indicated a significant effect due to stations ($\alpha = 0.05$). To determine which stations were different, an a posteriori test (Student Newman Keuls) was performed. The test showed that Station 3 was significantly different from all other stations ($\alpha = 0.05$). This can also be seen graphically in Figure 10. The data presented are geometric means from log-transformed data at the 95% confidence intervals. These data indicate quite clearly that the relative abundance at Station 3 was significantly lower than the levels at all other stations. Station.3 was found to have the highest cadmium sediment concentrations.

Only four benthic stations were used in 1974; Station 3, 6, and 27 were the same as in 1973, and Station 24 was a slight modification of the 1973 Station 25. Station 25 was modified during 1974 to ensure that it would undergo periodic

Table 9.   Foundry Cove Sediment Characterization

Station	Sand (%), 2.0–0.05 mm	Silt (%) 0.05–0.005 mm	Clay (%) <0.005 mm	Organic Matter (%)
3	89	5	6	11
6	87	5	7	6
24	91	5	4	12
27	89	5	6	6

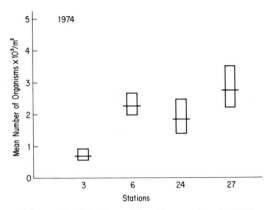

Figure 11.   Numbers of benthic organisms in 1974.

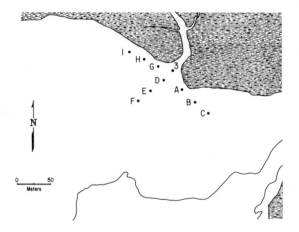

Figure 12.   Special benthic stations.

tidal exposure similar to that observed at Station 3. Benthic sampling was also reduced to once a month.

An ANOVA was run on the square-root-transformed number of organisms collected during the eight sampling dates from March 21 to October 24, 1974. A Student Newman Keuls test showed that Station 3 again was significantly different from all other stations, as can be seen graphically from Figure 11. The data presented are geometric means from log-transformed data, and the 95% confidence intervals. These data indicate quite clearly that the relative abundance at Station 3 was significantly lower than at all other stations. The fact that Station 3 was significantly lower than Station 24 as well as all other benthic

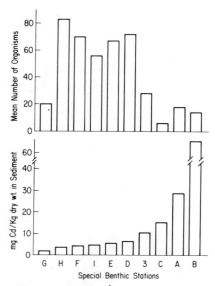

Figure 13.   Sediment cadmium and numbers of benthic organisms.

stations ($\alpha = 0.04$) in 1974 shows that tidal exposure, which both Stations 3 and 24 experience, is probably not responsible for this significant difference. As substrate conditions are also quite similar, only the contaminant concentrations remain as probable causes of the differences in population numbers.

A special series of benthic collections as made on July 2, 1974. Ten benthic stations were used, and three replicate samples were taken at each station. The samples were collected at various distances from Station 3 (Figure 12). Mud was also collected at each station for cadmium analysis.

The results of the special benthic sampling are presented in Figure 13. The purpose of these collections were to see what relationship, if any, could be found between decreasing cadmium contamination and numbers of benthic organisms per square meter. With a few exceptions an inverse relationship exists between cadmium concentration and mean number of organisms. As nickel concentrations generally parallel those of cadmium, this study cannot differentiate the effects of cadmium from those of the total contaminants.

7. PLANKTON POPULATIONS

Microzooplankton samples were collected from numerous stations in the Foundry Cove area from June through October 1973. It was of interest to determine whether there was a relationship between cadmium in the water and

Table 10. **Plankton in Foundry Cove**

Date	Time	Station	Total Cadmium (μg/l)	Soluble Cadmium	Total Organisms
6/28/73	0955		9.8	4.9	5,540
	0920	3	17.1	5	14,650
	1050	3	68.7	17.6	13,710
	1510	3	103.1	8.0	268,340
	1230	3	343.6	40.4	19,995
7/16/73	1005	3	24.2	3.0	20,650
	1045	3	25.6	9.5	8,042
9/27/73	1245	3	5.9	2.4	3,550
	1555	6	8.5	3.5	17,832
	1220	2	24.1	7.0	3,710
	1150	1	28.2	7.0	4,120
10/25/73	1240	6	5.4	2.4	2,095
	1200	3	7.9	4.9	1,400
	0825	WC[a]	8.4	3.5	1,765
	0835	HR(FC)[a]	8.4	3.5	2,508
	0855	HR(SC)[a]	8.9	4.0	2,650
	1035	21	9.0	4.0	2,895
	0800	6	13.5	5.0	2,362
	1135	2	14.0	5.0	1,688
	1210	1	16.0	8.5	1,140
	0745	3	25.6	7.5	4,340

[a] WC = the west cove, HR(FC) = Hudson River near Foundry Cove, HR(SC) = Hudson River near South Cove.

plankton counts. The procedure involved pouring 76 liters of water through a No. 20 (76-μm aperture) microplankton net, using a 7.6-liter bucket. Two replicate samples were collected at each sampling station. The plankton were preserved in a 3% formalin solution (Fennikoh et al., 1978).

Rotifers and copepods comprised between 75 and 99% of the total population in the samples (H. I. Hirshfield, unpublished results, this laboratory). The data for each sampling date and station are shown in Table 10 and include sampling location, sampling time, total cadmium in the water (suspended plus soluble cadmium), soluble cadmium, and total number of organisms. The data for each sampling date are listed in order of increasing cadmium concentration in the water. There is no apparent relationship between cadmium in the water and organism number. Collections made at stations closest to the outfall (1, 2, 3) do not show fewer total organisms than those from stations furthest from the outfall [21, WC, HR(FC), HR(SC)]. In particular, the samples taken on October 25, 1973, show fairly consistent total organism counts in spite of a wide variety of sample locations and bottom sediment cadmium concentrations.

The most sensitive species of plankton that were exposed in laboratory studies (Fennikoh et al., 1978) demonstrated a 96-hr LC_{50} of 85 μg Cd/l. The use of an application factor of 0.01 provides a presumable acceptable level of cadmium of 0.85 μg/l for chronic exposures. Periodic exposure to cadmium in Foundry Cove water would be greater than this calculated value of 0.85 μg/l.

8. CONCLUSIONS

The stability and bioavailability of the cadmium contamination at Foundry Cove are the significant parameters affecting both ecosystem and human health considerations. Measures of stability have been gained from the observed distribution of the contamination in the sediments over a number of years, as well as from data on the cadmium contained in Foundry Cove water. The bioavailability has been demonstrated by elevated cadmium concentrations in Foundry Cove biota and by laboratory and field experiments.

Evidence for an essentially insoluble deposit of cadmium contamination in the Foundry Cove area has been inferred from finding its chemical forms to be crystalline calcium-cadmium carbonate and organically bound cadmium. Further evidence was provided by the remarkably consistent spatial and quantitative distribution of cadmium in cove sediments during the 6 years of the study. The cadmium release to the water was shown to be at a rate insufficient to significantly affect the cadmium content of Hudson River water.

The low solubility and stability of the contamination, however, does not preclude its uptake by plants and animals in the cove. *Typha angustifolia* was shown to take up cadmium in proportion to the concentration in the sediments. An unusual multiple inflorescence was noted on plants in the area of the highest sediment contamination, although there was no demonstrable effect on seed germination.

Both vertebrate and invertebrate species captured in the cove were shown to have tissue elevations, particularly in the liver and kidney. (There was no evidence of biomagnification, however.) Laboratory experiments have shown that uptake can occur in simulated cove conditions and that ingestion of cadmium in food or sediments is the probable route of exposure. The only observed ecosystem effect was on the benthic invertebrate populations in the most highly contaminated region of the cove.

Human exposure is a possible result of the bioavailability of the cadmium contamination. As crabs have been found to be significantly contaminated, crabbing in the area is the only route of human exposure likely to be of significance.

Popular publications on edible wild plants suggest cattail rhizomes as a food source. Should people begin harvesting rhizomes from the Foundry Cove area, there would also be a likelihood of serious human exposure via this route.

ACKNOWLEDGMENTS

We wish to acknowledge the contributions of H. I. Hirshfield, J. O'Connor, G. Lauer, T. Waller, K. Fennikoh, G. Ré, T. Hernandez, T. Occhiogrosso, and B. Naumann to the program carried out in the study of Foundry Cove. The research has been supported by Grant AEN72-03571 A03, formerly GI37312, from the National Science Foundation, and is part of a center program from the National Institute of Environmental Health Sciences, ES00260, and the National Cancer Institute, CA 133343.

REFERENCES

Bondietti, E. A., Sweeton, F. H., Tamura, T., Perhac, R. M., Hulett, L. D., and Kneip, T. J. (1974). "Characterization of Cadmium and Nickel Contaminated Sediments from Foundry Cove, New York." In *Proceedings of the First Annual NSF Trace Contaminants Conference, Oak Ridge National Laboratory.* CONF-730802, p. 211.

Bower, P. M., Simpson, H. J., Williams, S. C., and Li, Y. H. (1978). "Heavy Metals in the Sediments of Foundry Cove, Cold Springs, New York," *Eviron. Sci. Technol.* **12**(6), 683–687.

Durum, W. H., Hem, J. D., and Heidel, S. G. (1970). *Reconnaissance of Selected Minor Elements in Surface Waters of the United States, October 1970.* Geological Survey Circular 643, U.S. Geological Survey, Washington D.C.

Fennikoh, K. B., Hirshfield, H. I., and Kneip, T. J. (1978). "Cadmium Toxicity in Planktonic Organisms of a Freshwater Food Web," *Environ. Res.* **15**, 357–367.

Fleisher, M., Sarofim, A. F., Fassett, D. W., Hammond, P., Shacklette, H. T., Nisbet, I. C. T., and Epstein, S. (1974). "Environmental Impact of Cadmium: A Review by the Panel on Hazardous Trace Substances," *Environ. Health Perspect.* May, p. 253.

Friberg, L., Piscator, M., Nordberg, G. F., and Kjellström, T. (1974). *Cadmium in the Environment,* 2nd ed. CRC Press, Cleveland, Ohio.

Gregor, H. P. (1973). "Exhibit A." Amended final judgment U.S.A. vs. Marathon Battery Co. et al., 70 Civ. 4110.

Hazen, R. E. and Kneip, T. J. (1976). "The Distribution of Cadmium in the Sediments of Foundry Cove." In *Hudson River Ecology.* Fourth Symposium on Hudson River Ecology, Bear Mountain, New York, March 1976.

Hem, J. D. (1972). "Chemistry and Occurrence of Cadmium and Zinc in Surface Water and Groundwater," *Water Resour. Res.* **8**(3), 661.

Kneip, T. J. (1978). "Effect of Cadmium in an Aquatic Environment." In *Cadmium 77: Edited Proceedings, of the First International Cadmium Conference, San Francisco, Jan. 31–Feb. 2, 1977.* Metal Bulletin Ltd., London, pp. 120–124.

Kneip, T. J., Hernandez, and Ré, G. (1974a). "Cadmium in an Aquatic Ecosystem: Transport and Distribution." In *Proceedings of the Second Annual NSF-RANN Trace Contaminants Conference, Asilomar, California,* pp. 279–283.

Kneip, T. J., Ré, G., Hernandez, T. (1974b). "Cadmium in an Aquatic Ecosystem: Distribution and Effects." In D. D. Hemphill, Ed., *Trace Substances in Environmental Health,* Vol. VIII. University of Missouri, Columbia, pp. 173–177.

Krattinger, K. (1975). "Genetic Mobility in *Typha., Aquatic Bot.* **1,** 57–70.

Moyer, B. R. and Budinger, T. F. (1974). *Cadmium Levels in the Shoreline Sediments of San Francisco Bay.* Report from Donner Laboratory and Lawrence Berkeley Laboratory, University of California, Berkeley, Califa.

National Marine Fisheries Service (1975). *First Interim Report on Microconstituent Resource Survey.* University of Maryland Campus, College, Park, Md.

12

TOXICITY AND ACCUMULATION OF CADMIUM IN MARINE AND ESTUARINE BIOTA

David J. H. Phillips

Fisheries Research Station, Aberdeen, Hong Kong

1.   INTRODUCTION

The first reports of the existence of cadmium in a marine organism concerned the liver of the scallop *Pecten maximus* (Fox and Ramage, 1930, 1931). Scattered reports of cadmium in marine and estuarine biota appeared in the literature thereafter until 1970, but the significance of the levels found was not fully appreciated. However, recognition of cadmium as the causal agent of itai-itai disease in the population of the Jintsu River basin in Japan (Kobayashi, 1970; Yamagata and Shigematsu, 1970) marked the beginning of a new era of concern for cadmium in marine foodstuffs. This concern was reinforced by the implication of cadmium in liver and kidney damage of occupationally exposed workers (Friberg et al., 1974) and in cardiovascular disease and hypertension (Schroeder, 1967).

In response to this growing body of data pointing to the extreme persistence of cadmium in human beings and to its toxic potential, the Joint Food and Agriculture Organization/World Health Organization Expert Committee on Food Additives (1972) suggested a provisional tolerable daily intake of cadmium of

Table 1. **Estimates of the Ingestion of Cadmium by Human Beings (μg Cd/individual·day) in Food**

All data are derived from total diet studies and should be considered "normal" means or ranges, with the exception of the Jintsu River population.

Country	Cadmium Intake	References
Czechoslovakia	60	Lener and Bibr (1970)
Japan, unpolluted	59–100	Yamagata and Shigematsu (1970)
Japan, Jintsu River basin	600	Yamagata and Shigematsu (1970)
Rumania	38–64	Rautu and Sporn (1970)
United Kingdom	15–30	Ministry of Agriculture, Fisheries, and Food (U.K.) (1973)
United States	4–60	Schroeder and Balassa (1961)
United States	25–> 200	Schroeder et al. (1967)

60 to 70 μg per individual (400 to 500 μg per individual per week). Estimates of the actual daily intake of cadmium by human populations vary according to author and country of origin (Table 1). However, it is evident that the average daily intake of the metal by large sections of some human populations may exceed the provisional FAO/WHO standard.

Seafoods contain high concentrations of cadmium compared to other components of the average diet considered in total diet studies. Recognition of this as a high-risk area has lead to a greatly increased research emphasis on cadmium in marine biota since 1970. Concomitantly with the concern generated at this time for the possible effects of the metal on public health, the toxic effects of cadmium on aquatic ecosystems were recognized. Most authors now estimate cadmium to be one of the five most toxic metals in the aquatic environment.

This chapter constitutes a review of the literature published to 1978 concerning the accumulation of cadmium by marine and coastal biota. Five major sections are presented. The first (Section 2) deals with cadmium in seawater and its biological availability. Section 3 concerns the toxicity of the metal to aquatic biota, and Section 4 discusses the data reported on the uptake of cadmium by marine and estuarine organisms. Section 5 presents a compilation of data on the concentrations of cadmium found in organisms of different phyla. Section 6 consists of speculation on the bases for the effects of different parameters on the accumulation and toxicity of cadmium as regards marine and coastal biota. The effects of variables concerning the organism itself or its ambient environment are emphasised where these are considered to be important factors in the determination of the response of the organism to cadmium in its immediate surroundings.

2. CADMIUM IN WATER AND ITS AVAILABILITY TO BIOTA

2.1. General

Many authors have reported data concerning the concentrations of cadmium found in marine or coastal waters. Reports dealing specifically with water chemistry often contain an estimate of the variability of the levels encountered with time or with location. Often, however, data are reported for single samples in time or space, particularly if studies of water chemistry are ancillary to the major objective of a study. Information that contains no estimate of the variation in cadmium levels in water should be interpreted cautiously, as the fluctuation of trace metal concentrations can be very large, on either a diurnal or a seasonal basis, at any one location (e.g., see Boyden, 1975; Fukai and Huynh-Ngoc, 1976; Grimshaw et al., 1976). In general, the variability of metal levels in water with time is greatest in locations of greatest pollution, perhaps reaching a maximum in estuarine areas. Factors contributing to this variation include the extent of freshwater runoff, the intermittent discharge of industrial effluents, hydrological factors such as tides, currents, and halocline or thermocline formation, and ecological factors, which include the effects of phytoplankton or detritus.

The analysis of water samples for trace metals is expensive and time-consuming and is also prone to both contamination and loss in samples (Martin et al., Chapter 5, this volume). Contamination may occur because of inadequate sample collection procedures, impurities of the analytical chemicals employed, leaching of metals from glassware or other equipment used, or a variety of other causes. Loss of cadmium from samples is most frequently due to adsorption of the metal onto the surfaces of equipment; however, incorrect maintenance of sample pH or instability of chelated complexes may also be important in some instances. Reports of high percentage recoveries in spiking or standard addition procedures do not necessarily imply the absence of contamination or loss from the sample, as the added salt may behave quite differently from the metal form or compound present in the original sample.

In addition to the above, the analysis of cadmium in seawater presents special problems not encountered in analytical procedures for other trace metals in seawater. Thus the atomic absorption technique, which is the one most frequently used, is subject to interference by sodium chloride. Similar problems have been encountered in the analysis of blood samples for cadmium in attempts to relate the levels of the metal in blood to the total body load or recent exposure to the metal (Friberg et al., 1974). The most common method presently used to analyze water samples for cadmium or other trace metals includes a preconcentration step using ammonium pyrrolidine dithiocarbamate (APDC) and methyl isobutyl ketone (MIBK), originally reported by Brooks et al. (1967). Other methods of preconcentration have been used, however, such as resin chelation (Riley and

Taylor, 1968); maintenance of column pH at the correct level is most important here if full recoveries are to be obtained. Such preconcentration techniques are used prior to analysis, most frequently by atomic absorption methods. More recent developments, such as anodic stripping voltametry (ASV) and polarography, show promise for the direct analysis of seawater without the need for preconcentration; elimination of this step helps to decrease problems connected with contamination, as less sample handling is necessary (See Raspor, Chapter 6, this volume).

2.2. Form and Biological Availability of Cadmium in Seawater

Trace metals exist in water as ions, complexes or chelates, or colloids, or adsorbed to organic and inorganic particulate material. The exact form of the metal may dictate not only its toxicity but also its availability to, and rate of uptake by, aquatic biota. Generalizations are difficult as each metal behaves differently from any other; in addition, very little information is available on which to base hypotheses.

Pollution of areas by trace metals is important only in terms of the biological availability of the metal. Thus the two facets of trace metal pollution of importance to human beings (the amounts of metals present in seafood they eat, and the toxic effects of trace metals in the ecosphere) both depend entirely on the biological availability of the metal.

Aquatic organisms may take up trace metals by one or more of three different routes (Phillips, 1977b). The first of these routes concerns the direct uptake of metals from solution in seawater. All organisms take up some trace metal directly from solution by adsorption to the body surface and/or by absorption across the body surface. Probably the whole body surface of the organism is involved in the uptake of metals, although in animals the gills are considered to predominate in the uptake process because of their large surface area and intimate contact with the respiratory water volume. The uptake process probably involves in all cases an initial adsorption of metal to a specific or nonspecific site on the body surface; specific sites may be membrane-bound polypeptides, although little information is available. If absorption occurs, the metal is translocated through the limiting surface membrane for transportation to the sequestration site; these processes of membrane translocation and sequestration are probably fairly specific, such as the storage of copper as granules in the wandering leucytes of oysters (Boyce and Herdman, 1898), or the similar granular sequestration of zinc and copper in the midgut tissues of barnacles (Walker et al., 1975; Walker, 1977). In biota that drink appreciable quantities of water, direct uptake of metals from the ingested water may occur across the gastrointestinal wall. It should be noted that the efficiency of uptake of a metal by this route may differ from that of the same metal taken up across the gills, as the metal form may be altered by the intestinal environment.

The second route of metal uptake concerns the ingestion of metals in food. This exposure route has not been studied as extensively as the direct uptake of metals from solution; however, most authors believe it accounts for the major proportion of the total body load of metals in aquatic organisms of higher trophic levels. The exact importance of the food and water routes of uptake must, however, differ according to the nature of the food ingested and the relative concentrations of metals in each component.

The ingestion of metals in inorganic particulates constitutes the third route of metal uptake. This process is presumably an important determinant of the total body load of metals only in filter-feeders, which may ingest large quantities of inorganic particulates. Amongst filter-feeding bivalves, the importance of this uptake route undoubtedly varies with species and with the metal considered (Boyden and Romeril, 1974). Oysters, in particular, are thought to obtain a large proportion of their total body load of metals via this route; Ayling (1974) has even suggested that concentration factors for trace metals in oysters should be based on metal levels in the surrounding sediments rather than those in the ambient waters, although this idea has met with little support.

It is evident, therefore, that the biological availability of a metal is a complex function of its uptake by one or more of three different exposure routes. The relation of the concentrations of trace metals found in water to the biological availability of the metals is uncertain, as the few data that exist are contradictory. The toxicity (and perhaps uptake also) of metals in gross particulates to aquatic organisms such as finfish is uncertain (e.g., compare Lloyd, 1960, and Herbert and Wakeford, 1964, with Sprague, 1964). However, most authors believe that the direct uptake of metals from water occurs at a faster rate with smaller particle sizes, that is, metals in gross inorganic particulates are relatively unavailable to biota (with the exception of filter-feeders), whereas metals in ionic form are highly available. The availability of metals in chelates or complexes may depend on the stability of the compound to dissociation, thus freeing the ion for uptake (Bryan, 1971); however, this is not always the case, and chelates may be absorbed intact in some instances at faster rates than the corresponding metal ion (George and Coombs, 1977a, 1977b).

Most authors studying the concentrations of trace metals in water employ a filtration step either during or after sample collection. This process eliminates all particles of greater than a given diameter (depending on the pore size of the filter used). The resulting fractions are then usually termed "soluble" and "particulate," and many authors imply that the former is highly available to biota, whereas the latter is unavailable. This tendency is so strong that most authors studying water chemistry ignore the particulate phase altogether, reporting data only for the soluble fraction. Although the distinction between the two phases may approximately parallel the availability of metals in each phase for *direct* uptake into biota, the total metal availability is determined in most cases largely by the uptake of metals from food, which depends in turn on the concentrations of metals in the particulate fraction. Thus the particulate fraction contains inorganic particulates, detritus, phytoplankton, and zooplankton, all of which are basic food items for higher organisms and therefore dictate the concentrations of metals passed upward through the trophic levels of the food web.

Despite this tendency of authors to ignore the metal concentrations found in the particulate fraction of seawater, some reports contain estimates of the relative concentrations of cadmium in solution and in particulates. Preston et al. (1972), in extensive studies of metals in coastal and offshore waters of the British Isles, reported figures of 16.7% and 20.0% for the ratio of particulate-associated cadmium (0.20-μ filter pore size) to total cadmium in shoreline samples. By contrast, the seven other trace metals studied exhibited much higher affinities for particulates, with ratios varying from 43.6% (zinc) to 99.0% (iron). This greater tendency of cadmium to remain in solution rather than adsorbing to particulates as do other trace metals has also been noted by other authors. For example, Dutton et al. (1973) found much higher concentrations of cadmium in solution than in particulates (0.20-μ filter pore size) for North Sea samples, although individual ratios are impossible to calculate from the data quoted by these authors. Nickel was also found mostly in solution, whereas Zn, Cu, and Mn were predominantly particulate associated. Bloom and Ayling (1977) reported a similar predominance of cadmium in solution in waters of the Derwent Estuary in Tasmania, Australia; pore sizes of 0.45 μ were used in this study.

A considerable body of data exists for trace metals in the waters of the Severn Estuary and Bristol Channel, United Kingdom. Boyden and Romeril (1974) found ratios of particulate-associated cadmium to total cadmium of <7.6 to 27.7% in waters of one site in the Bristol Channel sampled in November 1972, when silt loads were less than 305 mg/l. However, samples taken in January 1973 at the same site contained much higher silt loads and much higher percentages of the total cadmium in the particulate phase (75.0–86.6%). Calculation of the concentration of cadmium in these particulates (present author) gave levels of 0.51 to 5.96 μg/g; the major factor affecting these concentrations was the silt load, concentrations increasing markedly with increased particulate content of the water. Concentrations of cadmium reported for sediments (<204-μ fraction only) in the area were 2 to 3 μg/g dry weight, agreeing well with the mean levels of cadmium in the suspended particulates. The reason for the direct dependence of the cadmium levels in particulates on the total particulate load is not clear; possibly it represents the successively increasing delivery in the water column of highly contaminated sediments from upstream of the study site by scouring of the estuary bed during periods of high runoff.

Abdullah and Royle (1974), in a more extensive study of trace metals in solution (0.45-μ filter pore size) and in particulates of water from 44 stations in the Severn Estuary and Bristol Channel, found mean cadmium concentrations of 9.7 $\mu g/g$ (April) and 19.1 $\mu g/g$ (June) in particulates. These values are somewhat higher than those of Boyden and Romeril (1974), but differences in season, year, and exact location of sampling could account for the discrepancies between the two sets of results. More importantly, no relation between the concentration of cadmium in particulates and the total particulate load per unit volume of water was observed in the more extensive study. Thus, although the concentration of cadmium in particulates varies with the total particulate load at the upstream site of Boyden and Romeril (1974), the concentration in particulates of more open waters further down the estuary depends little on the extent of runoff or associated changes in total particulate content (Abdullah and Royle, 1974). The physicochemical equilibrium between cadmium in solution and cadmium in particulates thus appears fairly stable with distance down the Bristol Channel, once freshwater-saltwater mixing has attained a salinity of about 25‰. Loss of gross inorganic particulates kept in suspension by the turbulent river flow apparently occurs prior to the attainment of such a degree of mixing between the inflowing and receiving waters; desorption of cadmium from particulates into solution (reported to be a significant influence on the cadmium concentrations in estuaries elsewhere: see Cossa and Poulet, 1978) probably also occurs at lower salinities further upstream. The effects of phytoplankton on the concentrations of cadmium in solution and in the particulate-associated phase of seawater are considered in Section 2.3.

2.3.   Typical Concentrations of Cadmium in Seawater

Cadmium in Open Oceans

In comparison to coastal areas, relatively little information is available concerning the concentrations of trace metals in open oceans distant from the anthropogenic or other effects of land masses. Data reported for cadmium in solution are shown in Table 2; it should be noted that the present classification of a water body as "open ocean" is somewhat arbitrary, but generally corresponds to areas more than 100 km distant from the nearest land mass. Several of the values shown in Table 2 are estimates based on large amounts of data; the other values quoted represent direct experimental results. The overall mean concentrations fall within narrow limits (0.02 to 0.11 $\mu g/l$), although the ranges are more divergent (0.01 to 0.41 $\mu g/l$).

Conclusions concerning the abundance of cadmium in water in terms of location (different oceans or different offshore areas of the same ocean) should not be reached by comparing the results of different authors, as the methodol-

Table 2.   Ranges or Mean Concentrations of Cadmium in Open-Ocean Seawater (μg/l or parts per 10^9; salinity = 35‰)

All values represent estimates for filtered seawater and may be considered as "soluble" concentrations.

References	Water Body	Range	Mean
Goldberg (1965)	"Average seawater"	—	0.11
Turekian (1969)	"Natural"	—	0.05
Goldberg (1972)	"Baseline"	—	0.02
Preston et al. (1972)	Northeast Atlantic	0.01–0.41	0.04[a]
Knauer and Martin (1973)	Transect Hawaii-California	<0.02–0.06	—
Skinner and Turekian (1973)	"Natural"	—	0.11
Chester and Stoner (1974)	World oceans: "open ocean"	0.02–0.17	0.07[b]
Chester and Stoner (1974)	World oceans: "nearshore"	0.04–0.30	0.09[b]

[a] Geometric mean of five observations.

[b] "Open ocean" defined as >400 km from nearest land mass; "nearshore," as <400 km from nearest land mass. Arithmetic means of 38 and 51 sites, respectively.

ogies used differed somewhat. For example, pore sizes of the filters used varied from 0.2 μ (Preston et al., 1972) through 0.45 μ (Chester and Stoner, 1974) to 0.8 μ (Knauer and Martin, 1973). Such differences in techniques pervade research into trace metals; the need for standardization of sampling, preparative, and analytical procedures is obvious.

However, Chester and Stoner (1974) were able to produce some generalizations concerning regional differences in cadmium concentration. These authors studied surface waters from 38 "open-ocean" and 51 "nearshore" locations (all included here in the open-ocean category), the positions of which are shown in Figure 1. Concentrations of cadmium were slightly higher at sites less than 400 km from a land mass than at those more distant (Table 2), although in both sets of locations the largest number of samples centered around 0.06 to 0.08 μg/l (Figure 2). Regional differences found in cadmium concentrations are shown in Table 3; again, little gross variation was apparent, although nearshore samples from the waters bordering Japan and South Africa exhibited slightly elevated cadmium levels compared to other locations, the difference being approximately twofold. No conclusions were reached as to whether these local differences depended on anthropogenic changes or on natural differences in the weathering rate of cadmium from the local rocks or sediments.

In summary, presently available data concerning the concentrations of cadmium in solution in open-ocean waters distant from land masses suggest that few highly significant differences in connection with location exist; an average concentration could be taken as 0.05 to 0.10 μg/l at the present time. As the development of accurate analytical capability for cadmium in seawater is a rather

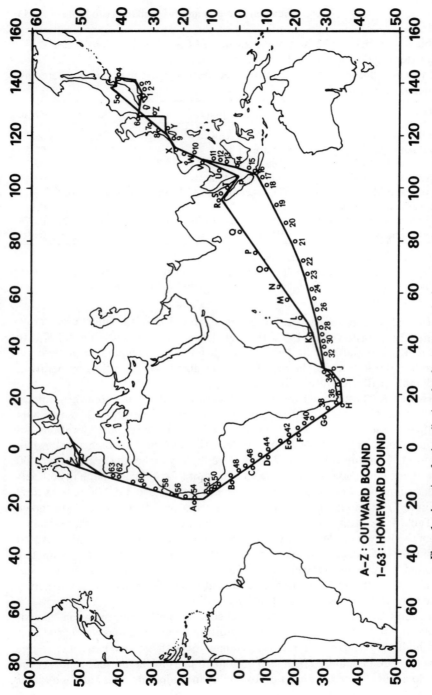

Figure 1. Locations for the collection of surface water samples for trace metal analysis in the studies of Chester and Stoner (1974).

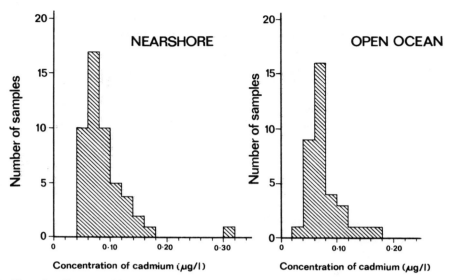

Figure 2. Concentrations of cadmium in 51 "nearshore" and 38 "open-ocean" locations. Water samples were taken in locations shown in Figure 1. (After Chester and Stoner, 1974.)

recent event, no conclusions are possible concerning possible changes in concentration with time. However, Dryssen et al. (1970) considered cadmium to be a local coastal, rather than a worldwide, pollutant; if this is so, few changes in the concentration of cadmium in open-ocean waters would be expected. More recent data have stressed the airborne route of cadmium introduction to the oceans; perhaps the categorization of cadmium as a coastal pollutant will change as more data become available on its precise mode of cycling in the ecosphere.

Cadmium in Offshore Waters

Offshore waters, loosely defined here as waters located within 100 km of a land mass but not showing gross evidence of cadmium pollution attributable to any certain source, may contain slightly higher concentrations of cadmium than those found in open oceans. Selected values from the literature are shown in Table 4. These data may be considered to be typical of most offshore areas of the world, although the higher values in the North Sea probably indicate a significant direct input of cadmium from industrial sources on the North Sea coasts.

As noted above, direct comparison of cadmium concentrations in different water masses is difficult as methodologies differ between authors (note differences in pore sizes and analytical techniques in Table 4). However, the work of

Table 3.   Mean Concentrations (μg/l or parts per 10^9) of Cadmium in Solution (0.45-μ pore size filtrate) in Surface Waters of Various World Regions [a]

Water Body and Region	Cadmium
"Nearshore waters"	
Northeast Atlantic	0.08
South African coast	0.09
Inland Sea (Japan)	0.11
South Japan coast	0.13
Java Sea	0.06
Malacca Straits	0.10
Sea of Japan	0.11
China Sea	0.08
Overall "nearshore" mean [b]	0.09
"Open-ocean waters"	
South Atlantic	0.07
Indian Ocean	0.07
Overall "open-ocean" mean [b]	0.07

[a] After Chester and Stoner (1974).
[b] Overall means calculated from original data for each station. Definition of "nearshore" and "open-ocean" waters as in Table 2, footnote [b].

certain authors permits tentative conclusions concerning the variation of cadmium levels with location and season in offshore waters.

Mullin and Riley (1956) reported as much as a threefold variation in the concentrations of soluble cadmium (0.45-μ pore size of filters) in waters of adjacent stations in the Irish Sea or English Channel. These authors speculated that such variations may be due to the adsorption of cadmium by detritus or by phytoplankton, thus removing the metal from solution. Riley and Taylor (1972) found similar variation in samples from one transect off Cap Blanc in the tropical northeast Atlantic, although samples from a transect further north exhibited relatively constant levels of cadmium. This report conforms to a familiar pattern in that the area showing greater variability in cadmium levels also exhibited a higher mean concentration of the metal; in general, the greatest variation in trace metal levels in water is found in the most polluted areas. This observation also holds true for mussels (*Mytilus edulis*) and has been ascribed in this case to differences in metabolic rates between different individuals (Schulz-Baldes, 1974; Phillips, 1976a, 1976b; see also Section 6). The explanation for the data concerning water may be that areas which exhibit high mean concentrations of trace metals are those in close proximity to land masses, where other pa-

Table 4. Selected Data for the Concentration of Cadmium in Solution in Seawater from Offshore Areas

Data for filtration and analytical techniques and for sampling extent and frequency are also shown.

References	Water Body Studied	Number of Stations	Fre- quency	Filter Pore Size (μ)	Preconcentration/ Analysis[a]	Concentration ($\mu g/l$) Range	Concentration ($\mu g/l$) Mean
Mullin and Riley (1956)	Irish Sea	12	2	[b]	Dithizone	0.03–0.25	0.11
	English Channel	6	3	[b]	extraction/S	0.02–0.26	0.11
Preston et al. (1972)	Five areas of U.K. coasts	52	2	0.20	APDC-chloroform/AA	<0.01–0.62	—
Riley and Taylor (1972)	Tropical northeast Atlantic	17	1	0.45	Ion exchange resin/AA	0.07–0.71	0.11
Windom and Smith (1972)	Continental shelf, southeast U.S.A.	57	1	0.45	Ion exchange resin/AA	0.02–0.23	0.09
Dutton et al. (1973)	North Sea	41	1	0.20	APDC-chloroform/AA	<0.10–1.60	—
Knauer and Martin (1973)	Monterey Bay, Calif.	1	28	0.80	Ion exchange resin/AA	0.02–0.51[c]	0.15[c]
Fukai and Huynh-Ngoc (1976)	Northwest Mediterranean	15	2	0.45	None/ASV	0.06–0.30	—

[a] S = spectrophotometry, AA = atomic absorption spectrophotometry, ASV = anodic stripping voltametry.

[b] Gross filtration with Whatman No. 1 only.

[c] Excludes two unusually high values, probably because of sample contamination.

rameters that may influence the concentrations of cadmium in solution (e.g., salinity, inorganic particulate load, phytoplankton productivity) are themselves much less stable than in open-ocean areas. The concerted effects of these extraneous factors may therefore cause the rapid changes in concentrations of cadmium observed in solution in areas directly affected by land masses. The effects of cadmium input from land masses on the concentrations of this metal found in offshore waters were emphasized by Windom and Smith (1972) in studies of continental shelf waters off the southeast United States. Two water masses were distinguished, differing in their contents of trace metals. The northern water mass was influenced by terrigenous sources of cadmium and exhibited patchy elevated concentrations of the metal; by contrast, the southern water mass was dominated by the Gulf Stream and contained relatively constant lower concentrations of cadmium and other metals.

Changes in the concentrations of cadmium in offshore waters with season are poorly documented. However, Mullin and Riley (1956) observed higher levels of the element in October to February than in April to July in the Irish Sea and English Channel. Such seasonal differences may be related either to an increased supply of cadmium to the offshore waters during periods of high winter runoff, or to changes in the cadmium content of phytoplankton. The exact effects of phytoplankton on the concentrations of metals in solution are a subject of some controversy; possibly the importance of the removal of metals from solution into phytoplankton varies according to the metal concerned. Thus Knauer and Martin (1973) observed decreased concentrations of cadmium in solution in Monterey Bay waters during periods of peak phytoplankton productivity, although the concentrations of Cu, Pb, Mn, and Zn were apparently unaffected by changes in phytoplankton abundance. Perhaps the concentration of cadmium in seawater is more easily (or more observably) affected by its adsorption to phytoplankton, as compared to other metals, because of its greater tendency to remain in the solution phase of seawater rather than adsorbing to inorganic particulates in the water column (see Section 2.2). Alternatively, cadmium may simply exhibit an unusually high affinity for adsorption sites on phytoplankters, although this may be expected to be species dependent.

In summary, the concentrations of cadmium found in offshore waters are, in general, slightly greater than those in open oceans, and are known to vary considerably with location and with season. Much of this variability is probably due to the effects of the nearby land masses, both in terms of cadmium influx per se and in terms of the effects of covarying parameters that influence the retention of cadmium in the solution phase of seawater.

Cadmium in Coastal or Estuarine Waters

The majority of the data published on cadmium in seawater concerns coastal or estuarine waters, many of which exhibit evidence of direct cadmium pollution, often attributable to known anthropogenic sources. Table 5 presents selected examples from the published literature. As previously noted, the concentrations of cadmium in coastal waters are highly variable compared to those found in waters more distant from land masses. Such variability may be ascribed to many factors, including the extent of freshwater runoff; the intermittent discharge of industrial or domestic effluents; tides, currents, and associated hydrological or oceanographical parameters; and ecological factors such as the abundance of phytoplankton or detritus. These factors act in concert with each other to produce continual short- and long-term fluctuations of the cadmium levels found in a coastal or estuarine water mass; the complexity of their interaction often leads to difficulty in studying single parameters in isolation. However, certain theoretical and practically observed variations may be discussed as examples.

The effects of season on the concentrations of cadmium found in seawater were alluded to in the preceding subsection. Two major seasonally dependent parameters may dictate the levels of cadmium in coastal or estuarine areas. The first of these is runoff from the adjacent land mass. The effects of changes in runoff on the supply of cadmium to estuarine or coastal environments depend on the nature of the cadmium source. Phillips (1978b) has discussed this aspect of pollutant supply as it relates to organochlorines; the theoretical basis for trace metals is similar. If the cadmium source in a catchment is industrial in nature, the total influx of cadmium per unit time is generally constant, or approximates constancy. In this instance an increase in runoff would lead to a greater immediate dilution of the primary effluent with the receiving river or estuarine waters; hence cadmium concentrations in the immediate area of the outfall would decrease. However, in a typical tidal estuary (especially if salinity stratification occurs to any extent), higher runoff leads to a more widespread decrease in salinity, that is, the effects of the fresh water extend further out to sea, estuarine mixing occurring at a site further downriver. In such a situation the availability of soluble cadmium to biota of different parts of the estuary will approximate that shown in Figure 3. A null point exists at the crossover of the two profiles where runoff does not affect the availability of cadmium. Upstream from this null point, the availability of cadmium is greater in low than in high runoff (less dilution of source effluent by receiving waters). Downstream from the null point, however, cadmium concentrations in solution are higher in high runoff periods because there is less dilution of fresh water with the receiving salt water. If Figure 3 were to accurately represent the availability of cadmium to biota, it is clear that any given species would exhibit different seasonal profiles for cadmium according to its exact position in the estuary. Evidence for such profiles will be considered in Section 5. Because anthropogenic sources of cadmium are predominantly industrial, this pattern may approximate the situation at least for levels of cadmium in solution in most polluted estuaries.

However, in some cases the source of cadmium may be nonindustrial, or at

Table 5. Selected Data for the Concentrations of Cadmium in Solution in Coastal or Estuarine Areas. Data for filtration and analytical techniques and for sampling extent and frequency are also shown.

References	Water Body Studied	Number of Stations	Frequency	Filter Pore Size (μ)	Preconcentration/ Analysis[a]	Concentration (μg/l) Range	Mean
Elderfield et al. (1971)	Conway Estuary, U.K.	16	1	0.45	Ion exchange resin/AA	0.16–2.12	0.67
Abdullah et al. (1972)	Liverpool Bay, U.K.	27	1	NQ[b]	Ion exchange resin/PP	0.14–0.74	0.27
Abdullah et al. (1972)	Cardigan Bay, U.K.	20	1	NQ	Ion exchange resin/PP	0.48–2.41	1.11
Abdullah et al. (1972)	Bristol Channel, U.K.	44	1	NQ	Ion exchange resin/PP	0.28–4.20	1.13
Butterworth et al. (1972)	Severn Estuary, U.K.	8	1	0.50	Ion exchange resin/PP	0.30–5.80	1.99
Preston et al. (1972)	Irish Sea, U.K.	20	1	0.20	APDC-chloroform/AA	0.03–1.43	0.43[c]
Dutton et al. (1973)	North Sea shore, U.K.	8	1	0.20	APDC-chloroform/AA	0.10–6.20	0.50[c]
Halcrow et al. (1973)	Firth of Clyde, U.K.	15	1	0.45	SDDC-chloroform/AA	0.20–1.20	0.50
Abdullah and Royle (1974)	Bristol Channel, U.K.	44	2	0.45	Ion exchange resin/PP	0.41–9.44	1.38–1.94[d]
Boyden and Romeril (1974)	Bristol Channel, U.K.	2	2	0.45	SDDC-chloroform/AA	1.00–1.30	1.15

Reference	Location				Method	Range	Mean
Stenner and Nickless (1974a)	Hardangerfjord, Norway	5	1	NQ	Ion exchange resin/AA	0.01–85.00	—
Stenner and Nickless (1974a)	Skjerstadfjord, Norway	4	1	NQ	Ion exchange resin/AA	0.01–0.03	—
Boyden (1975)	Poole Estuary, U.K.	1	37	0.45	SDDC-chloroform/AA	0.40–11.70	1.61
Boyden (1975)	Poole Estuary, U.K.	1	30	0.45	SDDC-chloroform/AA	0.70–20.50	5.70
Darracott and Watling (1975)	Poole Estuary, U.K.	1	8	NQ	NQ	0.20–1.50	0.80
Morris and Bale (1975)	Bristol Channel, U.K.	32	8	0.45	Ion exchange resin/AA	0.15–2.01	0.27–1.34[c,d]
Stenner and Nickless (1975)	Atlantic coast of Spain and Portugal	10	1	NQ	Ion exchange resin/AA	0.08–6.00	1.18
Foster (1976)	Afon Goch and Menai Straits, U.K.	NQ	NQ	NQ	NQ	0.20–3.90	—
Bloom and Ayling (1977)	Derwent Estuary, Australia	14	5	NQ	SDDC + APDC – MIBK/AA	0.30–5.00	1.90

[a] AA = atomic absorption spectrophotometry, PP = pulse polarography.
[b] NQ Not quoted.
[c] Geometric means.
[d] Means for different stations.

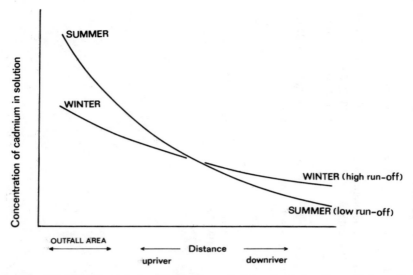

Figure 3.   Simplified theoretical profiles for the availability of cadmium from solution to biota of different areas in a tidal estuary under high and low runoff conditions, assuming a constant supply of cadmium from the catchment.

least may not be constant with time. An example might be a highly mineralized area from which cadmium is leached to the estuary in runoff water. In this case, higher runoff may be expected to leach the same amount of cadmium per unit volume from the substrate; thus final concentrations of cadmium would be similar in all periods, regardless of the extent of runoff. However, the amount of cadmium taken into suspension in particulates would increase with higher runoff because of the greater scouring action; most estuaries exhibit much higher loads of inorganic (and sometimes organic) particulates during periods of higher runoff. A greater total concentration of cadmium would therefore be found at any point in the estuary during periods of higher runoff in such a catchment (Figure 4); delivery of this metal to sites further downstream would also be enhanced under higher runoff conditions because of the slower dilution of the fresh water with the salt receiving waters. It should be noted that this example differs from the previous one in that total concentrations of cadmium are considered; the concentrations of soluble cadmium in Figure 4 may be very similar in all runoff conditions. In addition, it is important to emphasize that these profiles are highly simplified, taking no account of such phenomena as the desorption of cadmium from particulates into solution at the fresh water-salt water mixing boundary (Van der Weijden et al., 1977; Cossa and Poulet, 1978) and the adsorption of cadmium to humic acid.

The second major factor influencing the concentrations of cadmium in sea-

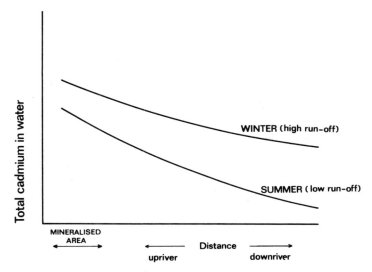

Figure 4. Simplified theoretical profiles for the total concentrations of cadmium in water of an estuary supplied with cadmium by leaching of the metal from exposed rocks or mineworkings.

water of coastal areas in a seasonally related fashion is the abundance and turnover of phytoplankton populations. The results of Knauer and Martin (1973) have already been discussed ("Cadmium in Offshore Water"); other authors have also reported reciprocal relationships between the concentrations of trace metals in solution and phytoplankton productivity, especially in times of major phytoplankton blooms (e.g., see Spencer and Brewer, 1969; Morris, 1971). The removal of cadmium from the solution phase of seawater into the phytoplankton-associated (a portion of the particulate) phase is probably an important mechanism dictating the levels of cadmium in solution at any one time; this mechanism is merely more obvious at times of peak phytoplankton productivity, as the effects of other parameters are overwhelmed.

A most extensive study of cadmium in both the soluble and particulate phases of seawater was reported by Abdullah and Royle (1974). These authors studied the levels of seven trace metals in solution and suspension in waters of the Bristol Channel and Severn Estuary, United Kingdom. Previous reports had already established the existence of gross contamination of both water and air of this area, especially by cadmium (Abdullah et al., 1972; Burkitt et al., 1972; Butterworth et al., 1972; Nickless et al., 1972; Peden et al., 1973). Abdullah and Royle (1974) performed two surveys of 44 sites located on 15 transects throughout the region, sampling in April and June 1971. Salinity profiles on the two occasions are shown in Figure 5A; no major differences in runoff were present, and the profiles are similar. The concentrations of cadmium in solution

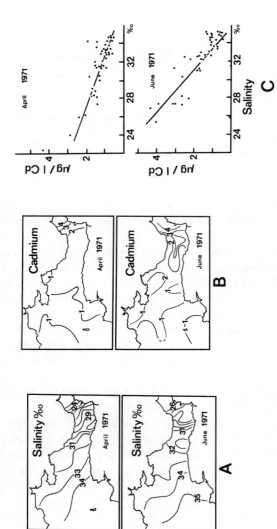

Figure 5. Water quality in the Bristol Channel, April and June 1971. (*A*) Salinity profiles for surface waters. (*B*) Concentrations of cadmium in solution. (*C*) Concentrations of cadmium plotted against salinity. (*D*) Relation of suspended solids to salinity. (*E*) Concentrations of cadmium in particulates. (*F*) Concentrations of cadmium in particulates plotted against distance down the channel. (*G*) Concentrations of cadmium in particulates plotted against total particulate load. (All data from Abdullah and Royle, 1974.)

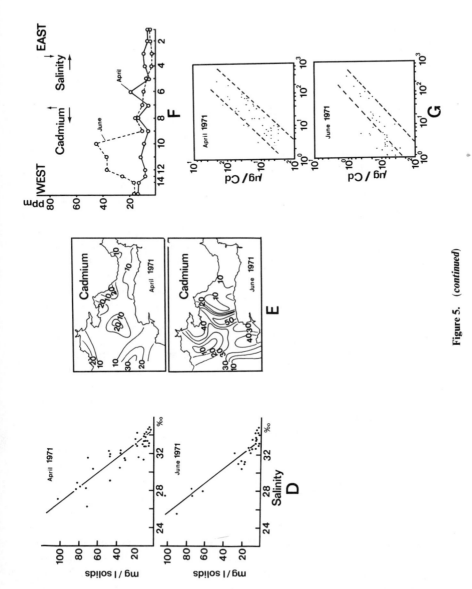

Figure 5. (*continued*)

445

are shown in Figure 5B (filter pore size was 0.45μ); these profiles were again similar, both revealing a general decrease in soluble cadmium in an east-west direction (downriver), with somewhat lower concentrations present in April than were found in June (means of 1.38 and 1.94 μg/l, and ranges of 0.55 to 4.29 and 0.41 to 9.44 μg/l, respectively). Profiles for soluble cadmium correlated to salinity, revealing a dependence of concentration on runoff, as expected (Figure 5C), with the June samples showing a greater slope due to the greater cadmium contamination in this month. The amount of particulates retained by a 0.45-μ filter at each station decreased markedly in a downriver direction, again correlating well with salinity (Figure 5D). Such profiles of particulate load are typical of many highly industrialized estuaries, and represent the rate of loss of particulates by settling out on diminution of the turbulent river flow, as well as the dilution of the turbid estuarine water with relatively clean marine receiving waters. In contrast to this pattern, the concentrations of cadmium found in the particulates exhibited a remarkable constancy throughout the study area in April (Figures 5E and 5F); means for the eastern and western halves of the region were 8.95 and 10.40 μg/g, respectively, at this time. The June profile was entirely different, however, exhibiting a large peak in the concentration of cadmium in particulates at transects located about two thirds of the way down the estuary (Figures 5E and 5F). This peak caused a large divergence in the mean particulate-associated concentrations of cadmium in the eastern and western halves of the area (9.24 and 28.50 μg/g, respectively). Comparison of these profiles for particulate-associated cadmium with profiles for particulate nitrogen (used as an indicator of phytoplankton biomass; see Abdullah et al., 1973) revealed the peaks in cadmium concentration in June to represent cadmium associated with phytoplankton. No such effect of phytoplankton was observed for the April results because of the relatively low levels of primary productivity at this time. The plots of cadmium concentration in particulates against total particulate load (Figure 5G) confirmed this hypothesis. A so-called positive anomaly seen in the samples from outer transects in June was caused by the adsorption of cadmium to phytoplankters, that is, the constant concentration of cadmium in total particulates was disturbed because of the higher affinity of cadmium for phytoplankton than for inorganic particulates.

The covariation of soluble cadmium concentrations and salinity in the Bristol Channel, noted by Abdullah and Royle (1974), was confirmed by Morris and Bale (1975) for the same area; agreement concerning the absolute concentrations of cadmium in solution and the concentration profile for cadmium in the Channel was very high between these authors. Increased levels of cadmium in solution with decreased salinity were also observed by Duinker and Nolting (1977) in studies of the Rhine Estuary and German Bight (Figure 6); the general behavior of cadmium in the German estuary appeared similar to that in the Severn Estuary and Bristol Channel. Unfortunately, Duinker and Nolting reported no

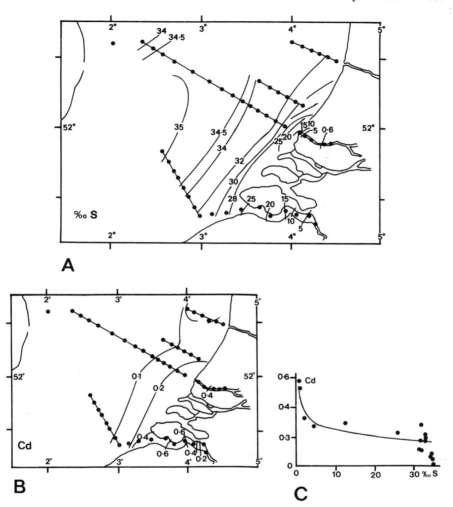

Figure 6. (*A*) Salinity profile for the Rhine Estuary and Southern Bight, February 1975. (*B*) Concentrations of cadmium in solution (0.45-μ filtrate) in the study area. (*C*) Relation between the concentrations of cadmium in solution and the ambient salinity (stations in the Rhine Estuary only). (All data after Duinker and Nolting, 1977.)

data for cadmium in particulates in the Rhine Estuary, so no comparison is possible for this phase.

The studies described above (Abdullah and Royle, 1974; Morris and Bale, 1975; Duinker and Nolting, 1977) concern areas with well-developed, relatively stable cadmium inputs and salinity profiles. Such areas are ideal for studying

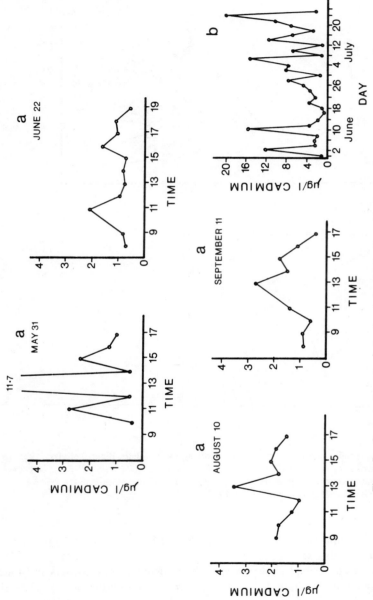

Figure 7. Concentrations of cadmium in solution (0.45-μ filtrate) in the Poole Estuary. (*a*) Samples taken hourly over 7 to 11 hr on four separate days. (*b*) Samples taken on alternate days over 2 months, June and July 1973. (After Boyden, 1975.)

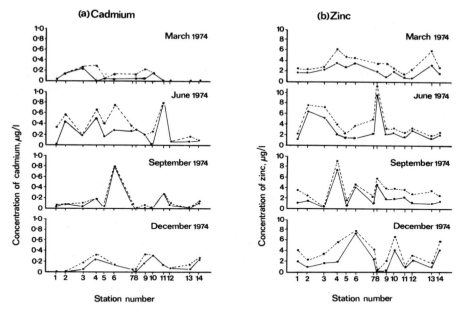

Figure 8.   Variation in the concentrations of (*a*) cadmium, and (*b*) zinc, in solution (0.45-μ filtrate), at 14 stations along the northwestern coast of the Mediterranean Sea; each station was sampled four times in 1974. Analysis by ASV at pH 8.0 (continuous lines) or pH 4.0 (dashed lines). (After Fukai and Huynh-Ngoc, 1976.)

cadmium cycling in estuaries, the stability of the cadmium profile permitting the use of water analysis as a means of identifying the pollution profile. However, stable profiles such as these are somewhat unusual; other authors have reported large fluctuations in the concentrations of cadmium observed with time at a single location. Boyden (1975) studied short-term fluctuations of cadmium levels in water of the Poole Estuary, U.K., near an oyster hatchery. Analyses performed on samples taken hourly over 11 hr on 4 separate days (Figure 7*a*) revealed considerable variation in the concentrations of cadmium in solution (0.45-μ filtrate), the daily ratio of maximum to minimum values varying between 3.5 and 29.25. These variations were probably related to tidal movements, amongst other factors, as the highest cadmium levels on 2 of the 4 days were observed at the time of second high water. The fluctuation of cadmium concentrations in water over 2 months at one site was also studied, water being sampled on alternate days. Concentrations varied from 0.70 to 20.50 μg/l (a factor of 29.3), consecutive samples often differing very considerably (Figure 7*b*). In such areas, where the supply of cadmium fluctuates considerably, the study of cadmium levels in water is of little value unless frequent samples are taken to establish representative mean concentrations at each study site. Areas of this nature are better studied using biological indication methods, which afford a time-integrated value for the relative metal pollution at each study site, at least if used correctly (Phillips, 1977b).

In addition to short-term variations in the concentrations of cadmium found in the water of any one estuarine or coastal site, differences exist in the seasonal variation in cadmium concentrations at different locations. Fukai and Huynh-Ngoc (1976) sampled water at 14 stations on the coast of the northwest Mediterranean Sea on four occasions in 1974. Profiles of cadmium concentration with location (Figure 8a) revealed gross differences in the seasonal variation of soluble (0.45-μ filtrate) cadmium levels at each site, at either pH 8 or pH 4. For example, water from Monaco (site 11) exhibited a seasonal maximum in cadmium levels in June; by contrast, water from San Remo (site 12, only about 30 km distant on a relatively open coastline) exhibited little variation in cadmium levels with season. Furthermore, these location-dependent differences in the seasonality of cadmium concentrations in solution were not matched by the seasonal variation in zinc concentrations (Figure 8b), showing that the changes were not due to gross factors such as runoff (which would affect both metals simultaneously), but were caused by local differences in metal supply. Differences such as these emphasize the need for multiple sampling with time or for a time-integrated method of measuring the abundance of metals if the relative contamination of different areas is to be accurately compared.

3. THE TOXICITY OF CADMIUM TO MARINE AND COASTAL BIOTA

3.1. General

The toxicity of cadmium to aquatic biota has been recognized fully only since 1970. Although it is difficult to rank trace metals on the basis of their degrees of toxicity to aquatic biota because of differences in the relative toxicities of metals to different test species, cadmium would certainly be ranked with Cu, Ag, and Hg amongst the most toxic metals for aquatic organisms.* Pollution by mercury has been markedly reduced in many areas of the world by increased restrictions on industrial effluents and other sources since the recognition of the toxic potential of methylmercury. Silver is a rather rare contaminant, at least in high concentrations, and pollution by this metal is probably restricted to small areas; Dryssen et al. (1970) considered it to be of relatively low toxic potential because of its restriction to river inputs and its low total load in most effluents.

* Such ranking of metals according to relative toxicities is generally performed by considering metal concentrations in milligrams per liter (part per million) terms. If molarities were used instead, metals of greater mass, such as cadmium and mercury, would appear to possess still higher relative toxicities in comparison to metals of less mass, such as copper.

Thus (as far as trace metals are concerned) the coastal and marine environments may be considered to be at greatest risk from pollution by cadmium and copper at the present time.

Many studies have been reported since 1970 concerning the toxicity of cadmium in seawater or brackish (estuarine) waters. For convenience, these reports are separated here into those dealing with the acute or lethal toxicity, and those including data on the sublethal effects, of the metal. The majority of the published data concerns the acute toxicity of cadmium, as such studies are easier to perform than are sublethal investigations. In addition, early attempts of environment protection bodies to produce meaningful water quality criteria for trace metals were based on the study of the acute toxicities of metals, using an application factor (arbitrary in nature, but generally 0.01 for trace metals) as a "safety margin" to protect the organism from sublethal effects also. Many authors have recognized the lack of any sound theoretical basis for this approach, and long-term studies of the sublethal effects of trace metals are becoming more numerous. Moreover, early tendencies of authors to study toxicants in isolation, using test organisms in optimal environments, are giving way to studies of the interactions of toxicants with each other and with natural stresses experienced by marine or coastal biota. The effects of ancillary parameters such as salinity and temperature on the toxicity of cadmium are most important and will be emphasized below where data exist; hypotheses explaining their mode of action will be discussed in Section 6.

3.2.   The Acute Toxicity of Cadmium

Finfish

Many of the data reported for the toxicity of cadmium to finfish concern the mummichog *Fundulus heteroclitus* (also termed the killifish by some authors; see Jackim et al., 1970). Results of acute toxicity bioassays of this species are tabulated in Table 6. Comparison of the results of different authors reveals discrepancies in the LC_{50} values (concentration of toxicant sufficient to kill 50% of the test organisms in a given time) quoted for 24-, 48-, or 96-hr tests. These differences in the results of different authors may be related to small variations in fish condition (age, sexual cycle, etc.) or in test parameters such as salinity and temperature.

The effects of salinity and temperature on the acute toxicity of cadmium to *F. heteroclitus* were studied by Eisler (1971). Complete results are shown in Figure 9. The fish used were acclimated to the test salinities and temperatures for 14 days prior to the addition of cadmium; the seawater used was artificial, after a recipe given by Zaroogian et al. (1969). Mummichogs were less resistant to cadmium at 20°C than at 5°C, by a factor of 1.7 to 3.0. The greatest tem-

Table 6. Acute Toxicity of Cadmium to Adults of the Mummichog *Fundulus heteroclitus* at Various Times in Differing Salinities and Temperatures

| References | Life Stage | LC$_{50}$ (ppm or mg/l) | | | | Test Conditions | | |
		24 hr	48 hr	96 hr	192 hr	Salinity (‰)	Temperature (°C)	Metal Form
Eisler (unpublished)[a]	Adult (?)	—	100	—	—	24	21	NQ[b]
Jackim et al. (1970)	Adult	—	—	~27	—	NQ	20–25	NQ
Eisler (1971)	Adult	140	—	49–55	—	20	20	Chloride
Eisler (1971)	Adult	—	—	—	15	5	20	Chloride
Eisler (1971)	Adult	—	—	—	27–28	35	20	Chloride
Eisler (1971)	Adult	—	—	—	75	15	5	Chloride
Eisler (1971)	Adult	—	—	—	50	35	5	Chloride
Voyer (1975)	Adult	175–>200	100	73	—	10	18	Chloride
Voyer (1975)	Adult	>200	175–>200	78	—	20	18	Chloride
Voyer (1975)	Adult	>50	40	30	—	30	20	Chloride
Middaugh and Dean (1977)	Adult	—	60	—	—	20	20	Chloride
Middaugh and Dean (1977)	Adult	—	43	—	—	30	20	Chloride

[a] Cited in Gardner and Yevich (1970).

[b] NQ = not quoted.

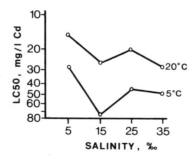

Figure 9. Concentrations of cadmium lethal to 50% of the test group in an exposure of 192 hr at four salinities and two temperatures. Each point represents results from 60 mummichogs (*Fundulus heteroclitus*). (After Eisler, 1971.)

perature effect was seen at the preferred salinity of this estuarine species (15 to 20‰). The effects of salinity (S) on LC_{50} values were similar in each thermal regime, fish exhibiting least resistance at the lowest salinity of 5‰ and greatest resistance at 15‰ S. The markedly decreased resistance of the fish acclimated to 5‰ S was suggested to be linked to the osmoregulatory stress to which they were exposed (Eisler, 1971).

Data from the studies of other authors concur with the conclusions mentioned above for the effects of salinity and temperature on cadmium toxicity to the mummichog. Thus Voyer (1975), in his studies of the effects of salinity and dissolved oxygen on the LC_{50} values for cadmium exhibited by this species, reported less susceptibility at 20‰ than at 10 or 30‰ S. No consistent effects of dissolved oxygen levels on the LC_{50} values were found, and temperature variations were too small to have significantly altered the toxicity pattern (Table 6). Middaugh and Dean (1977) also considered mummichog to be more susceptible to cadmium poisoning at 30‰ than at 20‰ S (Table 6); these data are highly comparable to those of Eisler (1971), test temperatures, fish size, and season of collection being similar in the two studies.

The results of studies on the acute toxicity of cadmium to teleosts other than the mummichog are shown in Table 7. The variation in LC_{50} values reported for different species is surprisingly small, perhaps indicating a similar mode and threshold of acute toxic action of cadmium in marine teleost species. Although freshwater finfish have been observed to be more susceptible to cadmium than are marine species (Pickering and Henderson, 1966; Ball, 1967; Pickering and Gast, 1972; Eaton, 1974; Benoit et al., 1976; Spehar, 1976; Pascoe and Mattey, 1977; McCarty et al., 1978), the exact sensitivity of freshwater fish depends to a great extent on the water hardness, and this distinction between the two environments does not seem to extend to estuarine teleosts when compared to marine species. Estuarine species may in fact be somewhat more resistant to

Table 7. Acute Toxicity of Cadmium to Various Finfish Species

Species	Common Name	LC_{50} (ppm or mg/l)					Test Conditions			References
		24 hr	48 hr	96 hr	120 hr	168 hr	Salinity (‰)	Temperature (°C)	Metal Form	
Fundulus majalis	Striped killifish	125	59	21	—	—	20	20	Chloride	Eisler (1971)
Cyprinodon variegatus	Sheepshead minnow	100	50	50	—	—	20	20	Chloride	Eisler (1971)
Agonus cataphractus	Bullhead	—	>33–<100	33	—	—	NQ[a]	15	Chloride	Portmann and Wilson (1971)
Aldrichetta forsteri	Yellow-eye mullet	—	—	—	14	16	~34	~19	Chloride	Negilski (1976)
Atherinasoma microstoma	Small-mouthed hardyhead	—	—	—	—	15–21	~34	~19	Chloride	Negilski (1976)
Menidia menidia	Atlantic silverside	—	12	—	—	—	20	20	Chloride	Middaugh and Dean (1977)
Menidia menidia	Atlantic silverside	—	13	—	—	—	30	20	Chloride	Middaugh and Dean (1977)

[a] NQ = not quoted.

Table 8.   Percentage Mortality (Nonemergence) of Larvae from Eggs of the Mummichog *Fundulus heteroclitus* and the Atlantic Silverside *Menidia menidia* Exposed to Cadmium for the First 48 Hr after Fertilization[a]

Species	Cadmium (mg/l)	Percentage Mortality	
		Salinity 20‰	Salinity 30‰
Fundulus heteroclitus	32.0	54	54
	10.0	40	54
	3.2	30	50
	1.0	27	47
	0.32	20	23
	Control	17	33
Menidia menidia	32.0	66	60
	10.0	52	38
	3.2	58	40
	1.0	36	36
	0.32	50	38
	Control	36	33

[a] After Middaugh and Dean (1977).

cadmium than are marine finfish; Eisler (1971) noted that the mummichog, compared to other fish, exhibits a high LC_{50} value, and this observation is confirmed by more recent results.

The preceding data concern only adult fish, or at least juvenile or preadult individuals. The life stage of the individuals tested is important, as the toxicity of most pollutants (including cadmium) varies according to the age of the organism used. In general, teleosts exhibit the greatest sensitivity to pollutants at the larval stage, both eggs and adults being less sensitive. Middaugh and Dean (1977) reported data for the acute toxicity of cadmium to eggs, larvae, and adults of the estuarine teleosts *Fundulus heteroclitus* and *Menidia menidia*. The results for the adult fish are included in Tables 6 and 7. Bioassays using eggs and larvae were performed at salinities of 20 and 30‰ and at 20°C. The eggs of both species were highly resistant to cadmium (Table 8); LC_{50} values were not given because of the high percentages of nonemergence from control eggs, but were certainly high and were probably greater than those for the adults of each species. The eggs of both these species are found in rigorous environments and are therefore presumably adapted to withstand natural stresses also. The low toxicity of cadmium to finfish eggs is probably due to fixation of the cadmium on the chorionic membrane. Rosenthal and Sperling (1974) reported high concentrations of cadmium in the chorion of developing eggs of the herring *Clupea harengus;* by contrast, very little cadmium was present in either embryos or

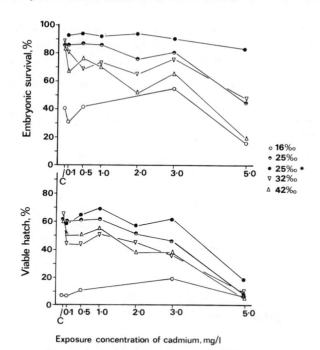

Figure 10. Embryonic survival (above) and viable hatch (below) of eggs and larvae of the Baltic flounder *Pleuronectes flesus,* maintained in different salinities at different concentrations of cadmium. *Fertilization occurred in cadmium-contaminated water. (After Westernhagen and Dethlefsen, 1975.)

newly emerged larvae. The sequestration of cadmium by adsorption to surface sites on the chorion thus appears to protect the embryo from the toxic effects of the metal, which may be elicited only if enough metal is present to pass through the limiting membrane.

A similar threshold effect was discovered by Westernhagen and Dethlefsen (1975), in studies of cadmium toxicity at four different salinities to the eggs of the Baltic flounder *Pleuronectes flesus.* No effects of cadmium were observed on hatching rates or viable hatch at exposure concentrations up to 1.0 mg/l. However, concentrations of 2.0 to 5.0 mg Cd/l markedly decreased both hatching rates and viable hatch (Figure 10). No consistent effects of salinity on the toxicity of the metal were discerned (a salinity of 16‰ eliciting deaths in the control group also), but the accumulation of cadmium in the eggs increased with decreasing salinity (see Section 4). By contrast, Westernhagen et al. (1974) had previously reported a clear increase in cadmium toxicity to eggs of the herring *Clupea harengus* with decreased salinity of the incubation medium, which correlated in this case to data on cadmium accumulation. The adsorption

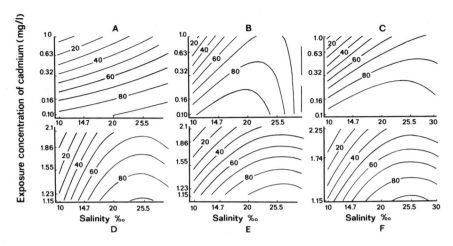

Figure 11. Response surfaces describing percentage viability of embryos of the winter flounder *Pseudopleuronectes americanus,* exposed to various combinations of cadmium and salinity. Surfaces *A, B, C* describe tests at 10°C; surfaces *D, E,* and *F* are for tests at 5°C. Each surface describes the responses of eggs from separate spawnings. (After Voyer et al., 1977.)

of cadmium to the chorion of the eggs was again suggested as central to the toxic action of the metal; Westernhagen and Dethlefsen (1975) noted that herring eggs adsorbed more metal than did flounder eggs and ascribed this to differences in the thickness and biochemical makeup of the two membranes.

Voyer et al. (1977) described investigations of the effects of cadmium, salinity, and temperature on viability of embryos of the winter flounder *Pseudopleuro-nectes americanus.* Concentrations of cadmium yielding 50% viable hatches (EC$_{50}$ values) varied with salinity, being lower at 10‰ than at 20 or 30‰, at both 5 and 10°C. Response surfaces generated from the data (Figure 11) clearly showed the greater susceptibility of embryos at the lowest salinity, also suggesting a slight increase in embryo susceptibility at salinities greater than 25‰. This finding agrees well with the conclusion of Rogers (1976) that salinities of 15 to 25‰ are optimal for the development of eggs of *P. americanus* at ambient temperatures greater than 3°C.

Finfish larvae in general exhibit a greater sensitivity to pollutants than do the eggs from which they emerge. This difference in response apparently depends on the rates of uptake of pollutants by each life stage. As noted above, the eggs of finfish are rather resistant to cadmium because of the protection afforded by the chorionic membrane. By contrast, larvae absorb pollutants directly from the water via the gills. Middaugh and Dean (1977) measured LC$_{50}$ values for cadmium toxicity to larvae of the mummichog *Fundulus heteroclitus* and the Atlantic silverside *Menidia menidia,* using larvae of different ages. Mummichog

Table 9.   Variation in the Toxicity of Cadmium Chloride to Larvae of the Mummichog *Fundulus heteroclitus* or of the Atlantic Silverside *Menidia menidia* with Age of the Larvae at Two Salinities and at 20°C[a]

Species	Larval Age (days)	Salinity (‰)	48-Hr LC_{50} (mg Cd/l)	95% Confidence Interval
Fundulus heteroclitus	1	20	16.2	12.7–21.2
	7	20	9.0	6.4–12.5
	14	20	32.0	24.6–41.6
Fundulus heteroclitus	1	30	23.0	19.2–27.6
	7	30	12.0	9.2–15.6
	14	30	7.8	5.6–10.3
Menidia menidia	1	20	3.8	2.7–5.3
	7	20	3.2	1.9–4.7
	14	20	2.2	1.6–2.9
Menidia menidia	1	30	5.6	4.0–7.8
	7	30	3.4	2.1–4.9
	14	30	1.6	1.1–2.4

[a] After Middaugh and Dean (1977).

larvae exhibited maximum sensitivity at 7 days after emergence in a salinity of 20‰, but were most sensitive at 14 days after emergence in 30‰ S (Table 9). The relative insensitivity of 1-day-old larvae to cadmium is probably related to the lack of gill function at this age; the gills do not become fully functional until the larvae attain an age of 4 days in this species. Atlantic silverside larvae exhibited increased sensitivity to cadmium with an increase in age (Table 9). In addition, young larvae of this species were more sensitive to cadmium at the lower salinity, agreeing with their survival rates at the two salinities in the absence of added toxicants (Middaugh and Lempesis, 1976).

    The interaction of pollutants with each other may be of great importance in determining the survival of biota, especially in situations such as estuaries where organisms are exposed to fluctuating levels of all pollutants simultaneously with their exposure to natural stresses. For finfish, only one report exists to date concerning the interaction of cadmium toxicity with the toxicities of other metals. Eisler and Gardner (1973) exposed the mummichog *Fundulus heteroclitus* to mixtures of cadmium, copper, and zinc chlorides in water of salinity 20‰ and temperature 20°C. Cadmium was added to produce final concentrations of 0.1, 1.0, and 10.0 mg/l in the test solutions. These levels alone are too low to produce any mortality in mummichogs (Eisler, 1971, and see Table 6). However, the addition of 10.0 mg Cd/l significantly increased the cumulative mortality over 96 hr in all Zn-Cu mixtures tested (except controls with no added zinc or copper),

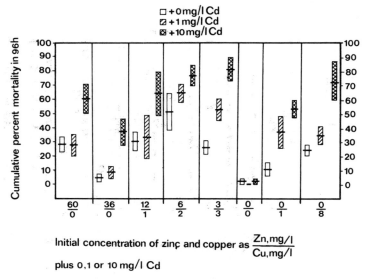

Figure 12. The effects of cadmium on the toxicity of zinc and copper alone and in combination to the mummichog *Fundulus heteroclitus* at 20°/oo salinity and 20°C. Each result represents data from five bioassays of nine fish each; means ± one standard deviation are plotted. Statistical significance was assumed between groups exhibiting no overlap of standard deviations. (After Eisler and Gardner, 1973.)

as shown in Figure 12. Even the comparatively low concentration of 1.0 mg Cd/l elicited significantly increased mortalities in the groups exposed to Zn-Cu combinations of 3:3, 0:1, and 0:8 mg/l. The implications of these data will be discussed in Section 6.

Molluscs

As noted by Eisler (1971), and since confirmed by the results of other authors, molluscs exhibit higher sensitivities to acute cadmium stress than do finfish. Table 10 shows, for various mollusc species, representative LC_{50} values of cadmium from the literature.

Certain problems exist concerning the use of some of the molluscs in acute toxicity bioassays. The greatest of these is accurate definition of the point of death of some species. For example, it is often difficult to accurately predict the death point of the blue mussel *Mytilus edulis,* or that of other bivalve molluscs. Ahsanullah (1976) attempted to overcome this problem by the use of a recovery period in clean seawater for any animal in a moribund condition. However, this solution is not altogether satisfactory, as animals that recover must be replaced in the toxicant tanks and their further resistance to the toxicant is presumably increased by the recovery period.

Table 10. Acute Toxicity of Cadmium to Various Species of Molluscs

| Species | Common Name | LC$_{50}$ (mg/l or ppm) | | | Test Conditions | | | References |
		24 hr	48 hr	96 hr	Salinity (‰)	Temperature (°C)	Metal Form	
Mya arenaria	Soft-shell clam	>200	50	2.2	20	20	Chloride	Eisler (1971)
Mytilus edulis	Blue mussel	>200	165	25	20	20	Chloride	Eisler (1971)
Nassarius obsoletus	Eastern mud snail	>200	125	10.5	20	20	Chloride	Eisler (1971)
Urosalpinx cinerea	Atlantic oyster drill	158	28	6.6	20	20	Chloride	Eisler (1971)
Cardium edule	Common cockle	—	10–33	2.0–3.3	NQ^a	15	Chloride	Portmann and Wilson (1971)
Mytilus edulis planulatus	Blue mussel	—	—	1.6	33	18.5	Chloride	Ahsanullah (1976)

^{a} NQ = not quoted.

In addition, some molluscs are able to reduce or eliminate contact of the tissues with the surrounding seawater when exposed to changes in the ambient water quality. Bivalve molluscs, for example, may adduct the valves tightly to eliminate contact with waters of fluctuating salinity. The blue mussel *M. edulis* has long been known to exhibit valve closure in adverse salinities (Cronklin and Krogh, 1938). Other lamellibranchs (perhaps all species) are also capable of valve closure, although the capacities of different species for total elimination of contact of the tissues with the ambient seawater vary considerably (Shumway, 1977). During periods of valve closure, these organisms become transient partial osmoregulators, the mantle cavity fluid exhibiting an osmoconcentration different from that of the surrounding water. However, the organism appears to sample the external environment by some mechanism, as the duration of valve closure in any one species is proportional to the extent and duration of the salinity change. Molluscs such as the limpet *Patella vulgata* and gastropods (*Littorina* spp.) are also capable of isolating the tissues from the external medium for short periods (Arnold, 1957; Avens and Sleigh, 1965).

Isolation of the internal tissues from the ambient environment not only protects against the stressful effects of salinity fluctuation, but also isolates the organism from toxicants present in the surrounding water. Davenport (1977) showed that relatively high concentrations of copper (0.5 mg/l) could elicit valve closure in the blue mussel *Mytilus edulis*. Phillips (unpublished data) has observed the same phenomenon in *M. edulis* exposed to Cd, Cu, or Zn; the concentration of each metal triggering valve closure appeared to correlate to its relative toxicity to this species. Although cadmium elicited valve closure at high concentrations (above about 5 mg/l), lower concentrations generally caused gaping, which was very rarely observed on exposure of mussels to other metals. Ahsanullah (1976) reported similar data, and this effect probably explains the discrepancy between the results of this author and those of Eisler (1971) for the acute toxicity of cadmium to the mussel. Such a conclusion is strengthened by the observation of very large differences in the LC_{50} values for 24 and 96 hr by Eisler (1971) for cadmium and *M. edulis;* the high concentrations of cadmium used by this author certainly elicited valve closure, which may be maintained for periods up to 96 hr by *M. edulis*. If the mussels used by Eisler (1971) opened their valves only after a long period of closure, the reported 96-hr LC_{50} in fact represents an LC_{50} for a much shorter period. The discrepancy between the toxicity data of the two authors cited thus appears much smaller, perhaps nonexistent.

The physiological effects of high concentrations of trace metals, such as those used in short-term toxicity testing, should therefore be considered as a possible interfering factor in the bioassay results. Although teleosts are known to exhibit trace metal avoidance in the environment, this is obviously impossible in laboratory bioassays. By contrast, sessile organisms that have evolved different avoidance methods such as valve closure may be highly sensitive to changes in the ambient water quality, presenting greater problems for successful toxicity testing. In these cases, longer term bioassays are necessary to overcome the interfering effects of the initial physiological response; the relative sensitivities of the various species should be based on incipient lethal levels (concentrations of toxicant required to kill 50% of the test organisms at time infinity), rather than on the results of short-term tests.

Few data are available concerning the variation in cadmium toxicity with the life stage of molluscs. Calabrese et al. (1973) estimated the LC_{50} for cadmium (test conditions were 25‰ salinity and 26°C in synthetic seawater) as 3.80 mg/l for embryos of the American oyster *Crassostrea virginica;* however, these authors state that the results for cadmium toxicity were highly variable, and no reason for this variability could be suggested. By analogy with telosts, it might be expected that molluscan species would exhibit greatest sensitivity to toxicants at the larval stage, eggs and adults being less sensitive. However, comparison of the results of Wisely and Blick (1967), Portmann (1970), Calabrese et al. (1973), and other authors concerning the toxicity of metals other than cadmium to different life cycle stages of bivalve molluscs suggests that this hypothesis may not be tenable. Wisely and Blick (1967) reported high values for the 2-hr LC_{50}s of Zn, Cu, and Hg, using larvae of the mussel *Mytilus edulis* and the oyster *Crassostrea commercialis.* It was suggested that the high resistance of these larvae was due to their ability to withdraw into the shells; thus the valve closure mechanism discussed above may be operative even at larval stages in some molluscs. It seems doubtful that larvae can maintain valve closure for periods as long as can the adult organism, as the requirements for oxygen and food are certainly greater at the larval stage. However, the lack of data for incipient lethal levels of toxicants to each life stage of molluscs permits no certain conclusion concerning the variation in sensitivity of these species with life stage.

Although the effects of salinity and temperature of the ambient water on the uptake of cadmium by molluscs have been studied by some authors (see Section 4), no data are available concerning the influence of these parameters on metal toxicity to members of this phylum. Similarly, no data are available concerning the effects of the interaction of metals with each other on the toxicity of any single metal. Eisler (1977) reported a study in which the toxicity of a mixture of six trace metals to the softshell clam *Mya arenaria* was evaluated. Cadmium was present as one of these six metals, but its concentration was very low (1.0 μg/l) and it probably contributed little to the overall toxic effect of the mixture. The effects of metal-metal and metal-salinity interactions are particularly important for organisms in their natural environment; the ability of molluscs to accumulate high concentrations of many metals also suggests that the study of such interactions would be most profitable.

Crustaceans

On the basis of extensive studies of cadmium toxicity to marine organisms, Eisler (1971) concluded that crustaceans were more sensitive to cadmium than were either molluscs or teleosts. A similar conclusion was reached by Ahsanullah (1976). Table 11 presents selected examples concerning the acute toxicity of cadmium to various crustaceans; comparison of the LC_{50} values quoted here with those in Table 10 (molluscs) or Tables 6 and 7 (finfish) confirms the generally greater sensitivity of crustaceans than of the other two groups. As noted above for molluscs, certain problems exist with respect to the use of crustaceans in acute bioassays for cadmium toxicity. The first of these concerns a protection mechanism similar to that found in bivalve molluscs. Thus certain species of barnacles are able to close their opercular valves, maintaining a constant internal environment under unfavorable conditions. Foster (1969) has demonstrated this phenomenon in *Balanus balanoides, B. balanus, B. crenatus, B. improvisus, Chthamalus stellatus,* and *Elminius modestus.* Again, decreases in ambient salinity appear most effective in eliciting this response; however, exposure to high concentrations of trace metals may also be effective, and this possibility should be kept in mind when interpreting the results of toxicity studies using barnacles.

The second major problem encountered during acute toxicity bioassays of crustaceans is cannibalism. Many crabs and some shrimp species exhibit cannibalism in holding tanks, and the incidence of this behavior may depend on the stresses placed on the organisms. Ahsanullah (1976) observed cannibalism in control animals, as well as in those exposed to zinc or cadmium, and suggested that the stress of captivity led to molting of the animals, which in turn increased the probability of attack by other individuals. Most authors find it necessary to separate individuals in order to avoid losses by cannibalism.

Molting of test organisms presents a further problem in acute or subacute bioassays of crustaceans, and this may affect the results of toxicity tests for cadmium more than those for other metals, as cadmium is highly concentrated in the exoskeleton of many crustaceans. In general, the concentrations of cadmium found in the exoskeleton are exceeded only by those in the viscera or hepatopancreas of most crustaceans (Eisler et al., 1972; Fowler and Benayoun, 1974; Nimmo et al., 1977; see also Section 4). Fowler and Benayoun (1974) observed that molting may be responsible for a large proportion of the total cadmium excretion in the benthic shrimp *Lysmata seticaudata;* the same is undoubtedly true for other species also. The high incidence of molting in crustaceans kept under laboratory conditions may therefore affect the results of toxicity tests using cadmium, as molting animals are able to lose a large amount of the cadmium taken up during the bioassay. Indeed, cadmium itself may affect

Table 11.   Acute Toxicity of Cadmium to Various Crustacean Species

Species	Common Name	LC$_{50}$ (mg/l or ppm)				Test Conditions			Reference
		24 hr	48 hr	96 hr	240 hr	Salinity ‰	Temperature (°C)	Metal Form	
Artemia salina	Brine shrimp	—	—	—	1.0	NQ[a]	20	Sulfate	Brown and Ahsanullah (1971)
Carcinus maenas	Green crab	100	16.6	4.1	—	20	20	Chloride	Eisler (1971)
Crangon septemspinosa	Sand shrimp	2.4	0.5	0.32	—	20	20	Chloride	Eisler (1971)
Pagarus longicarpus	Hermit crab	>200	3.7	0.32	—	20	20	Chloride	Eisler (1971)
Palaemonetes vulgaris	Grass shrimp	43	5.8	0.42	—	20	20	Chloride	Eisler (1971)
Crangon crangon	Brown shrimp	—	3.3–10	0.6–1.0	—	NQ	15	Chloride	Portmann and Wilson (1971)
Uca pugilator	Fiddler crab	—	28	10.4	3.5	20	30	Chloride	O'Hara (1973)
Uca pugilator	Fiddler crab	—	—	46.6	9.5	20	20	Chloride	O'Hara (1973)
Uca pugilator	Fiddler crab	—	—	—	42	20	10	Chloride	O'Hara (1973)
Palaemon sp.	Prawn	—	—	6.4–6.8	—	NQ	17	Chloride	Ahsanullah (1976)
Palaemonetes vulgaris	Grass shrimp	—	—	0.76	—	20	25	Chloride	Nimmo et al. (1977)
Penaeus duorarum	Pink shrimp	—	—	3.5	—	20	25	Chloride	Nimmo et al. (1977)
Paragrapsus gaimardii	Crab	—	—	24.1	—	17.5	19	Chloride	Sullivan (1977)
Paragrapsus gaimardii	Crab	—	—	61.5	—	17.5	5	Chloride	Sullivan (1977)

[a] NQ = not quoted.

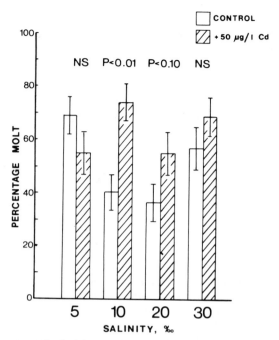

Figure 13.   Percentage molt of adult grass shrimp (*Palaemonetes pugio*), kept in static bioassay conditions for 21 days at 25°C and different salinities, with and without the addition of 50 μg cadmium chloride/l. NS = no significant difference between cadmium-treated and control groups. (After Vernberg et al., 1977.)

the incidence of molting in some species. Vernberg et al. (1977) observed significantly greater molting frequencies of grass shrimp (*Palaemonetes pugio*) exposed to 50 μg Cd/1 in water of 25°C and 10‰ salinity. However, no such effects were observed for the same species at salinities of 5, 20, or 30‰ (Figure 13). The exact effect of molting on the survival time of crustaceans exposed to high concentrations of cadmium is uncertain. If the effects of molting on the incidence of cannibalism (assuming individuals are separated during testing) are neglected, molting may have little influence on cadmium toxicity if the concentration of the metal in the target organ for lethality is not affected by the cadmium in the exoskeleton. Much of the cadmium present in the exoskeleton is probably adsorbed to the surface and is not available for exchange with other tissues. However, even if molting does not affect the acute toxicity of cadmium, the whole-body levels are certainly affected, as is the excretion rate for the metal in crustaceans (see Section 4).

Most of the data shown in Table 11 concerns the toxicity of cadmium added as the chloride salt ($CdCl_2 \cdot 2\frac{1}{2}H_2O$), though Brown and Ahsanullah (1971) used

Table 12.   Toxicity and Accumulation of Cadmium by Pink Shrimp (*Penaeus duorarum*) Exposed to One of Four Different Cadmium Salts[a]

Twenty individuals were tested in each solution for 7 days in flow-through conditions at 20°C and 20‰ salinity.

Toxicant	Exposure Concentration (mg/l)	Percentage Mortality	Whole-Body Cadmium (μg/g wet weight, mean $\pm$ 2SEM)[b]
Control	—	0	0.3 $\pm$ 0.2
Cadmium acetate	1.6	75	6.8 $\pm$ 1.4
Cadmium chloride	1.8	65	8.5 $\pm$ 4.4
Cadmium sulfate	1.6	55	10.5 $\pm$ 3.0
Cadmium nitrate	1.7	75	10.8 $\pm$ 3.2

[a] After Nimmo et al. (1977).
[b] SEM = standard error of the mean.

the sulfate rather than the chloride. As stated in Section 2.2, the form of the metal present may affect its toxicity greatly; however, whether the use of different simple salts in toxicity tests has any great influence on the final cadmium speciation in solution is uncertain and appears unlikely unless the added anion disturbs speciation by affecting the speciation of other seawater components. Nimmo et al. (1977) investigated this possibility, studying the toxicity of four cadmium salts to pink shrimp (*Penaeus duorarum*). The results (Table 12) indicated no significant difference in the toxicity or uptake of cadmium from each of the test solutions. Presumably the anions were not affecting the speciation of cadmium under the conditions of the bioassays, and did not contribute significantly to the toxic effect of the test solutions.

The effects of salinity and temperature on the toxicity of cadmium to crustaceans have been studied by several authors. O'Hara (1973a) reported extensive data for cadmium toxicity to the fiddler crab *Uca pugilator* at three temperatures (10, 20, 30°C) and three salinities (10, 20, 30‰). Results of these experiments (Table 13) indicated a dependence of the LC_{50} value on time of exposure in any experimental regime. At any given time and salinity, increases in water temperature increased the toxicity of cadmium (decreased the LC_{50} value). At any given time and temperature, increases in salinity decreased the toxicity of cadmium. Thus the crabs were most susceptible to cadmium at 30°C and 10‰ S (240-hr LC_{50} of 2.9 mg Cd/l) and least susceptible at 10°C and 30°/00 S (240-hr LC_{50} of 47.0 mg Cd/l). These data were also cited in Vernberg et al (1974); the uptake of cadmium by this species as influenced by salinity and temperature is discussed in Section 4.

Sullivan (1977) determined LC_{50} values for the toxicity of cadmium to the estuarine crab *Paragrapsus gaimardii*, using animals acclimated for 2 weeks

Table 13.   LC$_{50}$ Values (mg/l) for Cadmium Toxicity to the Adult Fiddler Crab *Uca pugilator* at Various Exposure Times, Salinities, and Water Temperatures (Cadmium Added as the Chloride Salt)[a]

Salinity (‰)	Exposure Time (hr)	LC$_{50}$ (mg Cd/l) 10°C	20°C	30°C
10	48	—	—	11.0
	96	—	32.2	6.8
	144	51.0	21.3	4.0
	192	28.5	18.0	3.0
	240	15.7	11.8	2.9
20	48	—	—	28.0
	96	—	46.6	10.4
	144	—	23.0	5.2
	192	52.0	16.5	3.7
	240	42.0	9.5	3.5
30	48	—	—	33.3
	96	—	37.0	23.3
	144	—	29.6	7.6
	192	—	21.0	6.5
	240	47.0	7.9	5.7

[a] After O'Hara (1973a).

to combinations of two temperatures (5 and 19°C) and four salinities (8.6, 17.5, 26.3, and 34.6‰ S). Within each temperature regime, decreased salinity caused decreases in the observed LC$_{50}$ values (Figure 14), the temperature effect being greatest at high salinity. These data therefore agree with those of O'Hara (1973a) reported above, maximum sensitivity occurring at high temperature and low salinity.

The effects of salinity on cadmium toxicity to adult grass shrimp (*Palaemonetes pugio*) were reported by Vernberg et al. (1977). Although no LC$_{50}$ values were quoted, as mortalities of 50% were not attained, static bioassays at 5, 10, 20, and 30‰ S revealed a clear dependence of mortality on salinity, higher mortalities being observed at lower salinities throughout the 21-day experiment (Figure 15). All studies were performed at 25°C; hence no data for the effects of temperature were noted.

Jones (1975) studied the effects of salinity and temperature on the mortality and osmoregulation of the marine isopods *Idotea baltica, I. neglecta, I. emarginata,* and *Eurydice pulchra*, and of the estuarine isopods *Jaera albifrons* (*sensu stricto*) and *J. nordmanni*. The marine isopods were studied at salinities of 34.0, 27.2, 20.4, and 13.6‰ (100, 80, 60, and 40% seawater); the estuarine species were exposed to 34.0, 17.0, 3.4, and 0.34‰ salinity (100, 50, 10, and 1% seawater). Both groups were tested at two temperatures (5 and 10°C) and two

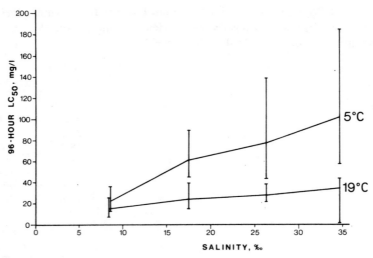

Figure 14.   Values for 96-hr LC_{50} and 95% fiducial limits for the toxicity of cadmium to the estuarine crab *Paragrapsus gaimardii* at two temperatures and four salinities. (After Sullivan, 1977.)

concentrations of cadmium (10 and 20 mg/l); the results are shown in Figure 16. At both concentrations of cadmium, decreased salinities led to greater mortalities for the marine species. For example, *Idotea baltica* exhibited an LT_{50} value (time to 50% death) in 10 mg Cd/l of more than 120 hr in 34‰ salinity; this LT_{50} diminished to only 34 hr in 13.6‰ S. Increased temperatures also caused a general increase in cadmium toxicity to each marine species. The estuarine species exhibited similar trends in regard to the effects of salinity and temperature on mortality rates, although in general these species were more resistant to cadmium than were the marine isopods. These six species thus appear to exhibit similar responses to changes of temperature or salinity in the presence of cadmium, and the response to each parameter is similar to that of the crabs and shrimp referred to above.

Information concerning the relative sensitivities of different life stages of crustacean species to acute cadmium stress is sparse. Brown and Ahsanullah (1971) reported higher sensitivities of larvae of the brine shrimp *Artemia salina* than of adults of the same species to zinc and copper salts; however, no data for cadmium were cited for larvae. More recently, Ahsanullah and Arnott (1978) determined 96-hr LC_{50} values for cadmium toxicity to adults and larvae of the crab *Paragrapsus quadridentatus*. Larval studies suggested an LC_{50} of 0.49 mg Cd/l at 17°C and 35‰ salinity, whereas adults of the same species exhibited a 168-hr LC_{50} of 14.0 mg Cd/l under similar bioassay conditions (Ahsanullah, 1976).

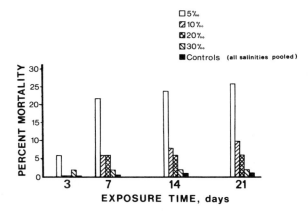

Figure 15.   Percentage mortality of adult grass shrimp (*Palaemonetes pugio*), exposed to cadmium in static tests at 25°C and four salinities. The exposure concentration was 50 μg/l, and 20 individuals per experimental group were used. (After Vernberg et al., 1977.)

Vernberg et al. (1974) kept zoeae of the fiddler crab *Uca pugilator* in water of 13 salinity-temperature combinations with or without the addition of 1 μg Cd/l. Even at this extremely low exposure concentration, the effect of cadmium could be observed in response surfaces generated from the results (Figure 17). Thus tolerance levels of the cadmium-exposed zoeae were narrowed compared to those of controls. Salinity tolerance was markedly narrowed, while temperature tolerance was shifted downward (toward lower temperatures). At 96 hr, for example, 80% survival of control larvae occurred in the ranges 21 to 28°C and 23 to 34‰ salinity; by contrast, cadmium-exposed larvae exhibited 80% survival in the range 18 to 23°C and 24 to 29‰ S. Salinity was the most important parameter affecting the survival of cadmium-treated larvae, whereas temperature was more important than salinity in determining the survival rates of control larvae. Effects of cadmium on the metabolism and activity of the larvae were also noted; these data are discussed in Section 3.3.

Rosenberg and Costlow (1976) studied the effects of salinity and of constant or cycling temperatures on the toxicity of cadmium to two species of estuarine crab, the blue crab *Callinectes sapidus* and the mud crab *Rhithropanopeus harrisii*. Both these species tolerate wide ranges of salinity and temperature during development (and as adults) in the absence of toxicants. However, 150 μg Cd/l was lethal to both species at a salinity of 10‰, and this concentration decreased survival at all salinities studied. Also, 50 μg Cd/l decreased the survival of zoeae of both species, these effects being much more apparent at lower salinities. The larval development of *R. harrisii* was also delayed by cadmium.

Information concerning the interaction of cadmium toxicity with the effects

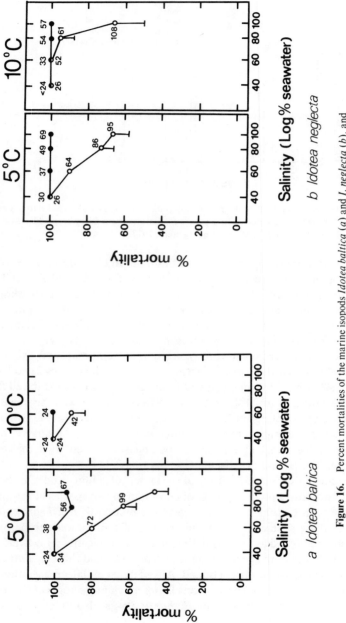

Figure 16. Percent mortalities of the marine isopods *Idotea baltica* (*a*) and *I. neglecta* (*b*), and the estuarine isopods *Jaera albifrons* (*c*) and *J. nordmanni* (*d*), exposed to cadmium chloride for 120 hr at 5 or 10°C. Closed circles represent 10 mg/l exposure concentrations; open circles, 20 mg/l. Vertical bars indicate one standard deviation about the mean; lack of bar lines indicates zero deviation. All data are means of three experiments; times to 50% death are also shown (in hr). (After Jones, 1975.)

470

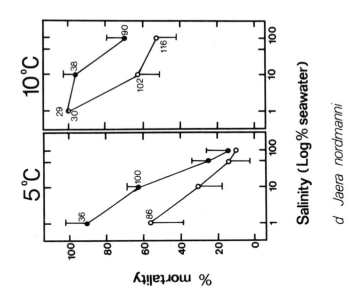

Salinity (Log% seawater)

d *Jaera nordmanni*

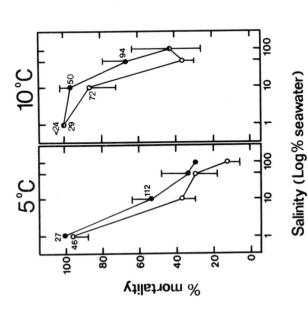

Salinity (Log% seawater)

c *Jaera albifrons*

Figure 16. (*continued*)

471

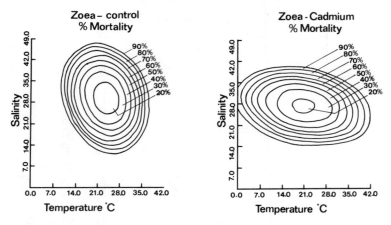

Figure 17.   Response surfaces describing percentage mortality of newly hatched zoeae of the fiddler crab *Uca pugilator* at different salinity-temperature combinations, with or without the addition of 1.0 µg cadmium chloride/l. Reponse surfaces were generated from data for 13 salinity-temperature combinations. (After Vernberg et al., 1974.)

of other pollutants is sparse. Bahner and Nimmo (1975) studied combinations of cadmium and malathion, cadmium and methoxychlor, and cadmium, methoxychlor, and Aroclor 1254, as well as a complex industrial waste containing both inorganic and organic pollutants. Comparison of LC_{50} values found for pink shrimp (*Penaeus duorarum*) for the individual components and the mixtures indicated little interaction between the metal and the pesticides; toxicities of the various components were independent. This conclusion was confirmed in a later paper (Nimmo and Bahner, 1976) using the same toxicant combinations and the same test species. The later work indicated an effect of methoxychlor on cadmium flux through the shrimp, however; this will be discussed in Section 4.

Annelid Worms and Echinoderms

The toxicities of cadmium to six species of annelids and two species of echinoderms are shown in Table 14. Comparison of the data quoted for 96-hr LC_{50} values suggests only small differences in the sensitivities of different annelid species; however, the long-term LC_{50} values for *Nereis diversicolor* indicate that this species is highly resistant to cadmium. No explanation for this species difference has been suggested; although *N. diversicolor* regulates its body loads of some trace metals, cadmium is not regulated (see Section 4) and resistance is not due to this method. Eisler (1971) suggested on the basis of his studies of cadmium toxicity to 13 species of marine organisms that annelids exhibited sensitivities similar to those of molluscs. Although this conclusion was based on data for only one species of annelid (*Nereis virens*), comparison of the data

Table 14.   Acute Toxicity of Cadmium to Annelid Worms and Echinoderms

Phylum	Species	LC$_{50}$ (mg/l)	Time (hr)	Salinity (‰)	Temperature (°C)	Metal Form	Reference
				Test Conditions			
Annelida	Ophryotrocha labronica	8.0	96	NQ[a]	20	Sulfate	Brown and Ahsanullah (1971)
	Ophryotrocha labronica	1.0	410	NQ	20	Sulfate	Brown and Ahsanullah (1971)
	Nereis virens	25.0	24	20	20	Chloride	Eisler (1971)
	Nereis virens	11.0	96	20	20	Chloride	Eisler (1971)
	Neanthes vaali	6.4	168	33	18	Chloride	Ahsanullah (1976)
	Nereis diversicolor	100.0	192	17.5[b]	NQ	Sulfate	Bryan (1976)
	Nereis diversicolor	10.0	816	17.5[b]	NQ	Sulfate	Bryan (1976)
	Ctenodrilus serratus	2.5–5.0	96	NQ	NQ	Chloride	Reisch and Carr (1978)
	Ophryotrocha diadema	2.5–5.0	96	NQ	NQ	Chloride	Reisch and Carr (1978)
Echinodermata	Asterias forbesi	12.0	24	20	20	Chloride	Eisler (1971)
	Asterias forbesi	0.8	96	20	20	Chloride	Eisler (1971)
	Patiriella exigua	>10	96	32	20	Chloride	Ahsanullah (1976)

[a] NQ = not quoted.
[b] Quoted as "50% seawater" in original.

presented in Tables 6, 7, 10, and 11 with those in Table 14 confirms the suggestion.

As previously noted for crustaceans, cannibalism is a problem in the use of annelid worms in toxicity tests. Ahsanullah (1976) noted fighting by *Neanthes vaali* in bioassay tanks, which resulted in some individuals being cut in two, as well as injuries leading to necrosis of the posterior portion of some intact worms.

These organisms should therefore be kept in separate compartments of the same bioassay tank, or in separate tanks, to avoid such problems.

The data reported by Eisler (1971) for cadmium toxicity to the common starfish *Asterias forbesi* suggest that echinoderms may be extremely sensitive to this metal. However, Ahsanullah (1976) found the asteroid *Patiriella exigua* to be rather insensitive to cadmium (Table 14), surviving exposure to 10 mg/l of the metal for 96 hr. Some difficulty was experienced by the latter author in accurately determining the point of death. The criterion used was failure to respond to mechanical stimulation of the tube-feet; however, this criterion was unsatisfactory, as animals judged to be dead later recovered in clean seawater. Later work by Ahsanullah (1976) suggested a 72-hr LC_{50} of between 25 and 50 mg Cd/l, a value similar to the ones exhibited by the more resistant teleosts. The differences between the results of Eisler (1971) and those of Ahsanullah (1976) may be based simply on the different species used by each author. However, Eisler (1971) gave no information concerning either his criterion of death or the use of a recovery period; conceivably the early values overestimate the toxicity of cadmium to *A. forbesi* if this animal reacts as does *P. exigua*. Evidently further studies are necessary to elucidate the real sensitivities of different echinoderm species to cadmium stress.

No data are available concerning the effects of natural variables or of other metals on the toxicity of cadmium to either annelids or echinoderms.

3.3.   The Sublethal Effects of Cadmium

As stated in Section 3.1, data concerning the sublethal effects of cadmium are much less common than those dealing with the acute lethal levels of this metal. Organisms are supposedly protected from the sublethal effects of a toxicant by maintaining its concentration in water at levels less than 0.01 of the 96-hr LC_{50} for the species. This factor of 0.01 is generally termed an application factor. As mentioned previously, its use has been criticized by many authors, as the magnitude of the factor is essentially arbitrary. Eisler (1971) calculated an application factor of 0.0018 for cadmium for the mummichog *Fundulus heteroclitus*, based on a 96-hr LC_{50} of 55 mg Cd/l and a maximum no-effect level of about 0.1 mg Cd/l. Negilski (1976) could find no incipient lethal level in bioassays, using either mullet (*Aldrichetta forsteri*) or hardyhead (*Atherinasoma microstoma*), suggesting that cadmium toxicity is a long-term phenomenon because of the slow accumulation of this metal; acceptable levels based on short-term acute tests would therefore be overestimated. Clearly there is no real substitute for studies of toxicants at the sublethal level; only from the results of sublethal studies can realistic water quality standards be devised, thus affording real protection to aquatic life from the toxic effects of cadmium and other pollutants.

Finfish

Gardner and Yevich (1970) exposed the mummichog *Fundulus heteroclitus* to relatively high concentrations of cadmium (50 mg/l, just less than the 96-hour LC$_{50}$ reported by Eisler in 1971) for periods between 0.5 and 48 hr. Pathological changes in three tissues were observed, at 1 hr (intestinal tract), 12 hr (kidney), and 20 hr (gill filaments) after the start of exposure. Eosinophils in the blood also increased in abundance on exposure of the fish to cadmium, the increase beginning at 4-hr exposure and reaching a plateau at 45% above the normal occurrence of these cells. These effects are similar in some respects to effects recorded for cadmium poisoning in mammals or in freshwater fish (Schweiger, 1957). Eisler (1971) cited unpublished studies of several authors on the histopathology of cadmium in both *F. heteroclitus* and the tautog *Tautoga onitis;* effects were similar to those cited above, although at some exposure concentrations cadmium also affected the heart rate of the fish. Possible target organs for cadmium in fish thus appear to be the heart, gills, intestinal mucosa, blood, and kidney. The precise effect elicited by the metal appears strongly dependent on the exposure dose and duration; for example, short-term acute dosage leads to respiratory problems (heart and gill effects), whereas long-term low dosage causes accumulation of cadmium in the kidney with associated nephritic damage (see also Eisler and Gardner, 1973). The histopathological effects are influenced by salinity, pH, and temperature of the ambient seawater in much the same way as these parameters affect the acute toxicity of cadmium (compare Gardner and Yevich, 1969, with Eisler, 1971, and others).

At the subcellular level, cadmium has been shown to affect the activities of enzymes in liver, kidneys, or gills of teleosts. Jackim et al. (1970) observed inhibition of four liver enzymes (acid phosphatase, catalase, ribonuclease, and xanthine oxidase) in *F. heteroclitus* after exposure of the fish to 27 mg Cd/l for 96 hrs; a fifth enzyme (alkaline phosphatase) was not significantly affected by cadmium. *In vivo* effects were often quite different from those produced by *in vitro* addition of the metal to enzyme solutions. In a later work, Jackim (1973) found slight stimulation of 5-aminolevulinate dehydrase activity in livers of mummichog exposed to 10 mg Cd/l for 14 days; Ag and Zn also stimulated the activity of this enzyme but Cu, Pb, and Hg caused inhibition (see also Jackim, 1974). MacInnes et al (1977) exposed the cunner *Tautogolabrus adspersus* for 30 or 60 days to 0.05 or 0.10 mg cadmium chloride/l. One liver enzyme (aspartate aminotransferase) was inhibited, whereas another (glucose-6-phosphate dehydrogenase) was induced. In addition, the respiratory rates of isolated gill tissue were decreased markedly at both exposure concentrations at 30 and 60 days, compared to those of control fish. Other studies concerning the effects of cadmium on enzymes in liver or other teleost tissues have also been reported (e.g., Gould and Karolus, 1974; Gould, 1977), although the precise relation of these *in vivo* effects (generally observed after exposure of the test organisms to high concentrations of cadmium compared to those present in solution in their natural environment) to the sublethal debilitating effects of cadmium in the field is uncertain.

In addition to its effects on enzyme activities in target organs of teleosts or in organs that accumulate high concentrations of cadmium, recent studies have suggested a possible link between cadmium and disturbances of steroid metabolism in fish and seals (Sangalang and O'Halloran, 1973; Sangalang and Freeman, 1974; Freeman and Sangalang, 1976, 1977). Present evidence is limited to freshwater teleosts, although marine species may exhibit similar effects.

In regard to teleost larvae, the studies of Voyer et al. (1977) and Westernhagen and Dethlefsen (1975) were cited in Section 3.2 ("Finfish"). Both groups observed sublethal effects of cadmium exposure in addition to the lethal action of the metal. Voyer et al. (1977) reported that rates of development of eggs of the winter flounder *Pseudopleuronectes americanus* varied greatly in different salinity-temperature treatments in the absence of cadmium. Addition of cadmium to developing eggs delayed development at 10°C but may have stimulated it at 5°C. High salinities also increased the rate of egg development in this species. Westernhagen and Dethlefsen (1975) could find no significant effects of cadmium on the growth of larvae of the Baltic flounder *Pleuronectes flesus*, although at salinities greater than 16‰ an increased incidence of bent and crippled larvae was apparent in cadmium-exposed specimens.

Middaugh et al. (1975) found an incipient lethal level of 0.2 to 0.3 mg Cd/l for larval spot (*Leiostomus xanthurus*). Larvae exposed to 0.5 to 0.8 mg Cd/l for 96 hr at 20°C exhibited a decreased postexposure tolerance to either thermal stress or low dissolved oxygen (DO) levels (1.6 mg DO/l), compared to controls that were not exposed to cadmium.

Molluscs

Very little information is available concerning the sublethal effects of cadmium on mollusc species. MacInnes and Thurberg (1973) reported studies in which the mud snail *Nassarius obsoletus* was exposed for 72 hr to cadmium or to a cadmium-copper mixture, both metals added as chlorides. Oxygen consumption rates were depressed by copper alone at all concentrations tested (0.1 to 2.0 mg/l); at concentrations greater than 2.0 mg/l the snails retracted into their shells. By contrast, cadmium caused slight increases in the oxygen consumption rates of the snails over the exposure range 0.5 to 4.0 mg/l; at 5.0 mg/l the snails retracted into their shells, and concentrations of 10 mg/l or greater were lethal. However, when present in combination, copper and cadmium decreased the oxygen consumption of mud snails below that observed in the presence of copper alone (Table 15). This is a most interesting example of metal interaction, the

Table 15.   Oxygen Consumption Rates (μl O/hr·g wet weight, means $\pm$ standard errors) of the Mud Snail *Nassarius obsoletus* Exposed to Cadmium and Copper Individually and in Combination[a]
The exposure period was 72 hr at 25‰ salinity and 20°C.

Metal Used	Exposure Concentration (mg/l)	Number of Tests	Oxygen Consumption
Control	—	11	30.5 ± 2.2
Copper	0.25	11	12.5 ± 2.6
Cadmium	1.00	11	40.4 ± 3.1
Copper + cadmium	0.25 + 1.00	11	6.4 ± 3.0

[a] After MacInnes and Thurberg (1973).

Table 16.   Concentrations of Cadmium, Glycolytic Rates, and Hemolymph Glucose Levels (means $\pm$ standard deviations) in Limpets (*Patella vulgata*) from Four Sites in the (Polluted) Bristol Channel and a (Unpolluted) Control Site in the English Channel[a]
Sites 1 to 4 run downstream in the Bristol Channel.

Location	Cadmium (μg/g dry weight)	Glycolytic Rate (μmol lactate/mg protein/hr)	Hemolymph Glucose (mg %)
1. Ladye Bay	537 ± 137	0.25 ± 0.04	8.6 ± 1.4
2. Weston-super-Mare	419 ± 25	0.27 ± 0.04	8.0 ± 1.2
3. Blue Anchor	257 ± 44	0.37 ± 0.055	6.2 ± 1.0
4. Combe Martin	116 ± 40	0.48 ± 0.04	4.3 ± 0.6
5. Sidmouth	27 ± 6	0.45 ± 0.095	5.1 ± 1.3

[a] After Shore et al. (1975).

more so in that it involves copper and cadmium; the acute toxicity interaction of Cu, Cd, and Zn reported by Eisler and Gardner (1973) was cited above. Hopefully, future studies will indicate the precise prevalence of metal-metal interactions in marine biota, and possibly the interaction of metals at the toxicity level will be correlated to their interaction at uptake.

Shore et al. (1975) studied limpets (*Patella vulgata*) from four sites in the Bristol Channel, United Kingdom, a region known to be severely polluted by cadmium from industrial sources (see Section 2.3: "Cadmium in Coastal or Estuarine Waters"). A control site in the English Channel was also studied. Rates of glycolysis were measured using homogenates of the digestive gland, and hemolymph glucose levels were ascertained, as well as the cadmium contents of the five samples. The results (Table 16) indicated that cadmium levels in

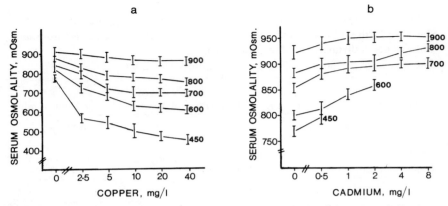

Figure 18. Effects of (*a*) copper and (*b*) cadmium on serum osmolality of the green crab *Carcinus maenas*. Means and standard errors are shown, six individuals being tested in each group. Note the difference in scales of both axes. (After Thurberg et al 1973.)

limpets increased with distance upstream in the Bristol Channel, agreeing with previous data of various authors (Sections 2.3, "Cadmium in Coastal and Estuarine Waters," and 4) on cadmium in water, air, and biota of this region, and confirming the results of Peden et al. (1973) for cadmium contents of the same species. Cadmium concentrations correlated well with both glycolytic rate and hemolymph glucose concentrations (Table 16). In addition, preliminary *in vitro* experiments were cited in which cadmium was observed to inhibit glycolysis in limpet homogenates. However, as stated by these authors, the existence of such correlations does not necessarily imply a cause and effect relationship. Numerous other parameters (e.g., salinity) covary with cadmium concentrations in the Bristol Channel, and the effects of such parameters on glycolysis in the limpet are unknown. Thus, although these results suggest a metabolic effect of cadmium on field populations of limpets, the conclusion is by no means certain until the effects of covarying parameters are fully ascertained.

Crustaceans

The literature concerning the sublethal effects of cadmium on crustaceans is somewhat more abundant than that for teleosts or molluscs; perhaps the extreme sensitivity of crustacean species to acute cadmium stress (see above) leads researchers to suspect a similar sensitivity in sublethal response.

Thurberg et al. (1973) studied the effects of cadmium and copper chlorides on oxygen consumption and osmoregulation of the estuarine crabs *Carcinus maenas* (green crab) and *Cancer irroratus* (rock crab). Exposure periods of 48 hr were used, in synthetic seawater (Zaroogian et al., 1969) at five salinities (17 to 32‰ S, quoted as 450 to 900 mosm in the original paper) and at 19 to 22°C.

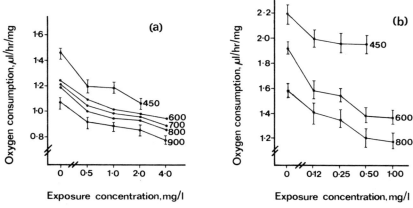

Figure 19. Effects of cadmium on oxygen consumption rates of isolated gill tissue from (*a*) the green crab *Carcinus maenas,* and (*b*) the rock crab *Cancer irroratus.* Means and standard errors are shown, six crabs per test group; standard errors for the three intermediate salinities in (*a*) are omitted for clarity. Note the difference in scale of both axes. (After Thurberg et al., 1973.)

Cancer irroratus was discovered to exhibit higher sensitivity to both metals than did *Carcinus maenas,* based on mortality data. Exposure concentrations were thus higher for the latter species.

Copper interfered with osmoregulation in both crab species, in each case causing a decrease in the osmolality of blood serum. The normally hyperosmotic serum thus became successively more isosmotic with increased exposure concentrations of copper. Figure 18*a* shows the data for *C. maenas.* By contrast, cadmium increased the osmolality of the blood serum of *C. maenas* (Figure 18*b*). Mortalities at the lower salinities forced early termination of experiments using higher concentrations of cadmium; this dependence of mortality rates on salinity of the ambient medium agrees with examples cited in Section 3.2 for crustaceans.

Oxygen consumption rates of isolated gill tissue from both species of crabs were affected by salinity, increasing with exposure to decreasing salinity. Exposure of either species to copper did not significantly affect its oxygen consumption. However, cadmium depressed gill oxygen consumption at all concentrations tested for each species (Figures 19*a* and *b*). It was suggested that these effects of cadmium on osmoregulation and oxygen consumption were due to deterioration of gill tissue caused by the metal, although no histological studies were presented. Similar effects of cadmium on oxygen consumption rates in crustaceans have been reported for gills of the adult mud crab *Eurypanopeus depressus* (Collier et al., 1973) and the grass shrimp *Palaemonetes pugio* (Vernberg et al., 1977), cadmium depressing respiration in both cases. By

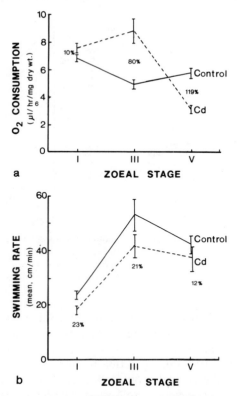

Figure 20.   Effects of exposure of larval fiddler crabs (*Uca pugilator*) to 1.0 μ cadmium chloride/l during development on (*a*) oxygen consumption and (*b*) swimming activity of zoeae of different stages. Percentage differences in cadmium-exposed response compared to control response are shown on each figure. (After Vernberg et al., 1974.)

contrast, Vernberg et al. (1974) observed that the effects of cadmium on oxygen consumption rates of larval fiddler crabs (*Uca pugilator*) varied with the zoeal stage. Thus exposure to 1 μg Cd/l during development caused stimulation of oxygen consumption rates of stage I and stage III larvae, whereas inhibition was recorded for stage V larvae (Figure 20*a*). Swimming activity was decreased, however, by exposure to cadmium at all stages of the larvae (Figure 20*b*).

Disruption of osmoregulation by cadmium was also observed by Jones (1975) in his work with six species of marine and estuarine isopods. Jones also noted a greater osmoregulation disturbance at lower salinities, agreeing with the data above for *Carcinus maenas* and with the data in Section 3.2 on the acute toxicity of cadmium.

The histological effects of cadmium on crustacean gills were investigated by

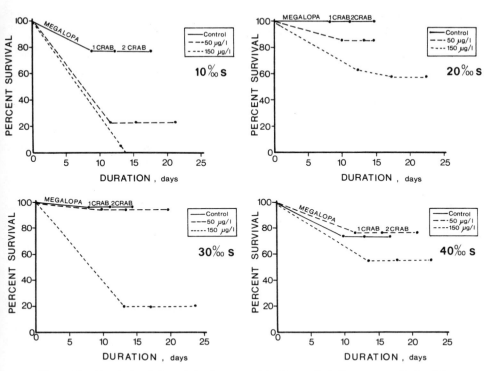

Figure 21. Survival and time of development of the blue crab *Callinectes sapidus* in different concentrations of cadmium at 25°C and at four salinities. Development from megalopa to third crab is shown in each case. (After Rosenberg and Costlow 1976.)

Nimmo et al. (1977) using pink shrimp (*Penaeus duorarum*), exposed to the metal in acute or subacute bioassays. Blackening of the branchial lamellae was frequently observed in cadmium-exposed shrimp, but never seen in control animals. Cross sections of the gills examined histologically revealed gill congestion, necrosis, and sloughing of individual lamellae, particularly near the distal ends of gill processes. Accumulation of hemocytes was also observed, resulting in distension and enlargement of gill processes. In addition, necrosis of the tissue led to infection of the gills and adjacent tissues by *Fusarium* molds in some cases. Similar blackening or melanization was occasionally observed in cuticular lesions of cadmium-exposed shrimp.

   Amongst other sublethal effects of cadmium on crustaceans, the effects of the metal on molting are possibly the most important. The data of Vernberg et al. (1977) and others were discussed in Section 3.2 ("Crustaceans") (also see Figure 13) as they related to cadmium excretion. Weis (1976) considered the effects of cadmium on regeneration of autotomized limbs in the fiddler crab *Uca*

Table 17.   Effects of Cadmium on Survival and Reproduction of the Polychaetes *Ctenodrilus serratus* and *Ophryotrocha diadema*.[a]
Forty individuals per concentration tested were added at time zero.

Added Cadmium Concentration (mg/l)	*Ctenodrilus serratus* Number at 96 hr	Number at 21 days	*Ophryotrocha diadema* Number at 96 hr	Number at 21 days
0 (control)	40	186	30	212
0.1	39	239[b]	40	267
0.5	40	197	39	218
1.0	40	183	40	95[c]
2.5	39	103[c]	37	34[c]
5.0	13	12[c]	11	9[c]
10.0	0	0[c]	0	0[c]

[a] After Reisch and Carr (1978).
[b] $P < .05$ for enhancement of reproduction.
[c] $P < .05$ for suppression of reproduction.

pugilator. Cadmium retarded regeneration rates at both 0.1 and 1.0 mg/l exposure concentrations; at 0.1 mg/l, cadmium was more effective than mercury. The relation of this effect to the environment is uncertain; probably any effect of this nature would be insignificant unless multiple autotomy as involved, as the loss of a single limb is not particularly debilitating in decapod crustacea. Finally, data concerning the effects of cadmium on intracellular enzymes of crustaceans are sparse, although Thurberg et al. (1977) observed inhibition of certain enzymes of the lobster *Homarus americanus* by cadmium, and Gould et al. (1976) reported similar data concerning the effect of cadmium on heart transaminase in the rock crab *Cancer irroratus*.

The studies of Rosenberg and Costlow (1976) on the effects of cadmium on larvae of the estuarine crabs *Callinectes sapidus* and *Rhithropanopeus harrisii* were mentioned in Section 3.2 ("Crustaceans"). Cadmium decreased survival and delayed development from megalopa to third crab in both species; data for *C. sapidus* are shown in Figure 21. Delay in the development of these species would presumably lead to greater predation at the vulnerable early life stages, selectively removing cadmium-exposed animals from the population. Elucidation of such effects in field conditions would be most difficult unless the populations were severely decimated.

Annelid Worms and Echinoderms

The only data known concerning the sublethal toxicity of cadmium to species belonging to either of these phyla are those of Reisch and Carr (1978) for the polychaetes *Ctenodrilus serratus* and *Ophyrotrocha diadema*. These species were exposed to cadmium chloride at six different concentrations, and the numbers of individuals alive at 96 hr and at 21 days were observed to determine both 96-hr LC_{50} values (see Table 14) and the effects of the metal on reproduction rates. The LC_{50} values were similar for both species, between 2.5 and 5.0 mg Cd/l. Reproduction was suppressed in *O. diadema* by concentrations of 1.0 mg Cd/l or greater. Reproduction in *C. serratus* was somewhat less sensitive, however, being affected at concentrations of 2.5 mg Cd/l or greater; additions of 0.1 mg Cd/l apparently enhanced reproduction in this species (Table 17). No explanation for such enhancement was suggested.

4. STUDIES ON THE UPTAKE OF CADMIUM BY MARINE AND COASTAL BIOTA

4.1. Finfish

Information from experimental studies on the uptake of cadmium by teleosts is restricted to a few reports dealing with cadmium uptake from solution. Eisler (1971) reported data for the whole-body content of cadmium in mummichogs (*Fundulus heteroclitus*) exposed to 11 concentrations of the metal (0 to 400 mg/l) for 264 hr at 20°C and 20‰ S. Both survivors and dead fish were analyzed. The results (Figure 22) showed, for both survivors and dead fish, increasing levels of cadmium in fish exposed to increasing concentrations of the metal in solution. Similar data showing a dependence of the cadmium concentration attained by surviving fish on both the concentration and the duration of exposure to the metal were also reported. On deliberate exposure of dead and live fish to cadmium in solution, Eisler (1971) also noted a much greater uptake of cadmium by dead fish (46 to 89 times that of live fish); this almost certainly represents nonspecific adsorption of the metal to the body surfaces of dead animals. The implication of this last observation is that reliance on the concentrations of metals recorded in dead animals as indices of exposure levels may lead to problems in the precise identification of the causative agent of fish kills in the field, and hence should be avoided.

In a later study using the same species, Eisler and Gardner (1973) investigated the interaction of Cd, Cu, and Zn. Toxicity results from this study were referred to in Section 3.2 ("Finfish"); sublethal concentrations of cadmium increased mortalities due to zinc or copper (or both) dramatically (see Figure 12). Interaction between the metals was also seen in uptake studies. Thus cadmium (at 1.0 or 10.0 mg/l) markedly inhibited the uptake of zinc from solution by mummichogs at high exposure concentrations of zinc (Figure 23a). Effects of cadmium on copper uptake were less dramatic and rather inconsistent (Figure 23b), but both zinc and copper altered whole-body cadmium uptake of the test

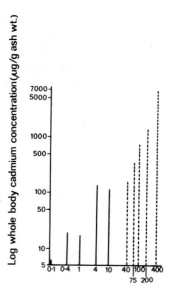

Figure 22.   Concentrations of cadmium (μg/g whole-body ash) in whole bodies of mummichogs (*Fundulus heteroclitus*) exposed to various concentrations of the metal in solution. Solid bars indicate survivors; broken bars indicate fish dying during exposure. Test conditions were 20°/oo salinity and 20°C; exposure time was 264 hrs. (After Eisler, 1971.)

fish (Figure 24). As started by the authors, the precise mechanism of this interaction between metals is unknown. Perhaps the metals interfere with each other at the site of uptake; however, speculation is unproductive, especially as no data for individual tissues were recorded. The effects of such metal-metal interactions on the uptake of each metal are most important, however, with respect to the use of organisms as indicators of the trace metal levels in their ambient environment (Phillips, 1977b).

Information concerning the uptake of cadmium into different tissues of the mummichog was published by Eisler (1974). These studies employed [115m]Cd, added with various concentrations of the stable cadmium isotope to the medium in which the test fish were maintained. Uptake curves for four tissues are shown in Figure 25. Gills and viscera accounted for most of the activity present, the high activities in the gills early in the experiment decreasing with time, concomitantly with increases in the activity present in the viscera. Shorter term experiments of 96-hr duration established the gastrointestinal tract, liver, and kidney as the tissues responsible for the high cadmium activity of viscera. These data may be interpreted as showing an initial accumulation of cadmium via the gills, followed by translocation of accumulated metal to sequestration sites in

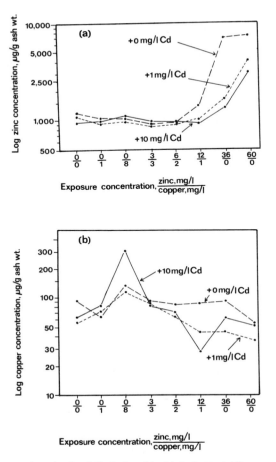

Figure 23. Concentrations (μg/g whole-body ash) of (*a*) zinc and (*b*) copper in mummichogs (*Fundulus heteroclitus*) exposed to various mixtures of Cd, Cu, and Zn for 96 hrs. (After Eisler and Gardner, 1973.)

the liver and kidney. The high cadmium content of the gastrointestinal tract is probably related to uptake of the metal via drinking. It should be remembered that tissue distributions for metals taken up from solution in laboratory experiments may not match those in organisms from the field, which probably absorb most of their total body load of metals from ingested food. In addition, the exposure concentrations used for metals in solution in laboratory uptake studies are generally far greater than those present (even in polluted areas) in the environment, although one study with exposure concentrations of 10 μg Cd/l has been reported (Eisler et al., 1972). Eisler (1974) also reported a loss of about

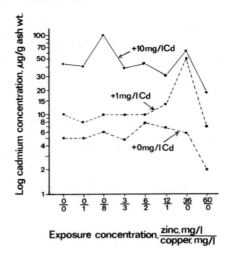

Figure 24. Concentrations of cadmium ($\mu g/g$ whole-body ash) in whole bodies of mummichogs (*Fundulus heteroclitus*) exposed to various mixtures of Cd, Cu and Zn for 96 hrs. (After Eisler and Gardner, 1973.)

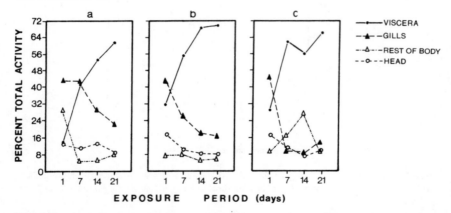

Figure 25. Percentage of total radioactivity in whole mummichog (*Fundulus heteroclitus*) found in selected tissues after exposure of the fish for 1, 7, 14, or 21 days to initial concentrations of 16,000 d/m·ml $^{115m}Cd(NO_3)_2$ with stable cadmium of concentration 0.3 $\mu g/l$ (*a*), 1.0 mg/l (*b*), or 10.0 mg/l (*c*). Each point represents the mean of three observations. (After Eisler, 1974.)

90% of whole-body radioactivity from *Fundulus heteroclitus* during 180 days in clean seawater following radiocadmium uptake. However, values obtained in laboratory experiments for half-lives of metals in organisms dosed previously via solution have been much criticized recently. Studies of metal excretion should be performed by the transplantation of organisms from one area to another in the field to obtain reasonably accurate estimates of metal retention by biota.

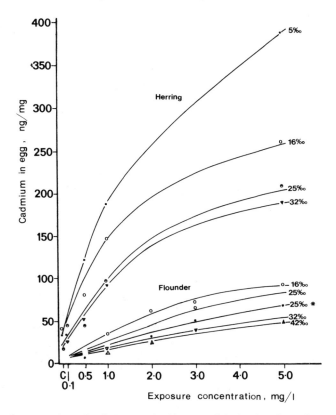

Figure 26.   Concentrations of cadmium attained by eggs of the herring *clupea harengus* and the Baltic flounder *Pleuronectes flesus* exposed to different concentrations of cadmium in various salinities. *Fertilization occurred in cadmium-contaminated water. (After Westernhagen and Dethlefsen, 1975.)

Apart from these studies of Eisler and co-workers, very little has been published concerning the uptake of cadmium by marine or estuarine teleosts. Middaugh et al. (1975) quoted data for cadmium concentrations present in larval spot (*Leiostomus xanthurus*) surviving a lethal bioassay. The use of survivors only to monitor cadmium levels in populations employed in bioassays is subject to some criticism, as the levels produced reflect those of a biased sample, that is, selecting survivors for analysis entails a sampling bias. However, in general it was evident that whole-body concentrations of cadmium in larval spot were proportional to the exposure concentration of the metal.

The relation of cadmium uptake by teleosts to ambient salinity has been little studied, despite the known effects of salinity on cadmium toxicity to this group. However, Westernhagen et al. (1974) and Westernhagen and Dethlefsen (1975) have reported data concerning the uptake of cadmium by eggs of the herring *Clupea harengus* and the Baltic flounder *Pleuronectes flesus*. These data are summarized in Figure 26; it can be seen that decreased salinities caused a generally greater net uptake of cadmium by the eggs of either species. The greater accumulation of cadmium at all test salinities by herring eggs than by flounder eggs correlated well with the species differences in susceptibility to cadmium, herring eggs exhibiting a much greater susceptibility to the metal than did flounder eggs. However, as mentioned in Section 3.2 ("Finfish"), the toxicity of cadmium to flounder eggs does not vary consistently with salinity, although herring eggs exhibit a clear increase in susceptibility to cadmium at lower salinities. The cause of these differences is uncertain, but they may be related to differences in the optimum salinities for development of the eggs of each species.

4.2. Molluscs

Eisler et al. (1972) exposed the oyster *Crassostrea virginica* and the scallop *Aquipecten irradians* for 21 days to flowing seawater of 30‰ salinity containing 10 µg added cadmium chloride/l. At the end of this exposure period, the whole soft parts of the cadmium-exposed individuals contained concentrations of the metal 4.52 (oyster) or 2.14 (scallop) times greater than those in control animals (wet weight basis). Similar data were reported by Pringle et al. (1968) for the soft shell clam *Mya arenaria* exposed to 50 to 100 µg Cd/l in solution. These simple laboratory tests amply demonstrate the ability of molluscs to accumulate cadmium rapidly from solution. Accumulation of the metal in the field has also been demonstrated, in animals transplanted from clean to cadmium-polluted environments for example, by Stenner and Nickless (1974a) for dog whelks (*Nucella lapillus*) and limpets (*Patella* sp.), and by Boyden (1975) for the oysters *Crassostrea gigas* and *Ostrea edulis*.

Each of the studies cited above reported data only for whole soft parts of the organisms concerned; hence no conclusions could be reached concerning the site of sequestration of the accumulated metal. However, Brooks and Rumsby (1967) analyzed six tissues of the oyster *Ostrea sinuata* after exposing this species to 50 mg Cd/l in solution for 100 hr. Concentrations of cadmium in tissues of the exposed oysters decreased in the order gills > heart > visceral mass > mantle > white muscle > striated muscle. Changes in the concentrations of elements other than cadmium were also recorded; however, these were probably an artifact of the small sample number and may in any case have been due to the excretion of gut contents during the exposure period. Further experiments with the radionuclide $^{115/115m}$Cd at much lower exposure concentrations confirmed the high concentrations of cadmium associated with the gills, visceral mass, and heart

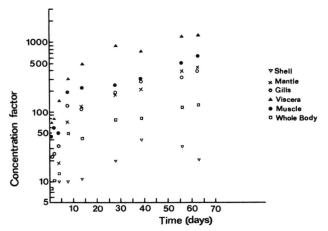

Figure 27. Uptake curves for [109]Cd in mussels (*Mytilus galloprovincialis*), describing uptake into whole body (including shell) or into selected tissues. Each point represents a mean for 2 to 4 individuals. (After Fowler and Benayoun, 1974.)

of this species, and suggested a decrease in concentration factor (ratio of cadmium concentration in the organism to that in the water) with increased exposure concentrations. Crude biochemical fractionation techniques were also used to study the binding of cadmium in heart tissue of the oyster; an appreciable percentage of the metal appeared to be protein-bound.

The distribution of cadmium absorbed from solution between the separate tissues of molluscs has also been reported by Fowler and Benayoun (1974) for the mussel *Mytilus galloprovincialis*, and by George and Coombs (1977a) for *Mytilus edulis*. Fowler and Benayoun (1974) exposed *M. galloprovincialis* to a mixture of [109]Cd and stable cadmium in solution. Equilibration between the animals and seawater was not attained over the 63-day exposure period (Figure 27). The radiotracer was found mainly in the viscera, muscle, mantle, and gills of the mussels, with much less occurring in the shell. This pattern was consistent even after 2 months' cleansing of the mussels in clean seawater containing no added cadmium (Table 18). Comparison of the data for the radiotracer with those reported for the stable isotope shows that the [109]Cd was not labeling the entire body pool of cadmium in the organism, that is, cadmium exchange rates were different for each tissue. The greatest anomaly was noted for the shell, which contained the highest stable cadmium concentration but exhibited the lowest concentration factor for [109]Cd. The shell of this mussel evidently contains much of its cadmium in a nonexchangeable matrix. Differences in the tissue distributions of [109]Cd and stable cadmium may also have been due to differences in the absorption route, as the radiotracer was accumulated from solution in the laboratory, whereas the stable isotope was taken up in the field (probably mainly

Table 18.   Mean Contents of ^{109}Cd in Separate Tissues of the Mussel *Mytilus galloprovincialis*, Expressed as a Percentage of the Total Body Burden after Varying Periods of Uptake and Loss of the Radiotracer, and Concentration of Stable Cadmium Found in Each Tissue

Tissue	Percent Total Body Wet Weight	Percent Total Body ^{109}Cd (after 63-day uptake)	Percent Total Body ^{109}Cd (after 2-month cleansing)	Concentration of Stable Cadmium ($\mu g/g$)
Shell	30.8	5.5	4.5	5.41
Mantle	6.5	14.8	18.1	1.26
Gills	4.3	14.5	18.3	1.21
Viscera	4.1	44.1	34.2	1.44
Muscle	1.5	10.2	11.6	2.08
Pallial fluid [b]	52.9	10.9	13.3	—

[a] After Fowler and Benayoun (1974).

[b] Determined by difference between whole animal and rest of tissues.

via food). George and Coombs (1977a) reported decreased concentrations of cadmium in tissues of *M. edulis* exposed for 20 days to 200 μg cadmium chloride in solution in the order kidney > viscera > gills > mantle > muscle. When allowance is made for the differences between the tissues studied by each group, these data agree well with those in Table 18 for *M. galloprovincialis*. Differences in the tissue distributions observed in these studies and in the work of Brooks and Rumsby (1967) probably depend on the dosage rate; higher levels of cadmium in gills would be expected to result from the more acute dosage regime of the latter authors.

Data concerning the effects of temperature or salinity on the net uptake of cadmium by molluscs have been published by several authors. Fowler and Benayoun (1974) could find no effect of temperature on ^{109}Cd uptake by whole mussels (*M. galloprovincialis*) exposed at high salinity for 26 days to this isotope. By contrast, ^{109}Cd uptake by the shrimp *Lysmata seticaudata* increased with higher temperatures (see Section 4.3). Phillips (1976a) kept mussels (*Mytilus edulis*) in waters of four different salinity-temperature combinations. The uptake of cadmium from solution into whole soft parts of the mussels (exposure concentration was 40 $\mu g/l$) was not affected by temperature at high salinity (35‰) but was increased by higher temperatures at low salinity (15‰); these results are shown in Table 19. Jackim et al. (1977) reported data for cadmium uptake by three species of bivalves exposed to 5 or 20 μg ^{109}Cd/l at 10 and 20°C and 30‰ salinity. Although *Mya arenaria* and *Mulinia lateralis* both exhibited significantly greater net uptake of ^{109}Cd at the higher temperature, no such effect

Table 19.   **Concentrations (μg/g wet weight, means $\pm$ standard deviations) of Stable Cadmium in Whole Soft Parts of the Mussel *Mytilus edulis*, exposed to 40 μg/l Cadmium Chloride in Solution at One of Four Salinity-Temperature Combinations**[a]

Each value represents 10 individuals; all mussels were acclimated for 21 days to their exposure regime before addition of the metal.

Sample	18°C, 35‰ S	18°C, 15‰ S	10°C, 35‰ S	10°C, 15‰ S
0-day (control)	0.47 ± 0.13	0.49 ± 0.07	0.51 ± 0.18	0.52 ± 0.07
6-day	1.01 ± 0.41	2.23 ± 0.40	0.99 ± 0.27	1.55 ± 0.34
14-day	1.22 ± 0.57	6.52 ± 2.04	1.42 ± 0.41	2.25 ± 0.66

[a] After Phillips (1976a).

Table 20.   **Concentrations (μg/g dry weight, means $\pm$ standard deviations) of Cadmium in Whole Soft Parts of Three Species of Bivalve Molluscs Exposed to One of Two Concentrations of ^{109}Cd in Solution at 10 or 20°C for 14 Days**[a]

Exposure salinity was 30‰; each value represents 10 individuals. Differences between values quoted for 10 and 20°C are statistically significant for all species except *Mytilus edulis*.

Species	5 μg Cd/l		20 μg Cd/l	
	10°C	20°C	10°C	20°C
Mya arenaria	2.2 ± 0.31	4.2 ± 0.25	16.8 ± 1.23	29.0 ± 0.94
Mytilus edulis	9.3 ± 2.48	9.4 ± 0.91	50.5 ± 8.57	60.3 ± 11.80
Mulinia lateralis	3.6 ± 0.21	8.9 ± 1.03	10.3 ± 0.27	20.46[b]

[a] After Jackim et al. (1977).
[b] No standard deviation quoted.

of temperature was noted for *Mytilus edulis* (Table 20). Present data on the effects of temperature on cadmium uptake by molluscs thus indicate that my-tilids may differ from other species in that no temperature effect is evident at higher salinities; at low salinities, however, a temperature effect is found. The basis for this species difference is unknown.

The effects of salinity on the uptake of cadmium by molluscs were studied by Phillips (1976a) and Jackim et al. (1977). As noted above, Phillips (1976a) used four salinity-temperature combinations; results are shown in Table 19. *Mytilus edulis* took up more cadmium at lower salinities; comparisons of the concentrations attained at the two temperatures were statistically significant at both 6 and 14 days. The salinity effect was more marked at higher tempera-tures, however. Jackim et al. (1977) published similar conclusions for three

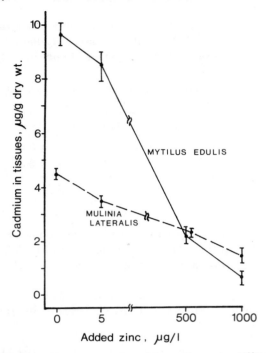

Figure 28.   The effect of ambient concentrations of zinc on the uptake of [109]Cd by whole soft parts of two bivalve species during an exposure period of 7 days. The exposure concentration of cadmium was 5 µg/l; test conditions were 28°/oo salinity and 10°C. Each value represents the mean and standard error for 10 individuals. (After Jackim et al, 1977.)

species of bivalve exposed to 10 or 20°C and 20 or 30‰ salinity for 21 days in the presence of 20 µg added cadmium chloride/l (Table 21).

The effects of other metals on the net uptake of cadmium by bivalve molluscs were also investigated in certain of the studies cited above. Fowler and Benayoun (1974) studied the effects of zinc on the uptake and elimination of [109]Cd by *Mytilus galloprovincialis.* Neither uptake nor elimination rates of [109]Cd were significantly affected by increasing the stable zinc concentration of the test medium from 9 to 30 or 100 µg/l. Phillips (1976a) exposed *M. edulis* to mixtures of four metals (Cd, Cu, Pb, and Zn) for 35 days. The net uptake of copper during this period was markedly affected by the concentrations of the other metals present. However, no effects of Cu, Pb, or Zn on the net uptake of cadmium were observed. By contrast to these negative reports, Jackim et al. (1977) found that increasing the ambient zinc concentration inhibited the uptake of [109]Cd from solution by both *M. edulis* and *Mulinia lateralis* (Figure 28). No reason for the disagreement between these authors has been suggested; this aspect requires further study.

Table 21.   Mean Concentrations (μg/g dry weight) of Cadmium in Whole Soft Parts of Three Bivalve Species Exposed to 20 μg/l Cadmium Chloride for 21 Days in One of Four Salinity-Temperature Regimes[a]
Each value represents a pooled sample of at least 10 individuals.

Species	10°C, 20‰ S	10°C, 30‰ S	20°C, 20‰ S	20°C, 30‰ S
Mytilus edulis	83.16	32.08	108.12	86.62
Mulinia lateralis	52.27	24.35	36.91	8.78
Nucula proxima	2.08	0.61	5.35	2.61

[a] After Jackim et al. (1977).

In summary, most studies performed to date on the uptake of cadmium by molluscs concern bivalve species. The tissue distribution of cadmium accumulated from solution is established for several species. In addition, metal uptake from solution is significantly affected, at least in some cases, by changes in the ambient salinity or water temperature, or by variations in the concentrations of other metals present.

4.3.   Crustaceans

Eisler et al. (1972) found increases in the concentration of cadmium in subadult lobster (*Homarus americanus*) exposed for 21 days to 10 μg cadmium chloride/l in solution. Whole lobsters exhibited a 41% increase in cadmium content over nonexposed individuals; changes in the tissues were 0% (viscera), 25% (muscle), 49% (exoskeleton), and 78% (gills). The tissue distribution of cadmium accumulated from solution or from food by the euphausiid *Meganyctiphanes norvegica* was reported by Benayoun et al. (1974). Concentrations were greatest in the visceral mass and were similar in eyes, exoskeleton, and muscle tissues. The percentage of the total body load of cadmium contributed by each tissue decreased in the order muscle > viscera, exoskeleton > eyes, hemolymph. These studies also indicated the greater importance of the food route for cadmium uptake than of the direct absorption route from solution; in addition, fecal pellets contained high cadmium concentrations and accounted for 84% of the total cadmium flux through this species.

O'Hara (1973a, 1973b) measured uptake of cadmium from solution by the fiddler crab *Uca pugilator*. Concentrations were greatest in the green gland, followed by gills, hepatopancreas, and muscle, and in general increased in proportion to the exposure concentration of cadmium used. The relative concentrations found in the gill and hepatopancreas were dependent on the salinity and temperature of the exposure medium; this aspect is discussed below. Hutcheson

Table 22.   Mean Concentrations of Cadmium (μg/g wet weight) in Tissues of the Blue Crab *Callinectes sapidus*, Exposed to 0, 0.1, 110, or 11,140 μg/l Cadmium Chloride in Solution for 8 Days at 20°C and 30‰ Salinity [a]

| Tissue | Exposure Solution Concentrations (μg/l) | | | |
	0 (Controls)	11,140	110	0.1
Claw	0.3	3.5	0.4	0.3
Carapace	2.4	14.4	2.8	2.8
Gills	0.4	56.6	1.0	0.3
Heart	0.2	3.2	0	0
Gastric mill	0.6	11.7	0.3	0.3
Hepatopancreas	3.1	41.1	1.4	1.5
Eyestalks	0.7	6.4	1.0	0.3
Green gland	BDL [b]	3.9	0	0
Supraesophageal ganglion	BDL	0	0	0

[a] After Hutcheson (1974).
[b] BDL = below detection limits (not quoted).

(1974) reported data for the cadmium contents of nine tissues in the blue crab *Callinectes sapidus* exposed to added cadmium chloride for 8 days at 0.1, 110, and 11,140 μg/l. Results are shown in Table 22. The tissue distribution of cadmium varied with the exposure concentration, although the carapace, gills, and hepatopancreas generally exhibited the greatest levels of the metal. The increased amounts of cadmium found in the gills at the highest exposure concentration are most noticeable; this effect of high exposure concentrations on gill metal levels was noted in Section 4.1. Wright (1977a) observed tissue distributions similar to these after exposure of the shore crab *Carcinus maenas* to cadmium in solution, concentrations decreasing in the order hemolymph > carapace > hepatopancreas > gill > muscle. Exposure concentrations for *C. maenas* were 1.15 or 2.3 mg/l, intermediate between the two highest concentrations used by Hutcheson (1974).

In the shrimp *Lysmata seticaudata,* Fowler and Benayoun (1974) found [109]Cd to be concentrated to the greatest extent by the viscera, followed by exoskeleton, muscle, and eyes in decreasing order. As noted in Section 4.2 for mussels, there was some evidence of differential labeling of the tissues when the [109]Cd concentrations were compared to those of the stable isotope in shrimp from the field. Nimmo et al. (1977) reported a similar tissue distribution for cadmium in pink shrimp (*Penaeus duorarum*), concentrations decreasing in the order hepatopancreas > exoskeleton > muscle > serum. These authors also noted mobilization of cadmium into muscle tissues on transferring pink shrimp to seawater with no added metal. In addition, the tendency for gill levels of the metal to increase more rapidly than the levels of other tissues with increases in the cadmium

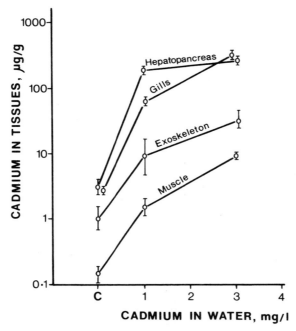

Figure 29. Concentrations of cadmium (μg/g wet weight) in tissues of the pink shrimp *Penaeus duorarum* exposed to various concentrations of cadmium chloride in solution for 96 hr at 20°/oo salinity and 25°C. (After Nimmo et al., 1977.)

exposure concentration was noted by Nimmo et al. (1977); these data are shown in Figure 29.

The effects of salinity and temperature on the uptake of cadmium from solution by crustaceans have been studied in several species. O'Hara (1973b) exposed fiddler crabs (*Uca pugilator*) to 10 mg cadmium chloride labeled with ^{109}Cd/l for 72 hr in six salinity-temperature regimes. Uptake of cadmium into gill and hepatopancreas was measured at the end of this exposure period. The results (Figure 30) indicated a general increase in cadmium uptake at low salinities and at the higher temperature. The temperature effect was greater at the lower salinity; this dependence of the magnitude of the temperature effect on salinity was also noted by Phillips (1976a) in studies of the mussel *Mytilus edulis* (see Section 4.2). In addition, the gill/hepatopancreas ratio of cadmium concentrations decreased wtih higher temperatures, suggesting not only greater net uptake of the metal at higher temperature but also more efficient translocation of cadmium from gills to hepatopancreas. It is interesting to compare these data for cadmium with those reported for mercury toxicity and uptake in the

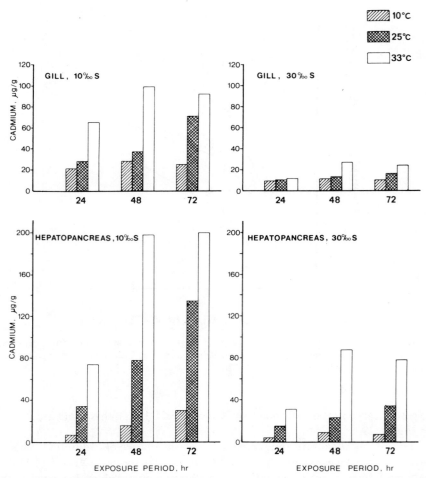

Figure 30. Concentrations of cadmium (means for four individuals per treatment) attained in gill and hepatopancreas of fiddler crabs (*Uca pugilator*) exposed to 10 mg cadmium chloride containing [109]Cd tracer/l for 72 hrs at six different salinity-temperature regimes. (After O'Hara, 1973b.)

same species. Vernberg and Vernberg (1972a, 1972b) reported a greater toxicity of mercury to *U. pugilator* at lower exposure temperatures. This unusual result was further studied by Vernberg and O'Hara (1972), who found that, although the total net uptake of mercury was similar at each temperature studied, translocation of the metal from gill to hepatopancreas was much slower at depressed temperatures. The ensuing faster buildup of mercury in the gills at lower temperatures was suggested as the cause of greater mortalities in these indi-

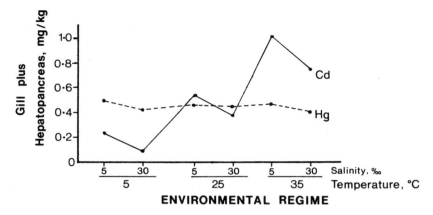

Figure 31.   Total concentrations of mercury and cadmium in gill plus hepatopancreas tissues of fiddler crabs (*Uca pugilator*) exposed for 72 hr to 0.18 mg Hg/l or 1.0 mg Cd/l in six salinity-temperature regimes. (After Vernberg et al., 1974.)

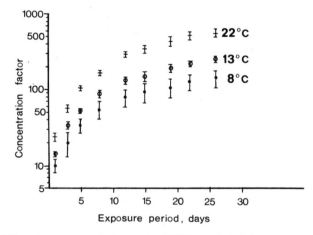

Figure 32.   Effect of temperature on the uptake of [109]Cd by whole shrimp (*Lysmata seticaudata*). Values shown represent means ± one standard deviation, six shrimp per sample. Experimental temperatures were 8°C (◆), 13°C (φ), and 22°C (+). (After Fowler and Benayoun, 1974.)

viduals. By contrast, the net uptake of cadmium is greater in *U. pugilator* at higher temperatures (see Figure 31). Although the translocation of cadmium from gill to hepatopancreas is temperature dependent, the higher rate of cadmium accumulation at higher temperature overrides the translocation effect to produce the observed direct dependence of toxicity on temperature (O'Hara, 1973a).

Hutcheson (1974) studied the effects of salinity and temperature on the uptake

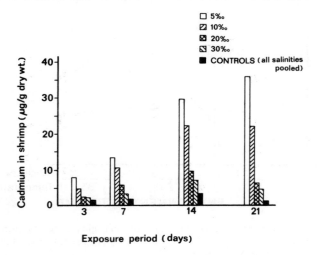

Figure 33.   Mean concentrations of cadmium attained by whole grass shrimp (*Palaemonetes pugio*) exposed to 50 μm cadmium chloride/l in solution at 25°C (static test). (After Vernberg et al., 1977.)

of stable cadmium chloride from solution by the blue crab *Callinectes sapidus*. The results of this study were very similar to those of O'Hara (1973a, 1973b) with *Uca pugilator*; the greatest uptake was noted at low salinities and high temperatures, and the temperature effect was greatest at the lower salinities. Mortalities were also observed in low-salinity/high-temperature treatments. Wright (1977a) reported that the uptake of cadmium from solution by the shore crab *Carcinus maenas* was greater at lower salinities. This profile was noted for whole body, hemolymph, gills, and carapace, but no salinity effect was observed for hepatopancreas or muscle cadmium levels. Uptake of cadmium by the gills, although initially dependent on salinity, apparently reached an equilibrium state of flux at about 48 days' exposure; the value was the same in animals from all salinities (quoted as 50 to 100% seawater, presumably 17.5 to 35‰ S) tested. This author also reported various data concerning ion interactions in *C. maenas;* these will be discussed in Section 6.

Data concerning temperature effects on cadmium uptake by shrimp were reported by Fowler and Benayoun (1974). *Lysmata seticaudata* exhibited a clear increase in uptake rate and final concentration factor for [109]Cd with temperature (Figure 32). The mean concentration factors for [109]Cd at day 26 were related to the exposure temperatures in an approximately linear fashion. The rate and incidence of molting increased with temperature also; activity loss with the molt depended on the duration of exposure of the molt. By contrast to the uptake data, the loss of radiocadmium by the same species after labeling for 41 days through

Table 23.   Mean Contents of [109]Cd in Separate Tissues of the Shrimp
Lysmata seticaudata, **Expressed as a Percentage of the Total Body Burden**[a]
Data for shrimp immediately after a 63-day accumulation period, and after
a 245-day cleansing period, are shown.

Tissue	Percent Total Body Wet Weight	Percent Total Body [109]Cd (after 63-day uptake)	Percent Total Body [109]Cd (after 245-day cleansing)
Exoskeleton	21.8	23.2	26.3
Viscera	6.6	21.7	21.9
Muscle	50.3	54.2	49.7
Eyes	1.4	0.3	0.6
Hemolymph[b]	19.9	0.6	1.5

[a] After Fowler and Benayoun (1974).
[b] Determined by difference between whole animal and rest of tissues.

food and water was independent of the temperature of the medium. Shrimp held
at 8 and 13°C were monitored for 245 days; at the end of this period, 55% of the
cadmium originally accumulated was still present. The loss curve for cadmium
was exponential after 30 days' cleansing and could be treated as loss from a single
compartment; a least squares analysis suggested a biological half-life of 307 to
378 days for this portion of the loss curve. Concentrations of [109]Cd in the tissues
after 250 days' cleansing were similar to those in the tissues of shrimp immediately after uptake (Table 23); thus the rates of radiocadmium loss from the
various tissues appeared approximately equal.

Vernberg et al. (1977) exposed the grass shrimp *Palaemonetes pugio* to 50
μg Cd/l at 5, 10, 20, and 30‰ salinity in static and flow-through tests. Profiles
for mortality were described in Section 3.2 ("Crustaceans," Figure 16), showing
increased percentage mortalities with decreased salinities. Profiles for cadmium
uptake (Figure 33) were very similar to those for mortalities, indicating a general
increase in the concentration of cadmium attained by whole shrimp with decreased salinities. The salinity effect appeared to be of greater magnitude at
higher temperatures; however, some of the results at 15°C showed a large scatter
between individuals, and these data should be treated with caution.

Published data concerning the effects of the interaction of pollutants on the
net uptake of cadmium by crustaceans are few. However, Nimmo and Bahner
(1976) suggested that methoxychlor (an organochlorine insecticide similar to
DDT) influences the accumulation or loss of cadmium from tissues of the pink
shrimp *Penaeus duorarum*. In one study of methoxychlor uptake, exposed
shrimp contained less cadmium in muscle than did controls. A second experiment
indicated depression of cadmium uptake into muscle in the presence of

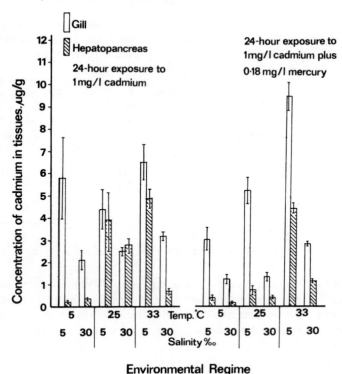

Figure 34. Concentrations of cadmium (mean ± one standard error, $\mu g/g$ wet weight) attained in gill and hepatopancreas of fiddler crabs (*Uca pugilator*) exposed to 1.0 mg cadmium chloride/l in solution with or without the addition of 0.18 mg mercuric chloride/l. Exposure time was 24 hr in all salinity-temperature regimes. (After Vernberg et al., 1974.)

methoxychlor, although PCB or copresence of methoxychlor + PCB appeared to have no effect on the cadmium levels attained by shrimp (Table 24). Vernberg et al. (1974) reported significant interaction between mercury and cadmium in the fiddler crab *Uca pugilator,* using gill and hepatopancreas for analyses of accumulated metals. The addition of 180 μg mercuric chloride/l significantly affected the concentrations of cadmium attained by each tissue after 24-hr exposure to 1.0 mg Cd/l (Figure 34). These effects of mercury on the cadmium level attained were complex, varying with salinity and temperature; they appeared to be based on changes both in total cadmium accumulated and in translocation of the metal between gill and hepatopancreas. In addition to this effect of mercury on cadmium uptake, cadmium was shown to influence the net uptake of mercury from solution (Figure 35). In this case the changes were somewhat more consistent, cadmium causing an increased concentration of mercury in gills and a generally decreased level in hepatopancreas. In addition,

Table 24.   Concentrations of Cadmium (μg/g wet weight $\pm$ 2 standard errors of the mean) Attained by Muscle of the Pink Shrimp *Penaeus duorarum*, Exposed to Mixtures of Cadmium Chloride, Methoxychlor, and the Polychlorinated Biphenyl Aroclor 1254[a]
All data refer to 10-day flow-through bioassays at 25°C and 20‰ salinity.

Measured Toxicant Concentration (μg/l)				Concentration
Cadmium	Aroclor 1254	Methoxychlor	Cadmium in Muscle	Factor
640	BG[b]	BG	15.58 ± 2.90	24
774	BG	1.1	9.90 ± 2.79	13
746	0.9	BG	13.99 ± 3.05	19
829	0.9	0.8	16.28 ± 5.44	20
BG	BG	BG	0.25 ± 0.10	—
BG	BG	1.0	0.25 ± 0.05	—
BG	0.7	BG	0.28 ± 0.10	—
BG	1.1	1.0	0.26 ± 0.07	—

[a] After Nimmo and Bahner (1976).
[b] BG = background concentrations only present (no addition).

the salinity effect noted for mercury uptake when this element was present alone was reversed by the addition of cadmium.

The effects of variables such as salinity and temperature on the net uptake of cadmium from solution by crustaceans can be seen from the data given above to parallel those exhibited by molluscs and teleosts in general. Thus high temperatures and low salinities generally lead to maximal uptake rates of cadmium in these organisms. The general applicability of this rule will be reexamined in Section 6, where the effects of other metals on the net uptake of cadmium will also be discussed.

4.4.   Other Organisms

Very few studies have been published on the uptake kinetics of cadmium in marine or estuarine organisms other than finfish, molluscs, or crustaceans. The only data of significance for other phyla concern the macroalga *Laminaria digitata* (Bryan, 1969) and the annelid *Nereis diversicolor* (Bryan and Hummerstone, 1973).

The net uptake of ^{65}Zn by *Laminaria digitata* was significantly affected by several parameters, including the copresence of other metals (Bryan, 1969). Thus the concentration of ^{65}Zn attained by the growing plant could be decreased by Cd, Cu, or Mn. The effects of cadmium and manganese were reversible; by contrast, those of copper appeared largely irreversible (Figure 36). No data were reported, however, concerning the uptake of cadmium itself by the alga.

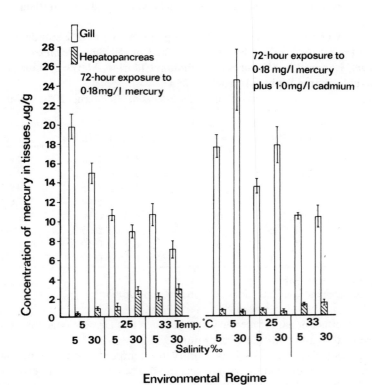

Figure 35. Concentrations of mercury (mean ± one standard error, μg/g wet weight) attained in gill and hepatopancreas of fiddler crabs (*Uca pugilator*) exposed to 0.18 mg mercuric chloride/l in solution with or without the addition of 1.0 mg cadmium chloride/l. Exposure time was 72 hr in all salinity-temperature regimes. (After Vernberg et al., 1974.)

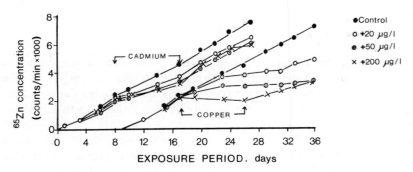

Figure 36. The effect of cadmium and copper additions on the net uptake of ^{65}Zn by the brown seaweed *Laminaria digitata* exposed to about 6 μg Zn/l in solution. Cadmium and copper were added as shown at concentrations of 20 (O), 50 (⊙), or 200 (x) μ/l. (After Bryan, 1969.)

502

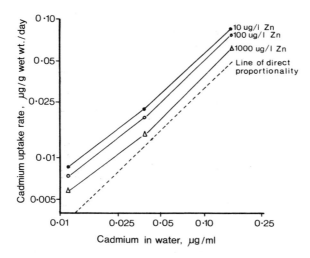

Figure 37. The effect of zinc on the relationship between cadmium absorption by the polychaete *Nereis diversicolor* and the exposure concentration of cadmium in solution. Each point represents a mean of four individuals; test salinity was 17.5°/oo. (After Bryan and Hummerstone, 1973.)

Bryan and Hummerstone (1973) studied the concentrations of zinc and cadmium present in the polychaete worm *Nereis diversicolor* from 26 locations of southwest England. The concentrations of zinc in the worm were strictly regulated; however, cadmium levels were not regulated and were approximately proportional to the concentrations of cadmium in the sediment in which the worms lived. The uptake of cadmium (as [115m]Cd) from solution by the polychaete was found to be significantly affected by the concentration of zinc present (Figure 37). Thus increasing the concentrations of zinc in the medium from 10 to 100 μg/l and from 10 to 1000 μg/l decreased the rate of cadmium absorption by average levels of 9% and 37%, respectively.

5.   CONCENTRATIONS OF CADMIUM FOUND IN MARINE ORGANISMS

5.1.   General

Most of the published data on cadmium in marine organisms concerns the concentrations of the metal found in organisms from various phyla taken directly from the environment. Such studies generally fall into one of two groups: studies examining metal concentrations to enforce public health regulations, and those using organisms to monitor the comparative pollution of different areas by trace

Table 25.   Concentration Factors for Cadmium (Cd in dry material/Cd in filtered seawater) in Various Marine Materials[a]

Material	Average Concentration Factor
Sediment	10^3
Plankton	10^4
Macroalgae	10^2-10^3 (10^4)[b]
Molluscs	10^3-10^4 (10^5)[b]
Crustacea	10^3
Finfish	10^2

[a] After Preston (1973b).
[b] Occasional values.

metals. Very many reports were published on this subject between 1970 and 1978. No single review could possibly attempt to cover all the published material comprehensively. The present chapter covers the reports considered to be of significant importance, either for their breadth of study or for their contribution to the development of accurate indicator techniques. The background to the use of biological indicator organisms, and critical reviews of the data published to 1977 concerning trace metals or organochlorines, have been published elsewhere (Phillips, 1977b, 1978b) and will not be repeated here.

Some authors have published estimates of average concentrations or concentration factors for cadmium in various marine phyla. Data from Preston (1973b) are shown in Table 25. Finfish are seen to exhibit the lowest concentration factor, and molluscs the highest. The concentration factor quoted for phytoplankton is based on very few data; however, it is evident that no food chain amplification occurs for cadmium. In fact, very little evidence exists for significant amplification of any trace metal up trophic levels with the possible exception of mercury; indeed, the evidence for food chain amplification of organochlorines is also a matter of controversy at present (Phillips, 1978b). Comparison of these average concentration factors with the relative sensitivities of phyla to the toxic effects of cadmium indicates to some extent the abilities of members of each phyla to sequester cadmium in the tissues without exhibiting overt toxic symptoms. Thus molluscs, which exhibit a greater concentration factor for cadmium than do many crustaceans, are nevertheless considerably less sensitive than crustaceans to the toxic effects of the metal (Section 3). This suggests that, in general, molluscs are able to tolerate higher concentrations of cadmium before the toxic effects of the body load of metal become apparent; possibly a different method of cadmium sequestration is involved in each phylum.

Table 26.   Geometric Mean Concentrations (μg/g dry weight in whole organisms, excluding shells of molluscs) of Cadmium in Different Groups of Marine Organisms Taken from Relatively Nonpolluted Areas[a]

Organism Group	Concentration (μg/g dry weight)
Phytoplankton	2
Zooplankton (copepods)	4
Macroalgae	0.5
Tunicates (mainly ascidians)	1
Coelenterates	1
Echinoderms	2
Decapod crustaceans	1
Bivalve molluscs[b]	2
Gastropod molluscs	6
Cephalopod molluscs	5
Finfish	0.2

[a] After Bryan (1976).
[b] Excludes high cadmium in oysters and scallops (see Table 31).

Estimates of the geometric mean concentrations of cadmium found in different groups of marine organisms were also published by Bryan (1976). These data do not include reports from areas of known contamination, but are mostly based on studies in coastal areas of the northern hemisphere and show a strong bias toward temperate rather than tropical species (Table 26). The agreement between the data in Tables 25 and 26 is good. However, it should be noted that the levels quoted here are general averages for different groups; certain species may exhibit concentrations of cadmium well outside the normal range noted for the phylum in general.

5.2.   Finfish

Finfish accumulate trace metals from solution and from food. The uptake of cadmium from solution may be considered as having two routes, the first being direct uptake across the body surface (especially the gills, as these exhibit a very large surface area and are in intimate contact with the respiratory flow) and the second being uptake across the gastrointestinal wall after drinking. The amounts of cadmium accumulated via each route are uncertain, and in any event depend largely on species and ambient conditions (Phillips, 1977b). Most authors believe

Table 27. Selected Data for the Concentrations of Cadmium (means or ranges, μg/g dry or wet weight) In Whole Individuals of Various Finfish Species

Species	Weight Basis	Cadmium Concentration		Study Area	Remarks	References
		Mean	Range			
Diaphus dumerili	Dry	0.73	—	Northwest Africa	Both species pelagic	Leatherland et al. (1973)
Hygophum macrochir	Dry	0.98	—	Northwest Africa	and migratory	Leatherland et al. (1973)
Platichthyes flesus	Dry	4.6	3.4–7.3	Severn Estuary, U.K.	Age and seasonal effects	Hardisty et al. (1974a)
Platichthyes flesus	Dry	4.7	4.0–5.2	Severn Estuary, U.K.	Age effects	Hardisty et al. (1974b)
Platichthyes flesus	Dry	1.4	1.1–1.7	North Devon, U.K.	Age effects	Hardisty et al. (1974b)
Ciliata mustela	Dry	8.1	—	Severn Estuary, U.K.	Concentrations correlated to diet	Hardisty et al. (1974b)
Gobius minutus	Dry	3.2	—	Severn Estuary, U.K.	Concentrations correlated to diet	Hardisty et al. (1974b)
Liza ramada	Dry	3.0	—	Severn Estuary, U.K.	Concentrations correlated to diet	Hardisty et al. (1974b)
Merlangus merlangus	Dry	6.2	—	Severn Estuary, U.K.	Concentrations correlated to diet	Hardisty et al. (1974b)
Trisopterus minutus	Dry	8.5	—	Severn Estuary, U.K.	Concentrations correlated to diet	Hardisty et al. (1974b)

Species						Reference
Brevoortia patronus	Wet	0.18	<0.05–0.43	Gulf of Mexico	Twice the concentration in muscle	National Oceanic and Atmospheric Administration (1975)
Brevoortia smithi	Wet	0.19	0.04–0.71	Gulf of Mexico		National Oceanic and Atmospheric Administration (1975)
Brevoortia tyrannus	Wet	0.19	0.06–0.43	North and South Atlantic	Four sites studied	National Oceanic and Atmospheric Administration (1975)
Engraulis mordax	Wet	0.26	0.02–0.57	California coast	Two sites studied	National Oceanic and Atmospheric Administration (1975)
Hemiramphus brasiliensis	Wet	0.24	0.06–0.43	North Atlantic		National Oceanic and Atmospheric Administration (1975)
Morone americana	Wet	0.12	0.09–0.17	North Atlantic	Headed and gutted	National Oceanic and Atmospheric Administration (1975)
Thaleichthys pacificus	Wet	0.10	0.01–0.21	Northwest Pacific		National Oceanic and Atmospheric Administration (1975)

Table 27. *Continued*

Species	Weight Basis	Cadmium Concentration		Study Area	Remarks	References
		Mean	Range			
Boreogadus saida	Dry	0.62	0.26–1.5	Baffin Island	No cadmium-size correlation	Bohn and McElroy (1976)
Hygophum hygomi	Wet	—	<0.07–0.11	Middle Atlantic Bight	Little difference between pelagic and demersal species	Greig et al. (1976)
Stephanolepsis hispidus	Wet	—	<0.13–0.14	Middle Atlantic Bight		Greig et al. (1976)
Synaphobranchus kaupi	Wet	0.12	—	Middle Atlantic Bight		Greig et al. (1976)
Merlangus merlangus	Dry	2.17	1.94–2.50	Severn Estuary, U.K.	Size and seasonal effects	Badsha and Sainsbury (1977)
Merluccius productus	Wet	0.12	0.07–0.15	West coast of U.S.A.	Size effect, correlated to diet	Cutshall et al. (1977)
Pomatoschistus minutus	Wet	0.24	0.20–0.28	Lower Medway Estuary, U.K.		Wharfe and Van den Broek (1977)
Sprattus sprattus	Wet	0.29	—	Lower Medway Estuary, U.K.		Wharfe and Van den Broek (1977)

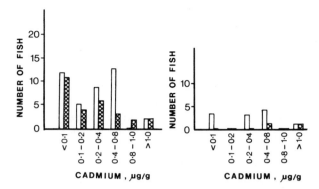

Figure 38. Concentrations of cadmium ($\mu g/g$ dry weight) in axial muscle of species of inshore (cross-hatched bars) and offshore (plain bars) finfish. The histogram at left refers to *Osteichthys*; that at right, to *Chondrichthys*. All samples were from the North Atlantic. (After Windom et al., 1973.)

that uptake of metals from food predominates; the significant effects of diet on the trace metal concentrations found in some finfish species (see below) support this hypothesis. However, the unusually great tendency of cadmium to remain in solution rather than adsorbing to inorganic particulates in estuarine areas (Section 2) would lead to a high availability of this metal to estuarine teleosts for direct uptake.

Phillips (1977b) has reviewed the literature concerning the ability of finfish to monitor ambient concentrations of metals, that is, to act as biological indicators of trace metals. It was concluded that teleosts are poor indicators of metals other than mercury. The levels of metals in muscle tissue, in particular, appear to depend far more on various ancillary perturbing factors (see below) than on the absolute ambient concentrations of the metals. Although organs other than muscle in teleosts may possess limited value as indicators, no study has yet been reported in which an indisputable correlation between levels of metals in the fish and those in the environment was demonstrated. Thus comparisons of finfish from inshore and offshore regions generally reveal few differences in cadmium concentrations present with distance from land (e.g., Ministry of Agriculture, Fisheries and Food, 1973; Preston, 1973a, 1973b; Windom et al., 1973); an example is shown in Figure 38. Similarly, comparisons of the levels of cadmium in finfish from polluted and nonpolluted areas have also failed to reveal significant differences between locations (Halcrow et al., 1973; Topping, 1973a; Eustace, 1974; Stenner and Nickless, 1974a). By contrast, some of these authors reported significant differences between the cadmium contents of molluscs and/or crustaceans taken from polluted and nonpolluted areas. It therefore appears that finfish regulate their concentrations of trace metals, at least in muscle tissue. In addition to regulating the muscle content of metals, most pelagic fish move considerable distances; the concentrations of trace metals found in organs other than muscle (even if these respond directly to ambient concentrations of available metal) thus represent at best a composite or average of available levels in both space and time. Unless the precise movements of each species or individual are known, such data are valueless as indices of differences between the extents of cadmium pollution in different areas. This fact has long been recognized by authors using finfish to monitor mercury pollution; the most successful studies have employed demersal species, which are generally territorial and are thus truly representative of the area in which they are caught (e.g., see Johnels et al., 1967; Dix et al., 1976).

Despite these criticisms of the use of finfish as indicators of pollution by trace metals other than mercury, a considerable body of published data exists on the concentrations of cadmium in these organisms. Tables 27 and 28 present selected data from the literature for the concentrations of cadmium in whole fish or in axial muscle of finfish, respectively. Even with the interference of species differences, it is noticeable that axial muscle in general contains lower concentrations of the metal than are found in whole fish. This is confirmed by studies on single species where both whole tissues and muscle alone have been analyzed (National Oceanic and Atmospheric Administration 1975; Bohn and McElroy, 1976; Cutshall et al., 1977). This pattern reflects the higher concentrations of cadmium found in tissues other than muscle, especially in liver and kidney; in addition, the lower concentrations present in muscle agree with the hypothesis of partial regulation of cadmium levels there. In this connection it is interesting that Stevens and Brown (1974) reported differences in the zinc and copper contents of different muscle sections along the body of the blue shark *Prionace glauca*. Such a difference has not been reported for cadmium in any species to date, but the possibility of significant differences in the cadmium contents of different portions of the musculature of finfish should not be ignored. Table 29 shows selected data for the tissue distribution of cadmium in finfish. Although there are certain exceptions, it is evident that the concentrations of cadmium in axial muscle are often exceeded by those in other tissues. Liver, kidney, and spleen appear to concentrate the metal in particular, concentrations being lower in general in the other tissues such as gills or gonads. This pattern is completely disobeyed by the liver and muscle samples of the demersal fish studied by Halcrow et al. (1973), however; the reasons are unknown but may be based on differences in the dominant uptake routes for cadmium in fish living in different environments.

As stated above, the concentrations of cadmium in finfish vary little, it at all, with changes in the ambient levels of the metal in water, but may vary considerably with parameters such as season, fish size, or the copresence of other pollutants. Hardisty et al. (1974a, 1974b) studied the variation of cadmium

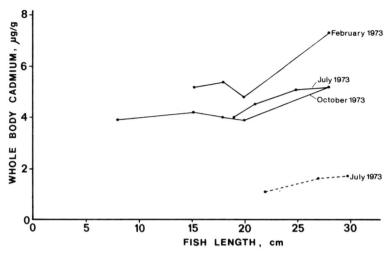

Figure 39.   Concentrations of cadmium (means, μg/g dry weight) in whole flounders (*Platichthyes flesus*) from Oldbury-on-Severn (continuous lines) or Barnstaple Bay (dashed lines) in relation to season and fish length. (Data for fish lengths were interpolated graphically from Hardisty et al., 1974b; data for October 1972 and February 1973 were derived from Hardisty et al., 1974a.)

concentrations in whole founder (*Platichthyes flesus*) from Oldbury-on-Severn in the Severn Estuary, United Kingdom. The results indicated some changes in cadmium levels with season, as well as a general correlation with age or size of the fish (Figure 39). In addition, the Oldbury samples exhibited higher levels of cadmium in all size groups than those present in the same species from a less polluted site. No data were reported for the individual tissues of the fish, although it is likely that changes in the cadmium levels of liver and kidney were mainly responsible for the differences between locations. These authors emphasized the uptake of cadmium in the diet as the most important factor determining the concentrations of the metal found in teleosts. Ingestion of cadmium-contaminated crustaceans in this polluted area would certainly lead to high body loads of the metal in predators; conceivably the increased concentrations found in older fish reflect selection of a higher proportion of crustaceans in the diet. Other mechanisms are possible however, and further research is necessary to distinguish between the effects of diet and those of other factors.

Data concerning the variation in cadmium concentration with age or size in finfish have been reported by other authors also. Eisler and La Roche (1972) cited unpublished results of Eisler indicating a tripling of cadmium concentration in tautogs (*Tautoga onitis*) with age. However, a later study by Mears and Eisler (1977) could find no correlation between body length and cadmium concentration in livers of either tautog (*T. onitis*) or blue fish (*Pomatomus saltatrix*).

Table 28. Selected Data for the Concentrations of Cadmium (means or ranges, $\mu g/g$ wet or dry weight) in Axial Muscle of Various Finfish Species

Species	Weight Basis	Cadmium Concentration ($\mu g/g$)		Study Area	Remarks	Reference
		Mean	Range			
8 species	Dry	—	0.6–1.0	Firth of Clyde, U.K.	All demersal species	Halcrow et al. (1973)
4 species	Dry	—	<0.005–0.13	Southern Norway	Species differences	Havre et al. (1973)
4 species	Wet	—	<0.03–0.12	Coasts of Scotland, U.K.	Herring exhibited highest levels	Topping (1973a)
35 species	Dry	—	<0.1–2.1	North Atlantic	*Chondrichthys* and *Osteichthys* similar	Windom et al. (1973)
8 species	Wet	—	0.001–0.024	North Island, New Zealand	Species differences	Brooks and Rumsey (1974)
31 species	Wet	—	<0.05–0.30	Derwent Estuary, Australia	No relation to cadmium pollution	Eustace (1974)
3 species	Dry	0.03	0.03	The Solent, U.K.	Similar levels in each species	Leatherland and Burton (1974)
5 species	Dry	—	<0.05	Southwest England, offshore	Relation to pollution?	Stevens and Brown (1974)

Species				Location	Relation to pollution?	Reference
3 species	Dry	—	0.4–0.9	Salcombe Estuary, U.K.		Stevens and Brown (1974)
Makaira indica	Wet	0.07	0.05–0.40	Queensland, Australia	Includes metal-size data	Mackay et al. (1975a)
10 species	Dry	—	0.08–3.20	Atlantic coast of Spain and Portugal	Some data below unquoted detection limits	Stenner and Nickless (1975)
Boreogadus saida	Dry	<0.5	—	Baffin Island	Compare with Table 27	Bohn and McElroy (1976)
4 species	Wet	—	<0.09–0.14	Middle Atlantic Bight	Mostly below detection limits	Greig et al. (1976)
9 species	Wet	—	0.01–0.10	Coasts of New South Wales, Australia	Very little species differences	Bebbington et al. (1977)
Merluccius productus	Wet	0.03	0.003–0.035[a]	West coast of U.S.A.	Compare with Table 27; size effect	Cutshall et al. (1977)
5 species	Wet	—	<0.1–0.1	New York Bight, Long Island Sound	Most samples below detection limits	Greig and Wenzloff (1977)
4 species	Wet	—	0.05–0.40	Lower Medway Estuary, U.K.	Differences between individuals noted	Wharfe and Van den Broek (1977)

[a] One anomalously high value (0.11 $\mu g/g$) reported also.

513

Table 29. Selected Data for the Concentrations of Cadmium (means or ranges, $\mu g/g$ wet or dry weight) in the Tissues of Various Species of Finfish

Species	Weight Basis	Muscle	Liver	Kidney	Gill	Gonad	Spleen	Brain	Stomach	Reference
Carcharinus falciformis	Dry	1.0	5.0	2.6	<0.2	<0.2	<0.2	<0.2	—	Windom et al. (1973)
Carcharinus milberti	Dry	—	<0.1	—	—	—	—	—	—	Windom et al. (1973)
Carcharinus obscurus	Dry	2.1	1.6	—	—	—	—	<0.1	—	Windom et al. (1973)
Rhinoptera bonasus	Dry	0.2	0.6	—	—	—	—	<0.1	0.4	Windom et al. (1973)
Raja eglanteria	Dry	0.6	<0.2	—	—	—	—	—	—	Windom et al. (1973)
Rhinoblatis lentiginous	Dry	0.4	1.4	—	—	—	—	—	0.4	Windom et al. (1973)
Sphyrna tiburo	Dry	0.4	0.9	—	—	0.7	0.6	—	0.9	Windom et al. (1973)
Sphyrna lewini	Dry	<0.1	<0.1	—	—	—	—	—	<0.1	Windom et al. (1973)
Squalus acanthius	Dry	0.4	1.0	—	—	—	1.4	—	3.7	Windom et al. (1973)
Arripis trutta	Wet	0.002	0.40	2.27	0.32	0.08	0.28	—	—	Brooks and Rumsey (1974)
Caranx lutescens	Wet	0.010	1.05	0.85	0.38	0.23	2.40	—	—	Brooks and Rumsey (1974)
Cheilodactylis macropterus	Wet	0.006	14.70	1.67	0.47	0.14	1.48	—	—	Brooks and Rumsey (1974)
Chrysophrys auratus	Wet	0.016	1.10	0.71	0.31	0.11	0.52	—	—	Brooks and Rumsey (1974)

Species										Reference
Latridopsis ciliaris	Wet	0.005	1.73	0.50	0.46	0.12	0.52	—	—	Brooks and Rumsey (1974)
Polyprion oxygeneios	Wet	0.005	12.15	5.35	0.19	0.17	0.50	—	—	Brooks and Rumsey (1974)
Seriola grandis	Wet	0.006	4.63	0.30	—	—	—	—	—	Brooks and Rumsey (1974)
Trigla kumu	Wet	0.015	4.23	0.12	0.14	0.14	0.73	—	—	Brooks and Rumsey (1974)
Norway pout	Dry	0.9	<0.1	—	—	—	—	—	—	Halcrow et al. (1973)
Cod	Dry	0.7	0.3	—	—	—	—	—	—	Halcrow et al. (1973)
Plaice	Dry	0.8	<0.1	—	—	—	—	—	—	Halcrow et al. (1973)
Whiting	Dry	0.9	<0.1	—	—	—	—	—	—	Halcrow et al. (1973)
Haddock	Dry	0.6	<0.1	—	—	—	—	—	—	Halcrow et al. (1973)
Saithe	Dry	0.7	<0.1	—	—	—	—	—	—	Halcrow et al. (1973)
Long rough dab	Dry	1.0	<0.1	—	—	—	—	—	—	Halcrow et al. (1973)
Flounder	Dry	1.0	<0.1	—	—	—	—	—	—	Halcrow et al. (1973)
Makaira indica	Wet	0.05–0.40	0.2–83.0	—	—	—	—	—	—	Mackay et al. (1975a)
Boreogadus saida	Dry	<0.5	0.68	—	—	—	—	—	—	Bohn and McElroy (1976)
Antimora rostrata	Wet	<0.12	0.34	—	—	—	—	—	—	Greig et al. (1976)
Halosauropsis macrochir	Wet	<0.12	—	—	—	—	—	—	—	Greig et al. (1976)

Table 29. *Continued*

Species	Weight Basis	Muscle	Liver	Kidney	Gill	Gonad	Spleen	Brain	Stomach	Reference
Nematonurus armatus	Wet	<0.10–0.14	1.21–1.33	—	—	—	—	—	—	Greig et al. (1976)
Seriola sp.	Wet	<0.10	0.24	—	—	—	—	—	—	Greig et al. (1976)
Clupea sprattus	Wet	0.24	—	—	—	—	—	—	—	Wright (1976)
Cyclopterus lumpus	Wet	0.12	1.08	1.83	—	—	—	—	0.6	Wright (1976)
Gadus morhua (young)[a]	Wet	1.35	1.72	—	1.25	—	—	—	0.65	Wright (1976)
Gadus morhua[a]	Wet	0.11	0.06	0.99	—	0.09	—	—	0.12	Wright (1976)
Gadus virens	Wet	0.64	—	4.20	3.23	—	—	—	2.10	Wright (1976)
Pleuronectes limanda	Wet	0.18	1.38	0.21	—	0.09	—	—	0.36	Wright (1976)
Pleuronectes platessa	Wet	1.44	2.91	—	—	—	—	—	3.96	Wright (1976)
Zoarces viviparus	Wet	0.29	0.42	—	—	1.63	—	—	1.07	Wright (1976)
Limanda ferruginea	Wet	<0.1–0.1	0.09–0.50	—	—	—	—	—	—	Greig and Wenzloff (1977)
Mustelus canis	Wet	<0.1	<0.1–<0.2	—	—	—	—	—	—	Greig and Wenzloff (1977)
Pseudopleuronectes americanus	Wet	<0.1	<0.1–0.8	—	—	—	—	—	—	Greig and Wenzloff (1977)
Urophycis chuss	Wet	<0.1	<0.3	—	—	—	—	—	—	Greig and Wenzloff (1977)
Urophycis tenuis	Wet	<0.1	<0.1	—	—	—	—	—	—	Greig and Wenzloff (1977)

[a] Difference in age and site of collection: see cited author for details.

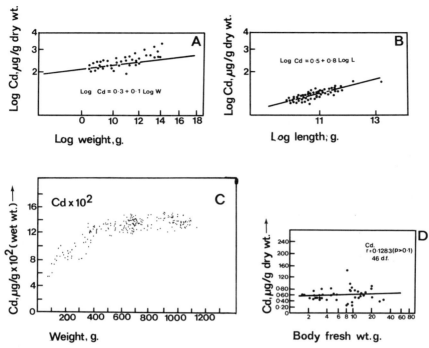

Figure 40.   The relationship of cadmium concentrations in whole fish to fish weight or length. (*A*) and (*B*) refer to whiting (*Merlangus merlangus*) from the Severn Estuary (Badsha and Sainsbury, 1977). (*C*) refers to Pacific hale (*Merluccius productus*) from the western coast of the United States. (Cutshall et al., 1977). (*D*) refers to Arctic cod (*Boreogadus saida*) from Baffin Island (Bohn and McElroy, 1976).

By contrast, tilefish (*Lopholatilus chamaeleonticeps*) exhibited not only significant increases in liver cadmium concentrations with greater body length, but also higher levels of the metal in females than in males. The species specificity of such differences is further emphasized by other reports; for instance, Mackay et al. (1975a) and Bohn and McElroy (1976) could find no cadmium-size correlations for black marlin (*Makaira indica*) or Arctic cod (*Boreogadus saida*), respectively, whereas both whiting (*Merlangus merlangus*) and Pacific hake (*Merluccius productus*) exhibited significant size-concentration relationships for cadmium (Badsha and Sainsbury, 1977; Cutshall et al., 1977). The results of the last three groups of authors are presented in Figure 40 for comparison.

The effects of coexisting pollutants on the concentration of cadmium found in organisms from the field may be significant in many cases, judging from their effects in laboratory uptake experiments (see Section 4). However, such effects would be difficult to identify in isolated studies, and even in monitoring programs

Table 30. Selected Data for the Concentrations of Cadmium (means or ranges, $\mu g/g$, wet or dry weight) Found in Gastropod Molluscs from Various Locations around the Coasts of the United Kingdom and Norway

| Species | Area | Weight Basis | Cadmium Concentration ($\mu g/g$) | | Remarks | Author |
			Mean	Range		
Buccinum undatum	Irish Sea	Dry	31.8	—	Shell contained 0.018 μg Cd/g	Mullin and Riley (1956)
Calliostoma zizyphinum	Irish Sea	Dry	3.1	—	Shell contained 0.049 μg Cd/g	Mullin and Riley (1956)
Gibbula umbilicalis	Irish Sea	Dry	0.8	—	Shell contained 0.029 μg Cd/g	Mullin and Riley (1956)
Littorina littoralis	Irish Sea	Dry	3.5	—	No data for shell	Mullin and Riley (1956)
Littorina littorea	Irish Sea	Dry	1.8	—	Shell contained 0.012 μg Cd/g	Mullin and Riley (1956)
Nucella lapillus	Irish Sea	Dry	37.9	—	Shell contained 0.082 μg Cd/g	Mullin and Riley (1956)
Patella vulgata	Irish Sea	Dry	16.4	—	No data for shell	Mullin and Riley (1956)
Buccinum undatum	The Solent	Dry	2.2	—	Cf. above	Segar et al. (1971)
Crepidula fornicata	The Solent	Dry	3.9	—		Segar et al. (1971)
Nucella lapillus	Isle of Man	Dry	73.0	—	Cf. above	Segar et al. (1971)
Patella vulgata	Isle of Man	Dry	31.0	—	Cf. above	Segar et al. (1971)
Littorina littorea	Severn Estuary	Dry	—	15–210	All species varied with location;	Butterworth et al. (1972)
Nucella lapillus	Severn Estuary	Dry	—	62–425	study area known to be polluted	Butterworth et al. (1972)

Species	Location		Mean	Range	Notes	Reference
Patella vulgata	Severn Estuary	Dry	—	30–550		Butterworth et al. (1972)
Buccinum undatum	Firth of Clyde	Dry	5.7	—	Sludge dumping area	Mackay et al. (1972)
Littorina littorea	Severn Estuary	Dry	—	8–75	Polluted area; compare to Butterworth et al. (1972)	Nickless et al. (1972)
Nucella lapillus	Severn Estuary	Dry	—	31–725		Nickless et al. (1972)
Patella vulgata	Severn Estuary	Dry	—	9–500		Nickless et al. (1972)
Patella vulgata	West Irish Sea	Dry	8.4	2.8–35	Geometric mean	Preston et al. (1972)
Patella vulgata	East Irish Sea	Dry	13.1	3.8–23	Geometric mean	Preston et al. (1972)
Patella vulgata	North Sea coast	Dry	4.4	2.9–7.1	Geometric mean	Dutton et al. (1973)
Buccinum undatum	Firth of Clyde	Dry	7.0	1.1–17.0	Location differences (sludge dumping)	Halcrow et al. (1973)
Nucella lapillus	Upper Bristol Channel	Wet	80.3	—	Cadmium of long half-life	Peden et al. (1973)
Nucella lapillus	Lower Bristol Channel	Wet	24.0	17–39	Relatively unpolluted	Peden et al. (1973)
Nucella lapillus	South Devon	Wet	2.2	—	Size effect reported	Peden et al. (1973)
Patella vulgata	Bristol Channel	Wet	66.1	10–118	Relatively unpolluted	Peden et al. (1973)
Patella vulgata	Devon	Wet	6.4	1–14		Peden et al. (1973)
Littorina littorea	Coasts of Scotland	Wet	0.2	0.03–0.5	Five locations, unpolluted	Topping (1973b)
Patella vulgata	Bristol Channel	Dry	114	82–145	Polluted area	Boyden and Romeril (1974)
Littorina littoralis	Severn Estuary	Dry	784	—	Severn Beach; highly polluted	Leatherland and Burton (1974)
Nucella lapillus	Portland	Dry	77.5	—		Leatherland and Burton (1974)
Patella vulgata	Dorset and South Wales	Dry	35.4	27–44	Two locations	Leatherland and Burton (1974)

Table 30. *Continued*

Species	Area	Weight Basis	Cadmium Concentration (μg/g)		Remarks	Author
			Mean	Range		
Nucella lapillus	Bristol Channel	Dry	780	500–1120	Polluted	Stenner and Nickless (1974b)
Nucella lapillus	Beer, Dorset	Dry	36	11–62	Unpolluted	Stenner and Nickless (1974b)
Patella vulgata	Bristol Channel	Dry	220	67–440	Polluted	Stenner and Nickless (1974b)
Patella vulgata	Beer, Dorset	Dry	11	3–28	Unpolluted	Stenner and Nickless (1974b)
Littorina littorea	Looe Estuary	Dry	1.4	0.5–2.6	Cadmium levels correlated to diet of each species	Bryan and Hummerstone (1977)
Nucella lapillus	Looe Estuary	Dry	12.8	5.5–16.0		Bryan and Hummerstone (1977)
Patella vulgata	Looe Estuary	Dry	8.6	3.3–21.5		Bryan and Hummerstone (1977)
Haliotis tuberculata	Guernsey	Dry	5.2	—	Tissue distribution of cadmium also reported	Bryan et al. (1977)
Littorina littorea	Southwest England	Dry	1.3	—		Bryan et al. (1977)
Patella vulgata	Southwest England	Dry	3.9	—		Bryan et al. (1977)
Patella vulgata	Trondheimsfjorden, Norway	Dry	10.0	2–22	Seven locations	Lande (1977)

of several sites. McDermott and Young (1974) studied Dover sole (species not named but believed to be *Microstomus pacificus*) from polluted areas near California sewage outfalls and from control areas off the California coast. The cadmium concentrations present in sediments were 150-fold greater in the polluted stations; however, the fish exhibited greater cadmium levels in liver (by a factor of 1.68) in control areas. Concentrations of the metal in the other tissues studied (muscle and gonads) were below the detection limits of 3.0 μg/g dry weight in fish from both sites. It was suggested that this reversal to the sediment profiles by the fish samples might possibly be caused by an interaction between DDT and cadmium (see Young and Jan, 1976), although the evidence is by no means conclusive. However, the significant depression of cadmium uptake in the shrimp *Penaeus duorarum* caused by methoxychlor (see Section 4.3) illustrates that such an interaction is possible. Further studies on this aspect would be most interesting, especially as the existence of pollutant interaction renders a species virtually useless as an indicator of ambient levels of any of the pollutants involved in the interaction (Phillips, 1977b, 1978b).

5.3. Molluscs

This chapter does not attempt totally comprehensive cover of the large amount of information published on cadmium in molluscs. However, most papers representing significant advances in knowledge in this area are considered. A detailed data bibliography on pollutant levels in bivalves has been published (Kidder, 1977); readers are referred to this document for more extensive comparative trace metal data for this group.

Concerning differences amongst the various taxonomic groups of the phylum, the large amount of available data permits reasonably accurate and meaningful comparisons without too much interference from other parameters such as location and season. Thus consideration of Tables 30, 31, and 32 shows that, in general, gastropod molluscs contain greater concentrations of cadmium than do lamellibranchs. Certain consistent differences may also be noted amongst the lamellibranchs, oysters and (especially) scallops exhibiting higher levels of cadmium than do mussels, cockles, clams, or quahaugs (compare Tables 31 and 32 and see Segar et al., 1971; Nielsen and Nathan, 1975; Bloom and Ayling, 1977). These differences between different groups represent the final result of variations in diet (dictating the amount of cadmium ingested) and in excretion rates between species. The difference between the cadmium taken up and that excreted per unit time is the amount that must be sequestered amongst the body tissues.

The effects of diet on the concentrations of metals found in different molluscs have been discussed by several authors. Differences in the cadmium contents

Table 31. Selected Data for the Concentrations of Cadmium (means or ranges, μg/g wet or dry weight) Found in Oysters and Scallops from Various Locations

Species	Area	Weight Basis	Cadmium Concentration (μg/g)		Remarks	Reference
			Mean	Range		
Ostrea sinuata	Tasman Bay, New Zealand	Dry	35	10–43	Compare to mussel data in Table 32	Brooks and Rumsby (1965)
Pecten novae-zelandiae	Tasman Bay, New Zealand	Dry	249	210–299		Brooks and Rumsby (1965)
Crassostrea gigas	Washington coast, U.S.A.	Wet	1.1	0.8–1.4	Seasonal variation	Pringle et al. (1968)
Crassostrea virginica	Coasts of U.S.A.	Wet	3.1	0.1–7.4	Location differences	Pringle et al. (1968)
Chlamys opercularis	Southwest England, U.K.	Dry	5.5	—	Tissue distribution data	Bryan (1973)
Ostrea edulis	Poole Harbour, U.K.	Dry	22.9	5.9–54		Boyden (1975)
Pecten maximus	Southwest England, U.K.	Dry	32.5	—	Tissue distribution data	Bryan (1973)
Chlamys septemradiata	Firth of Clyde, U.K.	Dry	13.5	9.5–17.5	Location differences (dumping)	Halcrow et al. (1973)
Crassostrea gigas/ Ostrea angasi	Tasmania, Australia	Wet	—	<2.0–19.8	Derwent and Tamar estuaries	Thrower and Eustace (1973a)
Crassostrea gigas	Tasmania, Australia	Wet	—	<2.0–31.7	Whole of Tasmania studied; claimed no species difference	Thrower and Eustace (1973b)
Ostrea angasi	Tasmania, Australia	Wet	—	<2.0–18.7		Thrower and Eustace (1973b)

Species	Location	Wet/Dry	Value	Range	Notes	Reference
Pecten maximus	Coasts of Scotland, U.K.	Wet	10.5	5.1–23.0	Cf. mussel data in Table 32	Topping (1973b)
Crassostrea gigas	Bristol Channel, U.K.	Dry	31.7	17–43	Introduced animals	Boyden and Romeril (1974)
Ostrea angasi	Derwent Estuary, Australia	Wet	10.7	—	Cf. mussel data in Table 32	Eustace (1974)
Crassostrea virginica	Northwest Atlantic	Dry	8.4	7.2–9.5	Fertilizer application effects	Valiela et al. (1974)
Crassostrea gigas	Poole Harbour, U.K.	Dry	4.6	—	Cf. data for mussels and cockles in Table 32	Boyden (1975)
Crassostrea virginica	Connecticut, U.S.A.	Dry	—	15.6–28.1	Two locations	Greig et al. (1975)
Crassostrea commercialis	New South Wales, Australia	Wet	0.2	0.1–1.0	Cadmium-size and Cadmium-metal correlations	Mackay et al. (1975b)
Ostrea angasi[a]	Port Phillip Bay, Australia	Dry	91.6	35–174	Incorrectly named[a]	Talbot et al. (1976)
Crassostrea gigas	Knysna Estuary, South Africa	Dry	3.7	—	Contains some previously unpublished data for other regions also	Watling and Watling (1976a)
Crassostrea margaritacea	Knysna Estuary, South Africa	Dry	2.5	—		Watling and Watling (1976a)
Ostrea edulis	Knysna Estuary, South Africa	Dry	3.1	—		Watling and Watling (1976a)
Crassostrea gigas	Saldanha Bay, South Africa	Dry	9.0	—	Cf. mussel data in Table 32	Watling and Watling (1976b)
Plagopecten magellanicus	Mid-Atlantic coast of U.S.A.	Dry	20.9	10.1–59.3	Location differences	Pesch et al. (1977)

[a] Incorrectly named "*Ostreidae angasi*" in original paper.

Table 32. Selected Data for the Concentrations of Cadmium (means or ranges, $\mu g/g$ wet or dry weight) Found in Bivalve Molluscs Other than Oysters and Scallops from Various Locations

Species	Area	Weight Basis	Cadmium Concentration ($\mu g/g$)		Remarks	Reference
			Mean	Range		
Mytilus edulis	Irish sea, U.K.	Dry	3.2	—	Shell contained 0.003 μg Cd/g	Mullin and Riley (1956)
Mytilus edulis aoteanus	Tasman Bay, New Zealand	Dry	<10	<10	Cf. Table 31	Brooks and Rumsby (1965)
Mercenaria mercenaria	Coasts of U.S.A.	Wet	0.2	0.1–0.7	Location differences	Pringle et al. (1968)
Mya arenaria	Coasts of U.S.A.	Wet	0.3	0.1–0.9	Location differences	Pringle et al. (1968)
Mytilus edulis	Bristol Channel, U.K.	Dry	17.9	4–60	Polluted area	Nickless et al. (1972)
Mytilus edulis	Coasts of Scotland, U.K.	Wet	0.4	0.08–2.0	Cf. Table 31	Topping (1973b)
Mytilus edulis	Derwent Estuary, Australia	Wet	5.5	—	Cf. Table 31	Eustace (1974)
Cerastoderma edule[a]	Southern coast of U.K.	Dry	5.8	4.9–6.7	Two locations	Leatherland and Burton (1974)
Mercenaria mercenaria	The Solent, U.K.	Dry	0.5	0.3–0.6	Age dependence	Leatherland and Burton (1974)

Species	Location		Value	Range	Comment	Reference
Cerastoderma edule	Hardangerfjord and Skjerstadfjord, Norway	Dry	10.4	4.5–19.3	Very marked location differences; very polluted area	Stenner and Nickless (1974a)
Mytilus edulis	Hardangerfjord and Skjerstadfjord, Norway	Dry	—	1.9–140		Stenner and Nickless (1974a)
Mercenaria mercenaria	Northwest Atlantic	Dry	1.8	1.3–2.2	Cf. data for oysters in Table 31	Valiela et al. (1974)
Modiolus demissus	Northwest Atlantic	Dry	4.3	2.0–7.1		Valiela et al. (1974)
Cerastoderma edule	Poole Harbour, U.K.	Dry	6.7	1.5–16.9	Location differences; polluted area	Boyden (1975)
Mytilus edulis	Poole Harbour, U.K.	Dry	34.6	3.7–65.4		Boyden (1975)
Cerastoderma edule	Atlantic coasts of Spain and Portugal	Dry	0.4	0.3–0.6	Some differences with location	Stenner and Nickless (1975)
Mytilus edulis	Atlantic coasts of Spain and Portugal	Dry	2.7	1.7–3.6		Stenner and Nickless (1975)
Mytilus galloprovincialis	Northwest Mediterranean	Dry	1.9	0.4–5.9	Seasonal data, 15 locations	Fowler and Oregioni (1976)
Mytilus edulis	Port Phillip Bay, Australia	Wet	—	0.5–18.2	Differences with location related to industrial discharges	Phillips (1976b)
Mytilus edulis	Western Port Bay, Australia	Wet	—	0.2–2.4		Phillips (1976b)
Mytilus edulis	Port Phillip Bay, Australia	Dry	21.8	4.2–83	Some location differences	Talbot et al. (1976)
Choromytilus meriodionalis	Saldanha Bay, South Africa	Dry	3.6	0.9–8.0	Possible size—cadmium relation	Watling and Watling (1976b)

Table 32. *Continued*

Species	Area	Weight Basis	Cadmium Concentration ($\mu g/g$)		Remarks	Reference
			Mean	Range		
Cerastoderma edule	Looe Estuary, U.K.	Dry	0.8	0.5–1.0	⎫ Attempted relation of cadmium contents to diet of each species	Bryan and Hummerstone (1977)
Macoma balthica	Looe Estuary, U.K.	Dry	0.7	0.2–0.8		Bryan and Hummerstone (1977)
Mytilus edulis	Looe Estuary, U.K.	Dry	1.8	0.8–2.6		Bryan and Hummerstone (1977)
Scrobicularia plana	Looe Estuary, U.K.	Dry	1.6	0.6–3.4	⎭	Bryan and Hummerstone (1977)
Mytilus edulis	Trondheimsfjorden, Norway	Dry	—	<1.0–5.0	Cf. Table 30	Lande (1977)
Mytilus edulis	Baltic waters	Dry	—	1.3–13.0	⎫ Levels related to salinity of each water mass	Phillips (1977c)
Mytilus edulis	Öresund and the Great Belt	Dry	—	0.6–4.2		Phillips (1977c)
Mytilus edulis	Kattegat and eastern Skagerrak	Dry	—	0.4–1.5	⎭	Phillips (1977c)
Mytilus edulis	Lower Medway Estuary, U.K.	Wet	1.0	0.5–1.5	Little polluted	Wharfe and Van den Broek (1977)
Pitar morrhuana	Rhode Island, U.S.A.	Dry	—	<0.1–3.3	Location differences	Eisler et al. (1978)

[a] Also known as *Cardium edule*.

of the various taxonomic groups in some cases exhibit a relationship with the feeding habits of each species (e.g., see Bryan and Hummerstone, 1977). Most gastropods are herbivorous (e.g., *Patella vulgata* and *Littorina littorea*), but some are carnivorous; for example, *Nucella (Thais) lapillus* feeds mainly on barnacles. The diet of these species may thus contain large amounts of cadmium (Phillips, 1977b), and some evidence exists for the amplification of cadmium levels from prey to gastropod predator (Butterworth et al., 1972), although this tendency is not very great. Lamellibranch molluscs are filter-feeders (e.g., the mussels *Mytilus edulis* and the cockle *Cerostoderma edule*) or deposit-feeders (e.g., *Macoma balthica* and *Scrobicularia plana*). The amounts of cadmium ingested by these organisms may depend to a large extent on the amount of suspended inorganic particulates or sediment ingested, as these components may contain very high concentrations of the metal. Ayling (1974) has suggested that this route is so important for oysters (*Crassostrea gigas*) that concentration factors for metals should be based on sediment values rather than on the concentrations of metals in solution. This proposal has received little support from other researchers; however, several authors have acknowledged the importance of this route of metal uptake in lamellibranchs (e.g., Raymont, 1972; Preston et al., 1972; Boyden and Romeril, 1974; Eisler et al., 1978), although quantification of its importance has not been seriously attempted.

Although a very loose relationship may exist between the concentrations of cadmium found in various species of molluscs and their respective feeding habits, the species differences undoubtedly depend to a large extent on the ability of each organism to excrete cadmium. Very little is known concerning the excretion of cadmium by molluscs, despite its central importance in determining the body load of the metal that must be sequestered amongst the tissues. The studies that have been reported (e.g., Fowler and Benayoun, 1974; see Section 4) suggest very slow excretion rates. In addition, loss of the metal on spawning appears to be quite minimal (Greig et al., 1975). Thus much of the ingested cadmium must be sequestered in the tissues, possibly in granular form (Bryan, 1973). This poor ability of molluscs to excrete ingested cadmium may have some bearing on the reported changes in cadmium concentration and total body load in molluscs with age or size (see below).

The distribution of cadmium amongst the various tissues of molluscs has been studied by several authors. Most of these reports concern bivalves, although one study has been done on the whelk *Buccinum undatum* (Mullin and Riley, 1956), and a second on the abalone *Haliotis tuberculata* (Bryan et al., 1977). Selected data are shown in Table 33. It can be seen that most species exhibit a very specific tissue distribution for cadmium. In general, both bivalves and gastropods contain the greatest concentrations of cadmium in the viscera, especially the digestive gland. The kidney is also often rich in cadmium, although organisms with paired kidneys, such as *H. tuberculata,* may exhibit quite different levels of cadmium (and other trace metals) in each kidney, which presumably relate to the different physiological functions of the left and right kidneys (see Bryan et al., 1977). In oysters the kidney and heart contain somewhat more cadmium than is found in the viscera (Brooks and Rumsby, 1965). Indeed, even in scallops, which exhibit the greatest concentrations in the viscera, the actual percentage of the total body load of cadmium found in these tissues varies considerably between species. Bryan (1973) quoted percentages of 89.9 and 1.7 for digestive gland and kidney of *Pecten maximus,* whilst comparable figures for *Chlamys opercularis* were 41.5 and 7.5, respectively. Although percentages of the total body load were not actually quoted by Young and Jan (1976) for tissues of the rock scallop *Hinites multirugosus,* the very high concentrations of cadmium in the digestive gland of this species (Table 33) suggest that this tissue contains a large proportion of the total body load of the metal.

Surprisingly, few data exist for the tissue distribution of cadmium in mussels. However, the work of Segar et al. (1971) on *Modiolus modiolus* and that of Fowler and Benayoun (1974) on *Mytilus galloprovincialis* (see Section 4) suggest a less asymmetric distribution of cadmium in mussel tissues than in the tissues of other bivalves. However, the general similarities exhibited by most lamellibranch and gastropod mollusc species suggest that most or all members of these groups sequester cadmium by a common mechanism. Although no specific data are presently available, it appears likely that these organisms store excess metals in granular form (see discussions in Bryan, 1973, and in Bryan et al., 1977) in much the same fashion as that described for zinc and copper storage in barnacles (Walker et al., 1975; Walker, 1977).

Only one report could be found citing a change in the tissue distribution of cadmium in a mollusc with an external variable. Bryan et al. (1977) published data concerning the variation in the cadmium concentrations found in the whole soft parts, foot, and viscera of the abalone *Haliotis tuberculata* with changes in the dry weight of the whole soft parts of the animal. These results are shown in Figure 41. Concentrations of cadmium increased in whole soft parts with increasing weight; a similar pattern was noted for viscera (which accounted for some 30% of the total dry weight of the tissues and for about 90% of the total body load of cadmium), but metal concentrations in the foot decreased with increasing whole tissue weight.

The effects of season on the concentrations of cadmium found in molluscs have been poorly studied in general. The best data available again concern the bivalves. Pringle et al. (1968) reported a gradual decrease in cadmium concentrations in the Pacific oyster *Crassostrea gigas* taken from two stations on the Washington coast in autumn, winter, and early summer of 1966–1967. Unfortunately these data barely cover one full year; thus separation of true recurrent seasonal effects and temporal changes in cadmium output by industries or other coastal sources is impossible. Frazier (1975, 1976) reported much more

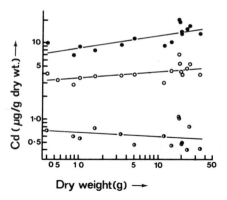

Figure 41. Concentrations of cadmium in whole soft parts (O), foot (◑), and viscera (●) of the abalone *Haliotis tuberculata* and their variation with dry tissue weight of the animal. Results for whole soft parts are simply additions of those for foot and viscera. (After Bryan et al., 1977.)

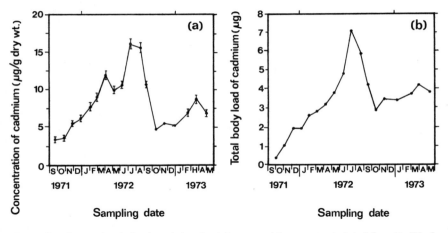

Figure 42. Seasonal variation in cadmium levels in oysters (*Crassostrea virginica*) from the Rhode River, Chesapeake Bay. (*a*) Concentrations of cadmium, means ± one standard error (lack of error bars indicates error smaller than symbol used for mean). (*b*) Total body loads of cadmium for the same oysters. (After Frazier, 1975.)

extensive studies on the American oyster *C. virginica,* taken from Chesapeake Bay. A well-defined seasonal peak in cadmium concentration was observed in summer, with much lower levels in winter (Figure 42*a*); this pattern coincided closely with the seasonal profile for the total body load of cadmium (Figure 42*b*). The changes in cadmium levels observed in the oyster soft parts did not correlate with shell deposition; thus loss of cadmium to the shell during periods of shell deposition was not a highly significant factor in the seasonality of the metal. It

Table 33. Selected Data for the Tissue Distribution of Cadmium, in Various Lamellibranch and Gastropod Molluscs (means or ranges, μg/g dry weight, except Topping, 1973b, and Young and Jan, 1976, which refer to wet weights)

Reference	Species	Shell	Gills	Mantle	Viscera	Muscle	Gonad	Foot	Kidney	Heart	Whole Soft Parts
LAMELLIBRANCHS											
Mullin and Riley (1956)	Chlamys opercularis	0.008	1.24	—	9.53–9.71[a]	0.87	2.21	—	—	—	—
	Pecten maximus[b]	—	12.5	4.1	111.0[b]	1.58	0.11–0.55	16.5	127.0	—	—
Brooks and Rumsby (1965)	Pecten novae-zelandiae	<20	<20	<20	2000	<20	<20	<20	<20	—	—
	Ostrea sinuata	<20	<20	207	61	<20–97	—	—	118	154	—
	Mytilus edulis aoteanus	<20	<20	<20	<20	<20	<20	<20	—	—	—
Brooks and Rumsby (1967)	Ostrea sinuata	—	<10	<10	14	<10	—	—	—	—	12
Segar et al. (1971)	Pecten maximus	0.04	3.1–17.0[c]		96	1.9	2.5	—	—	—	13
	Modiolus modiolus	0.03	7.0[d]		4.2	2.8	4.6	—	—	—	4.5–7.1

Bryan (1973)	*Pecten maximus*[e]	—	—	321[f]	—	—	—	79.0	—	32.5
	Chlamys opercularis	—	—	27[f]	—	—	—	41.0	—	5.5
Topping (1973b)	*Pecten maximus*	—	—	15.4	0.55–1.03	0.3–0.88	—	—	—	—
Young and Jan (1976)	*Hinites multirugosus*	—	—	520–540[f]	0.3–0.9[g]	2.6–5.4	—	—	—	—
GASTROPODS										
Mullin and Riley (1956)	*Buccinum undatum*	0.018	0.96	68.0[f]	—	11–24	0.29	17.7	2.32	31.8
Bryan et al. (1977)	*Haliotis tuberculata*	—	1.9	18.1[h]	0.24	—	0.71	4.1–43.0[i]	—	—

[a] Duplicate samples of mantle + viscera.

[b] Figures refer to purged animal; viscera values represent gut + gonad + digestive gland.

[c] Washed and unwashed samples of gill + mantle.

[d] Mantle + gills.

[e] Remaining soft parts of *P. maximus* (whole soft parts − digestive gland and kidney) contained 2.2 µg Cd/g.

[f] Digestive gland only.

[g] Adductor muscle only.

[h] Digestive gland contained 46 µg Cd/g.

[i] Lower value refers to left kidney; higher value, to right kidney.

is noticeable from Figures 42*a* and *b* that the total body burden of cadmium was greater in winter 1972 than in winter 1971; the change in cadmium concentration in the oysters from these two periods is, however, much less. This difference reflects the growth of the oysters studied during the spring and summer of 1972; this growth served to dilute preexisting cadmium and thus decrease metal concentrations, while body burdens remained unchanged.

Such a reciprocal relationship between seasonal changes in tissue weights of bivalves and seasonal variation in trace metal concentrations has been observed in mytilids also. Fowler and Oregioni (1976) and Phillips (1976a), studying *Mytilus galloprovincialis* and *M. edulis,* respectively, reported very similar conclusions concerning seasonality of metals in these species. These authors believe concentrations of metals depend mainly on the sexual state of the bivalve, which determines its tissue weight. Thus in postspawning periods the mussels are light and exhibit high concentrations of metals, whereas in the heavy prespawn condition their concentrations are much lower. These data suggest that most of the metal is present in tissues other than the gonad (or eggs), a hypothesis that is confirmed by the data shown in Table 33 and by the studies of Greig et al. (1975). Other factors also contribute to the seasonal change in the cadmium concentrations present in bivalves; the most important of these is the change in cadmium availability, which depends to a large extent on runoff if the organism is taken from a (polluted) coastal area. The relative dominance of the weight-based effect and the availability-based effect in determining the seasonal profile probably depends on the proximity of the bivalve sample to a river-borne source of cadmium, as discussed by Phillips (1976a, 1977b), although the pattern is complicated by the dependence of cadmium availability on the metal source (see Section 2.3, "Cadmium in Coastal or Estuarine Waters").

Size of the organism studied is also an important determinant of the cadmium concentrations present in molluscs. Data on the effects of size on cadmium concentrations exist for both bivalves and gastropod molluscs. The study of Bryan et al. (1977) on the effects of size on the tissue distribution of cadmium in the abalone *Haliotis tuberculata* was cited above, and will not be considered further here. Most of the data for other gastropod species concerns the limpet *Patella vulgata.* Nickless et al. (1972) first reported studies suggesting increased cadmium levels in limpets of greater tissue weight; however, these data were not extensive enough to support firm conclusions. Peden et al. (1973) reported increased levels of cadmium in limpets of greater size; this trend was followed in the whole soft parts of animals from four locations in Somerset, United Kingdom. By contrast, the concentrations of zinc present in the same limpets decreased with increasing size (Table 34). In addition, the shells of limpets from one station exhibited much lower levels (probably independent of size) of both cadmium and zinc than were found in the soft tissues.

Table 34. Mean Concentrations (μg/g wet weight) of Cadmium and Zinc in Whole Soft Parts or Shells of Limpets (*Patella vulgata*) of Various Sizes[a]
All locations are in Somerset, United Kingdom; no data for the number of individuals analyzed or for the definition of each size group were quoted.

Location	Cadmium (μg/g)			Zinc (μg/g)		
	Small	Medium	Large	Small	Medium	Large
Kilve (flesh)	23.2	25.8	37.6	76.4	61.2	51.2
Sand Bay (flesh)	56.4	86.4	81.6	77.2	64.0	47.6
Clevedon (flesh)	34.4	46.4	61.6	99.2	68.4	47.2
Watchet (flesh)	46.5	54.5	78.0	142.0	89.0	53.0
Watchet (shell)	2.7	3.3	3.3	8.9	7.5	8.9

[a] After Peden et al. (1973).

Boyden (1974, 1977) published extensive studies of metal-size relationships in six gastropod species taken from various locations around British coasts. Most of these data were treated by plotting log metal content against log tissue weight; each species-metal combination was then classified as to the regression slope produced. A slope of 1.0 indicates a slope of zero on a concentration-tissue weight plot, that is, no effect of size on concentration exists. A slope greater than 1.0 indicates an increase in cadmium concentration with increased tissue weight; conversely, a slope of less than 1.0 indicates a decrease in concentration with increased tissue weight. Most species-metal combinations produced regressions of slope of about 1.0; however, cadmium contents in three gastropod species (*Patella vulgata, P. intermedia,* and the whelk *Buccinum undatum*) were related to dry tissue weights by regressions of slope significantly greater than 1.0. By contrast, concentrations of cadmium in the winkle *Littorina littorea,* the opistobranch *Scaphander lignarius,* and the limpet *Crepidula fornicata* were independent of size (Table 35). Additional data for the carnivorous species *Nucella lapillus* exhibited too great a scatter for a regression to be fitted; this may have been due to the small size range studied and the differences in diet between the selected individuals.

The data published by Boyden (1977) for the limpet *Patella vulgata* agree well with the conclusions of Peden et al. (1973) in that cadmium concentrations increased with increasing tissue weights, whereas zinc concentrations decreased (above and see Tables 34 and 35). The data of Boyden (1977) for both contents and concentrations of the two elements are shown in Figure 43, using linear scales for both axes. The relationship between cadmium content (or concentration) and tissue weight in *P. vulgata* varied with season and with location. Thus in Table 35 it is evident that the regression coefficients for the five samples of this limpet differ somewhat from each other. Boyden (1977) ascribed this difference

Table 35. Regression Coefficients Relating the Cadmium Contents of the Whole Soft Parts of Six Species of Gastropod Molluscs to the Dry Weights of Their Tissues[a]

Slopes significantly different from 1.0 ($P < .05$) are marked with an asterisk.

Species	Number of Individuals	Size Range (g dry weight)	Location	Date	Regression Coefficient
Buccinum undatum	20	0.06–21.4	Looe Bay	4/17/74	1.18*
Crepidula fornicata	21	0.05–0.78	Poole Harbour	6/23/73	1.12
Littorina littorea	37	0.01–0.70	Restronguet Creek	4/9/74	0.97
Scaphander lignarius	20	0.09–3.80	Looe Bay	4/17/74	0.94
Patella intermedia	14	0.02–1.71	Helford Estuary	4/3/73	1.35*
Patella intermedia	16	Not quoted	Helford Estuary	1/10/74	1.49*
Patella vulgata	35	0.02–0.80	Portishead	3/8/73	2.05*
Patella vulgata	29	0.01–0.80	Portishead	8/13/73	1.98*
Patella vulgata	30	0.01–0.85	Portishead	1/11/74	1.96*
Patella vulgata	34	0.02–1.00	Portishead	5/5/74	1.70*
Patella vulgata	32	0.01–1.10	Watchet	3/18/74	1.37*

[a] After Boyden (1977).

534

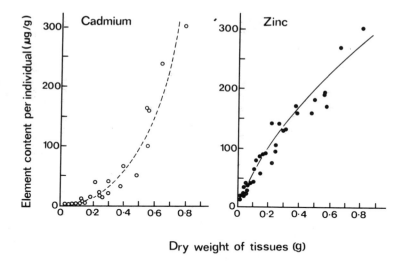

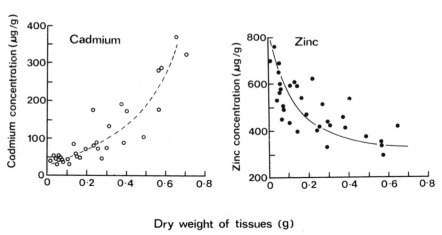

Figure 43. Total body content (above) or concentrations (below) of cadmium and zinc in limpets (*Patella vulgata*) of various dry tissue weights collected from Portishead in the Severn Estuary. Concentrations of the elements are based on dry weights. (After Boyden, 1977.)

to variation in the degree of cadmium pollution; hence limpets from Watchet (a relatively unpolluted location) exhibited lower regression coefficients than did those from Portishead in the polluted Severn Estuary. Possible explanations of this phenomenon will be discussed below.

Published information concerning the effects of body size (age, weight, length)

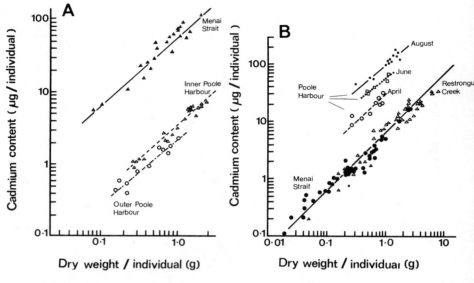

Figure 44.   (*A*) Regressions describing the relationship between cadmium content and tissue weight in mussels (*Mytilus edulis*) from three locations. (*B*) Similar regressions for oysters (*Ostrea edulis*) from three locations; oysters from Menai Strait and Restronguet Creek fitted the same regression line. (After Boyden, 1977.)

on the concentrations of cadmium found in bivalve molluscs in the field is available in several papers. Boyden (1974, 1977) has again dominated the literature to some extent; results from his extensive studies published in 1977 are shown in Table 36. Examples of the cadmium content-body weight regressions observed for the mussel *Mytilus edulis* and the oyster *Ostrea edulis* from several collection locations are shown in Figure 44. By contrast to the results cited above for *Patella vulgata,* each of the bivalve species studied exhibited slope constancy, that is, no significant differences in regression coefficients for any one species-metal combination were seen with changes in season or location. Boyden (1977) distinguished three groups of bivalves. The first group exhibited no significant change in cadmium concentration with changes in tissue weight (regression coefficient ≈1.0); examples of this group were *Mytilus edulis, Ostrea edulis,* and *Chlamys opercularis.* The second group, exemplified by *Crassostrea gigas* and the two clams *Mercenaria mercenaria* and *Venerupis decussata,* exhibited lower concentrations of cadmium with greater tissue weights (regression coefficients <1.0). The sole member of the third group was the scallop *Pecten maximus,* which exhibited a curved regression line for cadmium concentration (or content), although the curve occurred only for animals above 2.0 g in dry tissue weight. For scallops of less than 2.0 g dry weight, cadmium concentrations were independent of tissue weights (Figure 45).

Table 36. Regression Coefficients Relating the Cadmium Contents of the Whole Soft Parts of Seven Species of Bivalve Molluscs to the Dry Weights of Their Tissues[a]

Slopes significantly different from 1.0 are marked with an asterisk.

Group	Species	Number of Individuals	Size Range. (g dry weight)	Location	Date	Regression Coefficient
Mytilidae	*Mytilus edulis*	21	0.26–1.94	Poole Harbour, outer	2/22/74	0.97
	Mytilus edulis	17	0.18–1.30	Poole Harbour, outer	5/3/74	1.05
	Mytilus edulis	22	0.08–1.09	Poole Harbour, inner	2/22/74	1.02
	Mytilus edulis	20	0.17–1.03	Poole Harbour, inner	5/3/74	1.08
	Mytilus edulis	40	0.04–3.50	Restronguet Creek	10/31/74	0.95
Ostreidae	*Crassostrea gigas*	39	0.01–4.30	Menai Strait	3/22/74	0.85*
	Crassostrea gigas	22	0.02–4.00	Poole Harbour, outer	2/22/74	0.86*
	Crassostrea gigas	30	0.07–11.60	Restronguet Creek	10/31/74	0.85*
	Ostrea edulis	38	0.02–2.50	Menai Strait	3/22/74	0.94
	Ostrea edulis	24	0.34–6.63	Restronguet Creek	10/31/74	0.96
Pectinidae	*Chlamys opercularis*	20	0.20–3.70	Looe Bay	4/17/74	0.96
	Pecten maximus	37	0.16–8.30	Looe Bay	4/17/74	0.65*[b]
Veneridae	*Mercenaria mercenaria*	35	0.02–8.02	Southampton Water	1/24/74	0.81*
	Venerupis decussata	30	0.17–2.77	Poole Harbour	2/22/74	0.77*

[a] After Boyden (1977).
[b] Regression was curved.

537

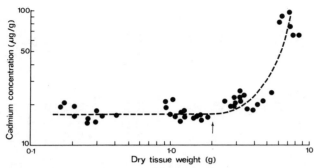

Figure 45.   Concentrations of cadmium in whole soft parts of the scallop *Pecten maximus* from the English Channel in relation to dry weight of the tissues. The vertical arrow marks the beginning of the deviation from a linear plot. (After Boyden, 1977.)

Other authors have also published information concerning the variation in cadmium concentration in bivalve species with tissue weight. Leatherland and Burton (1974) reported decreased concentrations of cadmium in older clams (*Mercenaria mercenaria*) from The Solent in England. Mackay et al. (1975b) studied Sydney rock oysters [*Crassostrea commercialis* (= *Saccostrea cuccullata*)] of three age groups (1.5, 2.5, and 3.5 years). The total content of Cd, Cu, or Zn in these oysters increased with increasing age, whereas the concentration of each metal decreased with age. Phillips (1976a, 1976b) observed a significant dependence of cadmium concentration in the mussel *Mytilus edulis* on wet tissue weight. In contrast to the results of Boyden (1977), these samples consistently exhibited lower concentrations of cadmium in larger individuals. This pattern was observed in mussels exposed to cadmium in solution in the laboratory, as well as in mussels from the field. In addition, the slope of the weight-concentration regression altered with season, again disagreeing with the data of Boyden (1977). No explanation for the differences between the results of these authors is available at present. Watling and Watling (1976b) reported concentration-weight relationships for several elements in the mussel *Choromytilus meridionalis* growing in Saldanha Bay, South Africa. Differences between the weight-concentration regressions exhibited by mussels from different locations were observed, and only 10 of the 24 regression slopes calculated fitted the relationships suggested by Boyden (1974, 1977). In addition, some elements exhibited differences between male and female mussels, although cadmium did not appear to be one of these. Bryan and Uysal (1978) reported a marked increase in cadmium concentrations with increased dry weight in the burrowing bivalve *Scrobicularia plana* from the Tamar Estuary; five other metals (Co, Cr, Ni, Pb, Zn) also increased in concentration, whereas concentrations of Cu, Mn, and Ag decreased with increased tissue weight. This profile for cadmium in *S. plana* is similar to that cited above for the scallop *Pecten maximus* (Boyden, 1977), although the latter organism exhibited a curved relationship. As was the case for *P. maximus,* a significant weight-concentration regression was observed in *S. plana* only for larger animals (greater than 0.006 g dry weight). These two bivalve species are the only examples studied to date that exhibit such profiles, although three species of gastropod molluscs are also known to show increased concentrations of cadmium with increased dry weights (Table 35).

The precise reasons for the frequent dependence of trace metal concentrations in molluscs on the size of the individual, and for the occasional differences between these relationships with season or location for one species, are a matter of speculation at present. The suggestions of Boyden (1974, 1977) concerning the effects of metabolic rates or surface areas on uptake and elimination rates of metals appear to possess limited value, as such mechanisms do not explain the differences between the regression slopes for one species from several locations, or indeed between metals in any one species. The patterns described above for *Pecten maximus* and *Scrobicularia plana* may conceivably be related to a change of preferred diet at a given size, thus explaining the lack of a relationship in individuals of less than a certain weight. If these organisms, on attaining a certain size, shifted from a low-cadmium to a high-cadmium diet, the profiles seen would be produced if one assumes a relatively long half-life of the metal in the scallop as compared to *S. plana.* For other organisms, the data of Phillips (1976a) showing significant weight-concentration regressions in mussels exposed to metals in solution in the laboratory may help in future identification of the parameters responsible for such an effect. These data are certainly fitted by assuming greater metabolic rates in smaller animals, which lead to greater rates of net uptake of the metal. However, the inconsistency of this effect and its dependence on season confuse this interpretation somewhat. One further possibility exists, which may correlate well with the changes in regression slopes seen for *Patella vulgata* from different locations. Boyden (1977) related these changes to differences in the ambient cadmium pollution. If the Portishead limpets were to exhibit slower growth rates than those from the less polluted Watchet site, the difference in regression slopes would be as reported. It is notable in this connection that Shore et al. (1975) considered limpets from the Severn Estuary to be severely affected by the high ambient concentrations of cadmium in this area (see Section 3.3, "Molluscs"). Further research is needed to confirm this hypothesis.

Both gastropod and lamellibranch molluscs are known to exhibit changes in cadmium concentration with sampling position on the shoreline (e.g., see Nickless et al., 1972; Peden et al., 1973; Nielsen, 1974; Phillips, 1976a; Bryan and Uysal, 1978). This variation is certainly caused by the differing exposure—largely due to stratification of estuarine waters—of animals from the various tidal zones to metals in both solution and food. No reports of such effects in animals from open coastal areas with efficient vertical mixing of water are known.

The effects of coexisting pollutants on the levels of cadmium found in molluscs

Table 37. Selected Data for the Concentrations of Cadmium (means or ranges, $\mu g/g$ wet or dry weight) in Crustaceans from Various Locations

Species	Area	Weight Basis	Cadmium Concentration ($\mu g/g$)		Tissue	Reference
			Mean	Range		
Balanus balanoides	Irish Sea, U.K.	Dry	0.15	—	Whole	Mullin and Riley (1956)
Cancer pagarus	Irish Sea, U.K.	Dry	0.15	—	Whole	Mullin and Riley (1956)
Corystes cassivelaunus	Irish Sea, U.K.	Dry	0.15	—	Whole	Mullin and Riley (1956)
Eupagarus sp.	Irish Sea, U.K.	Dry	1.31	—	Whole	Mullin and Riley (1956)
Crangon allmani	Firth of Clyde, U.K.	Dry	2.2	1.2–3.5	Whole	Halcrow et al. (1973)
Nephrops norvegicus	Firth of Clyde, U.K.	Dry	0.75	0.6–0.9	Whole	Halcrow et al. (1973)
Pandalus montagui	Firth of Clyde, U.K.	Dry	1.9	1.2–3.1	Whole	Halcrow et al. (1973)
Carcinus maenas	Severn Estuary, U.K.	Wet	22.4	14.3–33.1	Whole	Peden et al. (1973)
Cancer pagarus	North Devon, U.K.	Wet	5.0	—	Whole	Peden et al. (1973)
Cancer pagarus	Coasts of Scotland, U.K.	Wet	7.2	3.6–13.0	Edible tissue only	Topping (1973b)
Homarus vulgaris	Coasts of Scotland, U.K.	Wet	—	<0.03–0.09	Edible tissue only	Topping (1973b)
Nephrops norvegicus	Coasts of Scotland, U.K.	Wet	—	<0.03–0.1	Whole	Topping (1973b)
Palaemon elegans	Hamble, U.K.	Dry	0.31	—	Whole	Leatherland and Burton (1974)

Species	Location	Dry/Wet	Value	Range	Tissue	Reference
Carcinus maenas	Hardangerfjord, Norway	Dry	4.5	1.9–6.9	Whole	Stenner and Nickless (1974b)
Callinectes sapidus	South Atlantic	Wet	0.21	0.07–0.36	Edible tissue only	National Oceanic and Atmospheric Administration (1975)
Cancer magister	Northwest Pacific	Wet	0.12	<0.05–0.25	Edible tissue only	National Oceanic and Atmospheric Administration (1975)
Panulirus argus	South Atlantic	Wet	0.05	<0.05–0.13	Edible tissue only	National Oceanic and Atmospheric Administration (1975)
Penaeus aztecus	Gulf of Mexico	Wet	0.05	<0.05–0.16	Edible tissue only	National Oceanic and Atmospheric Administration (1975)
Penaeus duorarum	Gulf of Mexico	Wet	0.13	<0.05–0.25	Edible tissue only	National Oceanic and Atmospheric Administration (1975)
Penaeus setiferus	Gulf of Mexico	Wet	0.06	<0.05–0.48	Edible tissue only	National Oceanic and Atmospheric Administration (1975)
Penaeus setiferus	South Atlantic	Wet	0.05	<0.05–0.34	Edible tissue only	National Oceanic and Atmospheric Administration (1975)

Table 37. *Continued*

Species	Area	Weight Basis	Cadmium Concentration ($\mu g/g$)		Tissue	Reference
			Mean	Range		
Chthamalus stellatus		Dry	5.5	5.1–6.3	Whole	Stenner and Nickless (1975)
Balanus amphitrite	Atlantic coasts of Spain and Portugal	Dry	5.1	4.5–5.8	Whole	Stenner and Nickless (1975)
Balanus perforatus		Dry	11.4	10.8–12.1	Whole	Stenner and Nickless (1975)
Maia squinada		Dry	0.63	0.46–0.74	Edible tissue only	Stenner and Nickless (1975)
Nephrops norvegicus	Atlantic coasts of Spain and Portugal	Dry	2.00	0.70–3.30	Edible tissue only	Stenner and Nickless (1975)
Cancer pagarus	Northumberland coast, U.K.	Wet	0.36	—	Whole	Wright (1976)
Carcinus maenas	Northumberland coast, U.K.	Wet	0.98	—	Whole	Wright (1976)
Crangon vulgaris	Northumberland coast, U.K.	Wet	3.50	—	Whole	Wright (1976)
Leander serratus	Northumberland coast, U.K.	Wet	2.80	—	Whole	Wright (1976)
Carcinus maenas	Lower Medway Estuary, U.K.	Wet	1.22	0.7–3.0	Whole	Wharfe and Van den Broek (1977)
Crangon vulgaris	Lower Medway Estuary, U.K.	Wet	0.59	0.53–0.65	Whole	Wharfe and Van den Broek (1977)

are difficult to perceive in field studies, as noted previously. However, Young and Jan (1976) reported a possible suppression of cadmium uptake in the rock scallop *Hinites multirugosus* by coexisting DDT or other pesticides. This example is similar to that of McDermott and Young (1974) concerning Dover sole from the same study areas; once again, confirmation of the hypothesis would be a difficult undertaking.

The data cited in this section for the phylum Mollusca have emphasized the gastropod and lamellibranch molluscs, ignoring the cephalopods. The reason for this emphasis is the much greater amount of data published for the former groups; cephalopod molluscs have been comparatively little studied. Eustace (1974), in studies of a wide range of species from the Derwent Estuary in Tasmania, Australia, found <0.05 μg Cd/g wet weight in a squid (*Notodarius gouldi*) and an octopus, identified as *Octopus* sp. Leatherland and Burton (1974) reported 0.11 μg Cd/g in gills and 0.03 μg/g in the mantle (both by dry weight) of the cuttlefish *Sepia officinalis* from The Solent, United Kingdom. A National Oceanic and Atmospheric Administration (1975) report quoted cadmium concentrations (by wet weight for whole animals) of 0.18 to 0.34 μg/g for the short-finned squid *Illex illecebrosus* from the North Atlantic, <0.05 to 0.54 μg/g for the Pacific squid *Loligo opalescens* from California coasts, and 0.25 to 1.94 μg/g for the octopus *Polypus marmuratus* from Hawaii. Stenner and Nickless (1975) gave one figure of 8.0 μg/g by dry weight for cadmium in the legs of *Octopus vulgaris* from the Atlantic coast of Spain. In regard to the tissue distribution of cadmium in cephalopods, Martin and Flegal (1975) reported very high levels of the metal in livers of three squid species. Livers from *Loligo opalescens* contained 22.6 to 265.5 μg Cd/g by dry weight, whereas those of *Ommastrephes bartrami* contained 71 to 694 μg/g and 427 to 1106 μg/g was found in livers of *Symplectoteuthis oualaniensis*. The livers of squids thus appear comparable to the digestive glands of scallops in their ability to concentrate very large amounts of cadmium. Martin and Flegal also found a high correlation of cadmium levels with concentrations of silver and copper in squid livers; conceivably this reflects a common storage mechanism for these elements. No data concerning the effects of season, age, or other parameters on the concentrations of cadmium in cephalopod molluscs are known; further research in this area appears worthwhile.

5.4.   Crustaceans

Although crustaceans have been widely employed as test animals for cadmium bioassays and for studies of cadmium uptake in the laboratory, their use as indicators of cadmium pollution in the field is much less frequent. However, some studies of cadmium levels in crustaceans have been reported, the objective of

Table 38. Selected Data for the Concentrations of Cadmium (means or ranges, $\mu g/g$ dry weight) in Macroalgae from Various Locations

Species	Area	Cadmium Concentration ($\mu g/g$)		Remarks	References
		Mean	Range		
Chlorophyceae	Irish Sea, U.K.	0.86	—	One species	Mullin and Riley (1956)
Phaeophyceae	Irish Sea, U.K.	—	0.13–2.08	Variation with species; 5 species total	Mullin and Riley (1956)
Rhodophyceae	Irish Sea, U.K.	—	0.84–0.86	Two species	Mullin and Riley (1956)
Fucus vesiculosus	Severn Estuary, U.K.	—	15–220	Variation with location; polluted area	Butterworth et al. (1972)
Fucus vesiculosus	Severn Estuary, U.K.	—	2–40	Cf. Butterworth et al. (1972)	Nickless et al. (1972)
Fucus vesiculosus	Five areas around coasts of U.K.	—	0.05–21.0	1961 and 1970 data for 5 areas	Preston et al. (1972)
Porphyra umbilicalis	Coasts of Irish Sea	—	0.05–0.97	1970 data, 2 areas only	Preston et al. (1972)
Fucus vesiculosus	Bristol Channel, U.K.	13.8	3.8–25.6	Variation with location; polluted area	Fuge and James (1974)
Fucus vesiculosus	Caernarvon Bay, U.K.	2.4	0.9–4.3	Variation with season	Fuge and James (1974)
Ascophyllum nodosum	Hardangerfjord, Norway	6.5	0.7–16.0	Variation with location; polluted area	Haug et al. (1974)

Species	Location				Reference
Ascophyllum nodosum	Trondheimsfjord, Norway	—	<0.7–1.0	Most samples <0.7 µg Cd/g	Haug et al. (1974)
Halidrys siliquosa	Lee, U.K.	0.43	—	Variation with species	Leatherland and Burton (1974)
Laminaria digitata	Lee, U.K.	0.15	—		Leatherland and Burton (1974)
Laminaria saccharina	Lee, U.K.	0.35	—		Leatherland and Burton (1974)
Ascophyllum nodosum		—	1.0–11.5		Stenner and Nickless (1974b)
Enteromorpha sp.	All species from Hardangerfjord and Skjerstadfjord, Norway	—	0.7–13.0	Variation for each species with location: limited data for 7 other species also reported	Stenner and Nickless (1974b)
Fucus serratus		—	2.3–13.0		Stenner and Nickless (1974b)
Fucus vesiculosus		—	1.8–12.5		Stenner and Nickless (1974b)
Laminaria digitata		—	2.3–4.3		Stenner and Nickless (1974b)
Nereocystis luetkeana	Off Vancouver, B.C.	—	1.3–2.0	Fronds > stipes	Whyte and Englar (1974)
Ulva lactuca	Pole Harbour, U.K.	2.5	1.0–4.8	Variation with location	Boyden (1975)

Table 38. *Continued*

Species	Area	Cadmium Concentration (μg/g)		Remarks	References
		Mean	Range		
Enteromorpha sp.	Atlantic coasts of Spain and Portugal	—	0.8–7.4	Variation with location for each species; limited data for 5 other species also reported	Stenner and Nickless (1975)
Fucus sp.		2.4	1.7–3.2		Stenner and Nickless (1975)
Ulva lactuca		1.2	0.5–2.0		Stenner and Nickless (1975)
Fucus vesiculosus	Coasts of Cornwall, U.K.	2.2	0.9–8.5	Variation with location	Bradfield et al. (1976)
Ascophyllum nodosum	Two locations in North Wales, U.K.	1.65	1.5–1.8	Did not relate to differences in cadmium levels in water	Foster (1976)
Fucus vesiculosus		1.95	1.8–2.1		Foster (1976)
Fucus sp.[a]	Looe Estuary, U.K.	1.30	0.9–2.4	Variation with location; cf. other species	Bryan and Hummerstone (1977)

[a] *Fucus vesiculosus* mostly; *F. ceranoides* at 2 stations.

these generally being the protection of public health; some crabs and other crustacean species may contain large amounts of cadmium, even in the edible muscle tissues (cf. finfish in Section 5.2, which exhibit low levels of the metal in edible muscle, apparently due to metabolic regulation). Selected data from the literature concerning the concentrations of cadmium found in crustaceans in the field are shown in Table 37.

Some information on the tissue distribution of cadmium in crustaceans is available from metal uptake studies (see Section 4). In addition, several authors have reported tissue distributions of the metal in crustaceans collected from the field. In general, cadmium is distributed more evenly amongst the tissues of crabs than is the case for molluscs. Topping (1973b) reported higher levels of the metal in liver + gonad and in kidney of *Cancer pagarus* than in claw, body meat, or gills; however, the differences were only about one order of magnitude. Similar data were also reported by other authors for various species (Ministry of Agriculture, Fisheries, and Food, 1973; Wright, 1976). No data describing the effects of season, age, size, or other parameters on the concentrations of cadmium found in crustaceans are known.

5.5. Other Organisms

The literature contains scattered data alluding to the concentrations of cadmium found in organisms other than finfish, molluscs, or crustaceans. For example, some information on the metal levels in echinoderms has been reported (Mullin and Riley, 1956; Riley and Segar, 1970; Eustace, 1974; Leatherland and Burton, 1974; Stenner and Nickless, 1974b, 1975), as has data for annelids (e.g., Bryan and Hummerstone, 1973; Leatherland and Burton, 1974; Wharfe and Van den Broek, 1977). However, the use of these species as indicators of cadmium is suspect as no information is available on the effects of ancillary parameters that may affect the concentrations of the metal present in the animal. In addition, none of these species is important as a food source for human beings; thus they do not represent a significant hazard for public health even if they contain high levels of metals.

The only group of organisms other than those considered above with potential as indicators of cadmium pollution is the macroalgae. These organisms respond to trace metals in solution (Phillips, 1977b) and exhibit long time-integration of ambient soluble levels of metals. They have been used particularly in estuarine areas, where they may complement data from other indicators responding to different portions of the total trace metal load (e.g., bivalve or gastropod molluscs). In addition, the ability of macroalgae to concentrate cadmium from the ambient seawater is an important factor in the translocation of this element from solution into the food web. Table 38 presents selected data for the concentrations of cadmium reported in macroalgae from various locations. Most of these data are discussed in detail in the review of Phillips (1977b); therefore the use of macroalgae as indicators of cadmium will not be considered at length here.

The levels of cadmium encountered in macroalgae are quite high and represent a (dry weight-based) concentration factor of about 10^3 or 10^4 over water. Phytoplankton probably exhibit concentration factors similar to those found in macroalgae for cadmium. The high levels of the metal present in these primary producers account for the predominance of the food route of cadmium uptake in higher organisms, as the elevated levels of the metal in primary producers are transmitted directly through the food web via herbivores to the carnivores of higher trophic levels. This process may be most important in determining the exact environmental effect of a given discharge containing cadmium. Phillips (1977c, 1978a) has suggested that the phytoplankton productivity of a region dictates to a large extent the availability of cadmium (and other trace metals) to higher organisms of that region, as the discharged metal is taken up in the first instance primarily by phytoplankton, and the resultant concentrations are a function of the standing crop. Thus waters of low primary productivity may exhibit more efficient transmission of metals through the food chain, and the polluting capacity of a given amount of cadmium differs according to the area of discharge. One consequence of this theory is that nutrient enrichment of coastal waters (due, e.g., to sewage pollution) would tend to decrease the concentrations of metals transmitted through the food web because of greater metal dilution by the standing crop of phytoplankton. Improvement in water quality as regards nutrients might, therefore, coincide with greater problems with trace metal pollutants.

6.   PARAMETER EFFECTS ON CADMIUM TOXICITY AND ACCUMULATION: SPECULATION

6.1.   The Relation between Toxicity and Accumulation

Bryan (1971, 1976) published tables showing the factors that may influence the toxicity of trace metals in solution. The later version (Bryan, 1976) is shown as Table 39. Many of these factors have been alluded to in the present chapter and have been shown to affect either the toxicity of cadmium to marine biota or the accumulation of the metal by these organisms, or both. This section briefly explores hypotheses to account for the effects of such parameters in an attempt to understand the uptake of cadmium by marine organisms at the biochemical level.

Bryan (1971, 1976) has suggested that factors which influence the toxicity of metals in solution may do so primarily by affecting the rate of net uptake of

Table 39. Factors Influencing the Toxicity of Trace Metals in Solution to Marine and Estuarine Biota[a]

Form of metal in water

Soluble
{ Inorganic
  Organic
{ Ion
  Complex ion
  Chelate ion
  Molecule

Particulate
{ Colloidal
  Precipitated
  Adsorbed

Presence of other metals or toxicants
{ Joint interaction:
  No interaction:
  Antagonism:
{ More than additive
  Additive
  Less than additive

Factors influencing the physiology of organisms and possibly the form of the metal in water
{ Temperature
  pH
  Dissolved oxygen
  Light
  Salinity

Condition of organism and behavioral response
{ Stage in life history (egg, larva, etc.)
  Changes in life cycle (molting, reproduction, etc.)
  Age and size
  Sex
  Starvation
  Activity
  Additional protection (e.g., shell)
  Adaptation to metals
  Altered behavior

[a] After Bryan (1976).

549

these metals into the test organisms. Certain examples supporting this hypothesis have been reported (e.g., see O'Hara, 1973a, 1973b), although the available information is sparse. It should be noted that a particular factor need not influence toxicity by promoting whole-body uptake of an element (i.e., by increasing the total body load or concentration of a metal in a whole organism); as toxicity is based on the effect of the element on a target organ (or on target organs), a factor may influence toxicity merely by altering the concentration of metal in this organ. An excellent example of this effect is found in the studies of Vernberg and Vernberg (1972a, 1972b) and Vernberg and O'Hara (1972) on mercury toxicity to the fiddler crab *Uca pugilator*.

Theoretically, the above mechanism linking the toxicity of trace metals to their net uptake is not the sole possible explanation for the effects of parameters on trace metal toxicity. In fact, two possibilities exist:

1. Factors that influence toxicity do so via promoting or inhibiting metal uptake into the whole organism or into the target organ for toxicological effects (as above).
2. Factors that influence toxicity do so by decreasing the concentration threshold (in the whole organism or in the target organ) at which the toxic response (sublethal or lethal) is apparent. In this case no effect of the influencing factor on net uptake rates for the toxicant need exist.

The latter possibility leads to a discussion of the effects of multiple stresses on an organism. Fassett (1969) has represented the toxicity of chemicals to biota by a curve similar to a stress-strain engineering curve (Figure 46). The response of the organism depends on the interaction of stress (exposure to toxicant) with strain (which may be considered here as parameters affecting the toxicity of the chemical, e.g., salinity, temperature, organism condition). The distinction between the two alternatives listed above may be considered using this curve. Factors influencing metal uptake rates tend to compress or expand the horizontal axis (changes labeled *a* in Figure 46), changing the exposure stress (concentration) necessary to reach the limit of compensatory processes (or the concentration at which the toxic response becomes apparent). By contrast, factors influencing metal toxicities by altering the threshold level of the response shift the limiting line of compensatory changes in a horizontal direction along the stress axis (changes labeled *b* in Figure 46).

Distinguishing between these possibilities in practice is most difficult, not least because of the uncertainties concerning the target organ(s) responsible for the lethal or sublethal toxic effects of trace metals. However, no clear example of any factor interfering with metal toxicity by altering the response threshold of the organism for that metal is known. Thus most or all effects of parameters altering the toxicities of trace metals (including cadmium) appear to be mediated

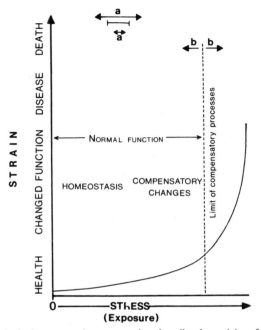

Figure 46. Hypothetical stress-strain curve used to describe the toxicity of a chemical to biota and the effects of interfering parameters on that toxicity. For explanation of changes (*a*) and (*b*), which are additions of the present author, please refer to text. (After Fassett, 1969.)

via their influence on the net uptake of these elements into the whole organism or a particular target organ. It is therefore important to consider possible mechanisms by which factors such as salinity and temperature may alter the uptake rates of cadmium in biota. An insight into these mechanisms may allow prediction of the occurrence of such effects in different species or phyla.

6.2. The Effects of Salinity

Both the toxicity of cadmium to biota (Section 3) and the accumulation of the metal by biota (Section 4) are significantly affected by the ambient salinity. In almost all the examples reported to date, a decrease in salinity leads to an increase in the toxicity, or the net uptake, of the metal. This general effect of salinity is specific for cadmium; for example, Phillips (1976a) found that, although the net uptake of cadmium by the mussel *Mytilus edulis* was greater at lower salinities, the net uptake of lead was less and that of zinc was unaffected under the same conditions. Zinc uptake by this species could, however, be increased by fluctuating salinities (Phillips, 1977a). These differences between the uptake kinetics of the various metals show that the observed salinity effects are not based on a simple physiological effect, for example, changes in the organism filtration rate. The salinity effects must be based on a specific change in the metal uptake process occurring at the body surface of the organism (mainly the gills), or in the excretion process for cadmium. The differences between the uptake kinetics of different cadmium species (George and Coombs, 1977a) are related to these effects of salinity in that each process must reflect a biochemical event occurring at the body surface during uptake.

George and Coombs (1977a) have suggested a mechanism for cadmium uptake in *Mytilus edulis* which involves a carrier ligand necessary to transport the metal across the limiting surface membrane into the organism. This scheme is highly reminiscent of the models for ion transport in bacteria, which generally involve a protein or lipoprotein molecule used to shuttle ions across the outer lipid membrane. Although the uptake of some trace metals by higher marine organisms may be based on a pinocytosis process (e.g., see George et al., 1976), this mechanism is probably more important for particulate-associated metals, such as iron or lead, than for metals found predominantly in the soluble phase, such as cadmium. As in bacteria, therefore, the transport of cadmium into organisms across the gills or gut wall may involve a specific carrier molecule; also by analogy with bacterial transport of ions, the transport of any one ion may be linked to that of another ion or molecule. The relationship of calcium to the toxicity of cadmium has been studied quite extensively in mammals (see the review by Friberg et al., 1974). Calcium appears to protect mammals against the toxic effects of cadmium to some extent, probably by serving to decrease cadmium absorption. This factor may also have been important in itai-itai disease; the available evidence shows that the group at highest risk consists of women aged 45 or more with several children. One effect of pregnancy is to serve as a drain on body calcium; in addition, women apparently absorb less calcium than men from ingested food (Friberg et al., 1974).

The effects of salinity on the toxicity of cadmium to marine biota and on the accumulation of the metal by these organisms may be based on a similar calcium-cadmium interaction. If the salinity of a water body is halved, its content of calcium will also be halved (neglecting for the present the concentrations of calcium in freshwater inflows). A given concentration of cadmium will therefore be present at a greater cadmium/calcium competition ratio. If the (hypothetical) carrier molecule serving to transport calcium across the body surface of the animal also transports the cadmium ion (acting as a substrate analogue), more cadmium will be absorbed at lower salinities. Such an effect would also explain the influence of salinity on the toxicity of cadmium, if toxicity depends on the amount of metal accumulated (Bryan, 1971).

The foregoing hypothesis predicts changes in the toxicity and accumulation of cadmium in direct proportion to the cadmium/calcium ratio. This is sufficient to explain the majority of reports on the effects of salinity on cadmium toxicity to, and/or cadmium accumulation by, biota. However, certain exceptions to this general pattern exist, in which the toxicity of cadmium is greatest at extreme salinities and least at intermediate ones. Such profiles have been reported for the adult mummichog *Fundulus heteroclitus* (Eisler, 1971; see Figure 9), as well as for eggs or larvae of this species (Middaugh and Dean, 1977) and of the winter flounder *Pseudopleuronectes americanus* (Voyer et al., 1977). None of these authors reported data for cadmium uptake at the various test salinities, and each of the test species inhabits estuarine environments, exhibiting a salinity preference of 15–20‰. In these cases, two processes may be operating together to produce the observed salinity effects on cadmium toxicity. First, the salinity effect noted above may be operative, increasing both the toxicity and the accumulation of the metal at lower salinities. Second, however, a combined stress phenomenon may exist whereby the extra energy used by organisms maintained at extreme (low or high) salinities serves to increase the toxicity of cadmium at these salinities; this effect may or may not be linked to parallel changes in the accumulation of the metal. Support for this hypothesis may be found in reports concerning the effects of salinity in each of these species in the absence of added toxicants; in each case, survival is greatest at intermediate salinities. The cadmium stress may be additive or even synergistic to the salinity effect under these conditions.

Wright (1977a, 1977b, 1977c) reported detailed studies of the relation between salinity, cadmium, and calcium, using the shore crab *Carcinus maenas*. Whole-body cadmium levels increased with decreased salinity, because of changes in the cadmium concentrations present in the hemolymph, gills, and carapace (levels of cadmium in hepatopancreas and muscle tissues were independent of salinity). These data were discussed in Section 4.3. The later papers (Wright, 1977b, 1977c) dealt mainly with cadmium in the hemolymph. The addition of calcium to the external medium generally caused a decrease in the cadmium accumulation by the tissues; this effect was most evident in the hemolymph but was also observed in other tissues and in whole-body concentrations. This represents powerful evidence for the hypothesis of a cadmium-calcium competition in marine animals, as suggested above. In addition, Wright (1977c) observed a rise in hemolymph calcium in intermolt or postmolt animals kept in cadmium-contaminated water, and suggested that this may be due to competition between the two elements for binding ("deposition") sites.

Further studies are needed on the relation between cadmium toxicity and accumulation and calcium; at present it appears that a competition between the two ions at uptake (either via solution across the gills, or via food across the gastrointestinal tract) may explain much of the salinity dependence observed to date in studies of the uptake and toxic effects of cadmium.

6.3. The Effects of Temperature

In general, as noted in Sections 3 and 4, the effect of higher temperature is to increase the toxicity and/or accumulation of cadmium in marine biota. Once again, this effect is specific for the metal; the uptake of other metals present at the same time may be unaffected by temperature increases (Phillips, 1976a). Indeed, the toxicity of organochlorine pesticides is often greater at lower temperatures (Phillips, 1978b), and a similar profile has been reported for mercury toxicity to the fiddler crab *Uca pugilator,* although this is rare for trace metals (Vernberg and O'Hara, 1972). If the effects of temperature on cadmium accumulation by biota were to be ascribed simply to changes in the metabolic rates or physiological processes (e.g., ventilation rate, feeding rate) of the organisms, one would expect all metals to be affected in a similar fashion (assuming that the uptake routes for the various metals are similar).

Cairns et al. (1975) suggested that the toxicity of cadmium to the mummichog *Fundulus heteroclitus* is greater at higher temperatures (data of Eisler, 1971; see Section 3.2) because the rate of drinking is greater. Although some evidence suggests that the uptake of cadmium from solution by drinking in this species is significant (e.g., see Gardner and Yevich, 1969, 1970), this mechanism is again based on a general physiological change and would therefore be coupled with higher toxicities of all metals with higher temperatures; there is no evidence for such a general effect of temperature. The present author thus favors specific effects as the basis for the influence of temperature on the toxicities of metals, operative probably at the site of metal uptake (gill or gastrointestinal wall), as suggested above for salinity effects. It should be noted in addition that parameters such as salinity and temperature may also affect excretion rates for metals in marine organisms. Such effects are at least as important as those occurring at the site of metal uptake, although research to date has been very sparse on this aspect.

6.4.   Interaction of Pollutants

Several examples have been cited in this chapter of pollutant interaction, either occurring at pollutant uptake (one component affecting the net uptake of another), or influencing pollutant toxicity (producing nonadditive effects). Such interaction of pollutants with each other is extremely important for two reasons. First, the number of toxicant combinations possible is almost infinite; thus little information is available on pollutant interactions in terms of the possible permutations. It follows that organisms are, and will remain, continually at risk from the effects of toxicants acting in combination, as no program of water quality management, however stringent, could predict the toxic effects involved. Second, the current trend toward the monitoring of pollutants by use of so-called indicator organisms (Phillips, 1977b, 1978b) is seriously threatened by the existence of pollutant interaction at the uptake stage. Thus, if an organism that is used as an indicator takes up one pollutant in accordance with the abundance of a second interfering toxicant, the profiles of contamination produced for the original component are meaningless in terms of its actual abundance.

Although pollutant interaction at uptake is not known (as yet) to be widespread, the isolated cases recorded to date should be a matter of concern. Thus Phillips (1976a, 1976b) suggested that the mussel *Mytilus edulis* should not be used as an indicator of copper in the marine environment, as the uptake of this element was affected by the presence of other metals. Jackim et al. (1977) later demonstrated an effect of zinc on cadmium uptake by the same species and by another bivalve, *Mulinia lateralis;* however, neither Phillips (1976a) nor Fowler and Benayoun (1974) could demonstrate this interaction in mytilids, and its existence requires confirmation. The uptake rates for cadmium exhibited by other species may be affected by pesticides (McDermott and Young, 1974; Nimmo and Bahner, 1976; Young and Jan, 1976), and the fiddler crab exhibits cadmium-mercury interaction of a complex nature (Vernberg et al., 1974). If the use of indicator organisms to monitor pollutants is to supplant the traditional techniques of direct measurement of these compounds in water or sediments, many more data on pollutant interaction in these organisms is needed. The tendency of many marine organisms to accumulate pollutants from seawater, long considered a disadvantage because of the potential hazard to human beings at the top of the trophic levels, may yet be turned into an advantage if these organisms can be used to monitor the coastal waters of the world, aiding in the elucidation and prediction of the insidious polluting effects of trace metals and other toxicants presently being discharged to all water masses.

REFERENCES

Abdullah, M. I. and Royle, L. G. (1974). "A Study of the Dissolved and Particulate Trace Elements in the Bristol Channel," *J. Mar. Biol. Assoc. U.K.* **54**, 581–597.

Abdullah, M. I., Royle, L. G., and Morris, A. W. (1972). "Heavy Metal Concentration in Coastal Waters," *Nature* **235**, 158–160.

Abdullah, M. I., Dunlop, H. M., and Gardner, D. (1973). "Chemical and Hydrographic Observations in the Bristol Channel during April and June, 1971," *J. Mar. Biol. Assoc. U.K.* **53**, 299–319.

Ahsanullah, M. (1976). "Acute Toxicity of Cadmium and Zinc to Seven Invertebrate Species from Western Port, Victoria," *Aust. J. Mar. Freshwater Res.* **27**, 187–196.

Ahsanullah, M. and Arnott, G. H. (1978). "Acute Toxicity of Copper, Cadmium and Zinc to Larvae of the crab *Paragrapsus quadridentatus* (H. Milne Edwards), and Implications for Water Quality," *Aust. J. Mar. Freshwater Res.* **29**, 1–8.

Arnold, D. C. (1957). "The Response of the Limpet, *Patella vulgata,* to Waters of Different Salinities," *J. Mar. Biol. Assoc. U.K.* **36,** 121–128.

Avens, A. C. and Sleigh, M. A. (1965). "Osmotic Balance in Gastropod Molluscs. I: Some Marine and Littoral Gastropods," *Comp. Biochem. Physiol.* **16,** 121–141.

Ayling, G. M. (1974). "Uptake of Cadmium, Zinc, Copper, Lead and Chromium in the Pacific Oyster, *Crasostrea gigas,* Grown in the Tamar River, Tasmania," *Water Res.* **8,** 729–738.

Badsha, K. S. and Sainsbury, M. (1977). "Uptake of Zinc, Lead and Cadmium by Young Whiting in the Severn Estuary," *Mar. Pollut. Bull.* **8**(7), 164–166.

Bahner, L. H. and Nimmo, D. R. (1975). "Methods to Assess Effects of Combinations of Toxicants, Salinity and Temperature on Estuarine Animals." In D. D. Hemphill, Ed., *Trace Substances in Environmental Health,* Vol. IX. University of Missouri, Columbia, pp. 169–177.

Ball, I. R. (1967). "The Toxicity of Cadmium to Rainbow Trout (*Salmo gairdnerii* Richardson)," *Water Res.* **1,** 805–806.

Baric, A. and Branica, M. (1967). "Polarography of Seawater," *J. Polarogr. Soc.* **13,** 4–8.

Bebbington, G. N., Mackay, N. J., Chvojka, R., Williams, R. J., Dunn, A., and Auty, E. H. (1977). "Heavy Metals, Selenium and Arsenic in Nine Species of Australian Commercial Fish," *Aust. J. Mar. Freshwater Res.* **28,** 277–286.

Benayoun, G., Fowler, S. W., and Oregioni, B. (1974). "Flux of Cadmium through Euphausiids," *Mar. Biol.* **27,** 205–212.

Benoit, D. A., Leonard, E. N., Christensen, G. M., and Fiandt, J. T. (1976). "Toxic Effects of Cadmium on Three Generations of Brook Trout (*Salvelinus fontinalis*)," *Trans. Am. Fish. Soc.* **105,** 550–560.

Bloom, H. and Ayling, G. M. (1977). "Heavy Metals in the Derwent Estuary," *Environ. Geol.* **2,** 3–22.

Bohn, A. and McElroy, R. O. (1976). "Trace Metals (As, Cd, Cu, Fe, and Zn) in Arctic Cod, *Boreogadus saida,* and Selected Zooplankton from Strathcona Sound, Northern Baffin Island," *J. Fish. Res. Board Can.* **33,** 2836–2840.

Boyce, R. and Herdman, W. A. (1898). "On a Green Leucocytosis in Oysters Associated with the Presence of Copper in Leucocytes," *Proc. R. Soc. London* **62,** 30–38.

Boyden, C. R. (1974). "Trace Element Content and Body Size in Molluscs," *Nature,* **251,** 311–314.

Boyden, C. R. (1975). "Distribution of Some Trace Metals in Poole Harbour, Dorset," *Mar. Pollut. Bull.* **6**(12), 180–187.

Boyden, C. R. (1977). "Effect of Size upon Metal Content of Shellfish," *J. Mar. Biol. Assoc. U.K.* **57,** 675–714.

Boyden, C. R. and Romeril, M. G. (1974). "A Trace Metal Problem in Pond Oyster Culture," *Mar. Pollut. Bull.* **5**(5), 74–78.

Bradfield, R. E. N., Kingsbury, R. W. S. M., and Rees, C. P. (1976). "An Assessment of the Pollution of Cornish coastal waters," *Mar. Pollut. Bull.* **7**(10), 187–193.

Brooks, B. R. and Rumsby, M. G. (1965). "The Biogeochemistry of Trace Element Uptake by Some New Zealand Bivalves," *Limnol. Oceanogr.* **10,** 521–527.

Brooks, R. R. and Rumsby, M. G. (1967). "Studies on the Uptake of Cadmium by the Oyster, *Ostrea sinuata* (Lamarck)," *Aust. J. Mar. Freshwater Res.* **15,** 53–61.

Brooks, R. R. and Rumsey, D. (1974). "Heavy Metals in Some New Zealand Commercial Sea Fishes," *N.Z. J. Mar. Freshwater Res.* **8,** 155–166.

Brooks, R. R., Presley, B. J., and Kaplan, I. R. (1967). "APDC-MIBK Extraction System for the Determination of Trace Elements in Saline Waters by Atomic-Absorption Spectrophotometry," *Talanta* **14,** 809–816.

Brown, B. and Ahsanullah, M. (1971). "Effect of Heavy Metals on Mortality and Growth," *Mar. Pollut. Bull.* **2**(12), 182–187.

Bryan, G. W. (1969). "The Absorption of Zinc and Other Metals by the Brown Seaweed *Laminaria digitata,*" *J. Mar. Biol. Assoc. U.K.* **49,** 225–243.

Bryan, G. W. (1971). "The Effects of Heavy Metals (Other than Mercury) on Marine and Estuarine Organisms," *Proc. R. Soc. London,* Series B, **177,** 389–410.

Bryan, G. W. (1973). "The Occurrence and Seasonal Variation of Trace Metals in the Scallops *Pecten maximus* (L.) and *Chlamys opercularis* (L.)," *J. Mar. Biol. Assoc. U.K.* **53,** 145–166.

Bryan, G. W. (1976). "Heavy Metal Contamination in the Sea." In R. Johnston, Ed., *Marine Pollution.* Academic Press, New York, pp. 185–302.

Bryan, G. W. and Hummerstone, L. G. (1973). "Adaptation of the Polychaete *Nereis diversicolor* to Estuarine Sediments Containing High Concentrations of Zinc and Cadmium," *J. Mar. Biol. Assoc. U.K.* **53,** 839–857.

Bryan, G. W. and Hummerstone, L. G. (1977). "Indicators of Heavy-Metal contamination in the Looe Estuary (Cornwall) with Particular Regard to Silver and Lead," *J. Mar. Biol. Assoc. U.K.* **57,** 75–92.

Bryan, G. W. and Uysal, H. (1978). "Heavy Metals in the Burrowing Bivalve *Scrobicularia plana* from the Tamar Estuary in Relation to Environmental Levels," *J. Mar. Biol. Assoc. U.K.* **58,** 89–108.

Bryan, G. W., Potts, G. W., and Forster, G. R. (1977). "Heavy Metals in the Gastropod Mollusc *Haliotis tuberculata* (L.)," *J. Mar. Biol. Assoc. U.K.* **57,** 379–390.

Burkitt, A., Lester, P., and Nickless, G. (1972). "Distribution of Heavy Metals in the Vicinity of an Industrial Complex," *Nature* **238,** 327–328.

Butterworth, J., Lester, P., and Nickless, G. (1972). "Distribution of Heavy Metals in the Severn Estuary," *Mar. Pollut. Bull.* **3**(5), 72–74.

Cairns, J., Heath, A. G., and Parker, B. C. (1975). "Temperature Influence on Chemical Toxicity to Aquatic Organisms," *J. Water Pollut. Control Fed.* **47,** 267–281.

Calabrese, A., Collier, R. S., Nelson, D. A., and MacInnes, J. R. (1973). "The Toxicity of Heavy Metals to Embryos of the American Oyster *Crassostrea virginica,*" *Mar. Biol.* **18,** 162–166.

Chester, R. and Stoner, J. H. (1974). "The Distribution of Zinc, Nickel, Manganese, Cadmium, Copper, and Iron in Some Surface Waters from the World Ocean," *Mar. Chem.* **2,** 17–32.

Collier, R. S., Miller, J. E., Dawson, M. A., and Thurberg, F. P. (1973). "Physiological Response of the Mud Crab, *Eurypanopeus depressus,* to Cadmium," *Bull. Environ. Contam. Toxicol.* **10,** 378–382.

Cossa, D. and Poulet, S. A. (1978). "Survey of Trace Metal Contents of Suspended Matter in the St. Lawrence Estuary and Saguenay Fjord," *J. Fish. Res. Board Canada* **35,** 338–345.

Cronklin, R. E. and Krogh, A. (1938). "A Note on the Osmotic Behaviour of *Eriocheir* in Concentrated and *Mytilus* in Diluted Sea Water," *Z. Vergl. Physiol.* **26,** 239–241.

Cutshall, N. H., Naidu, J. R., and Pearcy, W. G. (1977). "Zinc and Cadmium in the Pacific Hake *Merluccius productus* off the Western U.S. Coast," *Mar. Biol.* **44,** 195–201.

Darracott, A. and Watling, H. (1975). "The Use of Molluscs to Monitor Cadmium Levels in Estuaries and Coastal Marine Environments," *Trans. R. Soc. S. Afr.* **41,** 325–338.

Davenport, J. (1977). "A Study of the Effects of Copper Applied Continuously and Discontinuously to Specimens of *Mytilus edulis* (L.) Exposed to Steady and Fluctuating Salinity Levels," *J. Mar. Biol. Assoc. U.K.* **57,** 63–74.

Dix, T. G., Martin, A., Ayling, G. M., Wilson, K. C., and Ratkowsky, D. A. (1976). "Sand Flathead (*Platycephalus bassensis*), an Indicator Species for Mercury Pollution in Tasmanian Waters," *Mar. Pollut. Bull.* **6**(9), 142–144.

Dryssen, D., Patterson, C., Ui, J., and Weichart, G. F. (1970). In *Methods of Detection, Measurement and Monitoring of Pollutants in the Marine Environment.* Fish. Rep. Food and Agricultural Organization, 99, Suppl. 1, pp. 37–52.

Duinker, J. C. and Nolting, R. F. (1977). "Dissolved and Particulate Trace Metals in the Rhine Estuary and the Southern Bight," *Mar. Pollut. Bull.* **8**(3), 65–71.

Dutton, J. W. R., Jefferies, D. F., Folkard, A. R., and Jones, P. G. W. (1973). "Trace Metals in the North Sea," *Mar. Pollut. Bull.* **4**(9), 135–138.

Eaton, J. G. (1974). "Chronic Cadmium Toxicity to the Bluegill (*Lepomis macrochirus* Rafinesque)," *Trans. Am. Fish. Soc.* **103,** 729–735.

Eisler, R. (1971). "Cadmium Poisoning in *Fundulus heteroclitus* (Pisces: Cyprinodontidae) and Other Marine Organisms," *J. Fish. Res. Board Can.* **28,** 1225–1234.

Eisler, R. (1974). "Radiocadmium Exchange with Seawater by *Fundulus heteroclitus* (L.) (Pisces: Cyprinodontidae)," *J. Fish. Biol.* **6,** 601–612.

Eisler, R. (1977). "Toxicity Evaluation of a Complex Metal Mixture to the Softshell Clam *Mya arenaria*," *Mar. Biol.* **43,** 265–276.

Eisler, R. and Gardner, G. R. (1973). "Acute Toxicology to an Estuarine Teleost of Mixtures of Cadmium, Copper and Zinc Salts," *J. Fish. Biol.* **5,** 131–142.

Eisler, R. and LaRoche, G. (1972). "Elemental Composition of the Estuarine Teleost *Fundulus hereroclitus* (L)," *J. Exp. Mar. Biol. Ecol.* **9,** 29–42.

Eisler, R., Zaroogian, G. E., and Hennekey, R. J. (1972). "Cadmium Uptake by Marine Organisms," *J. Fish. Res. Board Can.* **29,** 1367–1369.

Eisler, R., Barry, M. M., Lapan, R. L., Telek, G., Davey, E. W., and Soper, A. E. (1978). "Metal Survey of the Marine Clam *Pitar morrhuana* Collected near a Rhode Island (USA) Electroplating Plant," *Mar. Biol.* **45**, 311–317.

Elderfield, H., Thornton, I., and Webb, J. S. (1971). "Heavy Metals and Oyster Culture in Wales," *Mar. Pollut. Bull.* **2**(3), 44–47.

Eustace, I. J. (1974). "Zinc, Cadmium, Copper and Manganese in Species of Finfish and Shellfish Caught in the Derwent Estuary, Tasmania," *Aust. J. Mar. Freshwater Res.* **25**, 209–220.

Fassett, D. W. (1969). "General Discussion, Session III." In M. W. Miller and G. G. Berg, Eds., *Chemical Fallout.* Charles C Thomas, Springfield, Ill., pp. 325–331.

Foster, B. A. (1969). "Responses and Acclimation to Salinity in the Adults of Some Balanomorph Barnacles," *Phil. Trans. R. Soc.* (B)**256**, 377–400.

Foster, P. (1976). "Concentrations and Concentration Factors of Heavy Metals in Brown Algae," *Environ. Pollut.* **10**, 45–53.

Fowler, S. W. and Benayoun, G. (1974). "Experimental Studies on Cadmium Flux through Marine Biota." In *Comparative Studies of Food and Environmental Contamination.* International Atomic Energy Agency, Vienna, pp. 159–178.

Fowler, S. W. and Oregioni, B. (1976). "Trace Metals in Mussels from the N.W. Mediterranean," *Mar. Pollut. Bull.* **7**(2), 26–29.

Fox, H. M. and Ramage, H. (1930). "Spectrographic Analysis of Animal Tissues," *Nature* **126**, 682–684.

Fox. H. M. and Ramage, H. (1931). "A Spectrographic Analysis of Animal Tissues," *Proc. R. Soc. London,* Series B, **108**, 157–173.

Frazier, J. M. (1975). "The Dynamics of Metals in the American Oyster, *Crassostrea virginica.* I: Seasonal Effects," *Chesapeake Sci.* **16**, 162–171.

Frazier, J. M. (1976). "The Dynamics of Metals in the American Oyster, *Crassostrea virginica.* II: Environmental Effects," *Chesapeake Sci.* **17**, 188–197.

Freeman, H. C. and Sangalang, G. (1976). "Changes in Steroid Hormone Metabolism as a Sensitive Method of Monitoring Pollutants and Contaminants." In *Proceedings of the Third Aquatic Toxicity Workshop, Halifax, Nova Scotia, Nov. 2–3, 1976,* pp. 123–132.

Freeman, H. C. and Sangalang, G. (1977). "A Study of the Effects of Methylmercury, Cadmium, Arsenic, Selenium, and a PCB (Aroclor 1254) on Adrenal and Testicular Steroidogenesis *in vitro,* by the Gray Seal *Halichoerus grypus*," *Arch. Environ. Contam. Toxicol.* **5**(3), 369–383.

Friberg, L., Piscator, M., Nordberg, G. F., and Kjellström, T. (1974). *Cadmium in the Environment,* 2nd. ed. CRC Press, Cleveland, Ohio.

Fuge, R. and James, K. H. (1974). "Trace Metal Concentrations in *Fucus* from the Bristol Channel," *Mar. Pollut. Bull.* **5**(1), 9–12.

Fukai, R. and Huynh-Ngoc, L. (1976). "Copper, Zinc and Cadmium in Coastal Waters of the North-West Mediterranean," *Mar. Pollut. Bull.* **7**(1), 9–13.

Gardner, G. R. and Yevich, P. P. (1969). "Toxicological Effects of Cadmium on *Fundulus heteroclitus* under Various Oxygen, pH, Salinity and Temperature Regimes," *Am. Zool.* **9**, 1096 (abstr.).

Gardner, G. R. and Yevich, P. P. (1970). "Histological and Hematological Responses of an Estuarine Teleost to Cadmium," *J. Fish. Res. Board Can.* **27**, 2185–2196.

George, S. G. and Coombs, T. L. (1977a). "The Effects of Chelating Agents on the Uptake and Accumulation of Cadmium by *Mytilus edulis*," *Mar. Biol.* **39**, 261–268.

George, S. G. and Coombs, T. L. (1977b). "Effects of High-Stability Iron-Complexes on the Kinetics of Iron Accumulation and Excretion in *Mytilus edulis* (L.)," *J. Exp. Mar. Biol. Ecol.* **28**, 133–140.

George, S. G., Pirie, B. J. S., and Coombs, T. L. (1976). "The Kinetics of Accumulation and Excretion of Ferric Hydroxide in *Mytilus edulis* (L.) and Its Distribution in the Tissues," *J. Exp. Mar. Biol. Ecol.* **23**, 71–84.

Goldberg, E. D. (1965). "Minor Elements in Sea Water." In J. P. Riley and G. Skirrow, Eds., *Chemical Oceanography,* Vol. 1. Academic Press, New York, pp. 163–196.

Goldberg, E. D., Ed. (1972). *Baseline Studies of Pollutants in the Marine Environment and Research Recommendations.* IDOE Baseline Conference, May 24–26, 1972, New York.

Gould, E. (1977). "Alteration of Enzymes in Winter Flounder, *Pseudopleuronectes americanus,* Exposed to Sublethal Amounts of Cadmium Chloride." In F. J. Vernberg, A. Calabrese, F. P. Thurberg, and W. B. Vernberg, Eds., *Physiological Responses of Marine Biota to Pollutants.* Academic Press, New York, pp. 209–224.

Gould, E. and Karolus, J. J. (1974). *Physiological Responses of the Cunner,* Tautogolabrus adspersus, *to Cadmium.* V: *Observations on the Biochemistry.* U.S. Department of Commerce, National Oceanic and Atmospheric Administration, Tech. Rep. NMFS SSRF 681, pp. 21–25.

Gould, E., Collier, R. S., Karolus, J. J., and Givens, S. A. (1976). "Heart Transaminase in the Rock Crab, *Cancer irroratus,* Exposed to Cadmium Salts," *Bull. Environ. Contam. Toxicol.* **15**, 635–643.

Greig, R. A. and Wenzloff, D. R. (1977). "Trace Metals in Finfish from the New York Bight and Long Island Sound," *Mar. Pollut. Bull.* **8**(9), 198–200.

Greig, R. A., Nelson, B. A., and Nelson, D. A. (1975). "Trace Metal Content in the American Oyster," *Mar. Pollut. Bull.* **6**(5), 72–73.

Greig, R. A., Wenzloff, D. R., and Pearce, J. B. (1976). "Distribution and Abundance of Heavy Metals in Finfish, Invertebrates and Sediments Collected at a Deepwater Disposal Site," *Mar. Pollut. Bull.* **7**(10), 185–187.

Grimshaw, D. L., Lewin, J., and Fuge, R. (1976). "Seasonal and Short-Term Variations in the Concentration and Supply of Dissolved Zinc to Polluted Aquatic Environments," *Environ. Pollut.* **11**, 1–7.

Halcrow, W., Mackay, D. W., and Thornton, I. (1973). "The Distribution of Trace Metals and Fauna in the Firth of Clyde in Relation to the Disposal of Sewage Sludge," *J. Mar. Biol. Assoc. U.K.* **53**, 721–739.

Hardisty, M. W., Huggins, R. J., Kartar, S., and Sainsbury, M. (1974a). "Ecological Implications of Heavy Metal in Fish from the Severn Estuary," *Mar. Pollut. Bull.* **5**(1), 12–15.

Hardisty, M. W., Kartar, S., and Sainsbury, M. (1974b). "Dietary Habits and Heavy Metal Concentrations in Fish from the Severn Estuary and Bristol Channel," *Mar. Pollut. Bull.* **5**(4), 61–63.

Hart, B. T. and Davies, S. H. R. (1977). "A New Dialysis-Ion Exchange Technique for Determining the Forms of Trace Metals in Water," *Aust. J. Mar. Freshwater Res.* **28**, 105–112.

Haug, A., Melsom, S., and Omang, S. (1974). "Estimation of Heavy Metal Pollution in Two Norwegian Fjord Areas by Analysis of the Brown Alga *Ascophyllum nodosum*," *Environ. Pollut.* **7**, 179–192.

Havre, G. N., Underdal, B., and Christiansen, C. (1973). "Cadmium Concentrations in Some Fish Species from a Coastal Area in Southern Norway," *Oikos* **24**, 155–157.

Herbert, D. W. M. and Wakeford, A. C. (1964). "The Susceptibility of Salmonid Fish to Poisons under Estuarine Conditions. I: Zinc Sulphate," *Int. J. Air Water Pollut.* **8**, 251–256.

Hutcheson, M. S. (1974). "The Effect of Temperature and Salinity on Cadmium Uptake by the Blue Crab, *Callinectes Sapidus*," *Chesapeake Sci.* **15**, 237–241.

IAEA/Rudjer Boskovic (1963–1966). Annual Reports: IAEA and the "Rudjer Boskovic" Institute Research Contract 201/RB, No. 201/R1/RB and No. 201/R2/RB, July 1963 to June 1966. Cited in Maljkovic and Branica (1971).

Jackim, E. (1973). "Influence of Lead and Other Metals on Fish δ-Aminolevulinate Dehydrase Activity," *J. Fish. Res. Board Can.* **30**, 560–562.

Jackim, E. (1974). "Enzyme Responses to Metals in Fish." In F. J. Vernberg and W. B. Vernberg, Eds., *Pollution and Physiology of Marine Organisms*. Academic Press, New York, pp. 59–65.

Jackim, E., Hamlin, J. M., and Sonis, S. (1970). "Effects of Metal Poisoning on Five Liver Enzymes in the Killifish (*Fundulus heteroclitus*)," *J. Fish. Res. Board Can.* **27**, 383–390.

Jackim, E., Morrison, G., and Steele, R. (1977). "Effects of Environmental Factors on Radiocadmium Uptake by Four Species of Marine Bivalves," *Mar. Biol.* **40**, 303–308.

Johnels, A. G., Westermark, T., Berg, W., Persson, P. I., and Sjöstrand, B. (1967). "Pike (*Esox lucius* L.) and Some Other Aquatic Organisms in Sweden as Indicators of Mercury Contamination in the Environment," *Oikos* **18**, 323–333.

Joint Food and Agriculture Organization/World Health Organization Expert Committee on Food Additives (1972). *Sixteenth Report: Evaluation of Certain Food Additives and the Contaminants Mercury, Lead and Cadmium*. FAO Nutrition Meetings Rep. Series 51, WHO Tech. Rep. Series 505.

Jones, M. B. (1975). "Synergistic Effects of Salinity, Temperature and Heavy Metals on Mortality and Osmoregulation in Marine and Estuarine Isopods (Crustacea)," *Mar. Biol.* **30**, 13–20.

Kidder, G. M. (1977). *Pollutant Levels in Bivalves: a Data Bibliography*. Scripps Institute of Oceanography publication under the mussel watch program, 1977.

Knauer, G. A. and Martin, J. H. (1973). "Seasonal Variations of Cadmium, Copper, Manganese, Lead, and Zinc in Water and Phytoplankton in Monterey Bay, California," *Limnol. Oceanogr.* **18,** 597–604.

Kobayashi, J. (1970). Relation between the "Itai-itai" Disease and the Pollution of River Water by Cadmium from a Mine. In *Proceedings of the International Water Pollution Research Conference,* Vol. 1, pp. 25–52.

Lande, E. (1977). "Heavy Metal Pollution in Trondheimsfjorden, Norway, and the Recorded Effects on the Fauna and Flora," *Environ. Pollut.* **12,** 187–198.

Leatherland, T. M. and Burton, J. D. (1974). "The Occurrence of Some Trace Metals in Coastal Organisms with Particular Reference to the Solent Region," *J. Mar. Biol. Assoc. U.K.* **54,** 457–468.

Leatherland, T. M., Burton, J. D., Culkin, F., McCartney, M. J., and Morris, R. J. (1973). "Concentrations of Some Trace Metals in Pelagic Organisms and of Mercury in Northeast Atlantic Ocean Water," *Deep Sea Res.* **20,** 679–685.

Lener, J. and Bibr, B. (1970). "Cadmium Content in Some Foodstuffs in Respect of Its Biological Effects," *Vitalstoffe* **15,** 139–167.

Lloyd, R. (1960). "The Toxicity of Zinc Sulphate to Rainbow Trout," *Ann. Appl. Biol.* **48,** 84–94.

MacInnes, J. R. and Thurberg, F. P. (1973). "Effects of Metals on the Behaviour and Oxygen Consumption of the Mud Snail," *Mar. Pollut. Bull.* **4**(12), 185–186.

MacInnes, J. R., Thurberg, F. P., Greig, R. A., and Gould, E. (1977). "Long-Term Cadmium Stress in the Cunner, *Tautogolabrus adspersus,*" *Fish. Bull. Fish. and Wildl. Serv. U.S.* **75,** 199–203.

Mackay, D. W., Halcrow, W., and Thornton, I. (1972). "Sludge Dumping in the Firth of Clyde," *Mar. Pollut. Bull.* **3**(1), 7–10.

Mackay, N. J., Kazacos, M. N., Williams, R. J., and Leedow, M. I. (1975a). "Selenium and Heavy Metals in Black Marlin," *Mar. Pollut. Bull.* **6**(4), 57–61.

Mackay, N. J., Williams, R. J., Kacprzac, J. L., Kazacos, M. N., Collins, A. J., and Auty, E. H. (1975b). "Heavy Metals in Cultivated Oysters (*Crassostrea commercialis = Saccostrea cucullata*) from the Estuaries of New South Wales," *Aust. J. Mar. Freshwater Res.* **26,** 31–46.

Maljković, D. and Branica, M. (1971). "Polarography of Seawater. II: Complex Formation of Cadmium with EDTA," *Limnol. Oceanogr.* **16,** 779–785.

Martin, J. H. and Flegal, A. R. (1975). "High Copper Concentrations in Squid Livers in Association with Elevated Levels of Silver, Cadmium and Zinc," *Mar. Biol.* **30,** 51–55.

McCarty, L. S., Henry, J. A. C., and Houston, A. H. (1978). "Toxicity of Cadmium to Goldfish, *Carassius auratus,* in Hard and Soft Water," *J. Fish. Res. Board Can.* **35,** 35–42.

McDermott, D. J. and Young, D. (1974). "Trace Metals in Flatfish around Outfalls." In *Annual Report, Southern California Coastal Water Research Project, 1974,* pp. 117–121.

Mears, H. C. and Eisler, R. (1977). "Trace Metals in Liver from Bluefish, Tautog and Tilefish in Relation to Body Length," *Chesapeake Sci.* **18**, 315–318.

Middaugh, D. P. and Dean, J. M. (1977). "Comparative Sensitivity of Eggs, Larvae and Adults of the Estuarine Teleosts, *Fundulus heteroclitus* and *Menidia menidia,* to Cadmium," *Bull. Environ. Contam. Toxicol.* **17**, 645–652.

Middaugh, D. P. and Lempesis, P. W. (1976). "Laboratory Spawning and Rearing of a Marine Fish, the Silverside *Menidia menidia menidia,*" *Mar. Biol.* **35**, 295–300.

Middaugh, D. P., Davis, W. R., and Yoakum, R. L. (1975). "The Response of Larval Fish, *Leiostomus xanthurus,* to Environmental Stress Following Sublethal Cadmium Exposure," *Contrib. Mar. Sci.* **19**, 13–19.

Ministry of Agriculture, Fisheries and Food (U.K.) (1973). *Survey of Cadmium in Food.* Fourth Report, Working Party on the Monitoring of Foodstuffs for Heavy Metals.

Morris, A. W. (1971). "Trace Metal Variations in Sea Water of the Menai Straits Caused by a Bloom of *Phaeocystis,*" *Nature* **233**, 427–428.

Morris, A. W. and Bale, A. J. (1975). "The Accumulation of Cadmium, Copper, Manganese and Zinc by *Fucus vesiculosus* in the Bristol Channel," *Estuarine Coastal Mar. Sci.* **3**, 153–163.

Mullin, J. B. and Riley, J. P. (1956). "The Occurrence of Cadmium in Seawater and in Marine Organisms and Sediments," *J. Mar. Res.* **15**, 103–122.

National Oceanic and Atmospheric Administration. (1975). *First Interim Report on Microconstituent Survey.* National Marine Fisheries Service, Maryland, Feb. 19, 1975.

Negilski, D. S. (1976). "Acute Toxicity of Zinc, Cadmium and Chromium to the Marine Fishes, Yellow-Eye Mullet (*Aldrichetta forsteri* C. & V.), and Small-Mouthed Hardyhead (*Atherinasoma microstoma* Whitley)," *Aust. J. Mar. Freshwater Res.* **27**, 137–149.

Nickless, G., Stenner, R., and Terrille, N. (1972). "Distribution of Cadmium, Lead and Zinc in the Bristol Channel," *Mar. Pollut. Bull.* **3**(12), 188–190.

Nielsen, S. A. (1974). "Vertical Concentration Gradients of Heavy Metals in Cultured Mussels," *N.Z. J. Mar. Freshwater Res.* **8**, 631–636.

Nielsen, S. A. and Nathan, A. (1975). "Heavy Metal Levels in New Zealand Molluscs," *N.Z. J. Mar. Freshwater Res.* **9**, 467–481.

Nimmo, D. W. R. and Bahner, L. H. (1976). "Metals, Pesticides and PCBs: Toxicities to Shrimp Singly and in Combination." In *Estuarine Processes,* Vol. 1; *Uses, Stresses, and Adaptation to the Estuary.* Academic Press, New York, pp. 523–532.

Nimmo, D. W. R., Lightner, D. V., and Bahner, L. H. (1977). "Effects of Cadmium on the Shrimps, *Penaeus duorarum, Palaemonetes pugio* and *Palaemonetes vulgaris.*" In F. J. Vernberg, A. Calabrese, F. P. Thurberg, and W. B. Vernberg, Eds., *Physiological Responses of Marine Biota to Pollutants.* Academic Press, New York, pp. 131–184.

O'Hara, J. (1973a). "The Influence of Temperature and Salinity on the Toxicity of Cadmium to the Fiddler Crab, *Uca pugilator*," *Fish. Bull.* **71**, 149–153.

O'Hara, J. (1973b). "Cadmium Uptake by Fiddler Crabs Exposed to Temperature and Salinity Stress," *J. Fish. Res. Board Can.* **30**, 846–848.

Pascoe, D. and Mattey, D. L. (1977). "Studies on the Toxicity of Cadmium to the Three-Spined Stickleback *Gasterosteus aculeatus* L.," *J. Fish Biol.* **11**, 207–215.

Peden, J. D., Crothers, J. H., Waterfall, C. E., and Beasley, J. (1973). "Heavy Metals in Somerset Marine Organisms," *Mar. Pollut. Bull.* **4**(1), 7–9.

Pesch, G., Reynolds, B., and Rogerson, P. (1977). "Trace Metals in Scallops from within and around Two Ocean Disposal Sites," *Mar. Pollut. Bull.* **8**(10), 224–228.

Phillips, D. J. H. (1976a). "The Common Mussel *Mytilus edulis* as an Indicator of Pollution by Zinc, Cadmium, Lead and Copper. I: Effects of Environmental Variables on Uptake of Metals," *Mar. Biol.* **38**, 59–69.

Phillips, D. J. H. (1976b). "The Common Mussel *Mytilus edulis* as an Indicator of Pollution by Zinc, Cadmium, Lead and Copper. II: Relationship of Metals in the Mussel to Those Discharged by Industry," *Mar. Biol.* **38**, 71–80.

Phillips, D. J. H. (1977a). "Effects of Salinity on the Net Uptake of Zinc by the Common Mussel *Mytilus edulis*," *Mar. Biol.* **41**, 79–88.

Phillips, D. J. H. (1977b). "The Use of Biological Indicator Organisms to Monitor Trace Metal Pollution in Marine and Estuarine Environments—a Review," *Environ. Pollut.* **13**, 281–317.

Phillips, D. J. H. (1977c). "The Common Mussel *Mytilus edulis* as an Indicator of Trace Metals in Scandinavian Waters. I: Zinc and Cadmium," *Mar. Biol.* **43**, 283–291.

Phillips, D. J. H. (1978a). "The Common Mussel *Mytilus edulis* as an Indicator of Trace Metals in Scandinavian Waters. II: Lead, Iron and Manganese," *Mar. Biol.* **46**, 147–156.

Phillips, D. J. H. (1978b). "The Use of Biological Indicator Organisms to Quantitate Organochlorine Pollutants in Aquatic Environments—a Review," *Environ. Pollut.,* **16**, 167–229.

Pickering, Q. H. and Gast, M. H. (1972). "Acute and Chronic Toxicity of Cadmium to the Fathead Minnow (*Pimephales promelas*)," *J. Fish. Res. Board Can.* **29**, 1099–1106.

Pickering, Q. H. and Henderson, C. (1966). "The Acute Toxicity of Some Heavy Metals to Different Species of Warm Water Fishes," *Int. J. Air Water Pollut.* **10**, 453–463.

Portmann, J. E. (1970). *Shellfish Inf. Leafl.* 19. Ministry of Agriculture, Fisheries and Food, London, U.K.

Portmann, J. E. and Wilson, K. W. (1971). *Shellfish Inf. Leafl.* 22. Ministry of Agriculture, Fisheries and Food, London, U.K.

Preston, A. (1973a). "Cadmium in the Marine Environment of the United Kingdom," *Mar. Pollut. Bull.* **4**(7), 105–107.

Preston, A. (1973b). "Heavy Metals in British Waters," *Nature,* **242,** 95–97.

Preston, A., Jefferies, D. F., Dutton, J. W. R., Harvey, B. R., and Steele, A. K. (1972). "British Isles Coastal Waters: the Concentrations of Selected Heavy Metals in Sea Water, Suspended Matter and Biological Indicators—a Pilot Survey," *Environ. Pollut.* **3,** 69–82.

Pringle, B. H., Hissong, D. E., Katz, E. L., and Mulawka, S. T. (1968). "Trace Metal Accumulation by Estuarine Mollusks," *J. Sanit. Eng. Div. Am. Soc. Civ. Eng.* **94,** 455–475.

Rautu, R. and Sporn, A. (1970). "Beitrage zur Bestimmung der Kadmiumzufuhr durch Lebensmittel," *Nahrung* **14,** 25–47.

Raymont, J. E. G. (1972). "Some Aspects of Pollution in Southampton Water," *Proc. R. Soc. London,* Series B, **180,** 451–468.

Reisch, D. J. and Carr, R. S. (1978). "The Effect of Heavy Metals on the Survival, Reproduction, Development, and Life Cycles for Two Species of Polychaetous Annelids," *Mar. Pollut. Bull.* **9**(1), 24–27.

Riley, J. P. and Segar, D. A. (1970). "The Distribution of the Major and Some Minor Elements in Marine Animals. I: Echinoderms and Coelenterates," *J. Mar. Biol. Assoc. U.K.* **50,** 721–730.

Riley, J. P. and Taylor, D. (1968). "Chelating Resins for the Concentration of Trace Elements from Sea Water and Their Analytical Use in Conjunction with Atomic Absorption Spectrophotometry," *Anal. Chim Acta* **10,** 479–485.

Riley, J. P. and Taylor, D. (1972). "The Concentrations of Cadmium, Copper, Iron, Manganese, Molybdenum, Nickel, Vanadium and Zinc in Part of the Tropical North-East Atlantic Ocean," *Deep Sea Res.* **19,** 307–317.

Rogers, C. A. (1976). "Effects of Temperature and Salinity on the Survival of Winter Flounder Embryos," *Fish. Bull.* **74,** 52–58.

Rosenberg, R. and Costlow, J. D. (1976). "Synergistic Effects of Cadmium and Salinity Combined with Constant and Cycling Temperatures on the Larval Development of Two Estuarine Crab Species," *Mar. Biol.* **38,** 291–303.

Rosenthal, H. and Sperling, K. R. (1974). "Effects of Cadmium on Development and Survival of Herring Eggs." In J. H. S. Blaxter, Ed., *The Early Life History of Fish.* Springer, Berlin, pp. 383–396.

Sangalang, G. B. and Freeman, H. C. (1974). "Effects of Sublethal Cadmium on Maturation and Testosterone and 11-Ketotestosterone Production *in vivo* in Brook Trout," *Biol. Reprod.* **11,** 429–433.

Sangalang, G. B. and O'Halloran, M. J. (1973). "Adverse Effects of Cadmium on Brook Trout Testes and *in vitro* Testicular Androgen Synthesis," *Biol. Reprod.* **9,** 394–402.

Schroeder, H. A. (1967). "Cadmium, Chromium, and Cardiovascular Disease," *Circulation* **35,** 570–595.

Schroeder, H. A. and Balassa, J. J. (1961). "Abnormal Trace Metals in Man: Cadmium," *J. Chronic Dis.* **14,** 236–259.

Schroeder, H. A., Nason, A. P., Tipton, I. H., and Balassa, J. J. (1967). "Essential Trace Metals in Man: Zinc. Relation to Environmental Cadmium," *J. Chronic Dis.* **20**, 179–210.

Schulz-Baldes, M. (1974). "Lead Uptake from Sea Water and Food, and Lead Loss in the Common Mussel *Mytilus edulis*," *Mar. Biol.* **25**, 177–193.

Schweiger, V. G. (1957). "Die Toxikologische Einwirkung von Schwermetallsalzen auf Fische und Fischnährtiere," *Arch. Fischereiwiss.* **8**, 54–78.

Segar, D. A., Collins, J. D., and Riley, J. P. (1971). "The Distribution of the Major and Some Minor Elements in Marine Animals. II: Molluscs," *J. Mar. Biol. Assoc. U.K.* **51**, 131–136.

Shore, R., Carney, G., and Stygall, T. (1975). "Cadmium Levels and Carbohydrate Metabolism in Limpets," *Mar. Pollut. Bull.* **6**(12), 187–189.

Shumway, S. E. (1977). "Effect of Salinity Fluctuation on the Osmotic Pressure and Na^+, Ca^{2+} and Mg^{2+} Ion Concentrations in the Haemolymph of Bivalve Molluscs," *Mar. Biol.* **41**, 153–177.

Skinner, B. J. and Turekian, K. K. (1973). *Man and the Ocean.* Prentice-Hall, Englewood Cliffs, N.J.

Spehar, R. L. (1976). "Cadmium and Zinc Toxicity to the Flagfish, *Jordanella floridae*," *J. Fish. Res. Board Can.* **33**, 1939–1945.

Spencer, D. W. and Brewer, P. G. (1969). "The Distribution of Copper, Zinc and Nickel in Sea Water of the Gulf of Maine and the Sargasso Sea," *Geochim. Cosmochim. Acta* **33**, 325–329.

Sprague, J. B. (1964). "Lethal Concentrations of Copper and Zinc for Young Atlantic Salmon," *J. Fish. Res. Board Can.* **21**, 17–26.

Stenner, R. D. and Nickless, G. (1974a). "Distribution of Some Heavy Metals in Organisms in Hardangerfjord and Skjerstadfjord, Norway," *Water, Air, Soil Pollut.* **3**, 279–291.

Stenner, R. D. and Nickless, G. (1974b). "Absorption of Cadmium, Copper and Zinc by Dog Whelks in the Bristol Channel," *Nature,* **247**, 198–199.

Stenner, R. D. and Nickless, G. (1975). "Heavy Metals in Organisms of the Atlantic Coast of S.W. Spain and Portugal," *Mar. Pollut. Bull.* **6**(6), 89–92.

Stevens, J. D. and Brown, B. E. (1974). "Occurrence of Heavy Metals in the Blue Shark *Prionace glauca* and Selected Pelagic Fish in the N.E. Atlantic Ocean," *Mar. Biol.* **26**, 287–293.

Sullivan, J. K. (1977). "Effects of Salinity and Temperature on the Acute Toxicity of Cadmium to the Estuarine Crab *Paragrapsus gaimardii* (Milne Edwards)," *Aust. J. Mar. Freshwater Res.* **28**, 739–743.

Talbot, V. W., Magee, R. J., and Hussain, M. (1976). "Cadmium in Port Phillip Bay Mussels," *Mar. Pollut. Bull.* **7**(5), 84–86.

Thrower, S. J. and Eustace, I. J. (1973a). "Heavy Metals in Tasmanian Oysters in 1972," *Aust. Fish.* **32**, 7–10.

Thrower, S. J. and Eustace, I. J. (1973b). "Heavy Metal Accumulation in Oysters Grown in Tasmanian Waters," *Food Technol. Aust.* **25**, 546–553.

Thurberg, F. P., Dawson, M. A., and Collier, R. S. (1973). "Effects of Copper and Cadmium on Osmoregulation and Oxygen Consumption in Two Species of Estuarine Crabs," *Mar. Biol.* **23,** 171–175.

Thurberg, F. P., Calabrese, A., Gould, E., Greig, R. A., Dawson, M. A., and Tucker, R. K. (1977). "Response of the Lobster, *Homarus americanus,* to Sublethal Levels of Cadmium and Mercury." In F. J. Vernberg, A. Calabrese, F. P. Thurberg, and W. B. Vernberg, Eds., *Physiological Responses of Marine Biota to Pollutants.* Academic Press, New York, pp. 185–198.

Topping, G. (1973a). "Heavy Metals in Fish from Scottish Waters," *Aquaculture* **1,** 373–377.

Topping, G. (1973b). "Heavy Metals in Shellfish from Scottish Waters," *Aquaculture* **1,** 379–384.

Turekian, K. K. (1969). In K. H. Wedepohl, Eds., *Handbook of Geochemistry,* Vol. 1. Springer Verlag, Berlin.

Valiela, I., Banus, M. D., and Teal, J. M. (1974). "Response of Salt Marsh Bivalves to Enrichment with Metal-Containing Sewage Sludge and Retention of Lead, Zinc, and Cadmium by Marsh Sediments," *Environ. Pollut.* **7,** 149–157.

Van der Weijden, C. H., Arnoldus, M. J. H. L., and Meurs, C. J. (1977). "Desorption of Metals from Suspended Material in the Rhine Estuary," *Neth. J. Sea Res.* **11,** 130–145.

Vernberg, W. B. and O'Hara, J. (1972). "Temperature-Salinity Stress and Mercury Uptake in the Fiddler Crab, *Uca pugilator*," *J. Fish. Res. Board Can.* **29,** 1491– 1494.

Vernberg, W. B. and Vernberg, F. J. (1972a). "The Synergistic Effects of Temperature, Salinity and Mercury on Survival and Metabolism of the Adult Fiddler Crab, *Uca pugilator,*" *Fish. Bull.* **70,** 415–420.

Vernberg, W. B. and Vernberg, F. J. (1972b). "Synergistic Effects of Temperature, Salinity and Mercury on Tissue Metabolism in the Fiddler Crab, *Uca pugilator.*" In F. H. Whitehead, Ed., *Physiological Ecology of Plants and Animals in Extreme Environments.* University of South Carolina Press, Columbia.

Vernberg, W. B., De Coursey, P. J., and O'Hara, J. (1974). "Multiple Environmental Factor Effects on Physiology and Behaviour of the Fiddler Crab, *Uca pugilator.*" In F. J. Vernberg and W. B. Vernberg, Eds., *Pollution and Physiology of Marine Organisms.* Academic Press, New York, pp. 381–425.

Vernberg, W. B., De Coursey, P. J., Kelly, M., and Johns, D. M. (1977). "Effects of Sublethal Concentrations of Cadmium on Adult *Palaemonetes pugio* under Static and Flow-Through Conditions," *Bull. Environ. Contam. Toxicol.* **17,** 16–24.

Voyer, R. A. (1975). "Effect of Dissolved Oxygen Concentration on the Acute Toxicity of Cadmium to the Mummichog, *Fundulus heteroclitus* (L.), at Various Salinities," *Trans. Am. Fish. Soc.* **104,** 129–134.

Voyer, R. A., Wentworth, C. E., Barry, E. P., and Hennekey, R. J. (1977). "Viability of Embryos of the Winter Flounder *Pseudopleuronectes americanus* Exposed to Combinations of Cadmium and Salinity at Selected Temperatures," *Mar. Biol.* **44,** 117–124.

Walker, G. (1977). " 'Copper' Granules in the Barnacle *Balanus balanoides*," *Mar. Biol.* **39,** 343–349.

Walker, G., Rainbow, P. S., Foster, P., and Holland, D. L. (1975). "Zinc Phosphate Granules in Tissue Surrounding the Midgut of the Barnacle *Balanus balanoides*," *Mar. Biol.* **33,** 161–166.

Watling, H. R. and Watling, R. J. (1976a). "Trace Metals in Oysters from the Knysna Estuary," *Mar. Pollut. Bull.* **7**(3), 45–48.

Watling, H. R. and Watling, R. J. (1976b). "Trace Metals in *Choromytilus meridionalis*," *Mar. Pollut. Bull.* **7**(5), 91–94.

Weis, J. S. (1976). "Effects of Mercury, Cadmium, and Lead Salts on Regeneration and Ecdysis in the Fiddler Crab, *Uca pugilator*," *Fish. Bull.* **74,** 464–467.

Westernhagen, H. Von and Dethlefsen, V. (1975). "Combined Effects of Cadmium and Salinity on Development and Survival of Flounder Eggs," *J. Mar. Biol. Assoc. U. K.* **55,** 945–957.

Westernhagen, H. Von, Rosenthal, H., and Sperling, K. R. (1974). "Combined Effects of Cadmium and Salinity on Development and Survival of Herring Eggs," *Helgoländer Wiss. Meeresunters.* **26,** 416–433.

Wharfe, J. R. and Van den Broek, W. L. F. (1977). "Heavy Metals in Macroinvertebrates and Fish from the Lower Medway Estuary, Kent," *Mar. Pollut. Bull.* **8**(2), 31–34.

Whyte, J. N. C. and Englar, R. J. (1974). *Elemental Composition of the Marine Alga* Nereocystis leutkeana *over the Growing Season.* Fisheries and Marine Service, Techn. Rep. 509.

Windom, H. L. and Smith, R. G. (1972). "Distribution of Cadmium, Cobalt, Nickel and Zinc in Southeastern United States Continental Shelf Waters," *Deep Sea Res.* **19,** 727–730.

Windom, H., Stickney, R., Smith, R., White, D., and Taylor, F. (1973). "Arsenic, Cadmium, Copper, Mercury, and Zinc in Some Species of North Atlantic Finfish," *J. Fish. Res. Board Can.* **30,** 275–279.

Wisely, B. and Blick, R. A. P. (1967). "Mortality of Marine Invertebrate Larvae in Mercury, Copper, and Zinc Solutions," *Aust. J. Mar. Freshwater Res.* **18,** 63–72.

Wright, D. A. (1976). "Heavy Metals in Animals from the North East Coast," *Mar. Pollut. Bull.* **7**(2), 36–38.

Wright, D. A. (1977a). "The Effect of Salinity on Cadmium Uptake by the Tissues of the Shore Crab *Carcinus maenas*," *J. Exp. Biol.* **67,** 137–146.

Wright, D. A. (1977b). "The Uptake of Cadmium into the Haemolymph of the Shore Crab *Carcinus maenas:* the Relationship with Copper and Other Divalent Cations," *J. Exp. Biol.* **67,** 147–161.

Wright, D. A. (1977c). "The Effect of Calcium on Cadmium Uptake by the Shore Crab *Carcinus maenas*," *J. Exp. Biol.* **67,** 163–173.

Yamagata, N. and Shigematsu, I. (1970). "Cadmium Pollution in Perspective," *Bull. Inst. Publ. Health* **19,** 1–27.

Young, D. R. and Jan, T.-K. (1976). "Metals in Scallops." In *Annual Report, Southern California Coastal Water Research Project, 1976,* pp. 117–121.

Zaroogian, G. E., Pesch, G., and Morrison, G. (1969). "Formulation of an Artificial Sea Water Suitable for Oyster Larvae Development," *Am. Zool.* **9,** 1144–1147.

Zirino, A. and Healy, M. L. (1972). "pH-Controlled Differential Voltammetry of Certain Trace Transition Elements in Natural Waters," *Environ. Sci. Technol.* **6,** 243–249.

Zirino, A. and Yamamoto, S. (1972). "A pH-Dependent Model for the Chemical Speciation of Copper, Zinc, Cadmium, and Lead in Seawater," *Limnol. Oceanogr.* **17,** 661–671.

13

CADMIUM TOXICITY
TO PHYTOPLANKTON
AND MICROORGANISMS

P. T. S. Wong

Canada Centre for Inland Waters, Burlington, Ontario, Canada

C. I. Mayfield

Department of Biology, University of Waterloo, Waterloo, Ontario, Canada

Y. K. Chau

Canada Centre for Inland Waters, Burlington, Ontario, Canada

1. INTRODUCTION

Cadmium is unique among chemicals in nature since it is always found in association with zinc. However, zinc is an essential trace element in living cells, whereas cadmium has no known useful biological function. In fact, cadmium is ranked among the most hazardous trace elements in the environment.

The principal uses of cadmium are in metal electroplating, in alloys, as a stabilizing material for plastics, and in batteries. Erosion and weathering of rocks and soil contribute far less cadmium to the aquatic environment than do human activities. It has been reported that mine waters contain up to 42 mg Cd/l (Fleischer et al., 1974). Tenny and Stanley (1967) examined more than 1000 samples of industrial wastes in the Chicago area and found that 1.4% contained more than 10 mg Cd/l and 0.3% contained more than 50 mg/l. The concentration of cadmium in natural waters, however, is usually less than 1 μg/l (Friberg et al., 1974); in the oceans it varies from 0.04 to 0.3 μg/l (Nordberg, 1974). The main inputs into aquatic systems are water from mine wastes, sewage, land leachates, atmospheric dust, and industrial effluents (Fleischer et al., 1974; Lagerwerff, 1967, 1971; Friberg et al., 1974).

Cadmium can exist in water as complexes with organic matter, as chelates, adsorbed onto organic material in the form of particulate matter or detritus, adsorbed onto inorganic matter, or in the form of the free ion. These forms may behave differently in terms of toxicity and availability to algae and microorganisms. In the case of microorganisms, the secondary decomposers in aquatic systems, the presence of cadmium in other living organisms, especially if these organisms accumulate the metal, is also of importance. Much of the microbial activity in fresh water is concentrated in the sediment (Wetzel, 1975), and so the levels and availability of cadmium in sediment are of primary importance in terms of toxic effects on processes carried out by sediment microorganisms. The complexing of metal ions with various organic molecules plays an important role in the toxicity of such metals to bacteria, as well as to other organisms. Little information is available regarding the effect of cadmium on aquatic biota, particularly at the primary producer level, as shown by the fact that the reviews published on cadmium toxicity in fresh water concentrate mainly on fish (Anonymous, 1977; Ray and Coffin, 1977).

2. TOXICITY

2.1. Algae

Studies on the effects of cadmium on algae have been primarily concerned with determining the toxic concentration of the ion in a single species. Very little is

Table 1.   Some Examples of Cadmium Toxicity Test Results in the Literature

Algal Species	Toxic Concentration (μg Cd/l)	Comments	References
Anacystis nidulans	10,000	40% growth inhibition in laboratory medium	Sparling (1968)
Asterionella formosa	10	Growth cessation in continuous culture	Conway (1978)
Chlorella sorokiniana	1,000	Growth inhibition in test tube bioassay	Moshe et al. (1972)
Gleocapsa apicola	10,000	60% growth inhibition in laboratory medium	Sparling (1968)
Lemna minor valdiviana	100	Plant dead	Hutchinson and Czyrska (1975)
Natural phytoplankton	100	40% primary production suppressed in brackish water	Pietilainen (1975)
Nostoc muscorum	10,000	30% growth inhibition in laboratory medium	Sparling (1968)
Merismopedia glauca f. *insignis*	10,000	10% growth inhibition in laboratory medium	Sparling (1968)
Myriophyllum spicatum L.	7,400 14,600	50% root weight inhibition 50% shoot weight inhibition	Stanley (1974)
Salvinia natans	50	Plant dead	Hutchinson and Czyrska (1975)
Selenastrum capricornutum	650	Laboratory study	Bartlett et al. (1974)
Scenedesmus obtusuisculus	50	Laboratory study with 2 light intensities	Monahan (1976)
Scenedesmus quadricauda	60	Laboratory study	Klass et al. (1974)

known about interactions between different metals, environmental factors, and algae in ecosystems. Some of the results of such studies are summarized in Table 1.

Klass et al. (1974) studied the effects of cadmium on algal growth. They found that as little as 6.1 ppb Cd had a significant inhibitory effect on the growth of *Scenedesmus quadricauda* in Bold's basal medium. Cadmium concentrations higher than 61 ppb severely inhibited growth. Gachter (1976) determined the effects of Hg, Cu, Zn, Cd, and Pb on algal photosynthesis in the eutrophic Lake of Alpnach and the mesotrophic Lake of Lucerne, Switzerland. Zinc and cadmium inhibited the photosynthetic activity most strongly in summer (May to September) in both lakes. A concentration of 5×10^{-7} mol Cd/l inhibited the mean photosynthesis for the year by 40% in the Lake of Alpnach and by 50% in the Lake of Lucerne. The order of toxicity for the metals studied was Hg > Cu > Cd > Zn > Pb, which coincides with the order of the electronegativity of the metals. Moshe et al. (1972) found that chromium was most toxic, followed by Cd, Cu, Ni, and Zn, when tested at 1 mg/l on *Chlorella sorokiniana*. Cadmium at 10^{-3} M was found to be nontoxic to several species of marine algae. However, cadmium was used in the form of $Cd(CH_3COO)_2$, and the diacetate could have modified the cadmium toxicity (Overnell, 1976).

In continuous culture apparatus, cadmium was shown to reduce the population growth of a freshwater diatom, *Asterionella formosa*, by an order of magnitude at the ambient level of about 2 µg Cd/l. At 10 µg Cd/l, growth of the diatom ceased. Fractionation experiments on cellular components showed that the cadmium was associated with the cell contents (Conway, 1978). Cadmium at 0.1 mg/l was found to be nontoxic to three marine organisms (*Phaeodactylum tricornutum, Chaetoceros galvestonensis,* and *Cyclotella nana*). However, the growth inhibition was observed to vary inversely with the concentration of nutrients available. Therefore toxicity must be defined in the context of a given nutrient level (Hannan and Patouillet, 1972). Berland et al. (1977) studied the sublethal effects of Hg, Cd, and Cu on the diatom *Skeletonema costatum* Cleve; the division rate was the first-affected and most sensitive parameter, more so than maximum growth yield, mean cell volume, particulate carbon, and nitrogen and [^{14}C]bicarbonate uptake. Cadmium (up to 100 ppb) increased the cell division rate and was then responsible for an obvious decrease. Exposure of the unicellular green alga *Chlorella vulgaris* to 10^{-3} M CdCl$_2$ resulted in decreased growth rates and survival rates of the alga, as shown by a decrease in the average number of autospores and an increase in the duration of the cell cycle (Anikeeva et al., 1975). In another study, addition of CdCl$_2$ (10^{-4} M) to the growth medium led to an increase in the lethal and mutagenic effect of chronic ultraviolet irradiation on *Chlorella vulgaris* (Kogan et al., 1975). Cadmium was found to increase the generation time (time for the algae to double once) of *Ankistrodesmus falcatus* from 20 to 50 hr when the concentration was increased from 0 to 1000 µg/l (Burnison et al., 1975). When cadmium was tested in the form of nitrate, carbonate, acetate, and chloride, the element was slightly more toxic as nitrate and acetate than as chloride or carbonate. Furthermore, the enzyme response (alkaline phosphatase activity) was more sensitive to cadmium than was the growth response of the algae.

Sparling (1968) studied the effect of cadmium in concentrations from 0.5 to 10 mg/l on four genera of blue-green algae. At 5 mg Cd/l the sensitivity of algae to the element was in the order *Gleocapsa > Nostoc > Merismopedia >*

Anacystis. At 10 mg Cd/l the order changed to *Gleocapsa* > *Anacytis* > *Nostoc* > *Merismopedia.* Hutchinson and Czyrska (1975) determined the relative toxicities of several metals to the floating aquatic weeds *Lemna minor* (valdiviana), the common duckweed, and *Salvinia natans,* a floating fern. *Salvinia* was more sensitive to cadmium than *Lemna.* The former was killed by 50 μg Cd/l, whereas the latter survived until 100 μg/l. When *Salvinia* and *Lemna* were grown together, they were more resistant to cadmium. This was explained by the lower concentrations in the plants when they were grown together. Different algal species responded differently to cadmium in a study by Burnison et al. (1975). For *Scenedesmus quadricauda* addition of 20 μg Cd/l decreased the primary productivity of the alga by 70%. For *Chlorella pyrenoidosa* 100 μg Cd/l was required to produce the same effect, and for *Ankistrodesmus falcatus* and *Chlorella vulgaris* cadmium levels as high as 1 mg/l were required.

Pietilainen (1975) compared the toxicities of lead and cadmium on natural phytoplankton communities in brackish water. Lead was found to be relatively nontoxic in low concentrations (0.3 mg/l or less), whereas an obvious inhibition of primary production occurred at the lowest cadmium concentration (0.1 mg/l). Mills and Colwell (1977) determined the effects of metals (Cd, Co, Cr, Pb, and Hg) on photosynthesis ($^{14}CO_2$ uptake) and nitrification of two pure algal cultures, *Dunaliella* sp. and *Chlorella* sp., and the nannoplankton in the surface water from Chesapeake beach. Addition of Cd, Cr, and Co up to 100 ppm did not inhibit $^{14}CO_2$ uptake, whereas Pb and Hg decreased the uptake of *Dunaliella.* On the other hand, *Chlorella* was inhibited by 25 ppm of Cd but not by 100 ppm. No apparent explanation for this interesting and reproducible observation was available. For natural populations of nannoplankton of Chesapeake Bay water, $^{14}CO_2$ uptake was inhibited by Cd, Co, and Hg but not by Cr or Pb.

2.2. Microorganisms

With respect to microorganisms and cadmium toxicity in fresh water, little is known concerning the toxic levels of cadmium in this environment. Many of the toxicological data on cadmium and microorganisms have been gathered from pure cultures of coliform bacteria, such as *Escherichia coli,* or other bacteria of medical importance (e.g., *Staphylococcus aureus*). Other studies have been performed on soil bacteria, such as members of the genus *Bacillus,* and on fungi from various sources. Data on the activities of microorganisms *in situ* in fresh water or in sediment are particularly lacking, with the exception of a few studies on model ecosystems and some research on bacteria freshly isolated from the aquatic ecosystem. Many of the laboratory studies have been concerned simply with determining the toxic level of cadmium under controlled conditions in rich media. Most have also concentrated on the effects of different cadmium levels on the growth rate of bacteria in batch culture. A few authors have attempted to examine the interrelationships between cadmium, microorganisms, and the physical and chemical features of different ecosystems, but data on the fresh-water system are again sparse. Therefore much of the following discussion on the toxicity of cadmium to freshwater microorganisms will be based primarily on information obtained from systems and microorganisms not associated with the freshwater milieu.

Various microorganisms have been studied in pure cultures with the addition of various levels of cadmium. Babich and Stotzky (1977a) tested a variety of different bacteria, including actinomycetes, and fungi for their sensitivity to cadmium. They also studied the effect of different pH levels in the broth cultures and agar plate cultures used for the bacteria and fungi, respectively. Wide variation was noted in the response of the fungi to cadmium; the yeasts tested were sensitive to levels as low as 0.1 μg/ml in the case of *Schizosaccharomyces octosporus*, with moderate growth, occurring at 500 μg/ml of *Rhodotorula* sp. The mycelial fungi could be grouped into three broad categories; the first group was capable of growth in the presence of up to 10 μg/ml but was inhibited by 100 μg/ml; the second group grew at 100 μg/ml but was inhibited by 1000 μg/ml; and the third group grew at 1000 μg/ml. Some common soil fungi, such as *Trichoderma viride* and *Rhizopus stolonifer*, were in the third group. No consistent correlation between genus and cadmium resistance was found. Some of these fungi also showed suppression of sporulation when in the presence of low cadmium levels. *Aspergillus niger* (in the second group in terms of growth response) showed a drastic reduction in spore production, to 35% of the control, when in the presence of 1 μg Cd/ml, even though this concentration did not significantly alter the mycelial growth of the fungus.

In the same study it was also found that the gram-negative eubacteria, such as *Enterobacter aerogenes*, *Proteus vulgaris*, and *Chromobacterium orangum*, were more tolerant of cadmium than were the gram-positive eubacteria, such as *Bacillus megaterium* and *Micrococcus agilis*. The range of cadmium levels necessary to produce growth inhibition ranged from 0.5 μg/ml with *Brevibacterium linens* and *Agrobacterium tumefaciens* to 10 μg/ml with *Bacillus cereus* and *Enterobacter aerogenes*. The members of the Actinomycetales that were tested were, in general, more tolerant of cadmium than the eubacteria.

In similar studies Doyle et al. (1975) found that *Escherichia coli* and *Bacillus cereus* were able to grow at 40 and 80 μg Cd/ml, but *Lactobacillus acidophilus*, *Staphylococcus aureus*, and *Streptococcus faecalis* were inhibited. In a measure of $^{14}CO_2$ production from ^{14}C-labeled glucose by *E. coli*, Zwarun (1973) concluded that the cells were totally inhibited by 12 μg/ml and that the threshold level for toxicity was 6 μg/ml. In a study on the effects of metals on glucose oxidation by the microorganisms in Chesapeake Bay and Colgate Creek, Mills and Colwell (1977) found that cadmium at a concentration of 10 ppm inhibited 84% and 59% of the glucose oxidation by water and sediment microorganisms, respectively. At 100 ppm 92% and 95% of the glucose oxidation activity was inhibited. Inhibition of bacteria from a relatively nonpolluted area was greater than that observed for the Colgate Creek samples, where the microorganisms were more metal resistant (Mills and Colwell, 1977). A heavy-metal-tolerant strain of *Pseudomonas* isolated from Chesapeake Bay sediment was able to convert cadmium in the presence of cobalamin to a volatile cadmium compound. Exposure of this volatile compound to mercury resulted in the formation of methylmercury, suggesting that the volatile compound was methylcadmium (Huey et al., 1975).

Thormann (1975), in a study of the effects of cadmium and lead on the heterotrophic bacteria of water from the estuary of the Weser River, found that all of the cadmium-sensitive bacteria isolated with a replica-plating technique were sensitive to 2 ppm Cd. This complete inhibition of growth was shown at a level of 0.2 ppm by the most sensitive strain tested. With lead-sensitive strains, all nine showed complete growth inhibition by 1 ppm, and the most sensitive strain was inhibited by 0.3 ppm. The $CdCl_2$ was more easily soluble in the brackish water than the $Pb(NO_3)_2$; therefore cadmium was more toxic than lead to the natural aerobic, heterotrophic bacteria of the estuary. In fact, more bacteria grew in the presence of high lead concentrations (200 to 400 ppm) than did in the presence of cadmium.

3. ENVIRONMENTAL FACTORS

There have been very few studies on the effects of cadmium toxicity as modified by environmental factors on algae and microorganisms in freshwater systems. Such factors should include temperature, pH, light, complexing capacity, water hardness, interaction with other metals, nutrient levels, oxygen concentration, and sediment composition. Only relatively few studies have been carried out on the effects of pH, light, and interactions with other metals on cadmium toxicity to algae.

In the case of algae the growth and morphogenic responses of *Scenedesmus obtusuisculus* to cadmium were investigated in laboratory studies on the effects of light intensities of 2000 and 4000 lux. The inhibition by cadmium appeared to be related to light intensity (Monohan, 1976). The survival rate of *Chlorella vulgaris* irradiated at a lower light intensity did not differ from that of the controls during growth in media with and without 10^{-4} M $CdCl_2$. At a higher light intensity the survival rate was reduced to 18% during 8 days of cultivation (Kogan et al., 1975). Studies on the effect of cadmium and its interaction with light (depth of water) with *in situ* techniques using polyethylene carboys as enclosures in northern Green Bay, Lake Michigan, showed that the effect of the metal was greater in light (3 to 5 m depth) than in darker areas (6 to 8 m depth) (Marshall et al., 1977).

Moshe et al. (1972) found that cadmium at 1 mg/l was toxic to algae in laboratory experiments but not in some experimental oxidation ponds. In fact, cadmium at 6 mg/l had no effect. The authors attributed the negative effects to the high pH (above 8) in the oxidation ponds, which caused metal ions to precipitate in the form of hydroxides. Moshe et al. (1972) concluded that oxidation ponds may serve as good receptors of industrial wastes as long as close control is maintained on the pH levels. The toxicity of cadmium to anaerobic digesters was demonstrated to be dependent on pH when the latter was greater than 7 and essentially independent of pH when it was less than 7. This effect was found to be caused by the insolubility of $CdCO_3$ (Mosey, 1974).

Kneip (1978) found that a combination of cadmium and nickel did not clearly show additive toxicity to an ecosystem consisting of fish, plankton, and benthic organisms. Mortality could be accounted for essentially as due to cadmium alone. Addition of five metals (Hg, Cu, Zn, Cd, and Pb) to lake water showed that the metals were more toxic in combination with other metals than when they were present alone in higher concentrations (Gachter, 1976). Moshe et al. (1972) observed no synergistic effect on *Chlorella sorokiniana* when Cd, Cu, Ni, and Cr were added together. Cadmium suppressed the stimulatory effect of zinc on the aquatic weeds *Lemna* and *Salvinia*. In fact, zinc accentuated cadmium toxicity (Hutchinson and Czyrska, 1975). Low concentrations (0.1 to 1.0 mg/l) of lead increased the toxicity of cadmium (0.1 mg/l) in a study by Pietilainen (1975). However, antagonism was apparent when the concentration of lead was greater than the concentration of cadmium. Synergism was observed in solutions where the concentration of cadmium was greater than the concentration of lead. Cadmium toxicity was found to be more pronounced when it was added to lake water than to a laboratory medium. Chemical analyses of the lake water revealed that it contained Pb, Cu, Zn, Ni, and Cd; therefore its higher toxicity was probably caused by the synergistic effects of the metals (Burnison et al., 1975).

Kneip (1978) studied the effects of the metals on the ecosystem in the Foundry Cove area, New York. No effect of cadmium was observed on the plankton, fish, and benthic organisms. In laboratory experiments a level of 1 μg Cd/l had no effect on organisms other than *Daphnia*. Serious effects appeared only when the cadmium concentration exceeded 10,000 mg Cd/kg sediment (on a dry weight basis).

The effects of environmental factors on cadmium toxicity to microorganisms have been examined mainly in the soil ecosystem. Examples of such studies in fresh water are rare, and the effects of many environmental factors important in the aquatic habitat have not been examined.

The toxicity of cadmium to fungi seems to be related to the pH of the environment. In broth cultures Babich and Stotzky (1977a) showed that alkaline pH increased the toxicity of 10 ppm Cd to *Aspergillus niger, Trichoderma viride,* and *Rhizopus stolonifer.* In another study by the same workers (Babich and Stotzky, 1977b) the influence of the clay minerals kaolinite and montmorillonite on cadmium toxicity to fungi was examined. It was apparent that both clay minerals, especially montmorillonite, protected the fungi against cadmium toxicity (at both lethal and inhibitory concentrations). The mechanism was ascribed to the cation exchange capacity of the clay minerals; the cadmium was exchanged with cations such as H, K, Na, Mg, and Ca on the clays. Since montmorillonite has a higher cation exchange capacity, it can afford greater protection than kaolinite. This conclusion was further reinforced by the observation that clay which was homoionic for cadmium was extremely toxic to the fungi when they were grown on nutrient agar, since the cadmium was then itself exchanged and was toxic to the fungi.

The effect of pH and clay minerals on the toxicity of cadmium to bacteria was also studied by Babich and Stotzky (1977a, 1977b). Results were similar to those obtained with fungi; alkaline pH levels increased the toxicity of cadmium at 10 ppm to the bacteria tested, except that the streptomycete *Streptomyces olivaceus* showed no increased susceptibility to the metal over the pH range 4 to 9. Again the presence of clay minerals decreased the toxic effect of cadmium. In most freshwater systems a large part of the cadmium is found in the bottom sediments, and the amount in the aqueous phase is low (Nordberg, 1974). In water with a low organic load the cadmium will be in the form of the Cd ion at normal pH levels. As the water becomes more acidic, the amount of Cd ion increases.

Mowat (1976) measured the toxicity of 11 common metals, including cadmium, using the standard biochemical oxygen demand test. The 5-day toxicities at 20 mg/l were in the order $Hg > Ag > Cr^{3+} > Al > Fe > Cu > Ni > Cd > Co > Cr^{4+} > Sn > Zn$. There was a trend toward lower toxicity with increasing amounts of suspended solids, and all of the metals examined showed decreases in toxicity at higher levels of suspended solids after 2 weeks of incubation.

The influence of other metal ions on the toxicity of cadmium was studied by Mitra et al. (1975). In the case of *Escherichia coli,* the addition of zinc shortened the lag phase in the growth of the bacteria when it was treated with cadmium. The addition of zinc for longer periods before the addition of cadmium shortened the lag phase even more when the cells were pretreated with zinc. In a similar manner the inhibition of *E. coli* by cadmium was relieved by the addition of 20 ppm Mg to a medium containing an inhibitory (2 ppm) concentration of cadmium. At a concentration of 6 ppm Cd, however, this alleviation by 20 ppm Mg was only partial (Abelson and Aldous, 1950).

4. MODE OF ACTION

Since cadmium is similar to zinc, it can displace the latter in many enzymes. Examples of enzymes inhibited by cadmium include alkaline phosphatase, alcohol dehydrogenase, carbonic anhydrase, dipeptidase, and aldolase (Vallee and Ulmer, 1972).

For the freshwater alga *Chlamydomonas reinhardii,* cadmium at 4×10^{-5} M was found to inhibit the light-induced oxygen evolution but had no effect on the Hill reaction and the modified Mehler reaction (Overnell, 1976). Silverberg (1976) determined whether ultrastructural changes in any of the cell organelles of three freshwater algae could be correlated with cadmium toxicity. He found that only the mitochondria exhibited morphological changes upon exposure to $CdCl_2$. Other organelles, such as the chloroplasts and endoplasmic reticulum, appeared normal. This observation suggests that the mitochondrion was the primary target for cadmium-associated cytotoxicity. Bazzaz and Govindjee (1974) showed that 0.5 mM $Cd(NO_3)_2$ caused complete inhibition of pigment system II reactions, in addition to changes in the concentration and composition of pigments in chloroplasts from maize. Monahan (1976) observed that cadmium at 50 $\mu g/l$ was a morphogenic factor in the growth of *Scenedesmus obtusuisculus.* Chlorosis was a common symptom of cadmium toxicity to two floating aquatic weeds (Hutchinson and Czyrska, 1975). It was suggested that cadmium affected the feedback control of the permease system that transported phosphate (or silicate) into the cells of *Asterionella formosa.* Cadmium was also shown to reduce the photosynthetic pigment content (Conway, 1978). Addition of cadmium to water from Chesapeake Bay completely inhibited nitrification as measured by loss of NH_4^+ and accumulation of NO_2^- (Mills and Colwell, 1977).

Cadmium was shown to interact with the phospholipid monolayer, implying that it may have a toxic effect on biological membranes. That this is indeed the case is shown by a kinetic analysis of the interactive effects of cadmium and nitrate on phytoplankton growth (Li, 1978). When cells of *Thalassiosira fluviatilis* were stressed by cadmium, not only was the maximum specific growth rate reduced, but also the half-saturation growth constant was increased, indicating interaction between cadmium and nitrate. Cadmium was more toxic at low than at high nitrate concentration, but the difference in severity diminished at higher cadmium levels. These results convinced Li (1978) that the algal membrane was the target of cadmium action. Prolonged cultivation of an alga in 10^{-3} M Cd medium resulted in genetic changes in *Chlorella vulgaris;* mutant clones, mainly mottled and dwarf, arose. These were unstable lethal sectorial clones (Anikeeva et al., 1975).

Studying the sortion and desorption of cadmium by the freshwater diatoms *Asterionella formosa* and *Fragilaria crotonensis,* Conway and Williams (1977) found that the cells contained approximately 2 ng Cd·$(mm^3)^{-1}$, even at the end of 2 months in a cadmium-free medium. This cadmium might represent the portion irreversibly tied up in metalloenzymes and complexed with amino acids, peptides, and proteins. The authors suggested that the <5% of the total cadmium sorbed by the cell was actively involved in the detrimental effects observed in the test population. In another study Cd, Cu, Hg, and Na_2SO_4 specifically altered some process of shoot elongation, upsetting the metabolic balance of Eurasian milfoil (*Myriophyllum spicatum* L.) (Stanley, 1974). Cells of *Ankistrodesmus falcatus, Scenedesmus quadricauda,* and *Chlorella pyrenoidosa* were cultured in medium with the addition of various levels of $CdCl_2$ for periods up to 9 days. Changes in ultrastructural morphology were first seen in cells treated with 30 μg Cd/l. These changes included dilation of the endoplasmic reticulum, mitochondrial vacuolation and granule accumulation, and loss of cell processes. Studies using X-ray energy microananalysis showed that cadmium was sequestered primarily in polyphosphate bodies within these algae. The mitochondrial changes suggest that cadmium may interfere with the production of ATP (Burnison et al., 1975).

Very little information is available on the toxicity of cadmium to freshwater microorganisms. Data are particularly lacking on the effects of the metal on activities of such microorganisms *in situ.*

Some bacteria have the ability to adjust to initially growth-inhibiting levels of cadmium. Mitra et al. (1975) showed that *Escherichia coli* can accommodate to cadmium. The cells showed an abnormal delay in the onset of proliferation when they were added to a glucose-inorganic salts medium containing 3 $\times$ 10^{-6} M Cd. It was found that the addition of 10^{-6} M Zn decreased the effect of cadmium in that it resulted in a shorter lag phase before growth occurred. No growth was seen with a 3 $\times$ 10^{-5} M concentration of cadmium. The accommodation of the cells to cadmium involved exclusion of the metal from the cell; in accommodated cells 56% of the cadmium was associated with the cell wall, 13% in the membranes, and 31% in the cytoplasm, whereas in the unaccommodated cells the corresponding values were 2%, 75%, and 23%. Cadmium resistance was found to begin decreasing in yeast after 5 passages and to be lost entirely after 17 transfers in cadmium-free medium (Ashida, 1965).

The penicillinase plasmid in *Staphylococcus aureus* contains genes conferring resistance to Cd ions. The resistant cells allow calcium, but not cadmium, to enter (Kondo et al., 1974). The effect of cadmium on *Escherichia coli* cells, measured by the production of $^{14}CO_2$ from ^{14}C-labeled glucose, was studied by Zwarun

(1973). The threshold level for toxicity was 6 mg Cd/l, with total inhibition occurring at 12 mg/l (Zwarun, 1973). *Escherichia coli* and *Bacillus cereus* were able to grow at 40 and 80 μg Cd/ml, respectively, but *Lactobacillus acidophilus, Staphylococcus aureus,* and *Streptococcus faecalis* were inhibited (Doyle et al., 1975). Novick and Roth (1968) showed that *E. coli* carried resistance factors to some inorganic ions, including Cd, in plasmids in the cell.

In many different types of organisms the reaction of cadmium with intracellular constituents involves the formation of a metallothionein with proteins. This has been shown also for a marine bacterium belonging to the genus *Vibrio.* The uptake rate of cadmium was increased by the presence of cysteine, which also acted to decrease the toxicity of the metal (Gauthier and Flatau, 1977).

5. CONCLUDING REMARKS

Despite the fact that algae and microorganisms are very important in the aquatic food chains, most of the studies on cadmium toxicity have been on fishes and some on invertebrates. Not much is known, however, of the effects on algae and microorganisms. The few such studies on cadmium toxicity have usually been carried out at concentrations above the ambient level present in water. It is essential, therefore, that more toxicity experiments be done by exposing these organisms for longer periods of time at lower concentrations so as to detect any sublethal effects of cadmium. The sensitivity of individual organisms to cadmium and the effects of environmental factors such as temperature, pH, complexing capacity, and water hardness are also important areas for more studies. The whole ecosystem approach, using simple food chains either in the laboratory or in the field, should be thoroughly investigated. Additional information in these areas will certainly provide a better understanding of the effects of cadmium on algae and microorganisms and in turn on the whole ecosystem.

REFERENCES

Abelson, P. H. and Aldous, E. (1950). "Ion Antagonism in Microorganisms: Interference of Normal Magnesium Metabolism by Nickel, Cobalt, Cadmium, Zinc and Manganese," *J. Bacteriol.* **60,** 401–413.

Anikeeva, I. D., Vaulina, E. N., and Kogan, I. G. (1975). "Effect of Cadmium Ions on *Chlorella.* I: Growth, Survival and Mutability of *Chlorella,"* *Genetika* (Moscow) **11,** 78–83. *Sov Genet* (English translation) **11,** 1543–1550.

Anonymous (1977). *Report on Cadmium in Freshwater Fish: Water Quality Criteria for European Freshwater Fish.* European Inland Fisheries Advisory Commission, Food and Agriculture Organization, Techn. Pap. 30, 21 pp.

Ashida, J. (1965). "Adaptation of Fungi to Metal Toxicants," *Ann. Rev. Phytopathol.* **3**, 153–174.

Babich, H. and Stotzky, G. (1977a). "Sensitivity of Various Bacteria, Including Actinomycetes, and Fungi to Cadmium and the Influence of pH on Sensitivity," *Appl. Environ. Microbiol.* **33**, 681–685.

Babich, H. and Stotzky, G. (1977b). "Reductions in the Toxicity of Cadmium to Microorganisms by Clay Minerals," *Appl. Environ. Microbiol.* **33**, 696–705.

Bartlett, L., Rabe, F. W., and Funk, W. H. (1974). "Effect of Copper, Zinc, and Cadmium on *Selenastrum capricornutum*," *Water Res.* **8**, 179–185.

Bazzaz, M. B. and Govindjee, (1974). "Effects of Cadmium Nitrate on Spectral Characteristics and Light Reactions of Chloroplasts," *Environ. Lett.* **6**, 1–12.

Berland, B. R., Bonin, D. J., Guerin-Ancey, O. J., Kapkov, V. I., and Arlhac, D. P. (1977). "Action of Sublethal Doses of Heavy Metals on the Growth Characteristics of the Diatom *Skeletonema costatum*," *Mar. Biol.* **42**, 17–30.

Bringman, G. and Kuhn, R. (1959). "The Toxic Effects of Waste Water on Aquatic Bacteria, Algae and Small Crustaceans," *Gesundheits-Ingenieur* **80**, 115–120.

Burnison, G., Wong, P. T. S., Chau, Y. K., and Silverberg, B. A. (1975). "Toxicity of Cadmium to Freshwater Algae," *Proc. Can. Fed. Biol. Soc. Winnipeg* **18**, 182.

Conway, H. L. (1978). "Sorption of Arsenic and Cadmium and Their Effects on Growth, Micronutrient Utilization, and Photosynthetic Pigment Composition of *Asterionella formosa*," *J. Fish. Res. Board Can.* **35**, 286–294.

Conway, H. L. and Williams, S. C. (1977). *Sorption and Desorption of Cadmium by* Asterionella formosa *and* Fragilaria crotonensis. Argonne National Laboratory, Radiological and Environmental Research Division Annual Report on Ecology, Argonne, Ill., pp. 51–53.

Doyle, J. J., Marshall, R. T., and Pfander, W. H. (1975). "Effects of Cadmium on the Growth and Uptake of Cadmium by Microorganisms," *Appl. Microbiol.* **29**, 562–564.

Fleischer, M. Sarofim, A. F., Fassett, D. W., Hammond, P., Shacklette, H. T., Nisbet, I. C. T., and Epstein, S. (1974). "Environmental Impact of Cadmium—a Review by the Panel on Hazardous Trace Substances," *Environ. Health Perspect.*, Environ. Issue 7, 253–323.

Friberg, L., Piscator, M., Nordberg, G. F., and Kjellstrom, T. (1974). *Cadmium in the Environment.* CRC Press, Cleveland, Ohio.

Gauthier, M. and Flatau, G. (1977). "Concentration and Method of Cadmium Fixation by a Marine *Vibrio*," *C. R. Acad. Sci.* (*Paris*) **285**, 817–820.

Gachter, R. (1976). "Untersuchungen uber die Beerinflussung der planktischer durch anorganische Metallsalze im eutrophen Alpnachersee und der mesotrophen Horwerbucht," *Schweiz. Z. Hydrol.* **38**, 97–119.

Huey, C. W., Brinckman, F. E., Iverson, W. P., and Grim, S. O. (1975). "Bacterial Volatilization of Cadmium." In *Proceedings of the International Conference on Heavy Metals in the Environment* (Abstr.), Toronto, Ont., C214-C216.

Hannan, P. J. and Patouillet, C. (1972). "Effect of Mercury on Algal Growth Rates," *Biotech. Bioeng.* **14,** 93–101.

Hutchinson, T. C. and Czyrska, H. (1975). "Heavy Metal Toxicity and Synergism to Floating Aquatic Weeds," *Verh. Int. Ver. Limnol.* **19,** 2102–2111.

Klass, E., Rowe, D. W., and Massaro, E. J. (1974). "The Effect of Cadmium on Population Growth of the Green Alga *Scenedesmus quadricauda,"* *Bull. Environ. Contam. Toxicol.* **12,** 442–445.

Kneip, T. J. (1978). "Effects of Cadmium in an Aquatic Environment." In *Proceedings of the 1st International Cadmium Conference, San Francisco, January 1977,* pp. 120–124.

Kogan, I. G., Anikeeva, I. D., and Vanlina, E. N. (1975). "Effect of Cadmium Ions on *Chlorella.* II: Modification of the UV Irradiation Effect," *Genetika* (Moscow), **11,** 84–87. *Sov. Genet.* (English translation) **11,** 1550–1553.

Kondo, I., Ishikawa, T., and Nakahara, H. (1978). "Mercury and Cadmium Resistances Mediated by the Penicillinase Plasmid in *Staphylococcus aureus,"* *J. Bacteriol.* **117,** 1–7.

Lagerwerff, J. V. (1967). "Heavy Metal Contamination of Soils." In N. C. Bradley, Ed., *Agriculture and the Quality of Our Environment.* American Association for the Advancement of Science, Washington, D.C.

Lagerwerff, J. V. (1971). "Uptake of Cadmium, Lead and Zinc by Radish from Soil and Water," *Soil Sci.* **111,** 129–133.

Li, W. K. W. (1978). "A Kinetic Analysis of the Interactive Effects of Cadmium and Nitrate on Phytoplankton Growth." In *Proceedings of the 41st Annual Meeting of the American Society on Limnology and Oceanography, University of Victoria, B.C.*

Marshall, J. S., Mellinger, D. L., and Saber, D. L. (1977). *Effects of Cadmium Enrichment on a Lake Michigan Plankton Community.* Annual Report, Argonne National Laboratory, Radiological and Environmental Research Division, Argonne, Ill., pp. 59–64.

Mills, A. L. and Colwell, R. R. (1977). "Microbiological Effects of Metal Ions in Chesapeake Bay Water and Sediment," *Bull. Environ. Contam. Toxicol.* **18,** 99–103.

Mitra, R. S., Gray, R. H., Chin, B., and Bernstein, I. A. (1975). "Molecular Mechanisms of Accommodation in *Escherichia coli* to toxic levels of Cd," *J. Bacteriol.* **121,** 1180–1188.

Monahan, T. J. (1976). "Effects of Cadmium on the Growth and Morphology of *Scenedesmus obtusuisculus,"* *J. Phycol.* **12** (Suppl.), 98.

Mosey, F. E. (1974). "The Toxicity of Cadmium to Anaerobic Digestion: Its Modification by Inorganic Ions," *Water Pollut. Control.* **43,** 584–598.

Moshe, M., Betzer, N., and Kott, Y. (1972). "Effect of Industrial Wastes on Oxidation Pond Performance," *Water Res.* **6,** 1165–1171.

Mowat, A. (1976). "Measurement of Metal Toxicity by Biochemical Oxygen Demand," *J. Water Pollut. Cont. Fed.* **48,** 853–866.

Nordberg, G. (1974). "Health Hazards of Environmental Cadmium Pollution," *Ambio* **3,** 55–66.

Novick, R. P. and Roth, C. (1968). "Plasmid-Linked Resistance to Inorganic Salts in *Staphylococcus aureus*," *J. Bacteriol.* **95,** 1335–1342.

Overnell, J. (1975). "The Effect of Some Heavy Metal Ions on Photosynthesis in a Freshwater Alga," *Pesticide Biochem. Physiol.* **5,** 19–26.

Overnell, J. (1976). "Inhibition of Marine Photosynthesis by Heavy Metals," *Mar. Biol.* **38,** 335–342.

Pietilainen, K. (1975). Synergistic and antagonistic of lead and cadmium on aquatic primary production. Intern. Conf. on Heavy Metals in the Environment. Symposium Proc., 2 (Pt. 2); 861–873.

Ray, S. and Coffin, J. (1977). *Ecological Effects of Cadmium Pollution in the Aquatic Environment.* Fisheries and Marine Service, Canada, Tech. Rep. 734, pp. 1–18.

Silverberg, B. A. (1976). "Cadmium-Induced Ultrastructural Changes in Mitochondria of Freshwater Green Algae," *Phycologia* **15,** 155–159.

Sparling, A. B. (1968). "Interactions between Blue-Green Algae and Heavy Metals." Ph.D. thesis, Washington University, St. Louis, Mo., 107 pp.

Stanley, R. A. (1974). "Toxicity of Heavy Metals and Salts to Eurasian Water Milfoil (*Myriophyllum spicatum L.*)," *Arch. Environ. Contam. Toxicol.* **2,** 331–341.

Tenny, A. M. and Stanley, G. H. (1967). *Application of Atomic Absorption Spectroscopy for Monitoring Selected Metals in an Industrial Waste.* Purdue Univ. Eng. Bull. Ext. Ser. 129.

Thorman, D. (1975). "Effects of Cadmium and Lead on the Indigenous Heterotrophic Bacterial Flora in the Brackish Estuarine Water of the Weser River," *Ver. Inst. Meeresforsch. Bremerhaven* **15,** 237–267.

Vallee, B. L. and Ulmer, D. D. (1972). "Biochemical Effects of Mercury, Cadmium and Lead," *Ann. Rev. Biochem.* **41,** 91–128.

Wetzel, R. G. (1975). *Limnology.* W. B. Saunders, Philadelphia.

Zwarun, A. A. (1973). "Tolerance of *Escherichia coli* to Cadmium," *J. Environ. Qual.* **2,** 353–355.

14

SOIL-PLANT-ANIMAL DISTRIBUTION OF CADMIUM IN THE ENVIRONMENT

R. P. Sharma

Utah State University, Logan, Utah

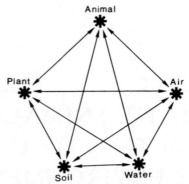

Figure 1. Movement of cadmium from one component of the ecosystem to another. An increase of cadmium concentration in any component will tend to cause its redistribution in all different phases.

1. INTRODUCTION

Cadmium is a relatively rare element and is uniformly distributed throughout the earth's crust. The metal has an appreciable vapor pressure and escapes in the atmosphere during smelting operations and in the processing of cadmium-containing alloys. The burning of coal and oil, cadmium-weighted plastics, and sewage sludge also contribute to atmospheric contamination by cadmium. Because of its coexistence with zinc and other metals in geological sources, relatively high emissions of cadmium are possible from zinc, copper, and lead smelting plants, from the flue dusts of zinc-blend roasting, and from the electrolytic purification of zinc.

Although cadmium is ubiquitous in the environment, its presence has not been associated with any essential biological function, and excessive amounts have been shown to be toxic to higher organisms. Cadmium has been found in air in concentrations of <0.01 to 0.35 μg/m^3 (U.S. Department of Health, Education, and Welfare, 1966), with levels usually highest in cities with considerable industrial activity. Cadmium levels in most soil are about 0.4 ppm (Fleisher et al., 1974). The soil obtained in the vicinity of smelters, however, may contain considerably higher amounts of cadmium (Dorn et al., 1975). Sewage sludge also contains higher amounts of cadmium, in concentrations ranging from 3 to over 3000 ppm (Somers, 1977).

2. CADMIUM AND THE BIOLOGICAL FOOD CHAIN

Cadmium is one of the elements that are selectively accumulated in various organs of higher animals. The primary reason for its differential accumulation is the presence of a specific metal-binding protein called metallothionein (Kagi and Valee, 1960). This protein, sometimes referred to as cadmium thionein, binds cadmium and other metals like Zn, Hg, and Cu. The presence of this protein in organs like liver and kidney, and its induction by exposure of animals to high amounts of metals that it binds, are responsible for the high levels of cadmium found in these organs.

An increase of cadmium in the components of an ecosystem will undoubtedly result in an increased consumption of this metal by animals. Animals may obtain the additional supply of cadmium directly from soil, air, water, or plants containing large amounts of this metal or by consuming the organs (e.g., liver and kidney) of animals that are exposed to this element. Figure 1 graphically illustrates the movement of cadmium in different components of the environment. Animals at the higher end of the food chain are exposed to larger amounts, particularly if they consume organs and tissues containing high levels of cadmium. It has been shown that the cadmium bound to proteins (as in metallothionein) is more selectively distributed after absorption than the metal available as a free chemical in a salt form (Cherian et al., 1978).

3. SOIL-PLANT RELATIONSHIP FOR CADMIUM

As indicated earlier, most soils contain very small amounts (<1 ppm) of cadmium (Allaway, 1968). Cadmium is added to soil by the use of fertilizers (Lee and Keeney, 1975), the disposal of sewage sludges on lands (Section 4), and in effluents from various industrial operations such as zinc and cadmium smelters. Plants grown on these soils, however, do not accumulate high amounts of cadmium until the soil cadmium level is very high.

There is a considerable lack of information regarding the chemistry of cadmium in different types of soils. The availability of cadmium is believed to be influenced by soil organic matter, clay content, pH, and redox potential. These factors ultimately determine the amount of soluble cadmium in soil and thus the level of the metal available to plants growing on it.

4. SEWAGE SLUDGE AS A SOURCE OF SOIL CADMIUM

Application of sewage sludge on farmlands has been suggested and practiced primarily for the value of sludge as a soil conditioner and fertilizer. It is also a practical method for the disposal of sludges. Nearly 25% of the sludge produced in the United States is currently applied to the land. In addition to the nutritive factors the sludge may add to the soil, it also contains some toxic elements, including cadmium, that are of current interest to scientists and legislators. A

Table 1.   Cadmium Concentrations in Various Types of Sludges

Type of Sludge	Cadmium (μg/g dry sludge)	Reference
From lagoons filled before 1969	190	McCalla et al. (1977)
From anaerobic digester, 1972	100	Peterson et al. (1973)
From aerobic digester, 1974	54	McCalla et al. (1977)
Sewage, anaerobic	160	Somers (1977)
	Range 3–3410	
Sewage, aerobic	135	Somers (1977)
	Range 5–2170	
Domestic	5	U.S. EPA (1976)
(from newer communities)		
Controlled municipal	25	Cheney et al. (1976)
(without excessive industrial		
waste source)		
Whole dry	87	U.S. EPA (1976)
	Range 0–1100	

report regarding the concern about heavy metals has been released by the U.S. Environmental Protection Agency (1976).

Table 1 lists the levels of cadmium observed in sewage sludges from different sources. The numbers are merely representative of what the levels might average and suggest that most of the cadmium in sludge is contributed by industrial processes. Domestic sludge has relatively low amounts of cadmium. A wide range of variability is apparent from the figures listed in Table 1.

It has been suggested that the contribution of sewage in increasing the cadium burden of soil is relatively minor as compared to other superphosphate fertilizers (Lee and Keeney, 1975). Moreover, the cadmium present in sludges may be in a bound form and hence not readily available to plants growing on soils treated with these sludges. Several experiments have been conducted to determine the bioavailability of cadmium from such soils. Table 2 illustrates some of the representative results from selected investigations. From the data in Table 2, it is clear that an increase in soil cadmium is related to the increase in cadmium levels in corn leaf. On the other hand, increasing amounts of cadmium in the grain can be produced only by applying high concentrations of cadmium to soil ($>$10 kg Cd/ha). Such high levels are unlikely in normal sludge disposal practices, especially if the sludge is not from industrial sources.

5. SOIL CADMIUM IN RELATION TO INDUSTRIAL ACTIVITY

Soils obtained from the vicinity of industrial operations, particularly smelting and other coal burning processes, generally have increased levels of cadmium.

Table 2.   Cadmium Concentrations in Various Parts of Corn Plants Grown on Soils Treated with Cadmium-Containing Sludges

Cadmium Applied (as sludge) (kg/ha)	Cadmium in Dry Leaf (ppm)	Cadmium in Dry Grain (ppm)	Reference
0	0.04	<0.04	Clapp et al. (1976)
1.3	0.12	<0.04	
2.6	0.19	<0.04	
5.1	0.25	<0.04	
0	0.05		Baker et al. (1976)
2.2	0.45		(1973 data)
4.5	1.07		
0		0.09	Kelling et al. (1976)
0.2		0.08	
0.53		0.10	
1.07		0.09	
2.15		0.09	
4.30		0.10	
0		0.10	Hinesly et al. (1976)
12.9		0.21	
25.7		0.57	
51.4		0.96	

Klein (1972) reported that soil samples obtained from industrial regions contained 1.5 times more cadmium, on an average, than those from residential areas. Levels of up to 69 μg Cd/g soil were found in rice fields of Japan where cadmium contamination was suspected from industrial sources (Yamamoto, 1972). Dorn et al. (1975) reported similar findings in soil samples obtained in the vicinity of lead smelters. Corresponding increases in cadmium in roots and aerial parts of plants were also reported in samples obtained from this industrial region.

6.   PLANT AND SOIL CADMIUM IN RELATION TO AUTOMOBILE EMISSIONS

Hemphill et al. (1973) reported that the level of cadmium decreased from 1.51 to <0.5 μg/g in grass samples obtained from 0 to 200 yards from a roadside. In similar findings reported by Lagerwerff and Specht (1971) the cadmium levels in plants were related to the traffic volume on the highway and also to the distance of the plants from the road. A significant decrease in soil cadmium levels

Table 3.   Cadmium Residues (μg/g fresh weight, 2 to 6 samples analyzed) in Usual Animal Rations[a]

Feed	Mean ± SE	Range
Hay	0.18 ± 0.04	0.14–0.33
Dairy grain ration	0.17 ± 0.04	0.12–0.23
Soybean meal	0.19 ± 0.01	0.18–0.20
Commercial hog ration	0.23 ± 0.04	0.18–0.30
Commercial chicken feed	0.32 ± 0.02	0.29–0.35

[a] R. P. Sharma and J. C. Street, unpublished data.

was reported with increasing distance from the road by Gish and Christensen (1973).

7.   UPTAKE OF CADMIUM BY PLANTS FROM SOIL

A number of factors influence the movement of cadmium from soil to plants. In addition to the soil factors mentioned above, crop species and varieties may have a major influence on the cadmium concentration in plant tissues. The growth rate of plants (and, indirectly, temperature and nutrients) and other available metals in soil may also be contributing factors for soil-plant cadmium movement. Increasing the soil pH has been shown to reduce the cadmium uptake by plants (Chaney et al., 1976; Hornick et al., 1976). Lettuce, tomato, eggplant, and cantaloupe showed an increase in their cadmium contents in various parts of the plant when grown on soil supplemented with cadmium, whereas no such accumulation of the metal was observed in broccoli, potato, or string beans (Giordano and Mays, 1977).

Considerable variation of cadmium levels in different tissues of plants has been observed. Corn grain contains a much smaller amount of cadmium than does the leaf (<15%), whereas the grain of soybean, wheat, oat, and sorghum may have cadmium levels closer to those in the levels of the plants (Chaney et al., 1976; Chaney and Giordano, 1976). The chemical nature of cadmium in plants is not well known.

8.   PLANT-SOIL-ANIMAL RELATIONSHIPS FOR CADMIUM

There is very little information regarding the natural plant-animal relationship for cadmium in the uncontaminated environment. It has been estimated that the daily cadmium uptake of most animals is relatively low and the cadmium

Table 4. Cadmium Levels in Selected Tissues of Animals

Tissue	Concentration (ppm)	Reference
Human liver	200, tissue ash	Tipton and Cook (1963)
Rat kidney (6 weeks old)	1.13 ± 0.22, dry	Fowler et al. (1975)
Rat kidney (12 weeks old)	0.83 ± 0.32, dry	Fowler et al. (1975)
Ewe liver	0.95 ± 0.37, dry	Mills and Delgarno (1972)
Lamb liver	0.14 ± 0.01, whole	Mills and Delgarno, (1972)
Lamb liver (at birth)	<0.14, whole	Mills and Delgarno (1972)
Lamb plasma (at birth)	0.21 ± 0.02	Mills and Delgarno (1972)
Calf blood	0.01 ± 0.0046	Lynch et al. (1976)
Cow muscle	0.054 ± 0.049, wet	Sharma et al. (1978)
Swine muscle	0.072 ± 0.004, wet	Sharma et al. (1978)
Swine kidney cortex	0.06, wet	Sharma et al. (1978)
Swine bone	0.44 ± 0.17, wet	Sharma et al. (1978)
Swine brain	0.14 ± 0.01, wet	Sharma et al. (1978)
Chicken muscle	0.06 ± 0.05, wet	Sharma et al (1978)

levels in animal organs gradually increase with age. The data presented in Table 3 suggest that most animal feeds contain less than 0.3 ppm Cd, although these values may be higher than those in most unprocessed human foods. A considerable amount of cadmium may be contributed to animal feed by the deposition of soil on plant tissues rather than the incorporation of cadmium in the plant itself. The values reported in Table 4 indicate the cadmium levels in different animal tissues. It is apparent that high amounts of this element are present primarily in liver and kidney of various animals.

The values reported for different animal tissues should be regarded with caution. The values may have been reported on an ash, dry, or wet tissue basis, and data from different sources are often difficult to compare. Advances in analytical technology have provided more accurate and sensitive means of metal detection, yet considerable variation has been observed between different laboratories and workers, sometimes using the same techniques. Most of the reports do not provide the justification for, or validity of, the analytical method employed, and large differences may thus be found.

Wild animals generally have a larger cadmium body burden than do domestic animals. This may depend on the cadmium content of their diet. Table 5 presents cadmium levels in various tissues of mule deer. Schroeder and Balassa (1961) have reported high cadmium levels in liver and kidney of deer, red squirrel, and rabbit. Cadmium levels as high as 2 ppm in liver (wet weight) and more than 17 ppm in kidney of squirrels were reported. The values in these animals were two- to ten-fold higher than the corresponding values for domestic animals reported by these workers.

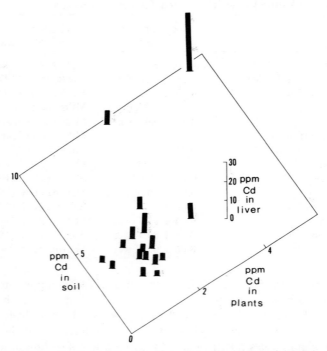

Figure 2. Relationship of cadmium in soil (surface, 0–5 cm), plants (aerial parts), and animal liver (from rock squirrels). Significant positive correlations between the three were noted and are given in Table 6. (Based on data from Sharma and Shupe, 1977.)

Sharma and Shupe (1977) reported the relationship of cadmium levels in soil, plant, and animals collected from 18 different regions. Soil samples were collected for 2 years, and during this period 332 plant specimens (representing 33 species) and 113 animals (109 rock squirrels, *Spermophilus varigatus*, and 4 pack rats, *Neotama cineren*) were obtained. The values of cadmium in samples from each region were averaged. Although there were considerable variations in cadmium values in different samples from the same location, highly significant correlations between the soil, plant, and animal cadmium levels were reported. The overall average cadmium contents ranged from 1 to 27 ppm in animal liver, 0.5 to 5 ppm in vegetation, and 1 to 10 ppm in surface soil. The relationships for soil-plant-animal cadmium values are shown graphically in Figure 2 with their corresponding correlations indicated in Table 6.

The results presented by Sharma and Shupe (1977) suggested that increasing amounts of cadmium in soil and plants are likely to increase the body burdens

Table 5.   Cadmium Levels (µg/g dry tissue) in Different Selected Tissues of Mule Deer (*Odocoileus hemionus*)[a]

Tissue	Number of Samples Averaged	Mean ± SE[b]	Range of Values[b]
Bone	21	5.58 ± 0.39	3.0–8.1
Kidney cortex	7	8.30 ± 1.80	4.8–18.2
Liver	10	2.78 ± 0.28	1.4–3.9
Lung	3	2.83 ± 0.07	2.7–2.9
Blood	4	0.10 ± 0.01	0.09–0.12

[a] R. P. Sharma et al., unpublished data.
[b] The relationship between liver and kidney of seven selected animals was significant at $p < 0.01$ ($r = 0.8725$). The regression equation was Cd in kidney cortex = 5.2233 × Cd in liver − 4.2120.

Table 6.   Linear Correlation Coefficients of the Mean Cadmium Concentrations in Different Components in the Environment[a]

	Vegetation	Soil[b]
Liver[c]	.7701*[e]	.6560*
Vegetation[d]		.8827*

[a] From Sharma and Shupe (1977). The data are shown graphically in Figure 2.
[b] Surface soil, 0 to 5 cm from surface.
[c] Liver samples from rock squirrels (*Spermophilus varigatus*).
[d] Plant samples representing 33 different species were obtained and analyzed without washing.
[e] Asterisk indicates significant correlation at $p < .01$.

of this element in animals. Higher significant correlations at the three values (Table 6) indicate that a slight increase of cadmium in soil or plants will provide a corresponding rise in cadmium intake by the animals and ultimately a tissue accumulation of the metal.

9.   CADMIUM ACCUMULATION IN INVERTEBRATES

Gish and Christensen (1973) reported the concentrations of various metals, including cadmium, in earthworms in relation to the cadmium residues in soil. The earthworms contained approximately 11 times more cadmium than was present in their surrounding soil (on a dry matter basis). When the samples were collected from 3 to 50 m away from the highway, the cadmium values in soil ranged from 0.66 to 1.59 ppm, whereas the values in earthworms were found to be from 3 to 14.4 ppm. The amounts of cadmium in the worms showed a significant positive correlation with soil cadmium and soil organic matter, and were also related to the moisture content in the worms. Since earthworms provide a source of food for many birds, as well as certain amphibians, reptiles, and mammals, the authors suggested that there may be an increased accumulation of cadmium and also possibly cadmium toxicity in predators eating the worms. This may be a possible mechanism of cadmium translocation from soil to higher animals.

10. CADMIUM LEVELS IN ANIMAL TISSUES WITH INCREASING CADMIUM EXPOSURE

The normal human intake of cadmium has been estimated to be 51.2 μg/day (Mahaffey et al., 1975), whereas the suggested limit for dietary intake ranges from 57.1 to 71.4 μg Cd/day (Food and Agriculture Organization/World Health Organization, 1972). Because of the selective accumulation of cadmium in organs like kidney and liver, the dietary intake of this element is not likely to increase with a rise of this metal in the environment, unless these organs are consumed. Several studies have been performed to relate the dietary exposure of cadmium to the residues of this metal in various edible tissues and food products from animals.

Baker et al. (1975) fed different levels of cadmium to laying hen and broiler chickens for 12 weeks. Little increase in cadmium residues was noted in muscle of broiler chickens and eggs, and a considerable increase in these two edible products was observed only when the dietary cadmium was 48 ppm. A dose-related increase in the liver and kidneys of cadmium-fed animals was observed, with kidney levels in laying hens approaching 306 ppm (dry matter).

Williams et al. (1976) fed sorghum and corn grown on soil with application of cadmium-containing sludges to meadow vole (dietary cadmium 2.76 and 1.09 ppm, dry matter, respectively) and reported that no accumulation of cadmium was observed in their muscle. The liver cadmium after a 40-day feeding trial was 1.86 and 0.43 ppm (dry matter) on sorghum and corn diets, respectively. The corresponding kidney cadmium levels were 2.84 and 0.42 ppm.

Doyle et al. (1974) added various levels of cadmium, up to 60 ppm, to the diet of lambs for 191 days. Cadmium levels in muscle and fat increased only at the highest dietary level for this metal, i.e., 60 ppm added to the diet. No increase in wool cadmium was observed at any cadmium level. A dose related increase, up to 276 ppm in liver and 469 ppm in kidney (wet basis), was observed at the 60-ppm dietary level of cadmium. Based on the total amount of cadmium in all tissues, it was estimated that 5.3% of the dietary cadmium was absorbed and retained in various organs.

Vogt et al. (1977) fed broiler chickens with various levels of dietary cadmium (1 to 80 ppm) for 4 and 7 weeks and found that kidney and liver accumulated cadmium up to 65 and 50 ppm, respectively. In breast and thigh muscles there was an increase in cadmium residues; the highest level of cadmium found in these tissues was about 0.6 ppm.

A feed-related increase in cadmium in mouse liver and kidney tissues was reported by Exon et al. (1977). After the withdrawal of cadmium-supplemented drinking water, a slight decrease in liver cadmium was noticed, the value being nearly 50% less in 180 days after stopping the cadmium feeding. On the other hand, no decrease in kidney cadmium levels was noticed for a period up to 180 days on a normal diet.

Sharma et al. (1978) have reported the relationship of dietary cadmium to the residues of this metal in different tissues of three food-producing species; cattle, swine, and chicken. Dairy cows were exposed to an equivalent of 2.4 and 11.3 ppm dietary Cd for 12 weeks. No accumulation of cadmium was observed in muscle, milk, or bone. Liver and kidney indicated a significant increase in cadmium levels at the highest exposure level (i.e., 11.3 ppm Cd in diet). At low cadmium intake (2.4 ppm) an increase in liver and kidney cadmium levels was observed, but the difference was not significant from the control group (0.2 ppm Cd in the diet). No depletion in bovine liver or kidney was observed for 12 weeks after the cadmium-supplemented diet was stopped.

Exposure of swine to cadmium-supplemented feed (2.4 and 10.1 ppm Cd vs. 0.2 ppm in control feed) provided results similar to those for dairy cows (Sharma et al., 1978). The feeding of cadmium for up to 6 months did not cause a significant rise in the cadmium residues in muscle, bone, or brain. In liver and kidney a dose- and time-related increase was observed, and the residues did not decline even when no cadmium was fed for 12 weeks after an initial 3-month exposure period.

Feeding of cadmium-supplemented diet to laying chickens did not produce any increase in egg cadmium levels even after a 6-month period (Sharma et al., 1978). After a 6-month exposure period the muscle cadmium increased to 0.14 ppm (fresh weight basis) when the dietary cadmium was 1.9 ppm (the control dietary cadmium was 0.3 ppm, and the corresponding value in control muscle tissue was 0.06 ppm), and to 0.26 ppm when the dietary metal level was 13.3 ppm. Similar increases were noted at intermediate feeding periods, but the values were not significantly different from those for controls. A time- and dose-related increase in liver and kidney cadmium was observed when the animals were given cadmium-supplemented diets. The residues in these organs did not decline appreciably within 7 weeks after exposure of the animals to cadmium for 6 weeks.

Table 7.   Liver and Kidney Concentrations of Cadmium in Various Species of Animals Given Different Levels of Dietary Cadmium and Their Relationships[a]

Species[b]	Cadmium Concentration (mean ± SE, μg/g fresh tisse)		Correlation Coefficient	Regression Equation
	In liver (x)	In kidney[c] (y)		
Chicken (47)	4.70 ± 0.98	24.31 ± 3.55	.9463*[d]	$y = 1.2846 + 5.4394x$
Swine (23)	3.01 ± 0.71	13.43 ± 3.10	.9253*	$y = 1.3457 + 4.0104x$
Cattle (12)	1.46 ± 0.38	6.31 ± 2.01	.8518*	$y = 0.3286 + 4.5427x$

[a] The animals were given various dietary levels of cadmium as cadmium chloride, ranging from 0.18 to 13.1 ppm Cd for different durations of 1 to 24 weeks. The relationship for each animal is indicated in Figures 3 to 5.
[b] The figures in parentheses indicate the number of values averaged in the table.
[c] Values are those in whole kidney of chicken, and in kidney cortex of swine and cattle.
[d] Asterisk indicates significant correlation at $p < .01$.

11.   ANIMAL BODY AS A FILTER FOR CADMIUM TRANSLOCATION IN FOOD CHAIN

The reports indicated above suggest that selected organs (namely, liver and kidney) accumulate cadmium preferentially when the levels of this element increase in the diet. The small increase in other tissues reduces the hazard of cadmium biomagnification in human beings and other animals consuming primarily the organs or foods where cadmium is not accumulated. Even after the exposure to cadmium is stopped, organs like liver and kidney do not show a depletion and sometimes even exhibit an increase in their cadmium contents (Sharma et al., 1978). This increase in cadmium suggests a redistribution of the metal from other tissues and probably an increasing binding capacity of liver and kidney for cadmium. The increase in binding capacity may result from the induction of metal-binding proteins in these organs.

The daily human consumption of cadmium is very close to the recommended maximum intake (Mahaffey et al., 1975) and perhaps will not be altered if tissues other than liver and kidney from cadmium-contaminated animals are consumed. Care should be exercised not to use liver and kidney from animals with diets that are exceedingly high in cadmium.

12.   DISTRIBUTION OF CADMIUM IN ANIMAL LIVER AND KIDNEY

In most normal animals the residues of cadmium in liver and kidney are comparable to each other (Schroeder and Balassa, 1961; Baker et al., 1975; Williams

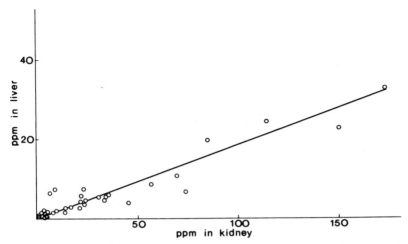

Figure 3. Relationship of chicken liver and kidney cadmium concentrations. There was no correlation when the liver and kidney cadmium values were below 2 and 5 ppm, respectively. A significant dependence of one on the other is apparent, and their correlation coefficient is listed in Table 7. All values are micrograms Cd per gram wet tissue. (Based on data from Sharma et al., 1978.)

et al., 1976; Verma et al., 1978). When the cadmium exposure of animals increases, however, the extent of cadmium accumulation in kidney is much higher than in liver. This difference may be due either to a greater inducibility of cadmium thionein in renal tissues than in hepatic ones or to a possible competition with other dietary metals in liver (Verma et al., 1978). In the studies reported by Sharma et al. (1978) the increase in renal cadmium was 4 to 5 times higher than in liver cadmium when the test animals were given cadmium-supplemented diets. The relationship between liver and kidney cadmium contents in cattle, swine, and chicken is shown in Figures 3 to 5. The increases in these two organs showed a significant positive correlation, and their overall correlation coefficients, together with the regression equations, are listed in Table 7. On the basis of these results it is suggested that a rise of cadmium in liver and kidney is expected when the animals are given high levels of this metal, although the increase in kidney is of considerably greater magnitude than that in liver.

13. DIETARY INTAKE OF CADMIUM FROM FOOD

Mahaffey et al. (1975) have surveyed the daily intake of cadmium in the U.S. population via common types of foods (Table 8). A large portion of daily cadmium is consumed through grains and cereals, potatoes, and fruits, and a smaller but considerable fraction via beverages, dairy products, meat and fish, and leafy

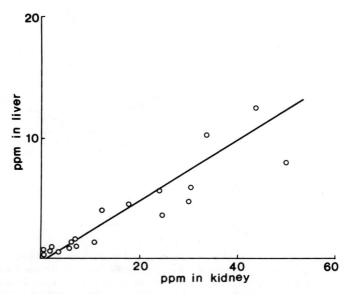

Figure 4. Relationship of cadmium concentrations in swine liver and kidney cortex. The correlation coefficient and regression equation are listed in Table 7. All values are micrograms Cd per gram wet tissue. (Based on data from Sharma et al., 1978.)

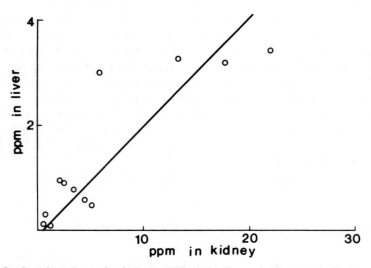

Figure 5. Interdependence of cadmium levels in bovine liver and kidney cortex. The correlation coefficient and regression equation are listed in Table 7. All values are micrograms Cd per gram wet tissue. (Based on data from Sharma, et al., 1978.)

600

Table 8.   Cadmium Levels in Selected Foods

Product	Concentration (ppm)	Reference
Sea food	0.79, wet Range 0.05–5.39	Schroeder et al. (1967)
Meats	0.88, whole Range 0.19–3.49	Schroeder et al. (1967)
Dairy products	0.27, whole Range 0.03–0.56	Schroeder et al. (1967)
Cereals and grains	0.16 Range 0.01–0.57	Schroeder et al. (1967)
Japanese cereals	0.07	Yamagata and Shigematsu (1970)
Meats (Japan)	0.032–0.052	Yamagata and Shigematsu (1970)
Fish (unprocessed)	0.015–0.027	Yamagata and Shigematsu (1970)
Clam (unprocessed)	0.16	Yamagata and Shigematsu (1970)
Clam (processed)	0.81	Yamagata and Shigematsu (1970)
Cuttle fish (unprocessed)	12–0.18, dry	Yamagata and Shigematsu (1970)
Cuttle fish (processed)	0.15–5.0, dry	Yamagata and Shigematsu, (1970)
Milk (California)	0.006 ± 0.004	Bruhn and Franke (1974)
Milk	0.014 ± 0.004	Sharma et al. (1978)
Eggs (yolk	0.07, whole	Sharma et al. (1978)

vegetables. Legumes, oils, fats, and sugar contribute a very small amount of cadmium. An illustration of cadmium levels in different types of foods that constitute our daily diet is provided in Table 8. Seafoods in general and certain meats were found to contain higher amounts of cadmium than did milk, eggs, and grains. The cadmium contents of different foods and commodities have been summarized from the market survey of The Food and Drug Administration (1975) and are listed in Table 9. The data presented in Tables 8 and 9 suggest that major contributors of dietary cadmium to human beings are seafoods and grains. One additional item evident from these tables is a consistently high cadmium level in processed foods as compared to the corresponding unprocessed materials. It is possible that processing provides a considerable amount of cadmium (and perhaps other heavy metal contaminants) in edible products.

14. CONCLUSIONS

The foregoing discussions are illustrative of the type and amount of information that is available on soil-plant-animal relationships for cadmium in the environment. Although cadmium is relatively rare (about 0.2 ppm) in the earth's

Table 9.   Cadmium Levels in Selected Foods (Food and Drug Administration, 1974 Compliance Program Evaluation)[a]

Commodity	Mean ± SD (ppm)	Range (ppm)
Carrots	0.051 ± 0.077	0–0.7
Potatoes	0.057 ± 0.139	0–0.36
Butter	0.032 ± 0.071	0–0.68
Eggs, whole	0.067 ± 0.072	0–1.48
Chicken, fryer, whole	0.039 ± 0.088	0–0.82
Bacon	0.040 ± 0.160	0–1.555
Frankfurters	0.042 ± 0.111	0–0.65
Raw liver, beef	0.183 ± 0.228	0–1.24
Ground beef	0.075 ± 0.122	0–2.56
Roast, chuck beef	0.035 ± 0.034	0–0.63
Rice, white milled	0.053 ± 0.226	0–2.16
Breakfast cereal	0.066 ± 0.267	0–2.62
Sugar	0.100 ± 0.709	0–6.16
Bread, white	0.036 ± 0.063	0–0.439
Tomato juice, canned	0.041 ± 0.059	0–0.468

[a] Food and Drug Administration (1975).

crust, the average human body burden of this element has been estimated to be nearly 30 mg, an amount larger than reported for several other trace metals that are more abundant in nautre (Schroeder, 1965). Part of it may reflect the physical characteristics of the metal (relatively volatile) and its biological behavior (selective accumulation in tissues), but much of it can be attributed to the commercial use of cadmium-containing products. The industrial uses of cadmium include the plating of other metals, pigments for glass and paint, alloys, and insecticides. Other uses of cadmium are dental amalgams, small-arms ammunition, and storage batteries. Cadmium is also a by-product of lead and zinc production.

Because of its toxic nature, cadmium has attained prominence in recent years. Increased intake of this element has been shown to cause public health hazards, such as the *itai-itai* disease in Japan (Yamagata and Shigamatsu, 1970). Its possible presence in sewage sludges, and the subsequent utilization of such sludges on land, are likely to increase the amount of cadmium in soils and possibly in plants and animals. More information on its distribution in the ecosystem is therefore desirable.

It has been shown that the concentrations of cadmium in some organs of animals (i.e., liver and kidney) are usually higher than the relative levels in the surrounding environment, particularly if the cadmium intake of the organism increases. More information is needed that will relate the levels of cadmium in the different components of the environment and will identify factors important in the translocation of this metal from one phase to another in the ecosystem.

REFERENCES

Allaway, W. H. (1968). "Agronomic Controls over the Environmental Cycling of Trace Elements," *Adv. Agronom.* **20**, 235–274.

Baker, D. E., Eshelman, R. M., and Leach, R. M., (1975). "Cadmium in Sludge Potentially Harmful When Applied to Crops," *Sci. Agric.* **22**, 14–15.

Baker, D. E., Amacher, M. C., and Doty, W. T. (1976). "Monitoring Sewage Sludges, Soils and Crops for Zinc and Cadmium." In *Land as a Waste Management Alternative.* Proceedings of the 8th Annual Waste Management Conference, Rochester, N.Y.

Bruhn, J. C. and Franke, A. A. (1974). "Lead and Cadmium in California Milk," *J. Dairy Sci.,* **57**, 588.

Chaney, R. L. and Giordano, P. M. (1976). "Micro-elements as Related to Plant Deficiencies and Toxicities." In, L. E. Elliott and F. J. Stevenson, Eds. *Soils for Management and Utilization of Organic Wastes and Waste Waters.* Soil Science Society of America, Madison, Wis.

Chaney, R. L., Hornick, S. B. and Simon, P. W. (1976). "Heavy Metal Relationships during Land Utilization of Sewage Sludge in the Northeast." In *Land as a Waste Management Alternative.* Proceedings of the 8th Annual Waste Management Conference, Rochester, N.Y.

Cherian, M. G., Goyer, R. A., and Valberg, L. S. (1978). "Comparative Studies on the Gastro-intestinal Absorption and Metabolism of Cadmium Chloride and Cadmium-Thionein in Mice," *Toxicol. Appl. Pharmacol.* **44**, 318.

Clapp, C. E., Dowdy, R. H., and Larson, W. E. (1976). In *Application of Sewage Sludge to Cropland: Appraisal of Potential Hazards of the Heavy Metals to Plants and Animals.* U.S. Environmental Protection Agency, Rep. MCD-33.

Dorn, C. R., Pierce, J. O., Chase, G. R., and Phillips, P. E. (1975). "Environmental Contamination by Lead, Cadmium, Zinc, and Copper in a New Lead Producing Area," *Environ. Res.* **9**, 159–172.

Doyle, J. J., Pfander, W. H., Grebing, S. E., and Pierce, J. O. (1974). "Effects of Dietary Cadmium on Growth, Cadmium Absorption, and Cadmium Tissue Levels in Growing Lambs," *J. Nutr.* **104**, 160–166.

Exon, J. H., Lamberton, J. G., and Koller, L. D. (1977). "Effect of Chronic Oral Cadmium Residues in Organs of Mice," *Bull. Environ. Contam. Toxicol.* **18**, 74–76.

Fleisher, M., Sarofin, A. F., Fassett, D. W., Hammond, P. B., Shacklette, H. T., Nisbet, C. T., and Epstein, S. (1974). "Environmental Impact of Cadmium," *Environ. Health Perspect.* **7**, 253–323.

Food and Agriculture Organization/World Health Organization (1972). Expert Committee on Food Additives, Rep. 16. Geneva.

Food and Drug Administration (1975). Compliance Program Evaluation, FY 1974, *Heavy Metals in Food Survey.* Bureau of Foods, U.S. Department of Health, Education, and Welfare.

Fowler, B. A., Jones, J. S., Brown, H. W., and Haseman, J. K. (1975). "The Morphologic Effects of Chronic Cadmium Administration on the Renal Vasculature of Rats Given Low and Normal Calcium Diets," *Toxicol. Appl. Pharmacol.* **34**, 233–252.

Giordano, P. M. and Mays, D. A. (1977). "Yield and Heavy Metal Content of Several Vegetable Species Grown in Soil Amended with Sewage Sludge." In H. Drucker and R. E. Wildung, Eds., *Biological Implications of Metals in the Environment.* Energy Research and Development Administration, pp. 417–425.

Gish, C. D. and Christensen, R. E. (1973). "Cadmium, Nickel, Lead and Zinc in Earthworms from Roadside Soil," *Environ. Sci. Technol.* **7**, 1060–1062.

Hemphill, D. D., Marienfeld, C. J., Reddy, R. S., Heidlage, W. D. and Pierce, J. O. (1973). "Toxic Heavy Metals in Vegetables and Forage Grasses in Missouri's Lead Belt," *J. Assoc. Offic. Anal. Chem.* **56**, 994–998.

Hinesly, T. D., Jones, R. L., Tyler, J. J., and Ziegler, E. L. (1976). "Soybean Yield Responses and Assimiliation of Zn and Cd from Sewage Sludge Amended Soil," *J. Water Pollut. Control Fed.* **48**, 2137–2152.

Hornick, S. G., Chaney, R. L., and Simon, P. W. (1976). In *Application of Sewage Sludge to Cropland: Appraisal of Potential Hazards of the Heavy Metals to Plants and Animals.* U.S. Environmental Protection Agency, Rep. MCD-33, p. 51.

Kagi, J. H. R. and Valee, B. L. (1960). "Metallothionein: a Cadmium and Zinc-Binding Protein from Equine Renal Cortex," *J. Biol. Chem.* **235**, 3460–3465.

Kelling, K. A., Keeney, D. R., Walsh, L. M., and Ryan, J. A. (1976). "A Field Study of the Agricultural Use of Sewage Sludge: Effect of Uptake and Extractability of Sludge Borne Metals," *J. Environ. Qual.* **6**, 352–358.

Klein, D. H. (1972). "Mercury and Other Metals in Urban Soils," *Environ. Sci. Technol.* **6**, 560–562.

Lagerwerff, J. V. and Specht, A. W. (1971). "Occurrence of Environmental Cadmium and Zinc and Their Uptake by Plants." In *Trace Substances in Environmental Health,* Vol. 4. University of Missouri, Columbia, pp. 85–98.

Lee, K. W. and Keeney, D. R. (1975). "Cadmium Additions to Wisconsin Soils by Commercial Fertilizer and Waste Water Sludge Applications," *Water, Air, Soil Pollut.* **5**, 109–112.

Lynch, G. P., Smith, D. F., Fisher, M., Pike, T. L., and Weinland, B. T. (1976). "Physiological Responses of Calves to Cadmium and Lead," *J. Anim. Sci.* **42**, 410–421.

Mahaffey, K. R., Corneliussen, P. E., Jelinek, C. F., and Fiorino J. A. (1975). "Heavy Metal Exposure from Foods," *Environ. Health Perspect.* **12**, 63–69.

McCalla, T. M., Peterson, J. R., and Lue-Hing, C. (1977). "Properties of Agricultural and Municipal Wastes and Waste Waters." In *Soils for Management of Organic Wastes and Waste Water.* American Society of Agronomists, Madison, Wis., pp. 9–44.

Mills, C. F. and Delgarno, A. G. (1972). "Copper and Zinc Status of Ewes and Lambs Receiving Increased Dietary Concentrations of Cadmium," *Nature* **239**, 171–173.

Peterson, J. R., Lue-Hing, C., and Denz, D. R. (1973). "Chemical and Biological Quality of Municipal Sludge." In W. L. Sopper and L. T. Kardos, Eds. *Recycling Treated Municipal Waste Water and Sludge Through Forest and Cropland*. Pennsylvania State University Press, University Park, Pa.

Schroeder, H. A. (1965). "The Biological Trace Elements," *J. Chronic Dis.* **18**, 217–228.

Schroeder, H. A. and Balassa, J. J. (1961). "Abnormal Trace Metals in Man: Cadmium," *J. Chronic Dis.* **14**, 236–258.

Schroeder, H. A., Nason, A. P., Tipton, L. H., and Balassa, J. J. (1967). "Essential Trace Metals in Man: Zinc Relation to Environmental Cadmium," *J. Chronic Dis.* **20**, 179–210.

Sharma, R. P. and Shupe, J. L. (1977). "Lead, Cadmium and Arsenic Residues in Animal Tissues in Relation to Those in Their Surrounding Habitat," *Sci. Total Environ.* **7**, 53–62.

Sharma, R. P., Street, J. C., Verma, M. P., and Shupe, J. L. (1979). "Cadmium Uptake from Feed and Its Distribution to Food Products of Livestock," *Environ. Health Perspect.*, **28**, 59–66.

Somers, L. E. (1977). "Chemical Composition of Sewage Sludge and Analysis of Their Potential Use as Fertilizers," *J. Environ. Qual.* **6**, 225–239.

Tipton, L. H. and Cook, M. J. (1963). "Trace Elements in Human Tissue. II. Adult Subjects from the United States," *Health Phys.* **9**, 103–145.

U.S. Department of Health, Education, and Welfare (1966). Cited by R. C. Dorn, *Cadmium, a Review*. FDA Rep. 3526-76F, June 1976.

U.S. Environmental Protection Agency (1976). *Application of Sewage Sludge to Cropland: Appraisal of Potential Hazards of the Heavy Metals to Plants and Animals*. Rep. MCD-33, 63 pp.

Verma, M. P., Sharma, R. P., and Street, J. C. (1978). "Hepatic and Renal Metallothionein (MT) Levels in Cows, Pigs, and Chickens Given Cadmium and Lead in Feed," *J. Am. Vet. Res.*, **12**, 1911–1915.

Vogt, H., Nezel, K., and Matthes, S. (1977). "Effect of Various Lead and Cadmium Levels in Broiler and Laying Rations on the Performance of the Birds and On the Residues in Tissues and Eggs," *Nutr. Metab.* **21** (Suppl. 1), 203–204.

Williams, P. H., Shenk, J. S., and Baker, D. E. (1976). In *Application of Sewage Sludge to Cropland: Appraisal of Potential Hazards of the Heavy Metals to Plants and Animals*. U.S. Environmental Protection Agency, Rep. MCD-33, p. 58.

Yamagata, N. and Shigematsu, I. (1970). "Cadmium Pollution in Perspective," *Bull. Jap. Inst. Pub. Health* **19**, 1–27.

Yamamoto, Y. (1972). *Present Status of Cadmium Environmental Pollution*. Japanese Association of Public Health, Kankyo Hoken, Rep. 11, 7 pp.

15

UPTAKE AND EFFECTS OF CADMIUM IN HIGHER PLANTS

Julie D. Jastrow

Division of Environmental Impact Studies, Argonne National Laboratory, Argonne, Illinois

David E. Koeppe

Departments of Agronomy and Forestry, University of Illinois, Urbana, Illinois

Although cadmium is not essential for plant growth, it is often readily taken up and accumulated by plants. Cadmium reaches plants, or the soil they grow on, by a variety of means, including water, air, pesticides, fertilizers, and solid wastes. A few studies have shown that some cadmium may be taken up through plant leaves (Little and Martin, 1972; Haghiri, 1973; Little, 1973). However, the predominant mode of cadmium uptake is through plant roots and is affected by numerous soil, plant, and environmental factors that will be discussed in this chapter.

1. CADMIUM UPTAKE—SOLUTION CULTURE STUDIES

The potential for cadmium uptake by plant roots and for eventual cadmium accumulation in aerial plant parts has been demonstrated by numerous studies in which various plant species were grown in cadmium-amended nutrient solution culture (Page et al., 1972; Turner, 1973; Cutler and Rains, 1974; Dijkshoorn et al., 1974; Carlson et al., 1975; Root et al., 1975; Jarvis et al., 1976; John, 1976; John and Van Laerhoven, 1976; Patel et al., 1976; Pettersson, 1976, 1977; Wallace et al., 1977). Turner (1973) found that some species can accumulate cadmium in their shoots when grown in solutions containing as little as 0.01 μg Cd/ml. Six vegetable species were grown in nutrient solution for 2 weeks and then transferred to solutions with added cadmium. After 2-weeks' growth in 0.01 μg Cd/ml, tomato tops accumulated 24.8 μg Cd per plant, giving a tissue concentration of 1.4 μg Cd/g dry weight. Lettuce tops accumulated similar concentrations after 5-weeks' exposure to 0.01 μg Cd/ml. Jarvis et al. (1976) measured the uptake of cadmium by 23 plant species exposed to nutrient solutions also amended with 0.01 μg Cd/ml. Fifty-five-day-old tomato and lettuce plants (among other species) were placed in the cadmium-amended solution for 3 days and then were removed to fresh nutrient solution for 4 days before harvesting. Three days after the cadmium addition, no cadmium was detected in the nutrient solution. Tomato and lettuce tops accumulated 5.6 μg Cd/g dry

Table 1. Uptake of Cadmium by Various Species from Flowing Solution[a]
Plants grown with 0.01 μg Cd/ml solution and harvested 7 days after the beginning of cadmium treatment.

	Cadmium Content (μg/g dry weight)	
Plant Species	Shoots	Fibrous Roots
Winter oat	6.6	44.2
Winter barley	2.6	48.8
Winter wheat	8.1	58.0
Maize	11.5	57.5
Perennial ryegrass	6.1	102.8
Cocksfoot	4.9	145.3
Meadow fescue	5.8	108.1
Lucerne	3.1	33.4
Tick bean	4.2	36.9
Sunflower	7.3	83.7
Kale	3.3	18.9
Fodder beet	21.1	151.1
Sorghum	17.5	59.6
Timothy	5.2	138.7
White clover	8.5	82.0
Red clover	8.3	61.8
Sainfoin	2.3	4.8
Parsnip	2.5	163.8
Carrot	4.1	57.5
Radish	1.8	33.6
Tomato	5.6	121.6
Lettuce	15.0	48.3
Watercress	21.0	77.5

[a] From Jarvis et al. (1976).

weight and 15.0 μg Cd/g dry weight, respectively. Cadmium accumulation data for the other 21 species assayed are presented in Table 1.

In experiments utilizing much higher cadmium concentrations than those of Turner (1973) and Jarvis et al. (1976), Root et al. (1975) observed tissue concentrations as high as 500 μg Cd/g dry weight after 14-day-old corn plants were transferred to nutrient solutions containing 40 μg Cd/ml for 12 days. John and Van Laerhoven (1976) transferred 2-week-old lettuce plants (var. Great Lakes 428) to nutrient solutions with 50 μg Cd/ml and after 5 weeks' exposure recorded tissue concentrations of 956.5 μg Cd/g dry weight.

2. PLANT FACTORS AFFECTING CADMIUM UPTAKE

2.1. Species and Variety Differences in Uptake

The results of a number of studies (Page et al., 1972; John, 1973; Turner, 1973) have shown that plant species exhibit genetic variability in cadmium uptake and accumulation. Table 1 illustrates the degree of this variability among 23 different species exposed to very low cadmium concentrations in a large batch solution culture, and Table 2 shows similar variability for 10 crop species grown in a greenhouse on Grenville loam soil with and without additions of $CdCl_2$.

Although the physiological mechanisms that affect species differences in cadmium uptake and accumulation are unknown at this time, experimental results generally point to genetic differences. Pettersson (1977) reported differences in the cadmium contents of seeds from 15 different barley cultivars grown on adjacent field plots. In a comparison of seeds from two wheat cultivars (also grown on adjacent field plots), Pettersson found significantly higher (averaging 1.5 times) cadmium concentrations in 'Starke' than in 'Holme'. Boggess and Koeppe (1977) grew 10 soybean varieties in a greenhouse for 3 weeks on a loamy sand soil amended with 1 μg Cd (as $CdCl_2$)/g dry soil. Mean tissue concentrations in the soybean shoots ranged from 12.5 to 47.4 μg Cd/g dry weight. Similar varietal differences were found in lettuce by John and Van Laerhoven (1976) when they compared nine lettuce varieties grown in cadmium-amended solutions for 3 to 5 weeks.

2.2. Plant Age

Few studies have examined the effects of plant age or developmental stage on cadmium uptake, and the studies that have measured cadmium uptake over time were not carried out through plant maturity and senescence. The study by Root et al. (1975) points to the general hypothesis that cadmium accumulation in plant tissues increases with time of treatment and with the growth medium concentration of cadmium when the plants are undergoing rapid vegetative growth.

Miller et al. (1977) harvested corn grown on Bloomfield loamy sand amended with 0, 2.5, and 5.0 μg Cd (as $CdCl_2$)/g dry soil at 10, 17, 24, and 31 days after emergence. In contrast to the nutrient solution experiments of Root et al. (1975), cadmium concentrations in the corn shoots decreased with time, whereas the total cadmium content (μg Cd per plant shoot) increased. Since it was determined that soil cadmium concentrations were not substantially reduced during the experiment, Miller et al. (1977) concluded that either the rate of cadmium uptake was not proportional to the rate of biomass production or the cadmium in the soil was becoming less available with time. Results of other greenhouse

Table 2. Cadmium (μg/g dry weight) in 10 Crop Species Grown with and without Addition of Cadmium in Grenville Loam[a,b]

Crop	Cadmium Added (μg/g soil)			Crop	Cadmium Added (μg/g soil)		
	0	2.5	5.0		0	2.5	5.0
Oats, grain	0.21	1.50	2.07	Lettuce, tops	0.66	7.72	10.36
Oats, straw	0.29	2.30	3.70	Lettuce, roots	0.40	2.96	5.60
Oats, roots	0.81	5.06	8.72	Carrots, tops	0.46	5.66	7.70
Soybeans, grain	0.29	1.88	2.51	Carrots, roots	0.24	2.53	2.65
Soybeans, vegetative part	0.71	3.95	4.88	Tobacco, leaves	0.49	5.41	11.57
Soybeans, roots	0.99	6.09	11.77	Tobacco, stems	0.28	2.42	4.80
Timothy, tops	0.21	1.04	1.41	Tobacco, roots	0.42	2.78	5.54
Timothy, roots	0.61	8.97	15.28	Potatoes, tops	0.58	3.46	7.35
Alfalfa, tops	0.28	1.34	1.72	Potatoes, tubers	0.18	0.89	1.09
Alfalfa, roots	0.52	7.99	10.51	Tomatoes, fruit	0.23	0.99	1.03
Corn, tops	0.22	1.84	2.68	Tomatoes, vegetative part	0.51	5.26	6.46
Corn, roots	0.73	10.47	17.02	Tomatoes, roots	0.59	5.08	10.35

[a] From MacLean (1976). [b] SE of mean: roots, ±0.05 for 0 cadmium level and ±2.25 for added cadmium levels; other plant parts, ±0.03 for 0 cadmium level and ±0.36 for added cadmium levels.

experiments with corn grown on the same soil used by Miller et al. (1977) but with higher cadmium amendments (up to 30 μg Cd/g dry soil) and longer growth periods (up to 60 days) showed similar trends in tissue concentration and total cadmium content over time (J. D. Jastrow, unpublished results).

In view of the toxic effects of cadmium in plants (see Section 4), there is probably an interaction between the rate of biomass production and the rate of cadmium uptake that is manifested over time. If enough cadmium is taken up by the plant to adversely affect growth and biomass production, then cadmium-induced changes in plant metabolism may in turn affect future cadmium uptake.

2.3. Translocation and Distribution of Cadmium within Plants

Wallace and Romney (1977) have tentatively classified many trace elements taken up by plants into three groupings based on their distribution between plant roots and shoots. Cadmium, along with Fe, Cu, Al, Co, and Mo, usually accumulates more in the roots than in the shoots; however, often moderate and sometimes large concentrations of these elements can be found in the shoots. In contrast Pb, Sn, Ti, Ag, Cr, Zr, V, and Ga usually accumulate in the roots with very little concentration in the shoots; whereas Zn, Mn, Ni, Li, and B are generally distributed somewhat uniformly between roots and shoots.

For all 23 species studied by Jarvis et al. (1976) in solution culture (Table 1), the tissue cadmium concentrations were higher in the fibrous roots than in the shoots. The storage "roots" of parsnip, carrot, and radish also had higher tissue concentrations of cadmium than the shoots; however, for each species the total cadmium content of the storage "roots" was less than that of the shoots. In the species where the stems were analyzed separately from the rest of the shoots, both the tissue concentrations and the total cadmium contents of the stems were lower than or equal to those of the rest of the shoots.

In solution culture studies using bush beans and relatively high cadmium concentrations, Wallace et al. (1977) reported higher cadmium concentrations in the roots than in the stems, with the lowest concentrations in the leaves. However, working with chrysanthemums and the same solution concentrations of cadmium as Wallace et al., Patel et al. (1976) found lower cadmium concentrations in the stems than in the leaves (both lower than the root cadmium concentrations).

Haghiri (1973) followed [115m]Cd movement from nutrient solution into 4-week-old soybeans. The percentage of applied [115m]Cd was higher in the stems than in the leaves, and at maturity the percentage of applied [115m]Cd was higher in the pods than in the seeds. Interestingly, in concurrent experiments designed to measure foliar absorption and translocation, when the same dose of ^{115m}Cd was applied to the surface of the first trifoliate leaf above the third branch, the percentage of applied ^{115m}Cd in the various plant parts followed the same concentration order as in the root application. However, for each plant part examined, the percentages of cadmium were much lower following foliar application than after root application.

In other soybean experiments (Cunningham et al., 1975) cadmium concentrations followed this sequence: roots > stems > leaves. Using autoradiography in addition to quantitative cadmium analysis, it was observed that cadmium movement from the roots to the stem was not severely impeded at solution concentrations up to 0.3 μg Cd/ml, but that movement from the stems and petioles into the leaves was restricted by localized blocking of vascular tissues along major veins in the leaves and at the stem nodes. A report by Brisson et al. (1977) suggests that phenolic compounds may be largely responsible for this blockage.

The data presented in Table 2 illustrate that the distribution pattern of cadmium (μg Cd/g dry weight) varies between crop species. In lettuce, carrots, and tobacco the tissue concentrations of cadmium were greater in the aerial plant parts than in the roots. The reverse was true for the other species. Cadmium concentrations were lower in the grain of oats and soybeans, the tubers of potatoes, and the fruit of tomatoes than in the other parts of each species.

John (1973) compared the cadmium concentrations in various tissues of eight food crops grown to maturity in a controlled environment chamber on a silt loam soil amended with higher rates of applied cadmium (40 and 200 μg Cd/g soil) than were used by MacLean (1976). These data are generally similar except that John reported overall higher tissue concentrations and higher concentrations in lettuce roots than in the leaves. There were relatively low cadmium concentrations in the below-ground storage organs of carrot, radish, and potato, possibly because these organs are not directly involved in nutrient absorption from soil.

Mitchell and Fretz (1977) found more cadmium (μg Cd/g dry weight) in the roots than in the leaves of 2- or 3-year-old seedlings of red maple and white pine grown for varying times in both soil and solution culture amended with varying concentrations of cadmium. Norway spruce seedlings, grown in solution culture, also exhibited the same pattern of cadmium distribution.

The distribution and the cycling of cadmium in a mixed deciduous forest in eastern Tennessee have been studied by Van Hook et al. (1977). Although cadmium and other trace element concentrations in this area were relatively low, cadmium concentrations in trees generally followed this pattern: roots > foliage > branch > bole.

2.4. Mechanisms of Cadmium Uptake and Translocation

Cutler and Rains (1974) investigated the mechanism of cadmium uptake by excised barley roots from solution culture. On the basis of short-term experiments they hypothesize that cadmium uptake by barley roots involves three mechanisms. The first is exchange adsorption, in which cadmium is reversibly bound to exchange sites on the root and can be readily exchanged by desorption solutions (solutions containing a large excess of another transition type metal cation, e.g., Zn, Cu, Hg) or calcium solutions. The second proposed mechanism is an irreversible, nonmetabolic binding or sequestering to a limited number of sites on the cell wall or other cellular macromolecules. The third mechanism, diffusion, accounts for movement across cell membranes, which is necessary for translocation from roots into the above-ground portions of the plant. Evidence for the third mechanism being diffusion, and not a metabolic process, comes from the linear sorption of cadmium with temperature changes that normally affect metabolic processes. Although cadmium uptake was inhibited by anaerobic conditions and the presence of a metabolic inhibitor, 2,4-dinitrophenol, these results were explained by the fact that both anaerobic conditions and 2,4-dinitrophenol are known to reduce membrane permeability.

Similar results were obtained in short-term solution culture experiments conducted by Jarvis et al. (1976). Cadmium uptake was depressed in 10- to 15-week-old perennial ryegrass roots by the presence of Ca^{2+}, Mn^{2+}, or Zn^{2+} in the solution, leading to the suggestion that calcium (Mn^{2+} or Zn^{2+}) may inhibit cadmium uptake by competing for exchange sites on the root surface and by an effect on cell membranes. These investigators also found that cadmium could be readily released from ryegrass roots by desorption solutions containing Ca^{2+}, Mn^{2+}, or Zn^{2+} and that cadmium uptake was increased when roots were killed by immersion in boiling water prior to placement in cadmium solutions. The latter observation may provide support for the irreversible binding mechanism proposed by Cutler and Rains (1974), or it may simply reflect the destruction of membrane integrity.

The rate of plant water uptake and loss (transpiration) affects the movement of essential plant nutrients, and also that of cadmium. Jaakkola and Ylaranta (1976) grew radish plants in ^{115m}Cd-amended synthetic soils and harvested them at different age levels. Autoradiographic analysis showed that much of the cadmium was present in pointlike accumulations in the leaf veins. The older leaves contained more cadmium than the younger ones, but some of the cadmium in the older leaves was distributed between the points of accumulation. Although the roots contained relatively high concentrations of cadmium, the swollen stem base (a nutrient storage organ) accumulated very little of the metal. Thus the authors suggested that cadmium translocation within the plant occurs upwards via the transpiration stream. Carlson et al. (1975) observed 2 to 3 times greater cadmium accumulation in sunflower than in corn, a fact that they attributed to the higher (2 times) rate of transpiration per unit leaf area in sunflower. Since the effects of cadmium on stomatal opening were determined to be approximately equal for sunflower and corn, it was concluded that cadmium may have an inhibitory effect on water movement through corn roots and into the leaves that is not present in sunflower.

On the basis of experiments with soybeans grown in solution culture at very low cadmium concentrations (0.838×10^{-3} μg Cd/ml), Cunningham et al. (1975) suggested that long-distance transport of cadmium in soybeans is carried out by a number of processes, including (a) carrier mechanisms similar to those proposed by Tiffin (1972) for zinc, (b) cation exchange phenomena with the cell walls lining the conducting vessels, (c) diffusion of cadmium from the roots to the stem, caused by cadmium adsorption in the stem, and (d) control by ongoing metabolism. The authors based their proposal of the last process on data indicating that the total cadmium content of the mature leaves reached a constant level relatively early in leaf development (approximately 40% of eventual dry weight). Since transpiration continues as a leaf develops, Cunningham et al. concluded that the absence of a net increase in leaf cadmium demonstrates that cadmium movement cannot be dictated solely by passive diffusion with the transpiration stream.

3.   EDAPHIC FACTORS AFFECTING CADMIUM UPTAKE

3.1.   Cadmium Concentration in the Rooting Medium

Numerous investigators have observed that cadmium uptake and accumulation by plants is usually positively correlated with substrate concentrations of cadmium below those where severe toxicity occurs. For example, Root et al. (1975) observed increased cadmium concentrations in the shoots and roots of corn as the cadmium concentration in solution culture was increased from 0 to 40 μg Cd/ml. John (1972) grew radishes for 5 weeks in both limed and unlimed silty clay loam soil amended with $CdCl_2$ to give cadmium concentrations of 0.5 to 100 μg Cd/g dry soil. For both the limed and unlimed soil, cadmium concentrations in radish tops and roots increased with increasing soil cadmium concentration.

Depending on soil and environmental conditions and on the susceptibility to cadmium toxicity of the plant species or variety, tissue cadmium concentrations do not always increase with increasing substrate cadmium concentration. In solution culture studies with nine lettuce varieties exposed to varying concentrations of cadmium for 5 weeks, John and Van Laerhoven (1976) observed increased tissue concentrations of cadmium with increased solution cadmium concentrations up to 10 μg Cd/ml. When the plants were exposed to 50 μg Cd/ml, only one variety had a significantly increased cadmium tissue concentration, seven varieties exhibited tissue concentrations that were not significantly different from those produced by the 10 μg Cd/ml treatment, and one variety had a significantly lower tissue concentration. Carlson and Bazzaz (1977) grew 2- to 3-year-old saplings of American sycamore on a silty clay loam soil amended with 0 to 100 μg Cd (as $CdCl_2$)/g dry soil. They observed increased cadmium concentrations in the foliage with increasing soil cadmium up to and including 50 μg Cd/g soil; no further increase in tissue concentration was observed at 100 μg Cd/g soil. The concentration of cadmium in new stems did not increase after soil concentrations exceeded 25 μg Cd/g soil.

The factors that cause a leveling off or even a decrease in tissue cadmium concentrations with increasing substrate concentrations of cadmium are not easily discerned. Studies of micronutrient uptake and theories of active uptake mechanisms indicate that this "plateau" phenomenon may occur when a limiting rate of uptake is reached and the transport system becomes saturated (Moore, 1972). Although Cutler and Rains (1974) proposed a passive mechanism of cadmium uptake by excised barley roots in solution culture, further experimentation is necessary before active uptake can be ruled out. Additionally, there is evidence that more than one mechanism of uptake can occur, depending on the substrate concentration of a given ion, the plant tissues involved, and other factors (Moore, 1972; Cunningham et al., 1975). Furthermore, as cadmium is accumulated in plant tissue, it may have a myriad of direct and indirect effects on plant functions, thereby modifying further cadmium uptake.

Results of experiments by Miller et al. (1976) indicate that better correlations of plant uptake with soil cadmium concentrations may be obtained when the capacity for a soil to sorb cadmium is considered in addition to the total cadmium concentration in the soil. In their experiments soybeans grown for 4 weeks on nine Illinois agricultural soils amended with $CdCl_2$ accumulated varying concentrations of cadmium, depending on soil type; however, tissue concentrations generally increased with increasing soil cadmium concentration. After determining the cadmium sorptive capacities of the nine soils, cadmium uptake by soybeans was found to be positively correlated with the degree of saturation of each soil's cadmium sorptive capacity at each concentration of added cadmium (the amount of added cadmium divided by the cadmium sorptive capacity of the soil reflects the degree of saturation of that soil).

3.2. Cadmium Availability

Plants do not take up and accumulate cadmium from soils as readily as from solution culture because the Cd ions are not as readily available in soil. The availability of soil cadmium and its uptake and accumulation by plants are affected by many soil factors, including pH, cation exchange capacity, organic matter content, soil moisture and temperature, phosphorus level, and the concentrations of other cations or trace elements in the soil.

Several investigators have attempted to determine the best method of extracting the available fraction of cadmium from soils by correlating extracted cadmium with plant uptake of this element on the same soils. John et al. (1972) amended 30 surface soils from southwestern British Columbia with 100 μg Cd (as $CdCl_2$)/g dry soil and correlated cadmium uptake by radish and lettuce plants grown on these soils with the cadmium extracted by 1 N NH_4OAc, 1 N HCl, and 1 N HNO_3. Miller et al. (1976) amended their nine Illinois soils with the cadmium levels described previously and correlated cadmium uptake by soybeans with cadmium extracted by Bray P_1 reagent (0.025 N HCl + 0.03 N NH_4F), Bray P_2 reagent (0.1 N HCl + 0.03 N NH_4F) (Bray and Kurtz, 1945), 0.1 N EDTA, and 2 N $MgCl_2$. In a similar experiment Symeonides and McRae (1977) collected 25 surface soils from different soil series throughout Kent, England. Each soil was amended with 50 μg and 100 μg Cd (as $CdCl_2$)/g dry soil, and cadmium uptake by radish was correlated with cadmium extracted by a 1 N NH_4OAc-HOAc solution buffered to pH 7.0, 5% HOAc at pH 2.5, 1 N NH_4NO_3 unbuffered, 0.05 N EDTA (diammonium salt), 1 N HCl, concentrated HF (for total cadmium), and hot concentrated HNO_3. For all extractants the weight of soil/volume of extractant ratio and the length of time of extraction are important. The individual papers should be consulted for procedures. The results of these experiments are summarized in Table 3. It appears that for the soils and plants tested the 1 N NH_4NO_3 extraction gives the best correlation, and Bray P_1 reagent the second best correlation, with plant tissue concentration (μg Cd/g dry weight). However, further experimentation, including field testing, is suggested. Also, the data of Symeonides and McRae show that, although tissue concentrations (in the shoots) correlated well with the cadmium concentrations extracted, other plant parameters such as root tissue concentrations and plant yield were not as well correlated.

3.3. Cation Exchange Capacity

Several investigators have reported decreased cadmium uptake with increasing soil cation exchange capacity (CEC), presumably due to the greater capacity of the higher CEC soils to adsorb Cd ions (Williams and David, 1973; Haghiri, 1974; Miller et al., 1976). Haghiri (1974) reported an inverse relationship between cadmium concentrations in oat shoots and the CEC of the soil, based on greenhouse pot experiments. He also measured decreased exchangeable cadmium (as determined by extraction with 1 N NH_4OAc) with increasing CEC of the soil. Working with relatively low concentrations of added cadmium in the

Table 3. Correlation Coefficients (r) of Plant Cadmium Concentrations and Concentrations of Extracted Soil Cadmium

Reference	Extractant	Plant Tissue		
		Radish		Lettuce
		Tops	Roots	Tops
John et al. (1972)	1 N NH$_4$OAc	.36*[a]	.19	.54**[a]
	1 N HCl	-.15	-.09	.01
	1 N HNO$_3$	-.22	-.13	-.14
Miller et al. (1976)			Soybean Shoots	
	Bray P$_1$		.865**	
	Bray P$_2$		.689*	
	0.1 N EDTA		.700*	
	2 N MgCl$_2$		.650*	
Symeonides and McRae (1977)			Radish Tops	
	1 N NH$_4$OAc-HOAc (pH 7.0)		.50***[a]	
	5% HOAc (pH 2.5)		.36**	
	1 N NH$_4$NO$_3$ (unbuffered)		.97***	
	0.05 N EDTA		.29	
	1 N HCl		.19	
	Conc. HF		.04	
	Hot conc. HNO$_3$		.17	

[a] * = Significant at 5% level, ** = significant at 1% level, *** = significant at 0.1% level.

soil (~0.1 μg Cd/g dry soil), Williams and David (1973) observed a fivefold difference in the percentage of added cadmium taken up by oats on two different soils. Uptake on a third soil was intermediate, and the authors attributed the differences largely to the variation in cation exchange properties of the soils (CEC ranged from 5.3 to 27.9 mequiv/100 g). Miller et al. (1976) found significant negative correlations of soybean tissue cadmium concentrations with soil CEC and the interaction of CEC with pH.

3.4.   Soil Organic Matter Content

The work of John et al. (1972), Haghiri (1974), MacLean (1976), and others suggests that plant uptake of cadmium is inversely related to the percentage of soil organic matter, again probably because of the increased capacity of the soil to adsorb cadmium with increasing organic matter content. In laboratory experiments, Andersson and Nilsson (1974) and Andersson (1977) found that organic soils adsorb cadmium more effectively than mineral soils, particularly under acidic conditions. The results of Haghiri (1974), Levi-Minzi et al. (1976), and Singh and Sekhon (1977) indicate that the effect of organic matter on cadmium availability is primarily due to the high CEC of the soil organic colloids, rather than their chelating ability. Experiments conducted by Jaakkola and Ylaranta (1976) indicate that the quantity of humic and fulvic acids present in organic matter may be directly proportional to its cadmium adsorptive ability. Thus both the quantity and the composition of the organic matter in a soil may affect cadmium adsorption.

3.5.   pH

In experiments with soils of varying pH values, John et al. (1972) and Miller et al. (1976) found that decreasing soil pH was associated with increasing concentrations of cadmium in radish, lettuce, and soybeans. Working with a soil amended with cadmium at varying rates (0 to 100 μg Cd/g dry soil), John (1972) found that an application of lime which increased the pH of the soil from 4.1 to 5.5 resulted in decreased cadmium concentrations in radish tops and roots. Williams and David (1976) amended a soil [pH = 5.4, <1 μg Cd/g soil (1 N HCl extraction)] with $CaSO_4$ to pH 5.1 and with $CaCO_3$ or $MgCO_3$ to pH 6.0 and 6.8. They observed that cadmium uptake by subterranean clover increased when the pH was lowered and decreased when the pH was raised. Other investigators have observed decreased cadmium uptake by subterranean clover, lettuce, and corn with additions of lime or $CaCO_3$ to the soil (Williams and David, 1973; MacLean, 1976; Wallace et al., 1977).

There is evidence that the availability of soil cadmium increases with decreasing soil pH, probably because of the increased solubility of many cadmium compounds, such as hydroxides, carbonates, and phosphates, at the lower pH values (Andersson and Nilsson, 1974; Santillan-Medrano and Jurinak, 1975). Additionally, Miller et al. (1976) have suggested that the increased solubility of other ions (e.g., Fe^{3+}, Mn^{2+}, Zn^{2+}, and H^+) at low soil pH may cause a greater competition for exchange sites than occurs in neutral to alkaline soils, thereby resulting in increased concentrations of cadmium in the soil solution.

3.6.   Soil Phosphorus

The effects of soil phosphorus concentrations on cadmium uptake are somewhat more complex. MacLean (1976) observed decreased cadmium accumulation by lettuce with the addition of phosphate to acidic soils, whereas cadmium accumulation was relatively unchanged by phosphate additions to more neutral soils. However, for nine soils Miller et al. (1976) found that cadmium accumulation by soybeans increased with increasing concentrations of available phosphorus. Williams and David (1976, 1977) suggested that the effects of phosphate addition on cadmium uptake vary from soil to soil and according to plant species. However, they concluded that on phosphorus-deficient soils the addition of phosphate may decrease cadmium accumulation via a dilution effect caused by large increases in plant yield; but when the soil phosphorus supply is adequate, phosphate additions apparently increase both cadmium accumulation and plant yield.

3.7.   Other Cations

The effects of other cations on the uptake of cadmium appear to be somewhat varied, depending on soil type and characteristics, plant species, and the concentrations of the various cations, including cadmium. Haghiri (1976) reported that additions of potassium or calcium to a silt loam soil suppressed cadmium uptake by soybeans and that, irrespective of soil pH levels, potassium had a greater effect than calcium. John (1976) observed a reduction in cadmium uptake by hydroponically grown lettuce and oats when potassium concentrations in the solution were increased. Short-term uptake (4 hr) of cadmium (0.25 μg Cd/ml in solution) by perennial ryegrass in solution culture was depressed by additions to the solution of Ca^{2+}, Mn^{2+}, and Zn^{2+}, both singly and in combinations of calcium with the other two ions (Jarvis et al., 1976). It is suspected that these ions (particularly calcium because its atomic radius is similar to that of cadmium) may compete with the Cd ions for exchange sites on the root.

Andersson (1976) reported that increased cadmium concentrations in wheat grain occurred with increasing rates of fertilization. Since cadmium impurities in the fertilizers were very low, Andersson attributed the increased cadmium uptake to increased salt concentrations (from the fertilizers), which led to intense ion exchange, thereby releasing cadmium already in the soil to the soil solution. Working with oats and lettuce in solution culture, John (1976) observed that both a complexity of nutrient element interrelationships in the substrate and the resultant element balance within the plant affect cadmium tissue concentrations. For example, John reported that solution cadmium concentrations and the resultant plant cadmium concentrations can affect the plant tissue concentrations of elements such as P, Fe, Mn, Al, and Ca, and that changes in the tissue concentrations of these elements may, in turn, affect subsequent cadmium uptake.

Haghiri (1974) grew soybeans for 16 weeks on a silty clay loam soil amended with 10 μg Cd/g dry soil and varying concentrations of zinc [as $Zn(NO_3)_2$], from 0 to 400 μg Zn/g dry soil. Cadmium concentrations were increased in the soybean shoots when added zinc ranged from 5 to 50 μg/g soil, but cadmium tissue concentrations were decreased relative to the control at soil zinc concentrations from 100 to 400 μg/g. Williams and David (1976) reported increased cadmium concentrations in oat tops when 5 μg Zn/g dry soil was added to a soil with low cadmium concentrations. MacLean (1976), however, found that additions of zinc to a cadmium-amended soil had no appreciable effect on cadmium concentrations in lettuce tops.

Working with corn, Hassett et al. (1976) and Miller et al. (1977) compared the effects of lead and cadmium added to a soil in combination with the effects of either cadmium or lead additions alone. Additions of lead (≥ 100 μg Pb/g dry soil) along with cadmium (as low as 2.5 μg Cd/g dry soil) tended to increase both the tissue concentrations and the total cadmium uptake of corn shoots and roots relative to the uptake and tissue concentrations of plants grown on soil amended with cadmium alone. Carlson and Bazzaz (1977) reported similar effects of added lead and cadmium on the cadmium concentrations in the foliage and new stems of American sycamore saplings.

3.8. Other Factors

In pot experiments (soil amended with 10 μg Cd/g dry soil) where soil temperatures were controlled at 15.5, 21.1, 26.6, and 32.2°C, cadmium concentrations in 16-week-old soybean shoots increased with rising temperatures. With the exception of the 32.2°C treatment, plant dry weight also increased as temperatures rose (Haghiri, 1974).

The availability of soil water may also affect cadmium uptake and accumulation. When corn was grown in cadmium-amended nutrient solution (0 to 40 μg Cd/ml) and subjected to various levels of water stress induced by additions of polyethylene glycol (an osmoticant) to the solution, cadmium uptake was significantly reduced by increasing water stress (J. D. Jastrow, unpublished results).

Williams and David (1977) studied the effect of vertical cadmium distribution on uptake by subterranean clover. In pot experiments, cadmium concentrations in plant tops were substantially lower when the element was mixed only into the surface 2 cm of the soil, as compared to mixture throughout the soil.

Many recent studies have determined that cadmium is taken up by plants after reaching soils through amendment with sewage sludge. This uptake is dependent on soil, plant, and environmental factors, as delineated previously in this chapter. The picture is complicated, however, by changes in these factors (e.g., CEC) that result from the addition of sludge itself. A substantial literature dealing specifically with the topic of soil amendment with sewage sludge is growing rapidly and will not be covered here. Data on plant cadmium uptake can be extrapolated from salt-amended soils to sludge-amended soils only if the soil variables affecting cadmium uptake (measured after the addition of sludge) are taken into consideration.

4.   EFFECTS OF CADMIUM

Studies on the effects of cadmium generally fall into two categories: (*a*) those relating to the general vegetative and/or reproductive growth of entire plants, and (*b*) studies of more specific physiological activities, both *in vivo* and *in vitro*.

4.1.   Gross Parameters

Probably because of the ease of making the measurements, vegetative yields of a number of plant species treated with cadmium have been reported. With the notable exception of reports of cadmium stimulation of radish (John, 1972; Turner, 1973) and lettuce (Turner, 1973; John and Van Laerhoven, 1976) at relatively low concentrations, virtually all other studies where cadmium was the clearly defined variable demonstrate that this element effects a reduction in growth. Growth reductions have been observed in plants grown in either soil or solution culture, the severity of toxicity for a given species showing a positive correlation with both the concentration of cadmium in the measured plant part and the duration of treatment. Table 4 is a summary of a number of studies where vegetative yield and other measures of plant growth were determined to be specifically affected by cadmium. In most instances these studies produced a range of toxicity (yield reduction), with the range including the effect (as % of control) at the lowest and the highest cadmium concentrations. Although the table cannot be used to precisely compare the effects of cadmium on different species, it does suggest that cadmium in above-ground tissues is toxic and that species and varietal differences in cadmium susceptibility are related to uptake, or the lack of it, and not to an *in vivo* tolerance of certain tissues (Boggess and Koeppe, 1977). Boggess et al. (1978) found a wide range of cadmium uptake in soybean cultivars that was correlated with the continuum of cadmium toxicity exhibited by these plants.

Unlike lead, cadmium is not sequestered in large, amorphous deposits in tissues. One of the first indications in soybeans of cadmium toxicity is a reddish purple coloration of the veins near the blade-petiole junction. Recent reports indicate that this may be due to a phenolic complex (Brisson et al., 1977). Whether cadmium is associated with this phenolic complex is not known at this time, although Jaakkola and Ylaranta (1976) have reported that cadmium is concentrated in the veins of radish, and Cunningham et al. (1975) found that in soybeans the movement of cadmium was restricted to the major veinal areas, a restriction that coincided with the deposition of anthocyanin pigments. The localization of cadmium in the area of the pulvinus of soybeans is probably associated with the decreased movement of soybean leaves in response to light (S. F. Boggess, unpublished results). Cadmium is often found in animal tissue as "point sources," possibly being associated with metallothioneins (Cherian and Goyer, 1978). In any event it might be hypothesized that cadmium complexes serve to keep the transported cadmium from reaching concentrations toxic to enzyme systems, although it is likely that the continued presence of these complexes in the vascular tissue may in themselves have an adverse effect on the physiological activities of the plant.

4.2. Specific Effects

The effects of cadmium on the more specific physiological activities of plants, other than growth, are divided here into studies of *in vivo* processes and studies on *in vitro* organelles or enzymes.

The *in vivo* effects collated in Table 5 were generally obtained when physiological processes or organic components were analyzed after a period of cadmium treatment to the roots of an intact plant. In virtually all studies, except those of Lee et al. (1976) with soybeans, cadmium was observed to effect a (toxic) reduction in the processes assayed. Unfortunately the cadmium concentrations of the assayed tissues were determined in only several instances. These data, like the growth experiments previously discussed, lead to the conclusion that cad-

Table 4. The Effects of Cadmium on General Parameters of Plant Growth

Species	Parameter Measured	Effect (% of control)[a]	Effective Concentration (Growth Medium or Tissue Analysis)	Cadmium Source	Growth Medium	Cadmium Concentration in Medium (μg Cd/g growth medium)	Reference
Acer saccharinum (silver maple)	Leaf dry weight	−48 to −95		$CdCl_2$	Sand and Hoagland's solution	5–20	Lamoreaux and Chaney (1977)
	Stem dry weight	−56 to −96					
	Root dry weight	−47 to −93					
	Height	−40 to −88					
	Leaf length	−14 to −48					
Beta vulgaris (beetroot)	Veg. yield	+4 to −53	0.9–2.5 μg/g dry wt	$CdCl_2$	Nutrient solution	0.01–1.0	Turner (1973)
Beta vulgaris (swiss chard)	Veg. yield	−38 to −53	0.6–150 μg/g dry wt	$CdCl_2$	Nutrient solution	0.01–1.0	Turner (1973)
Cynodon dactylon (bermudagrass)	Veg. yield	−25	43 μg/g dry wt	Sludge + $CdSO_4$	Domino silt loam	145	Bingham et al. (1976a)
Daucus carota (carrot)	Veg. yield	−24 to −72	1.5–2.2 μg/g dry wt	$CdCl_2$	Nutrient solution	0.01–1.0	Turner (1973)
Festuca elatior (tall fescue)	Veg. yield	−25	37 μg/g dry wt	Sludge + $CdSO_4$	Domino silt loam	95	Bingham et al. (1976a)

Species	Parameter	Response		Cd form	Medium	Conc.	Reference
Glycine max (soybean)	Veg. yield	−26 to −91	18–900 μM		Sand and vermiculite	18–900 μM	Huang et al. (1974)
	Nodule weight	−27 to −98					
	Pod weight	−35 to −98					
	Root weight	−6 to −80					
	Shoot weight	+1 to −78					
	Leaf weight	−44 to −99					
Glycine max cv. Columbus (soybean)	Chlorosis				Solution	1.35 μM	Lee et al. (1976)
	Epinasty						
	Growth reduction						
	Leaf abscission						
Glycine max cv. Williams (soybean)	Growth (height)	−64		CdCl₂	Sand and Hoaglands solution	4	Chaney et al. (1977)
	Dry weight	−94					
Lactuca sativa (many varieties) (lettuce)	Top dry weight	+4 to −88		CdCl₂	Nutrient solution	0.1–50	John and Van Laerhoven (1976)
	Root dry weight	+17 to −71					
Lactuca sp. (lettuce)	Veg. yield	+26 to +18	1.5–24.3 μg/g dry wt	CdCl₂	Nutrient solution	0.01–0.1	Turner (1973)
Lolium sp. (ryegrass)	Top growth	−25 to −90	15–500 μg/g dry wt	CdSO₄	Solution	1.0–82	Dijkshoorn et al. (1974)

Table 4. *Continued*

Species	Parameter Measured	Effect (% of control)[a]	Effective Concentration (Growth Medium or Tissue Analysis)	Cadmium Source	Growth Medium	Cadmium Concentration in Medium (μg Cd/g growth medium)	Reference
Lycopersicum esculentum (tomato)	Veg. yield	+2 to −53	1.4–158 μg/g dry wt	$CdCl_2$	Nutrient solution	0.01–1.0	Turner (1973)
Medicago sativa (alfalfa)	Veg. yield	−25	24 μg/g dry wt	Sludge + $CdSO_4$	Domino silt loam	30	Bingham et al. (1976a)
Oryza sativa var. Colusa (rice)	Grain yield	−25	1.8 μg/g leaf tissue	Sludge + $CdSO_4$	Domino silt loam (flooded)	320	Bingham et al. (1976b)
	Grain yield	−25	1.8 μg/g leaf tissue		Domino silt loam (nonflooded)	17	
Phaseolus vulgaris cv. Red Kidney (bean)	Embryo growth	+4 to −80		$CdCl_2$		2–10	Imai and Siegel (1973)
	Embryo greening	0 to −80					
	Embryo curvature	0 to −65					

Species	Parameter	Response[a]	Plant concentration	Metal form	Medium	Metal concentration	Reference
Platanus occidentalis (American sycamore)	Foliage biomass Stem diameter New stem weight Root weight	−20 to −50 −10 to −62 −12 to −45 −10 to −30		CdCl₂	Drummer silty clay loam soil	10–100	Carlson and Bazzaz (1977)
Raphanus sativus (radish)	Top yield	+32 to −37		Inorganic salt	Hjorth silty clay loam	0.5–100	John (1972)
Raphanus sp. (radish)	Roots Veg. yield	+9 to −61 +207 to +85	0.7–144.2 μg/g dry wt	CdCl₂	Nutrient solution	0.01–1.0	Turner (1973)
Sorgum halepense var. Sudanese Hitche (sudangrass)	Veg. yield	−25	9 μg/g dry wt	Sludge + CdSO₄	Domino silt loam	15	Bingham et al. (1976a)
Trifolium repens (white clover)	Veg. yield	−25	17 μg/g dry wt	Sludge + CdSO₄	Domino silt loam	40	Bingham et al. (1976a)
Trifolium sp. (subterranean clover)	Top yield	−3 to −93		CdCl₂	Red podzolic soil	0.2–20	Williams and David (1977)
Typha latifolia (broadleaf cattail)	Root yield Shoot biomass Root dry weight Root length	−13 to −97 −31 to −74 −10 to −86 0 to −60		Smelting complex	Palmerton soil	[Zn] = 5000 [Pb] = 435 [Cd] = 73	McNaughton et al. (1974)

[a] Minus (−) = less than the control; plus (+) = more than the control.

627

Table 5. Effects of Cadmium on Plant Physiological Processes after Whole-Plant Exposure

Species	Process Affected	Effect (% of control)[a]	Effective Concentration	Cadmium Source	Growth Medium	Cadmium Concentration in Medium	Notes	Reference
Avena sativa (oats)	Respiration	−35 to −49		$CdSO_4$	Solution	1 mM	2–24 hr Excised roots treatment	Keck (1978)
	ATP levels	−26 to −81					2–24 hr treatment	
	ATPase	−75					Mg^{2+} + K^+ stimulated	
	K^+ influx	−50					2 hr treatment	
Crepis capillaris	Increased ratio of chromosome to chromatid aberrations					0.1–0.01 M	Treatment for 1 hr	Ruposhev (1976)
Crepis capillaris	Seed ethylenimine	−35 to −75		$Cd(NO_3)_2$				Ruposhev and Garina (1977)
Glycine max (soybean)	Nodule acetylene reduction	−71 to −99	18–900 μM	$CdCl_2$	Sand and vermiculite	18–900 μM		Huang et al. (1974)
	Photosynthesis	−17 to −100						
	Nodule NH_4^+	−2 to −99						

Species	Parameter	% Change	−50% at	Cd compound	Growth media	Concentration	Treatment	Reference
Glycine max cv. Columbus (soybean)	Acid phosphatase	+115		$CdSO_4$	Solution	1.35 μM		Lee et al. (1976)
	Carbonic anhydrase	−18						
	DNase	+300						
	Malate dehydrogenase	+69						
	Peroxidase	+405						
	Respiration	+53						
	RNase	+226						
Helianthus annuus (sunflower)	Net photosynthesis	−10 to −80	−50% at 96 $\mu g/g$ leaf tissue	Cd salt	Vermiculite and Hoagland's solution	2–200 $\mu g/ml$	Excised tops 4–5 day treatment in Cd solution	Bazzaz et al. (1974)
	Stomatal opening	−10 to −65			Solution	1–1000 μM	Epidermal peels 8-hr treatment	
Helianthus annuus (sunflower)	Net photosynthesis	−10 to −70	−50% at 340 $\mu g/g$ leaf tissue	$CdCl_2$	Hoagland's nutrient solution	1–10 $\mu g/ml$	4–7 day treatment	Carlson et al. (1975)
Lactuca sativa (lettuce)	Chlorophyll content	−83		$CdSO_4$	Artificial rooting media		Plus ozone treatment	Czuba and Ormrod (1974)
	Carotenoid content	−80					Plus ozone treatment	
Lepidium sativum (cress)	Chlorophyll content	−36		$CdSO_4$	Artificial rooting media		Plus ozone treatment	Czuba and Ormrod (1974)
	Carotenoid content	−34					Plus ozone treatment	

Table 5. *Continued*

Species	Process Affected	Effect (% of control)[a]	Effective Concentration	Cadmium Source	Growth Medium	Cadmium Concentration in Medium	Notes	Reference
Phaseolus vulgaris cv. Red Kidney (bean)	Greening (chlorophyll biosynthesis)	−80	10 μg/ml	$CdCl_2$	Solution	10 μg/ml	Excised embryos	Imai and Siegal (1973)
Platanus occidentalis (American sycamore)	Photosynthesis	−5 to −25	Most inhibition at 20 μg Cd/g soil	$CdCl_2$	Drummer silty clay loam soil	10–100 μg/g soil		Carlson and Bazzaz (1977)
	Transpiration	−5 to −24	Most inhibition at 20 μg Cd/g soil					
Zea mays (corn)	Micronutrient concentration:			$CdCl_2$	Bloomfield sandy loam	2.5–5.0 μg/g soil	24-day treatment	Walker et al. (1977)
	B	−3 to −19						
	Cu	+8 to −44						
	Mn	+5 to −33						
	Zn	−10 to −9						
Zea mays (corn)	Net photosynthesis	0 to −70	−50% at 160 μg/g leaf tissue	$CdCl_2$	Nutrient solution	1–10 μg/ml		Carlson et al. (1975)

[a] Minus (−) = less than the control; plus (+) = more than the control.

mium has a significant toxic effect on the physiological processes of whole plants.

Some of the most definitive of the studies of isolated organelles and enzyme systems presented in Table 6 were done with the energy transfer mechanisms of isolated chloroplasts and mitochondria. In both instances cadmium was found to inhibit electron transfer and, in a separate manner, the synthesis of ATP. These experiments and others suggest that cadmium has an interactive role with plant membranes. Additionally these data are in agreement with reports of an interaction of cadmium with polythiols. In both the mitochondrial and chloroplast studies the effects of cadmium were reversed upon the addition of compounds that protect the thiol groups (Miller et al., 1973; Lucero et al., 1976). Cadmium is also known to complex with certain proteins, to substitute for other metals in metalloenzymes, and to have a high affinity for metallothioneins (a group of low molecular weight cytoplasmic metalloproteins with high thiol contents). While most specific binding studies have been done with organic compounds isolated from animal tissues, a few similar studies, such as those with mitochondria from both plants and animals, suggest that the location and extent of cellular binding of cadmium are comparable in plant and animal tissues.

5. SUMMARY

The references included in this brief review of cadmium in plants will probably be seen in retrospect as a mere introduction to a rapidly expanding literature on cadmium and other trace elements in the plant environment. At this point we do know enough to be concerned about the uptake of cadmium from localized sites of high impact. It is clear that cadmium is readily accumulated by many plants, the accumulation being affected by plant age and by most plant environmental conditions, but most prominently by the soil factors known to influence the availability and uptake of most nutrients (cation exchange capacity, pH, organic matter, moisture, microbial activity, and other ions). Plant species (and varieties) differ, mostly in their uptake patterns and probably not in tissue resistance, in their responses to cadmium. However, the distribution of accumulated cadmium within plants generally follows this order: roots > stems and leaves > fruits, grain, seeds, or nutrient storage organs. Cadmium accumulated in plant tissues affects, increasingly with time, a number of plant physiological processes. The majority of the reports here concern two categories of research: (a) a number of studies showing general cadmium-effected reductions in growth, and (b) a lesser number of experiments showing that cadmium has a toxic influence on most physiological processes studied. Unfortunately in the latter category there is little precision as to which processes are more affected by cadmium at the lowest concentrations.

Table 6. Effects of Cadmium on Isolated Enzymes or Organelles from Plant Tissues

Species	Process Affected	Effect (% of control)[a]	Effective Concentration	Cadmium Source	Notes	Reference
Nicotiana tabacu cv. Sabolesky (tobacco)	PEP carboxylase	−19	0.5 mM		Isolated enzyme	Leblova and Mares (1975)
Pisum sativum var. hortense (pea)	PEP carboxylase	−5	0.5 mM		Isolated enzyme	Leblova and Mares (1975)
Spinacea oleracea (spinach)	Photosynthesis $^{14}CO_2$ fixation	−77	10 μM	$CdCl_2$	Isolated chloroplasts	Hampp et al. (1976)
	Hill reaction	−20	500 μM			
Spinacea oleracea (spinach)	Photophosphorylation ATP synthesis	−60	100 μM	$CdCl_2$	Isolated chloroplasts	Lucero et al. (1976)
	Electron transport	−43			H_2O to methylviologen	
Spinacea oleracea (spinach)	Glycolipid biosynthesis	+2 to −89	0.01 mM– 0.5 mM	$CdCl_2$	Isolated chloroplasts	Mudd et al. (1971)

Spinacea oleracea (spinach)	DCPIP photoreduction	−100	0.5 mM	Cd(NO₃)₂	Isolated chloroplasts fixed with gluteraldehyde	Zilinskas and Govindgee (1976)
	Hill reaction	−55				
Spinacea oleracea cv. verbeterd breedblad (spinach)	Photosynthesis PS II Electron transport	−75	9 mM	Cd(NO₃)₂	Isolated chloroplasts	Van Duijendijk-Matteoli and Desmet (1975)
	Fluorescence	−50	10 mM		Inhibits on donor side of complex	
Spinacea oleracea (spinach)	Photosynthesis PS II photoreaction	−90	0.7 mM	Cd(NO₃)₂	Isolated chloroplasts 25 min incubation time	Li and Miles (1975)
Zea mays var. Konsky (corn)	PEP carboxylase	−22	0.5 mM		Isolated enzyme	Leblova and Mares (1975)
Zea mays WF9 × M14 (corn)	Mitochondrial NADH oxidation	+120 to −80	0.03–1 mM	CdCl₂	Isolated mitochondria	Miller et al. (1973)
	Succinate oxidation	+15 to −80	0.0001– 0.1 mM			
	Phosphorylation (ADP/O ratio)	−100	0.25 mM			

^{a} Minus (−) = less than the control; plus (+) = more than the control.

Although many specific questions relating to cadmium uptake and effects in plants are unanswered at this time, there are enough data to warrant words of caution to those involved with the growth of plants in localized areas of high cadmium contamination. Cadmium is readily taken up and accumulated by many plants, it gets into food chains via plant uptake, and it has profound toxic effects on plants when accumulated over a period of time. Further details regarding cadmium-resistant plants remain to be elucidated, as do the many complex interactions of soil and environmental variables with cadmium uptake. Since areas of high cadmium contamination are likely to occur in conjunction with high concentrations of other toxic trace elements, the interactions and combined effects of cadmium along with other trace elements are important areas requiring further investigation. Until these observations are made for very specific situations, caution should be used in the disposal of cadmium-containing wastes on lands where vegetation of any sort is to be grown.

ACKNOWLEDGMENTS

The research and literature search were supported in part by funds from the Illinois Agricultural Experiment Station, Urbana (D.E.K.). The submitted manuscript has been authored by a contractor of the U.S. Government (J.D.J.) under Contract W-31-109-ENG-38. Accordingly, the U.S. Government retains a nonexclusive, royalty-free license to publish or reproduce the published form of this contribution, or allow others to do so, for U.S. Government purposes.

REFERENCES

Andersson, A. (1976). "On the Influence of Manure and Fertilizers on the Distribution and Amounts of Plant-Available Cd in Soils," *Swed. J. Agric. Res.* **6**, 27–36.

Andersson, A. (1977). "Heavy Metals in Swedish Soils: On Their Retention, Distribution and Amounts," *Swed. J. Agric. Res.* **7**, 7–20.

Andersson, A. and Nilsson, K. O. (1974). "Influence of Lime and Soil pH on Cd Availability to Plants," *Ambio* **3**, 198–200.

Bazzaz, F. A., Carlson, R. W., and Rolfe, G. L. (1974). "The Effect of Heavy Metals on Plants. I: Inhibition of Gas Exchange in Sunflower by Pb, Cd, Ni and Tl," *Environ. Pollut.* **7**, 241–246.

Bingham, F. T., Page, A. L., Mahler, R. J., and Ganje, T. J. (1976a). "Yield and Cadmium Accumulation of Forage Species in Relation to Cadmium Content of Sludge-Amended Soil," *J. Environ. Qual.* **5**, 57–60.

Bingham, F. T., Page, A. L., Mahler, R. J., and Ganje, T. J. (1976b). "Cadmium Availability to Rice in Sludge-Amended Soil under 'Flood' and 'Nonflood' Culture," *Soil Sci. Soc. Am. J.* **40**, 715–719.

Boggess, S. F. and Koeppe, D. E. (1977). "Differences in the Susceptibility of Soybean Varieties to Soil Cadmium." In R. C. Loehr, Ed., *Food, Fertilizer and Agricultural Residues*. Proceedings of the 1977 Cornell Agricultural Waste Management Conference. Ann Arbor Science Publishers, Ann Arbor, Mich., pp. 229–238.

Boggess, S. F., Willavize, S., and Koeppe, D. E. (1978). "Differential Response of Soybean Varieties to Soil Cadmium," *Agron. J.* **70**, 756–760.

Bray, R. H. and Kurtz, L. T. (1945). "Determination of Total, Organic, and Available Forms of Phosphorus in Soils," *Soil Sci.* **59**, 39–45.

Brisson, J. D., Peterson, R. L., Robb, J., Rauser, W. E., and Ellis, B. E. (1977). "Correlated Phenolic Histochemistry Using Light, Transmission and Scanning Electron Microscopy, with Examples Taken from Phytopathological Problems," *Scanning Electron Microsc.* **2**, 667–676.

Carlson, R. W. and Bazzaz, F. A. (1977). "Growth Reduction in American Sycamore (*Platanus occidentalis* L.) Caused by Pb-Cd Interaction," *Environ. Pollut.* **12**, 243–253.

Carlson, R. W., Bazzaz, F. A., and Rolfe, G. L. (1975). "The Effect of Heavy Metals on Plants. II: Net Photosynthesis and Transpiration of Whole Corn and Sunflower Plants Treated with Pb, Cd, Ni, and Tl," *Environ. Res.* **10**, 113–120.

Chaney, W. R., Strickland, R. C., and Lamoreaux, R. J. (1977). "Phytotoxicity of Cadmium Inhibited by Lime," *Plant Soil* **47**, 275–278.

Cherian, M. G. and Goyer, R. A. (1978). "Metallothioneins and Their Role in Metabolism and Toxicity of Metals," *Life Sci.* **23**, 1–10.

Cunningham, L. M., Collins, F. W., and Hutchinson, T. C. (1975). "Physiological and Biochemical Aspects of Cadmium Toxicity in Soybean. I: Toxicity Symptoms and Autoradiographic Distribution of Cd in Roots, Stems and Leaves." In *Symposium Proceedings of the International Conference on Heavy Metals in the Environment, Toronto, Canada*, pp. 97–120.

Cutler, J. M. and Rains, D. W., (1974). "Characterization of Cadmium Uptake by Plant Tissue," *Plant Physiol.* **54**, 67–71.

Czuba, M. and Ormrod, D. P. (1974). "Effects of Cadmium and Zinc on Ozone Induced Phytoxicity in Cress and Lettuce," *Can. J. Bot.* **52**, 645–649.

Dijkshoorn, W., Lampe, J. E. M., and Kowsoleea, A. R. (1974). "Tolerance of Ryegrass to Cadmium Accumulation," *Neth. J. Agric. Sci.* **22**, 66–71.

Haghiri, F. (1973). "Cadmium Uptake by Plants," *J. Environ. Qual.* **2**, 93–96.

Haghiri, F. (1974). "Plant Uptake of Cadmium as Influenced by Cation Exchange Capacity, Organic Matter, Zinc, and Soil Temperature," *J. Environ. Qual.* **3**, 180–183.

Haghiri, F. (1976). "Release of Cadmium from Clays and Plant Uptake of Cadmium from Soil as Affected by Potassium and Calcium Amendments," *J. Environ. Qual.* **5**, 395–397.

Hampp, R., Beulich, K., and Ziegler, H. (1976). "Effects of Zinc and Cadmium on Photosynthetic CO_2-Fixation and Hill Activity of Isolated Spinach Chloroplasts," *Z. Pflanzenphysiol.* **77**, 336–344.

Hassett, J. J., Miller, J. E., and Koeppe, D. E. (1976). "Interaction of Lead and Cadmium on Maize Root Growth and Uptake of Lead and Cadmium by Roots," *Environ. Pollut.* **11**, 297–302.

Huang, C. Y., Bazzaz, F. A., and Vanderhoef, L. N. (1974). "The Inhibition of Soybean Metabolism by Cadmium and Lead," *Plant Physiol.* **54**, 122–124.

Imai, I. and Siegel, S. M. (1973). "A Specific Response to Toxic Cadmium Levels in Red Kidney Bean Embryos," *Physiol. Plant.* **29**, 118–120.

Jaakkola, A. and Ylaranta, T. (1976). "The Role of the Quality of Soil Organic Matter in Cadmium Accumulation in Plants," *J. Sci. Agric. Soc. Finl.* **48**, 415–425.

Jarvis, S. C., Jones, L. H. P., and Hopper, M. J. (1976). "Cadmium Uptake from Solution by Plants and Its Transport from Roots to Shoots," *Plant Soil* **44**, 179–191.

John, M. K. (1972). "Uptake of Soil-Applied Cadmium and Its Distribution in Radishes," *Can. J. Plant Sci.* **52**, 715–719.

John, M. K. (1973). "Cadmium Uptake by Eight Food Crops as Influenced by Various Soil Levels of Cadmium," *Environ. Pollut.* **4**, 7–15.

John, M. K. (1976). "Interrelationships between Plant Cadmium and Uptake of Some Other Elements from Culture Solutions by Oats and Lettuce," *Environ. Pollut.* **11**, 85–95.

John, M. K. and Van Laerhoven, C. J. (1976). "Differential Effects of Cadmium on Lettuce Varieties," *Environ. Pollut.* **10**, 163–173.

John, M. K., Van Laerhoven, C. J., and Chuah, H. H. (1972). "Factors Affecting Plant Uptake and Phytotoxicity of Cadmium Added to Soils," *Environ. Sci. Technol.* **6**, 1005–1009.

Keck, R. W. (1978). "Cadmium Alteration of Root Physiology and Potassium Ion Fluxes," *Plant Physiol.* **62**, 94–96.

Lamoreaux, R. J. and Chaney, W. R. (1977). "Growth and Water Movement in Silver Maple Seedlings Affected by Cadmium," *J. Environ. Qual.* **6**, 201–205.

Leblova, S. and Mares, J. (1975). "Thermally Stable Phosphoenolpyruvate Carboxylase from Pea, Tobacco and Maize Green Leaves," *Photosynthetica* **9**, 177–184.

Lee, K. C., Cunningham, B. A., Paulsen, G. M., Liang, G. H., and Moore, R. B. (1976). "Effects of Cadmium on Respiration Rate and Activities of Several Enzymes in Soybean Seedlings," *Physiol. Plant.* **36**, 4–6.

Levi-Minzi, R., Soldatini, G. F., and Riffaldi, R. (1976). "Cadmium Adsorption by Soils," *J. Soil Sci.* **27**, 10–15.

Li, E. H. and Miles, C. D. (1975). "Effects of Cadmium on Photoreaction II of Chloroplasts," *Plant Sci. Lett.* **5**, 33–40.

Little, P. (1973). "A Study of Heavy Metal Contamination of Leaf Surfaces," *Environ. Pollut.* **5**, 159–172.

Little, P. and Martin, M. H. (1972). "A Survey of Zinc, Lead, and Cadmium in Soil and Natural Vegetation around a Smelting Complex," *Environ. Pollut.* **3**, 241–254.

Lucero, H. A., Andreo, C. S., and Vallejos, R. H. (1976). "Sulphydryl Groups in Photosynthetic Energy Conservation. III: Inhibition of Photophosphorylation in Spinach Chloroplasts by $CdCl_2$," *Plant Sci. Lett.* **6**, 309–313.

MacLean, A. J. (1976). "Cadmium in Different Plant Species and Its Availability in Soils as Influenced by Organic Matter and Additions of Lime, P, Cd and Zn," *Can. J. Soil Sci.* **56,** 129–138.

McNaughton, S. J., Folsom, T. C., Lee, T., Park, F., Price, C., Roeder, D., Schmitz, J., and Stockwell, C. (1974). "Heavy Metal Tolerance in *Typha latifolia* without the Evolution of Tolerant Races," *Ecology* **55,** 1163–1165.

Miller, J. E., Hassett, J. J., and Koeppe, D. E. (1976). "Uptake of Cadmium by Soybeans as Influenced by Soil Cation Exchange Capacity, pH, and Available Phosphorus," *J. Environ. Qual.* **5,** 157–160.

Miller, J. E., Hassett, J. J., and Koeppe, D. E. (1977). "Interactions of Lead and Cadmium on Metal Uptake and Growth of Corn Plants," *J. Environ. Qual.* **6,** 18–20.

Miller, R. J., Bittell, J. E., and Koeppe, D. E. (1973). "The Effect of Cadmium on Electron and Energy Transfer Reactions in Corn Mitochondria," *Physiol. Plant.* **28,** 166–171.

Mitchell, C. D. and Fretz, T. A. (1977). "Cadmium and Zinc Toxicity in White Pine, Red Maple, and Norway Spruce," *J. Am. Soc. Hortic. Sci.* **102,** 81–84.

Moore, D. P. (1972). "Mechanisms of Micronutrient Uptake by Plants." In J. J. Mortvedt, P. M. Giordano, and W. L. Lindsay, Eds., *Micronutrients in Agriculture.* Soil Science Society of America, Madison, Wis., pp. 171–198.

Mudd, J. B., McManus, T. T., Ongun, A., and McCullogh, T. E. (1971). "Inhibition of Glycolipid Biosynthesis in Chloroplasts by Ozone and Sulfhydryl Reagents," *Plant Physiol.* **48,** 335–339.

Page, A. L., Bingham, F. T., and Nelson, C. (1972). "Cadmium Absorption and Growth of Various Plant Species as Influenced by Solution Cadmium Concentration," *J. Environ. Qual.* **1,** 288–291.

Patel, P. M., Wallace, A., and Mueller, R. T. (1976). "Some Effects of Copper, Cobalt, Cadmium, Zinc, Nickel, and Chromium on Growth and Mineral Element Concentration in Chrysanthemum," *J. Am. Soc. Hortic. Sci.* **101,** 553–556.

Pettersson, O. (1976). "Heavy-Metal Ion Uptake by Plants from Nutrient Solutions with Metal Ion, Plant Species and Growth Period Variations," *Plant Soil* **45,** 445–459.

Pettersson, O. (1977). "Differences in Cadmium Uptake between Plant Species and Cultivars," *Swed. J. Agric. Res.* **7,** 21–24.

Root, R. A., Miller, R. J., and Koeppe, D. E. (1975). "Uptake of Cadmium—its Toxicity, and Effect on the Iron Ratio in Hydroponically Grown Corn," *J. Environ. Qual.* **4,** 473–476.

Ruposhev, A. R. (1976). "Cytogenetic Effect of Heavy Metal Ions on *Crepis capillaris* Seeds," *Genetika* **12,** 37–43.

Ruposhev, A. R. and Garina, K. P. (1977). "Modification of the Mutagenic Effect of Ethylenimine by Cadmium Nitrate in *Crepis capillaris,*" *Genetika* **13,** 32–36.

Santillan-Medrano, J. and Jurinak, J. J. (1975). "The Chemistry of Lead and Cadmium in Soil: Solid Phase Formation," *Soil Sci. Soc. Am. Proc.* **39,** 851–856.

Singh, B. and Sekhon, G. S. (1977). "Adsorption, Desorption and Solubility Relationships of Lead and Cadmium in Some Alkaline Soils," *J. Soil Sci.* **28,** 271–275.

Symeonides, C. and McRae, S. G. (1977). "The Assessment of Plant-Available Cadmium in Soils," *J. Environ. Qual.* **6,** 120–123.

Tiffin, L. O. (1972). "Translocation of Micronutrients in Plants." In J. J. Mortvedt, P. M. Giordano, and W. L. Lindsay, Eds., *Micronutrients in Agriculture.* Soil Science Society of America, Madison, Wis., pp. 199–229.

Turner, M. A. (1973). "Effect of Cadmium Treatment on Cadmium and Zinc Uptake by Selected Vegetable Species," *J. Environ. Qual.* **2,** 118–119.

Van Duijvendijk-Matteoli, M. A. and Desmet, G. M. (1975). "On the Inhibitory Action of Cadmium on the Donor Side of Photosystem II in Isolated Chloroplasts," *Biochim. Biophys. Acta* **408,** 164–169.

Van Hook, R. I., Harris, W. F., and Henderson, G. S. (1977). "Cadmium, Lead, and Zinc Distributions and Cycling in a Mixed Deciduous Forest," *Ambio* **6,** 281–286.

Walker, W. M., Miller, J. E., and Hassett, J. J. (1977). "Effect of Lead and Cadmium upon the Boron, Copper, Manganese and Zinc Concentration of Young Corn Plants," *Commun. Soil Sci. Plant Anal.* **8,** 57–66.

Wallace, A. and Romney, E. M. (1977). "Roots of Higher Plants as a Barrier to Translocation of Some Metals to Shoots of Plants." In H. Drucker and R. E. Wildung, Chairmen, *Biological Implications of Metals in the Environment.* Proceedings of the Fifteenth Annual Hanford Life Sciences Symposium, Richland, Wash., 1975. Technical Information Center, Energy Research and Development Administration, CONF-750929. pp. 370–379.

Wallace, A., Romney, E. M., Alexander, G. V., Soufi, S. M., and Patel, P. M. (1977). "Some Interactions in Plants among Cadmium, Other Heavy Metals, and Chelating Agents," *Agron. J.* **69,** 18–20.

Williams, C. H. and David, D. J. (1973). "The Effect of Superphosphate on the Cadmium Content of Soils and Plants," *Aust. J. Soil. Res.* **11,** 43–56.

Williams, C. H. and David, D. J. (1976). "The Accumulation in Soil of Cadmium Residues from Phosphate Fertilizers and Their Effect on the Cadmium Content of Plants," *Soil Sci.* **121,** 86–93.

Williams, C. H. and David, D. J. (1977). "Some Effects of the Distribution of Cadmium and Phosphate in the Root Zone on the Cadmium Content of Plants," *Aust. J. Soil Res.* **15,** 59–68.

Zilinskas, B. A. and Govindjee (1976). "Stabilization by Glutaraldehyde Fixation of Chloroplast Membranes against Inhibitors of Oxygen Evolution," *Z. Pflanzenphysiol.* **77,** 302–314.

16

BIOCHEMICAL ASPECTS OF CADMIUM IN PLANTS

W. H. O. Ernst

Biologisch Laboratorium, Vrije Universiteit, Amsterdam, The Netherlands

1. INTRODUCTION

Preceding chapters have considered the uptake, transport, and distribution of cadmium by respiration in plants. This chapter will cover the biochemical aspects of this heavy metal in regard to the general principles affecting plants investigated up to now: bacteria, yeast, algae, and higher plants. The toxic and sometimes deleterious effects of cadmium in these organisms demand an accurate estimation of the compartmental concentrations of this heavy metal, because the degree of inhibition or stimulation of metabolic processes is also a function of the metal concentration (Ernst, 1976). Not only the individual enzymic reactions, but also the multiple reactions catalyzed by very different types of or-

Table 1.   Compartmentation of Cadmium in the Leaves of Trees Growing in the Vicinity of a Cadmium Smelter [a]

	Tilia platyphyllos	*Aesculus hippocastanum*
Total amount (mmol/kg dry matter)	0.79	1.21
Distribution (%)		
Cell wall	72.8	87.6
Plasm, vacuole sap	19.2	4.4
Plastids, nuclei	8.0	8.0
Mitochondria, ribosomes	0.01	0.01

[a] After Ernst (1972).

ganelles, especially mitochondria and chloroplasts, are sensitive to added metals.

2.   COMPARTMENTATION OF CADMIUM IN CELLS

Cell fraction studies are useful in shedding some light on the subcellular distribution of cadmium, especially if we want to know the mechanisms whereby plants can combat toxic levels of heavy metals. As far as bacteria are concerned, work on cadmium-tolerant strains of *Klebsiella rhinoscleromatis* growing in a medium with $\leq 8.9 \mu M$ Cd has shown that in the stationary phase 94% of the cadmium was eliminated from the cells by washing with a chelating agent, for example, ethylenediaminetetraacetic acid (EDTA) (Masao and Minami, 1978). In cadmium-resistant strains of *Pseudomonas* sp. more or less the same amount of this heavy metal was deposited on the cell surface, so that most of it remained outside the physiologically active sites of the bacterial cell. The incorporated Cd^{2+} was distributed in equal amounts in the 140,000 g supernatant and in the precipitation of the 5400 g fraction (Tokuyama and Asano, 1976). Also in the yeast *Candida utilis* surface binding of heavy metals took place (Failla et al., 1976).

A discontinuous distribution of Cd^{2+} is also probable within the cells of higher plants. In the leaves of *Tilia platyphyllos* and *Aesculus hippocastanum,* both trees growing in the vicinity of a cadmium smelter, more than 70% of the total cadmium was associated with the cell wall (Table 1), where the cation exchange sites consisted mainly of the exposed polygalacturonic acid carboxylate groups. In the brown alga *Laminaria digitata* the polysaccharide alginate should have the same ability for metal binding (Bryan, 1971). The remaining cadmium in the cells of the plants mentioned above was distributed in the cytosol, the vacuole sap, and the cell organelles (Ernst, 1972).

The exclusion of cadmium from the susceptible sites of the cell may prevent severe problems for the metabolism of plants. However, if the capacity of the cell wall is saturated, the cell is not able to render all the metal ions into a metabolicially innocuous form. As presented in Table 1, within the cell the maximum cadmium will be found in the cytoplasm and the vacuolar system. The chemical form of this cadmium is uncertain. Part of it is bound to macromolecules, as in *Crithidia fasciculata,* a trypanosomid flagellate, and in *Anabaena nidulans,* a blue-green alga (MacLean et al., 1972). The formation of a cadmium metallothionein, as indicated from animal tissues (Kagi and Vallee, 1960; Shaikh and Lucis, 1971), is dubious for plants (MacLean et al., 1972; Failla et al., 1976). The cadmium content in the cell organelles varies with their functions and their possibilities for ion uptake. Up to 50 or even 97 μmol Cd^{2+}/kg dry weight can be present in chloroplasts. In mitochondria only 0.012 μM Cd was found on analysis (Ernst, 1972). This is in good agreement with absorption experiments, where mitochondria of corn bound 0.058 μmol Cd/mg protein (Bittell et al., 1974) and those of beans about 1 μM (Wilson and Hopkinson, 1972).

3. INTERACTION OF CADMIUM WITH THE DIFFERENT COMPARTMENTS

3.1. Cytosol

Despite the great amount of research on the toxicological effects of cadmium in plants (cf. Jastrow and Koeppe, this volume, chapter 15), reports of the biochemical basis of these interactions are very limited. It is difficult to localize the reaction of cadmium in organized systems and to identify the protein most sensitive to this heavy metal. As to the effect of cadmium on the metabolism of plants growing in cadmium-contaminated medium (i.e., *in vivo*), this element either activates or inhibits a large number of enzyme systems *in vivo* (Lee et al., 1976). Leaves of seedlings of *Glycine max,* cultivated in the presence of 1.35 μM Cd, showed increased activity of hydrolytic enzymes and peroxidases (Figure 1) and decreased activity of carbonic anhydrase. The involvement of cadmium with all essential metabolic pathways suggests that not all effects of cadmium on enzyme systems are fundamental changes due to the heavy metal, but may also be interpreted as stimulation of one process (e.g., senescence) or as inhibition of another process, the energy supply by photosynthesis, with further physiological chain reactions.

A stimulation of cytosol enzymes *in vivo* by cadmium may indicate that these enzyme systems are cadmium dependent, replacing more efficiently another metal in the active center of the enzyme, or that the enzyme is protected against the heavy metal ion by complexes of Cd^{2+} with amino acids and peptides (Vallee

% activity

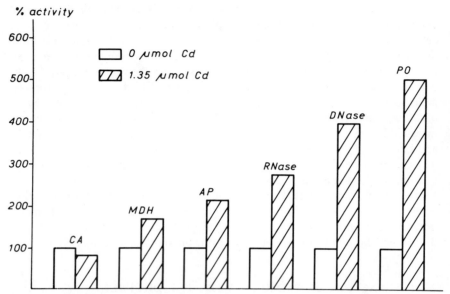

Figure 1.   Activities of enzymes in leaf blades of soybean seedlings 10 days after cadmium treatment. Values (in g fresh weight) of control experiments (= 100%): malate dehydrogenase (MDH), 230 μmol NADH/hr; ribonuclease (RNase), 3 μmol nucleotide/hr; desoxyribonuclease (DNase), 1.2 μmol nucleotide/hr; acid phosphatase (AP), 190 μmol p-nitrophenol/hr; peroxidase (PO), 880 μmol H_2O_2 consumed/hr; carbonic anhydrase (CA), 0.28 (velocity constant). (After Lee et al., 1976.)

and Ulmer, 1972), thus altering metabolic pathways. In *in vitro* experiments, some of these possibilities have been investigated. One of the enzymes of CO_2 metabolism, phosphoenolpyruvate (PEP) carboxylase, catalyzing nonautotrophic CO_2 fixation in plants, is normally dependent on manganese. Addition of cadmium to the enzyme *in vitro* lowers the maximum activity by nearly 70%, whereas zinc reduces the rate of the reaction by 98% (Walker, 1957; Leblova and Mares, 1975). On the contrary, a further metabolic step, reduction of the formed oxalacetate to malate by the cytosol malate dehydrogenase (MDH), is inhibited to a greater degree by cadmium than by zinc (Mathys, 1975). *In vitro,* cadmium concentration up to 10 μM did not alter the activity of MDH, whereas 500 μM Cd or 1200 μM Zn lowered the activity by 50% (Figure 2). However, *in vivo* an excess of cadmium (1.35 μM) stimulated the MDH (Lee et al., 1976). Replacement of zinc by cadmium seems improbable because MDH is not a metalloenzyme (Scrutton, 1973), and also an excess of ammonia has the same effect on the activity of this enzyme *in vivo* (Wakiuchi et al., 1971).

A further example of an unspecific stimulation of an enzyme *in vivo* by Cd^{2+} concerns peroxidase, which has the highest cadmium tolerance of the enzymes

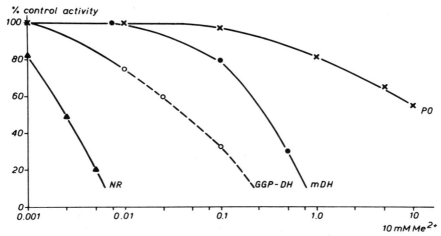

Figure 2. The effect of cadmium in the incubation medium on the activity of nitrate reductase (NR), glucose-6-phosphate dehydrogenase (G6P-DH), malate dehydrogenase (MDH), and peroxidase (PO) of leaves from *Silene cucubalus*. 100% of NR = 2.8 μmol NO_2/g · hr; 100% of G6P-DH = 19,9 μmol glucose-6-phosphate/g · hr; 100% of MDH = 3.050 μmol oxaloacetic acid/g · hr; 100% of PO = 72.000 OD at 420 nm/g · min. (After Ernst et al., 1974; Mathys, 1975; Ernst, 1976.)

so far tested (Ernst, 1976), followed by isocitrate dehydrogenase (Mathys, 1975). To halve the maximum peroxidase activity, 10 mM Cd is necessary. Therefore the stimulation of this enzyme by Cd^{2+} should be considered as strong indication of enhanced senescence of plant tissue (Lee et al., 1976).

The remarkable affinity of cadmium for SH groups will inhibit many enzymes. In this way it has been shown (Ernst et al., 1974; Mathys, 1975) that *in vitro* nitrate reductase, which catalyzes the initial step for the amino groups if plants are fed with nitrate, is one of the enzymes most affected by cadmium (Figure 2). As little as 2.5 μM Cd^{2+} inhibited the activity of the enzyme by 50%, and only less than 0.5 μM did not interact. Also a direct effect of cadmium on nitrate uptake was found in corn seedlings (Volk and Jackson, 1973). In situations where legumes are cultivated without nitrogen sources and the nitrogen supply has to come from the nitrogen fixation of symbiotic N_2 = fixing bacteria, 18 μM Cd^{2+} depressed the nitrogenase activity *in vivo* by 30 to 70% (Huang et al., 1974). But this effect is not specific for cadmium; other heavy metals, such as Zn, Ni, U, and Hg, show similar behavior (Wilson and Reisenauer, 1970; Vesper and Weidensaul, 1975). At times, however, the molecular aspects of this toxicity are unknown despite some *in vitro* experiments with nitrogenase isolated from the blue-green alga *Anabaena cylindrica* (Haystead and Stewart, 1972). One possibility may be a change in the active center of the leghemoglobin, which is the O_2-scavenging site, to produce a localized anaerobiosis for the O_2-sensitive

Table 2.   Effect of a Surplus of Cadmium on the Iron and Cadmium Contents (mmol Me^{2+}/kg dry weight) and Subcellular Distribution of Cadmium in Green and Chlorotic Leaves of *Cardaminopsis halleri* and *Agrostis tenuis*, Sampled in a Cadmium-Polluted Mine Area [a]

	Agrostis tenuis		*Cardaminopsis halleri*	
	Green	Chlorotic	Green	Chlorotic
Total amount of:				
Fe	1.4	1.4	3.8	2.2
Cd	0.01	0.03	0.08	0.50
Distribution of cadmium (%)				
Cell wall	45.6	31.6	59.4	56.9
Chloroplasts, mitochondria	13.8	9.3	17.3	10.2
Plasm, vacuole sap, microbodies	40.6	59.1	23.3	32.9

[a] After Mathys (1972).

nitrogenase system. Like myoglobin, leghemoglobin is able to undergo reversible oxygenation (Appleby, 1969), which can be affected by divalent metal ions, such Cd (Rifkind, 1973).

Replacement of zinc by cadmium is possible in the metalloenzyme alkaline phosphatase, this process taking place in *Escherichia coli,* but cadmium phosphatase is enzymatically inactive (Lazdunski et al., 1969; Applebury et al., 1970). Although completely inactive as a catalyst, cadmium phosphatase binds phosphate ions quite effectively and forms substantial amounts of phosphorylenzyme. Apparently the reason for the inactivity of the enzyme is reluctance to effect the necessary dephosphorylating step at alkaline pH. Interaction of cadmium with other enzymes is also known (enolase: Ernst, 1975, and Mathys, 1975; glucose-6-phosphate dehydrogenase: Mathys, 1975; flavin phosphate synthetase: Tachibana et al., 1969; synthesis of ascorbic acid: Zamfirescu and Tacu, 1970).

3.2.   Chloroplasts

An excess of cadmium may affect the metabolism in such a manner that plants become chlorotic. Sometimes the increase in cadmium is negatively correlated with the amount of iron (e.g., in the crucifere *Cardaminopsis halleri*), sometimes not (e.g., in *Agrostis tenuis*) (see Table 2). Lack of chlorophyll caused a reduction in photosynthesis as commonly analyzed in plants under cadmium treatments (Sempio et al., 1971; Nordeen, 1972; Zingmark, 1972; Auerbach et al., 1973; Bazzaz et al., 1974a, 1974b; Bazzaz and Govindjee, 1974; Carlson

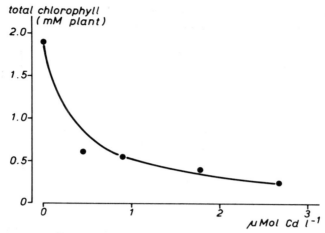

Figure 3.   Effect of Cd^{2+} treatment on total chlorophyll of leaves of *Glycine max* after 21 days. (After Cunningham et al., 1975.)

et al., 1975; Huang et al., 1974; Overnell, 1975; Root et al., 1975; Nakani and Korsak, 1976; Peneda-Saraiva, 1976; Carlson and Bazzaz, 1977).

Cadmium can affect photosynthesis in several ways. It may influence biosynthesis of biomembranes and of the photosynthetic pigments, especially chlorophyll. It may inactivate enzymes by binding SH-groups necessary for catalytic activity or substitute other divalent cations in metalloenzymes and interact with the electron transport in chloroplasts.

With regard to the first aspect, glycolipid biosynthesis in spinach chloroplasts, especially the incorporation of UDP-galactose into trigalactosyl diglyceride, is inhibited by 10 μM Cd (Mudd et al., 1971). As for the effect on chlorophyll synthesis, iron is required for the synthesis of δ-aminolaevulinate, the first intermediate. Cunningham et al. (1975) have shown that the chlorophyll concentrations of all plants of *Glycine max* being treated with 4.4 to 2.7 μM Cd^{2+} fell substantially below the level in the control (Figure 3). Cadmium is also known to degradate chlorophyll (Pakshina and Krasnovskii, 1975), especially after fumigation of plants with ozone (Czuba and Ormrod, 1974).

The second aspect, the interaction with enzymes of CO_2 fixation, may be possible, One effect may be due to the functional consequences of substituting cadmium for zinc in zinc metalloenzymes, for example, carbonic anhydrase (Cedeno-Malonado and Asencio, 1976). In soybean leaves treated with 1.35 μM Cd, the activity of carbonic anhydrase is lowered by 18%. Comparison of the stability constant and the activity of the zinc metalloenzyme with the corresponding values for the metallocarbonic anhydrases (Coleman, 1973) shows that cadmium is not an effective catalyst of carbonic anhydrases at the active

Table 3.   Metal Activation of Spinach Phosphoribulokinase with a Metal Concentration of 5 mmol Me^{2+}/l^a

Metal Added	ADP Formed (μm)	Inhibition (%)
Mg^{2+}	0.252	0
Mn^{2+}	0.203	19.4
Ca^{2+}	0.162	35.7
Co^{2+}	0.048	81.0
Cd^{2+}	0.024	90.5

a After Hurwitz et al. (1956).

site of these enzymes in human beings. Biochemical research on plant carbonic anhydrases has not thoroughly been carried out with regard to Cd^{2+}.

A further effect is due to the high affinity of cadmium to SH groups. At least phosphoribulokinase and ribulose-1,5-diphosphate carboxylase are known to be sulfhydryl enzymes. *In vitro* the replacement of magnesium, normally the activator, by the same amount (5 mM) of another divalent metal gives a less active phosphoribulokinase (Hurwitz et al., 1956) or, by 10 μM Cd^{2+}, an inactive ribulose diphosphate carboxylase (Jyung and Camp, 1976). The Cd^{2+} ion inhibited phosphoribulokinase by more than 90% (Table 3). Therefore it may be expected that CO_2 fixation is severely affected. With 10 μM Cd the CO_2 fixation of *Chlorella pyrenoidosa* was inhibited by 62% (Hart and Scaife, 1977); nearly the same inhibition (74%) was found for isolated spinach chloroplasts (Hampp et al., 1976). In the presence of 10 μM Zn, however, the effect of cadmium was less strong (53% inhibition) in spinach chloroplasts. In cadmium-treated soybeans as much as 450 μM external Cd reduces CO_2 fixation by only 34% (Huang et al., 1974), indicating that because of the internal compartmentation a large percentage of cadmium never reaches the chloroplasts.

As a third aspect, it may be assumed that cadmium influences the electron flow in photosystems I and/or II. However, no inhibition was observed on recording the electron flow in photosystem I (Li and Miles, 1975). On comparing the redox potentials (Cd: -402 mV, photosystem II: -150 mV; Seel, 1960), it is evident that cadmium cannot act as an artificial electron acceptor for photosystem II. Results of experiments with chloroplasts of maize (Bazzaz and Govindjee, 1974) and spinach (Li and Miles, 1975; Lucero et al., 1976) have demonstrated an inhibition of photosystem II due to Cd^{2+}. The cadmium action appears to take place at the oxygen-evolving site. Inhibition at this site may occur by a substitution of manganese by cadmium, as has already been found for zinc (Singh and Steenberg, 1974). In spinach chloroplasts, exogenous manganese, the electron donor for the oxidation of water, could not suppress the effect of cadmium, but could partially reverse cadmium inhibition (Van Duijvendijk-Matteoli and Desmet, 1975). Using hydroxylamine, which donates electrons to photosystem II, made it possible to bypass the effect of Cd^{2+}, so that Cd^{2+} had to act before Y_2, the second electron donor of photosystem II (Van Duijvendijk-Matteoli and Desmet, 1975). Since Y_2 is localized on the inner surface of the thylacoid membrane of the grana, cadmium, like other divalent cations, has to penetrate the membranes to act. In this way the toxicity of cadmium is dependent on its chemical status, which may change on the way from the cell wall through the cytosol to the chloroplasts. Although O_2 evolution is lower in cadmium-grown cells of *Chlorella pyrenoidosa* ($10 \mu M$ Cd^{2+}), the degree of inhibition is small (20%) in comparison to that of CO_2 fixation (Hart and Scaife, 1977).

The severe reduction in net photosynthesis (up to 76%) of up to 4.5 mM cadmium-treated detached sunflower and corn leaves (Bazzaz et al., 1974a, 1974b) may therefore be the result of all three aspects of cadmium interaction in photosynthesis.

3.3. Mitochondria

The effect of cadmium on the respiration of plants depends on their carbohydrate metabolism. C-heterotrophic organisms, for example, the yeasts *Candida utilis* (Lobyreva and Ruban, 1973), *Saccharomyces cerevisiae* (Subik and Kolarov, 1970; Lindegren and Lindegren 1973), and *Endomycopsis lipolytica* (De Troostembergh and Nyns, 1974), react with an inhibition of respiration after cadmium treatment. In the presence of a surplus of cystein, however, cadmium, as well as other heavy metals, can have a beneficial effect on the respiration of *Candida utilis* (Lobyreva and Ruban, 1973). Cadmium treatments on C-autotrophic species, for example, soybean seedlings, stimulated the respiration rate by 154% (Lee et al., 1976), an effect that should be due to a demand for ATP production through oxidative phosphorylation because photophosphorylation is reduced. This indirect effect of cadmium can act in several ways on mitochondria. The Cd^{2+} stimulated the rate of NADH oxidation at low concentrations (10 to 50 μM Cd), followed by inhibition at higher concentrations: 100 μM for isolated corn mitochondria (Miller et al., 1973), 60 μM for isolated bean mitochondria (Wilson and Hopkinson, 1972). Treatment with 250 μM Cd^{2+} completely inhibited the phosphorylation associated with succinate oxidation. Since the number of binding sites is the same for cadmium as for other heavy metals, which have the same inhibiting effect on NADH oxidation (Miller et al., 1973; Bittell et al., 1974), the site of cadmium action may be on the electron transport chain. Work with artificial electron-donating systems indicates that Cd^{2+} should inhibit the electron transfer at the terminal oxidase (Koeppe, 1977).

Moreover, cadmium caused pronounced swelling of mitochondria (Bittell et al., 1974), this effect being morphologically analyzed by Silverberg (1976). The presence of cadmium gives rise to morphological alterations within this cell organelle. In mitochondria of the green algae *Chlorella pyrenoidosa, Scenedesmus quadricauda,* and *Ankistrodesmus falcatus,* the ultrastructural changes are the presence of intramitochondrial inclusions containing cadmium, mitochondrial vacuolation, and the degeneration of well-developed cristae, as previously observed in cadmium-treated mitochondria of *Saccharomyces cerevisiae* (Lindegren and Lindegren, 1973). The disappearance of the cristae can definitely cause respiratory-deficient variants, as found in *Saccharomyces elipsoideus* (Nakamura, 1961).

3.4. Nucleus

Like a number of other metals, cadmium markedly affects the properties of DNA. When cultures of *Escherichia coli* were grown in the presence of 3 μM Cd^{2+}, loss of viability of the cells was accompanied by considerable single-strand breakage of the DNA, with no detectable increase in double-strand breaks (Mitra and Bernstein, 1978). Another effect was found in *Euglena gracilis,* where Cd^{2+} inhibited cell division, presumably by disrupting the processes critical to karyokinesis and cytokinesis (Falchuk et al., 1975a, 1975b). The DNA content of these blocked cells is 3 to 4 times greater than that of normal cells. In higher plants (e.g., *Crepis capillaris*) treatment of the seeds with 1 to 100 mM Cd for 1 hr increases the number of structural chromosome aberrations, as compared to controls (Ruposhev and Garina, 1976). This strong implication of cadmium with DNA makes it quite likely that cadmium resistance can evolve in bacteria (Uchida et al., 1973; Massao and Minami, 1978) and in higher plants (Coughtrey and Martin, 1977; Simon 1977). However, in some cases prevention of a direct confrontation of the nuclei with cadmium is possible by subcellular compartmentation.

REFERENCES

Applebury, M. L., Johnson, B. P., and Coleman, J. E. (1970). "Phosphate Binding to Alkaline Phosphatase," *J. Biol. Chem.* **245,** 4968–4976.

Appleby, C. A. (1969). "The Separation and Properties of Low-Spin (Haemochrome) and Native, High-Spin Forms of Leghaemoglobin from Soybean Nodule Extracts," *Biochim. Biophys. Acta* **189,** 267–279.

Auerbach, S., Pruefer, P., and Weise, G. (1973). "Gasstoffwechselphysiologische Schädigungskriterien bei submersen Makrophyten vom Typ *Fontinalis antipyretica* L. unter Einwirkung von Schwermetallen and Phenol," *Int. Rev. Ges. Hydrobiol.* **58,** 19–32.

Bazzaz, M. B. and Govindjee,   (1974). "Effects of Cadmium Nitrate on Spectral Characteristics and Light Reactions of Chloroplasts," *Environ. Lett.* **6**, 1–12.

Bazzaz, F. A., Carlson, W. R., and Rolfe, G. L. (1974a). "The Effect of Heavy Metals on Plants. 1: Inhibition of Gas Exchange in Sunflower by Lead, Cadmium, Nickel and Thallium," *Environ. Pollut.* **7**, 241–246.

Bazzaz, F. A., Rolfe, G. L., and Carlson, R. W. (1974b). "Effect of Cadmium on Photosynthesis and Transpiration of Excised Leaves of Corn and Sunflower," *Physiol. Plant.* **32**, 373–376.

Bittell, J. E., Koeppe, D. E., and Miller, R. J. (1974). "Sorption of heavy metal Cations by Corn Mitochondria and the Effects on Electron and Energy Transfer Reactions," *Physiol. Plant.* **30**, 226–230.

Bryan, G. W. (1971). "The Effects of Heavy Metals (Other than Mercury) on Marine and Estuarine Organisms," *Proc. R. Soc. London* **177**, 389–410.

Carlson, R. W. and Bazzaz, F. A. (1977). "Growth Reduction in American Sycamore (*Platanus occidentalis* L.) Caused by Pb-Cd Interaction," *Environ Pollut.* **12**, 243–253.

Carlson, R. W., Bazzaz, F. A., and Rolfe, G. L. (1975). "The Effect of Heavy Metals on Plants. 2: Net Photosynthesis and Transpiration of Whole Corn and Sunflower Plants Treated with Lead, Cadmium, Nickel and Thallium," *Environ. Res.* **10**, 113–120.

Cedeno-Malonado, A. and Asencio, C. I. (1976). "Carbonic Anhydrase, an Alternate Site for Cadmium Inhibition of Photosynthesis," *Plant Physiol.* **57** (5 Supply), 7.

Coleman, J. E. (1973). "Carbonic Anhydrase." In G. L. Eichhorn, Ed., *Inorganic Biochemistry,* Vol. 1, Elsevier, Amsterdam, London, and New York, pp. 488–548.

Coughtrey, P. J. and Martin, M. H. (1977). "Cadmium Tolerance of *Holcus lanatus* from a Site Contaminated by Aerial Fallout," *New Phytol.* **79**, 273–280.

Cunningham, L. M., Collins, F. W., and Hutchinson, T. C. (1975). "Physiological and Biochemical Aspects of Cadmium Toxicity in Soybean. I: Toxicity Symptoms and Autoradiographic Distribution of Cd in Roots, Stems and Leaves." In *Symposium Proceedings of the International Conference on Heavy Metals in the Environment, Toronto, Ont., 1975,* Vol. 2, pp. 97–120.

Czuba, M. and Ormrod, D. P. (1974). "Effects of Cadmium and Zinc on Ozone Induced Phytotoxicity in Cress and Lettuce," *Can. J. Bot.* **52**, 645–649.

De Troostembergh, J. C. and Nyns, E. J. (1974). "Inhibition by Cadmium Ion of the Endogenous Respiration of *Endomycopsis lipolytica," Arch. Int. Physiol. Biochem.* **82**, 785.

Ernst, W. (1972). "Zink- and Cadmium-Immissionen auf Böden und Pflanzen in der Umgebung einer Zinkhütte," *Ber. Dtsch. Bot. Ges.* **85**, 295–300.

Ernst, W. (1975). "Physiology of Heavy Metal Resistance in Plants." In *Symposium Proceedings of the International Conference on Heavy Metals in the Environment, Toronto, Ont., 1975,* Vol. 2, pp. 121–136.

Ernst, W. (1976). "Physiological and Biochemical Aspects of Metal Tolerance," In T. A. Mansfield, Ed., *Effects of Air Pollutants on Plants.* Society for Experimentals Biology, Seminar Series Vol. 1, Cambridge University Press, pp. 115–133.

Ernst, W., Mathys, W., Salaske, J., and Janiesch, P. (1974). "Aspekte von Schwerme-tallbelastungen in Westfalen," *Abh. Landesmus. Naturk, Münster Westf.* **36**(2), 1–30.

Failla, M. L., Benedict, C. D., and Weiberg, E. D. (1976). "Accumulation and Storage of Zinc Ion by *Candida utilis*," *J. Gen. Microbiol.* **94**, 23–36.

Falchuk, K. H., Fawcett, D. W., and Vallee, B. L. (1975a). "Competitive Antagonism of Cadmium and Zinc in the Morphology and Cell Division of *Euglena gracilis*," *J. Submicrosc. Cytol.* **7**, 139–152.

Falchuk, K. H., Krishan, A., and Vallee, B. L. (1975b). "DNA Distribution in the Cell Cycle of *Euglena gracilis:* Cytofluorometry of Zinc Deficient Cells," *Biochemistry* **14**, 3439–3444.

Hampp, R., Beulich, K., and Ziegler, H. (1976). "Effects of Zinc and Cadmium on Photosynthetic Carbon Dioxide Fixation and Hill Activity of Isolated Spinach Chloroplasts," *Z. Pflanzenphysiol.* **77**, 336–344.

Hart, B. A. and Scaife, B. D. (1977). "Toxicity and Bioaccumulation of Cadmium in *Chlorella pyrenoidosa*," *Environ. Res.* **14**, 401–413.

Haystead, A. and Stewart, W. D. P. (1972). "Characteristics of the Nitrogenase System of the Blue-Green alga *Anabaena cylindrica*," *Arch. Mikrobiol.* **82**, 325–336.

Huang, C. Y., Bazzaz, F. A., and Vanderhoef, L. N. (1974). "The Inhibition of Soybean Metabolism by Cadmium and Lead," *Plant Physiol.* **54**, 122–124.

Hurwitz, J., Weissbach, A., Horecker, B. L., and Smyniotis, P. Z. (1956). "Spinach Phosphoribulokinase," *J. Biol. Chem.* **218**, 769–783.

Jyung, W. H. and Camp, M. E. (1976). "The Effect of Zinc on the Formation of Ribulose Diphosphate Carboxylase in *Phaseolus vulgaris*," *Physiol. Plant.* **36**, 350–355.

Kagi, J. H. R. and Vallee, B. L.(1960). "Metallothionein: A Cadmium and Zinc Containing Protein from Equine Renal Cortex," *J. Biol. Chem.* **235**, 225–236.

Koeppe, D. E. (1977). "The Uptake, Distribution and Effect of Cadmium and Lead in Plants," *Sci. Total Environ.* **7**, 197–206.

Lazdunski, C., Petitclerc, C., and Lazdunski, M. (1969). *Eur. J. Biochem.* **8**, 519.

Leblova, S. and Mares, J. (1975). "Thermally Stable Phosphenol Pyruvate Carboxylase from Pea, Tobacco and Maize Green Leaves," *Photosynthetica* **9**, 177–184.

Lee, K. C., Cunningham, B. A., Paulsen, G. M., Liang, G. H., and Moore, R. B. (1976). "Effects of Cadmium on Respiration Rate and Activities of General Enzymes in Soybean Seedlings," *Physiol. Plant.* **36**, 4–6.

Li, E. H. and Miles, C. D. (1975). "Effects of Cadmium on Photoreaction II of Chloroplasts," *Plant Sci. Lett.* **5**, 33–40.

Lindegren, C. C. and Lindegren, G. (1973). "Mitochondrial Modification and Respiratory Deficiency in the Yeast Cell Caused by Cadmium Poisoning," *Mut. Res. Environ. Mutag. Relat. Subj.* **21**, 315–322.

Lobyreva, L. V. and Ruban, E. L. (1973). "Effect of Metallic Ions on the Respiration Activity of the Yeast *Candida utilis*," *Izv. Akad. Nauk SSSR*, Ser. Biol., **1**, 128–132.

Lucero, H. A., Andreo, C. S., and Vallejos, R. H. (1976). "Sulfhydryl Groups in Photosynthetic Energy Conservation. 3. Inhibition of Photophosphorylation in Spinach Chloroplasts by Cadmium Chloride," *Plant Sci. Lett.* **6**, 309–313.

MacLean, F. I., Lucis, O., Shaikh, Z., and Jansy, E. R. (1972). "The Uptake and Subcellular Distribution of Cd and Zn in Microorganisms," *Fed. Proc.* **31**, 699A.

Masao, O. and Minami, K. (1978). "Isolation and Identification of Cadmium Ion-Tolerant Microorganisms and Accumulation of Cadmiun Ion by the Cell," *Hakkado Kogaku Kaishi* **56**, 1–8.

Mathys, W. (1972). "Physiologische Untersuchungen der Zinkresistenz von *Agrostis tenuis*—Populationen," M. Sc. Thesis, University of Münster, 71 pp.

Mathys, W. (1975). "Enzymes of Heavy-Metal-Resistant and Non-Resistant Populations of *Silene cucubalus* and Their Interaction with Some Heavy Metals *in vitro* and *in vivo*," *Physiol. Plant.* **33**, 161–165.

Miller, R. J., Bittell, J. E., and Koeppe, D. E. (1973). "The Effect of Cadmium on Electron and Energy Transfer Reactions in Corn Mitochondria," *Physiol. Plant.* **28**, 166–171.

Mitra, R. S. and Bernstein, I. A. (1978). "Single-Strand Breakage in DNA of *Escherichia coli* Exposed to Cadmium," *J. Bacteriol.* **133**, 75–80.

Mudd, J. B., McManus, T. T., Ongun, A., and McCullogh, T. E. (1971). "Inhibition of Glycolipid Biosynthesis in Chloroplasts by Ozone and Sulfhydryl Reagents," *Plant Physiol.* **48**, 335–339.

Nakani, D. V. and W. N. Korsak, (1976). "Effects of Zinc, Chromium and Cadmium on the Intensity of Photosynthesis in Short-Term Experiments," *Biol. Nauki* (Moscow) **19**, 84–86.

Nakamura, H. (1961). "Adaptation of Yeast to Cadmium. IV: Production of Respiratory-Deficient Variant by Cadmium, "*Mem. Konan Univ. Sci.,* Ser. 5, 111–115.

Nordeen, S. (1972). "Effects of Cadmium on Respiration and Photosynthesis in Intact Algae and Spinach Leaf," *Proc. Nebr. Acad. Sci. Affil. Soc.* **82**, 60–61.

Overnell, J. (1975). "The Effect of Some Heavy Metal Ions on Photosynthesis in a Fresh Water Alga," *Pest. Biochem. Physiol.* **5**, 19–26.

Pakshina, E. V. and Krasnovskii, A. A. (1975). "Pheophytinization of Zinc and Cadmium Derivatives of Chlorophyll and Its Analogs: Effect of Light," *Dokl. Akad. Nauk SSSR,* Ser. Biol., **224**, 1216–1219.

Peneda-Saraiva, M. C. (1976). "Use of a Nannoplanktonic Alga as a Test Organism in Marine Molysmology: Some Reactions of *Dunaliella bioculata* to Gamma Irradiation and to Contamination by Chromium and Cadmium," *Rev. Int. Oceanogr. Med.* **43**, 111–115.

Rifkind, J. M. (1973). "Hemoglobin and Myoglobin," In G. L. Eichhorn, Ed., *Inorganic Biochemistry,* Vol. 2. Elsevier, Amsterdam, London, and New York, pp. 832–901.

Root, R. A. Miller, R. J., and Keoppe, D. E. (1975). "Uptake of Cadmium: Its Toxicity and Effect on the Iron Ratio in Hydroponically Crown Corn," *J. Environ. Qual.* **4**, 473–476.

Ruposhev, A. R. and Garina, K. P. (1976). "Mutagenic Effect of Cadmium Salts," *Tsitol. Genet.* **10**, 437–439.

Scrutton, M. C. (1973). "Metal Enzymes." In G. L. Eichhorn, Ed., *Inorganic Biochemistry,* Vol. 1. Elsevier, Amsterdam, London, and New York, pp. 381–437.

Seel, F. (1960). *Grundlagen der analytischen Chemie.* Verlag Chemie, Weinheim/ Bergstr.

Sempio, C., Raggi, V., Barberini, B., and Draoli, R. (1971). "Action of Cadmium on the Resistance of Frassineto Wheat-M to Powdery Mildew," *Phytopathol. Z.* **70**, 281–294.

Shaikh, Z. A. and Lucis, O. J.(1971). "The Nature of Biosynthesis of Cadmium Binding Proteins," *Fed. Proc.* **30**, 238.

Silverberg, B. A. (1976). "Cadmium-Induced Ultrastructural Changes in Mitochondria of Freshwater Green Algae," *Phycologia* **15**, 155–159.

Simon, E. (1977). "Cadmium Tolerance in Populations of *Agrostis tenuis* and *Festuca ovina,*" *Nature* **165**, 328–330.

Singh, B. R. and Steenberg, K. (1974). "Plant Response to Micronutrients: Interaction between Manganese and Zinc in Maize and Barley Plants," *Plant Soil* **40**, 655–667.

Subik, J. and Kolarov, J. (1970). "Metabolism of Calcium and Effect of Divalent Cations on Respiratory Activity of Yeast Mitochondria," *Folia Microbiol.* **15**, 448–458.

Tachibana, S., Takai, Y., and Siode, J. (1969). "Influence of Iron in the Effect of Cadmium upon the GTP Dependent Flavinphosphate Synthetase Activity of *Rhizopus javanicus,*" *Vitamins* (Kyoto) **40**, 46–49.

Tokuyama, T. and Asano, K. (1976). "Studies on Cadmium-Ion-Resistant Bacteria. II: Cultural Characteristics and Distribution of Intracellular Cadmium Ion of Cadmium-Ion-Resistant Bacteria," *Bull. Coll. Agric. Vet. Med, Nihon Univ.* **33**, 186–199.

Uchida, Y., Saito, A., Kaziwara, H., and Enomoto, H. (1973). "Cadmium-Resistant Microorganism. I: Isolation of Cadmium-Resistant Bacteria and the Uptake of Cadmium by the Organism," *Saga Daigaku Nogaku Iho* **35**, 15–24.

Valle, B. L. and Ulmer, D. D. (1972). "Biochemical Effects of Mercury, Cadmium and Lead," *Ann. Rev. Biochem.* **41**, 91–128:

Van Duijvendijk-Matteoli, M. A., and Desmet, G. M. (1975). "On the Inhibitory Action of Cadmium on the Donor Side of Photosystem II in Isolated Chloroplasts," *Biochem. Biophys. Acta* **408**, 164–169.

Vesper, S. J. and Weidensaul, T. C. (1975). "Effects of Cadmium, Nickel, Copper and Zinc on Nitrogen Fixation by Soybeans," *Proc. Am. Phytopathol. Soc.* **2**, 92.

Volk, R. J. and Jackson, W. A. (1973). "Mercury and Cadmium Interaction with Nitrate Absorption by Illuminated Corn Seedlings," *Environ. Health Perspect.* **4**, 103–104.

Wakiuchi, N., Matsumoto, H., and Takahashi, E. (1971). "Changes of Some Enzyme Activity of Cucumber during Ammonium Toxicity," *Physiol. Plant.* **24**, 248–253.

Walker, D. A. (1957). "Physiological Studies on Acid Metabolism. 4: Phosphoenolpyruvic Carboxylase Activity in Extracts of Crassulacean Plants," *Biochem. J.* **67,** 73–83.

Wilson, D. O. and Reisenauer, H. M. (1970). "Effects of Some Heavy Metals on the Cobalt Nutrition of *Rhizobium meliloti,*" *Plant Soil* **32,** 81–80.

Wilson, R. and Hopkinson, J. (1972). "The Effects of Cadmium on the Metabolism of Bean Mitochondria," *Plant Physiol.* **49,** (Suppl.), 11.

Zamfirescu, N. and Tacu, F. (1970). "Impact of Some Microelements upon the Synthesis of the Ascorbic Acid in Maize," *Lucr. Stiint. Inst. Cercet. Zooteh.* **13,** 215–227.

Zingmark, R. G. (1972). "Effect of Mercury and Cadmium on the Photosynthetic Rates of Atlantic Oceanic and Estuarine Phytoplankton Populations," *J. Phycol.* **8,** 9.

17

CADMIUM IN FOREST ECOSYSTEMS

William E. Sopper

Sonja N. Kerr

Institute for Research on Land and Water Resources, The Pennsylvania State University, University Park, Pennsylvania

1. INTRODUCTION

Metals such as cadmium may, when released into the environment, constitute a potential hazard to human health and natural biological systems. The concentration of cadmium naturally encountered in the environment is relatively low. Production and consumption of cadmium is increasing, however; consequently this element is becoming a potentially hazardous pollutant in the environment.

Cadmium toxicity in plants is well established; however, plants vary greatly in tolerance to the metal and in capacity to accumulate it in relation to substrate concentration. Most research regarding the phytotoxicity of cadmium has been conducted on agronomic crops. By contrast, relatively little has been done in relation to the effects of cadmium on forests, particularly individual tree species. A forest ecosystem can be viewed as a nutrient element-conserving system, controlled by climatic constraints and nutrient availability. Nutrient cycling mechanisms in forests collectively constitute a closed cycle in which all processes are critical for the system to be maintained. Forest productivity depends on the recycling of nutrients through consumers, predators, decomposers, and higher plants. Disruption of any of these processes can seriously affect the entire ecosystem through both accelerated nutrient losses and reduced mineralization rates.

Cadmium may be introduced into forest ecosystems by various pathways. Rainfall may deposit cadmium in forests as an indirect result of air pollution from smelters and metal reprocessing plants. A more direct application may occur as a result of the current interest in disposal of municipal wastewater and sludge in forest ecosystems. Disposal of these municipal wastes in forested areas seems to be a viable alternative to applications on agricultural cropland, thus limiting the inherent risk of introducing trace metals, particularly cadmium, into the human food chain. Forest vegetation is not harvested as an edible crop, and the only means of food chain transfer of trace metals would be through foraging wildlife and through leaching into groundwater.

In light of the continuing enrichment of cadmium in the environment, further information is needed concerning the effects of this trace metal on trees. Furthermore, since trees generally remain in the environment for considerably longer periods of time than do agricultural crops, the potential for accumulation or magnification of cadmium is great.

Most of the existing information on cadmium in forest ecosystems is derived from studies of trees on or near areas affected by industrial pollution. For instance, Buchauer (1973) investigated the metal contents of six tree species growing near a zinc smelter and found cadmium as high as 70 ppm in foliage of black tupelo (*Nyssa sylvatica*). Smith (1973) analyzed leaf and twig tissue of six woody plants growing in an urban environment for metal contamination and found a range of cadmium concentrations from 0.5 to 4.1 ppm. Smith considered the above range as "normal" amount of cadmium for the analyzed woody plants growing in the New Haven, Connecticut, ecosystem. Lamoreaux and Chaney (1977) studied the effect of cadmium chloride on the dry weight accumulation, height growth, and relative water-conductivity of stems of silver maple (*Acer saccharinum* L.). Seeds were planted in pots of white silica and treated with 0, 5, 10, or 20 ppm $CdCl_2 \cdot 2\frac{1}{2}H_2O$. These authors reported that the dry weights of seedling leaves, stems, and roots were all significantly reduced by the cadmium treatments. Seedling height growth was also severely reduced and was strongly correlated with cadmium level. The relative conductivity of excised stem sections was also significantly reduced by the cadmium treatments. The reduced water conductivity was attributed to a reduction in the amount of xylem tissue, reduced size of vessels and tracheids, and blockage of existing xylem elements by cellular debris or gums.

A literature survey by Shacklette (1972) showed cadmium concentrations in deciduous foliage sampled in polluted areas to be in the range from 4.0 to 17.0 ppm, compared with cadmium levels of similar vegetation from normal environments of 0.1 to 2.4 ppm. Cadmium concentrations in coniferous foliage were 0.05 to 1.0 ppm in polluted areas and 0.1 to 0.9 ppm in normal areas. Selected Norway spruce (*Picea abies* L.) forest sites in central Sweden subjected to industrial pollution had cadmium levels of 0.4 to 1.0 ppm in spruce needles, compared to 0.2 to 0.4 ppm in spruce foliage sampled from nonpolluted areas (Tyler, 1972).

Similarly, Parker et al. (1978) investigated trace metal distributions in forested ecosystems in urban and rural northwestern Indiana. The urban area had been exposed to industrial contamination for about 100 years. The levels of cadmium were significantly higher in soils and vegetation of the urban forest than in the rural forest 67 km away. Cadmium concentration in the top 2.5 cm of soil in the urban forest averaged 10 ppm, whereas that of the rural forest was 0.2 ppm. In addition, surface litter concentrations of cadmium averaged 4.7 ppm, over 3 times the value for the rural forest litter (1.2 ppm). It was estimated that the total annual input of cadmium on the urban forest was 8.2 g/ha, in comparison to 6.7 g/ha on the rural forest. Parker et al. reported that more than 95% of the total amount of cadmium in the urban forest ecosystem was found in the upper 25 cm of soil. The remaining amount was in the surface litter and in the plant biomass. The above-ground biomass contained less than 1% of the total cadmium on the site. Thus these authors concluded that forest vegetation will not be effective as sinks for heavy metals or as pumps to remove metals from contaminated sites. However, the living vegetation is important in that it is responsible for the development of the surface litter and organic matter in the soil. Any activity that would remove these organic sinks would allow more rapid movement of trace metals into the groundwater.

Tyler (1972) and Rhuling and Tyler (1973) reported that heavy metal deposition on the forest floor litter can result in inhibition of decomposer activity. These inhibitory effects are due to the toxicity of heavy metals to microorganisms responsible for decomposition. Tyler (1972) hypothesized that the consequences of decreased decomposition would be an increase in the amount of litter accumulation that would eventually bind up available plant nutrients and limit primary production, which, in turn, would reduce the total site productivity. The results of studies by Jordan and Lechevalier (1975) and Jackson and Watson (1977) support this theory. The latter investigated the disruption of nutrient pools in forested ecosystems located near a lead smelter in Missouri. The annual deposition rate of particulate cadmium was 0.72 g/m^2. At a distance of 1.2 to 2.0 km from the smelter stack, Jackson and Watson found no adverse effects on the soil-litter biota. However, at a distance of 0.4 to 0.8 km, they observed a depletion of soil and litter nutrient pools with evidence of depressed decomposer communities and nutrient translocation. Most of these studies were conducted near smelters or metal reprocessing plants, where metal contaminations were extremely high.

Published information concerning the impact of heavy metals at more moderate concentrations is limited. Some insight is provided, however, by the results of a laboratory study reported by Chaney et al. (1978). They studied litter decomposition in two black oak forests, one of which was impacted with cadmium and zinc from an industrial complex. Microcosms containing litter and mineral soil were collected from each forest and returned to the laboratory for measurements of carbon dioxide evolution. Measurements indicated a lower decomposition rate for the microcosm from the forest area impacted with trace metals. Microcosms from the unimpacted control forest were also treated with 0.1 and 10 ppm $CdCl_2$. The high concentration of $CdCl_2$ (10 ppm) similarly reduced the respiration rate, indicating inhibition of decomposition. However, respiration rates in the microcosms treated with the low concentration of $CdCl_2$ did not decline over time as did those of normal untreated microcosms. This suggests the possibility of stimulation of respiration at low metal levels. Similar results were obtained by Bond et al. (1976), who noted a stimulation of soil and litter respiration at low levels of cadmium and an inhibition at high levels.

Although many studies have been conducted on trace metal accumulations in forest soils, few have addressed the questions of residence time and leaching and transport of metals to groundwater. It is well known that the extractability of most trace metals in soils is greatly dependent on the acidity of the system. Lowering the pH of the extractant increases the exchange. The pH of precipitation in many parts of the world has been decreased markedly during the last two decades, and the possibility cannot be ignored that this acidification may affect the leaching and residence time of many trace metals in the soil. The effect of precipitation pH was investigated by Tyler (1978). He estimated that the current mean pH of precipitation reaching the ground in a spruce forest in Sweden was 4.2. Samples of the organic spruce forest soil were treated with artificial rainwater acidified to pH 4.2, 3.2, and 2.8. The amounts of trace metals released from the mor soils (40-mm depth) increased with decreasing pH. Approximately 85% of the total cadmium content was released at pH 2.8. Tyler also estimated the average residence time for heavy metals. He estimated that the number of years for a 10% decrease in the total concentration of cadmium in the mor horizon through leaching would be 1.7 years with a mean precipitation pH of 2.8, 4 to 5 years at a pH of 3.2, and 20 years at a pH of 4.2.

Sidle and Kardos (1977) also studied the transport of trace metals in a sludge-treated hardwood forest. Anaerobically digested sewage sludge was applied in the forest at 12.7 and 27 metric tons/ha. The high treatment applied 0.253 kg Cd/ha. Measurements of cadmium in soil water at the 120-cm depth indicated that only 6.6% of the cadmium applied in the sludge was leached.

Most of the studies reviewed here have been short-term ones. Few long-term studies on the fate of trace metals in forest ecosystems have been conducted. At The Pennsylvania State University treated municipal wastewater has been spray irrigated in several forest ecosystems since 1963 (16 years). A continual monitoring program has evaluated the environmental effects on soil, water, and vegetation. The results of these studies will now be discussed to provide some insight into the fate of cadmium.

2. PROJECT DESCRIPTION

The Pennsylvania State University's Wastewater Renovation and Conservation Project includes several forested sites irrigated with wastewater. An abandoned old field planted with white spruce (*Picea glauca* Muench Voss.) seedlings in 1955 has been irrigated since 1963. The irrigation rate is 5.0 cm sewage effluent/week from April through November of each year. An adjacent old field plot on the same Hublersburg clay loam soil (Typic Hapludalf) has been maintained as an unirrigated control. A mixed hardwood area, designated as old gamelands, has been irrigated with wastewater at rates varying from 5.0 to 15.0 cm/week since 1964. Sewage effluent alone was applied in the old gamelands over the entire span of the project except during 1971 and 1972, when approximately one half of the irrigations consisted of a mixture of sewage effluent and sludge. The soil in the old gamelands is a Morrison loamy sand (Ultic Hapludalf).

Wastewater used for irrigation was obtained from the University Park sewage treatment plant. The wastewater received secondary treatment either by activated sludge or trickling filter processes, followed by chlorination. Sludge, which was injected into the effluent at a ratio of 1 part sludge to 13 parts effluent during selected irrigation periods (1971–1972 in old gamelands), had been anaerobically digested.

A composite wastewater sample was collected during each irrigation cycle and subsequently analyzed for cadmium. Wastewater samples were prepared for analysis by digesting a 50-ml liquid sample with 5 ml conc. HNO_3 and 2 ml conc. $HClO_4$ in a micro-Kjeldahl flask. The digested samples were diluted to

Table 1.   Mean Heavy Metal Content of Wastewater

Element	Effluent (mg/l)	Effluent-Sludge Mixture (mg/l)
Cu	0.068	0.501
Zn	0.197	0.730
Cd	0.003	0.005
Pb	0.140	0.299
Ni	0.050	0.075
Co	0.040	0.057
Cr	0.022	0.121

a final volume of 50 ml and analyzed for cadmium in an atomic absorption spectrophotometer equipped with a graphite furnace and deuterium arc background corrector.

Foliar samples were collected to obtain an estimate of the average cadmium levels in vegetation. The foliar samples were prepared for analysis by digesting 2.000 g of dry plant tissue in 10 ml conc. HNO_3 and 4 ml conc. $HClO_4$ in a micro-Kjeldahl flask. The digested samples were diluted and analyzed for cadmium by the procedures outlined for wastewater samples.

Soils in the old field area were sampled with an Oakfield probe at depth intervals of 0 to 5, 5 to 10, 10 to 15, and 15 to 30 cm in 1978 and at depth intervals of 0 to 30 cm in 1963, 1965, 1967, 1971, 1976, and 1978. The old gamelands soils were sampled at intervals of 0 to 5, 5 to 10, 10 to 15, and 15 to 30 cm in 1975 and in the surface 30 cm only in 1966. Soils were air/dried, and 10.000 g was extracted twice with 0.1 N HCl, using an extract volume soil weight ratio of 2.5:1. For each extraction the suspension was shaken for 15 min and then centrifuged at 2500 rpm (1775 g). The supernatant liquid was decanted into a volumetric flask, and the supernatant liquids were combined and diluted to a final volume of 50 ml with 0.1 N HCl. The soil extracts were then analyzed for cadmium by the same procedure as for the wastewater samples.

Soil percolate water was collected from suction lysimeters installed at the 120-cm depth in both the irrigated and the unirrigated areas. Samples were acidified with concentrated HNO_3 at the time of collection and analyzed for cadmium by the procedures outlined for wastewater samples.

3.   RESULTS AND DISCUSSION

3.1.   Wastewater Quality

Concentrations of heavy metals in the sewage effluent and the effluent-sludge mixture are shown in Table 1. Values for all heavy metals represent average

Table 2.   Mean Annual Trace Metal Loading Applied with 5.0 cm of Irrigation per Week

Element	Effluent[a] (kg/ha)	Effluent-Sludge Mixture[b] (kg/ha)
Cu	0.9	6.0
Zn	2.0	7.7
Cd	0.1	0.1
Pb	0.3	1.5
Ni	0.3	0.9
Co	0.2	0.5
Cr	0.5	1.3

[a] Irrigated only during the growing season.
[b] Irrigated year round.

concentrations of composite monthly samples taken for all irrigated areas. Average concentrations of cadmium in the effluent and the effluent-sludge mixture are below the U.S. Public Health Service mandatory limit of 0.01 mg/l for drinking water (McKee and Wolf, 1963). Concentrations of the other heavy metals are within the ranges commonly encountered for municipal sewage effluents (Menzies and Chaney, 1974). The mean annual trace metal loading applied with the 5.0-cm irrigation per week is given in Table 2.

3.2.   Old Field Area

The old field area received a total of 2344 cm of effluent containing no sludge from 1963 to 1978. This corresponded to a total cadmium application of 0.69 kg/ha over the 16-year period (Table 1).

Vegetation sampled in 1978 for cadmium analysis included foliage from white spruce, wild strawberry (*Fragaria virginia* Duchesne), and goldenrod (*Solidago* spp. Ait.). Wild strawberry and goldenrod were the most abundant ground vegetation common to both the treated and control areas. The cadmium levels of white spruce and wild strawberry foliage collected in the treated and control areas did not differ significantly (Table 3). All species had higher concentrations of cadmium in the control than in the treated area. This was caused by an effective dilution of cadmium concentration in the treated areas, since the individual plants were much larger in the irrigated than in the control area. In previous studies by Sopper and Kardos (1973) and Sidle and Sopper (1976) conducted in the old field area, results indicated that the total biomass of herbaceous vegetation produced in the irrigated area was approximately 3 times greater than in the unirrigated control area. In both the treated and control areas goldenrod had higher levels of cadmium than did white spruce or wild strawberry.

Table 3. Cadmium Concentrations (μg/g) in Plant Samples from Old Field Area (1978)

Plant	White Spruce	Wild Strawberry	Goldenrod
Treated	0.07	0.07	0.19
Control	0.24	0.31	1.50

Table 4. Cadmium Concentrations (μg/g) in Soils from Old Field Area (1978)

	Soil Depth (cm)			
Soil	0–5	5–10	10–15	15–30
Treated	0.15	0.08	0.05	0.05
Control	0.19	0.06	0.04	0.01

Table 5. Cadmium Concentrations (μg/g) in Soils from Old Field Area (0 to 30 cm depth) over Time [a]

Soil	1963	1965	1967	1971	1976	1978
Treated	0.04 AB	0.04 A	0.07 BC	0.05 AB	0.03 A	0.08 C
Control	0.05 A	0.05 A	0.03 A	0.06 A	0.07 A	0.09 A

[a] Means within a given treatment followed by the same letter are not significantly different at the 5% level.

Cadmium levels in all treated species examined in the old field area are below the suggested tolerance levels for plant tissue (3.00 μg/g) as reported by Melsted (1973).

The concentrations of 0.1 N HCl-extractable soil cadmium in the old field area did not differ significantly between treated and control areas at all depths in 1978 (Table 4). Cadmium levels in both the treated and control areas were highest in the 0 to 5 cm depth. The decrease in cadmium with depth is due to the native distribution of the element in the soil profile rather than to the effluent irrigation. Average soil cadmium concentrations prior to effluent irrigation (1963) were 0.042 and 0.005 μg/g in the 0 to 30 and 30 to 60 cm depths of the treated area, and 0.050 and 0.012 μg/g in the corresponding depths of the control area. These data confirm that there was no measurable effect from the effluent irrigation on the 0.1 N HCl-extractable cadmium in the soil.

Soil samples were collected from the 0 to 30 cm depth over the entire 16-year period of wastewater irrigation. These samples were extracted with 0.1 N HCl and analyzed for cadmium. The results of these analyses are given in Table 5. There was a significant increase in the cadmium concentration of the 0 to 30 cm soil from 1963 to 1978 because of the wastewater irrigation. However, this

Table 6.   Concentrations (mg/l) of Heavy Metals in the Soil Percolate Water

Location	Cu	Zn	Cr	Pb	Co	Cd	Ni
Unirrigated	0.01	0.22	0.013	0.062	<0.001	<0.001	<0.001
Irrigated[a]	0.01	0.16	0.013	0.038	<0.001	<0.001	<0.001

[a] Irrigated at 5 cm/week during the growing season for 16 years.

increase (0.08 μg Cd/g) is still below the soil cadmium concentration (0.09 μg/g) found in the control area. All of the values reported are within the normal range (0.01 to 0.70 μg/g) of cadmium concentrations found in unirrigated Pennsylvania soils.

Soil percolate water samples were collected at the 120-cm depth during 1978 and analyzed for cadmium, as well as other trace metals. The results of the analyses indicated that the cadmium concentrations for both the irrigated and unirrigated soil percolate water were below the detection limit of 0.001 mg/l. Concentrations of Cu, Zn, Cr, Co, and Ni, as well as Cd, were all below the U.S. Public Health Service drinking water standards (Table 6).

3.3.   Old Gamelands Area

A mixed hardwood forest, known as the old gamelands, received a total application of 1813 cm of effluent and 245 cm of an effluent-sludge mixture from 1964 to 1974. This represented a total cadmium loading of 0.61 kg/ha over the 11-year period (Tables 1 and 2). Cadmium contributions from the effluent and effluent-sludge mixtures were 0.49 and 0.12 kg/ha, respectively.

Vegetation sampled in the old gamelands for cadmium analysis consisted of two hardwood species, white oak (*Quercus alba* L.) and red maple, and a predominant ground vegetation common to both treated and control areas, wild sarsaparilla (*Aralia nudicaulis* L.). The cadmium concentrations of the red maples sampled in the treated and control areas did not differ significantly (Table 7). However, there was a significant difference in the cadmium concentrations of the white oak and wild sarsaparilla between the treated and control areas. In both cases the foliar cadmium concentration was significantly higher in the control area. As stated before, this was caused by an effective dilution of cadmium concentration in the treated vegetation since the individual plants were larger and more abundant. Red maple foliage had higher cadmium levels than either white oak or wild sarsaparilla in both the treated and control areas. Cadmium levels in white oak and red maple were within ranges reported for deciduous foliage sampled from normal environments (Shacklette, 1972).

Table 7.   **Cadmium Concentrations (μg/g) in Plants from Old Gamelands Area (1975)** [a]

Plant	White Oak	Red Maple	Wild Sarsaparilla
Treated	0.09 A	0.30 A	0.15 A
Control	0.13 B	0.24 A	0.23 B

[a] Means within a given species followed by the same letter are not significantly different at the 5% level.

Table 8.   **Cadmium Concentrations (μg/g) in Soils from Old Gamelands Area (1975)** [a]

Soil	Soil Depth (cm)			
	0–5	5–10	10–15	15–30
Treated	0.37 A	0.09 A	0.04 A	0.01 A
Control	0.25 B	0.07 A	0.05 A	0.04 A

[a] Means within a given depth followed by the same letter are not significantly different at the 5% level.

Extractable soil cadmium levels in the old gamelands (1975) are shown in Table 8. Soil extracts at the 0 to 5 cm depth had significantly higher cadmium concentrations in the treated area than in the unirrigated control area. Soil extracts at the 5 to 10, 10 to 15, and 15 to 30 cm depths showed no treatment effects. Soil cadmium concentrations at the 0 to 5 cm depth were significantly higher than at the 5 to 10, 10 to 15, or 15 to 30 cm depths in both treated and control areas. High concentrations of cadmium in the surface layer of soil (0 to 5 cm) are primarily due to the higher amount of organic matter at this depth, as well as the input of cadmium through wastewater irrigation in the treated area. Organic matter content exceeded 5 % in the 0 to 5 cm soil depth and ranged from 1.8 to 1.9% at the 5 to 15 cm soil depth. The cation exchange capacity of the soil ranged from 12.0 to 14.8 mequiv/100 g soil at the 0 to 5 cm depth and from 7.6 to 11.0 mequiv/100 g soil at the 5 to 15 cm depth. The organic matter increases the cation exchange capacity of the soil, as well as providing additional chelation of heavy metals in this surface organic horizon. The average soil cadmium concentration sampled in 1966 (0 to 30 cm) in the treated area of the old gamelands was 0.028 μg/g.

4. SUMMARY

Data have been presented showing levels of cadmium in forest vegetation and soils subjected to wastewater irrigation for 16 years. In the old field areas white spruce, goldenrod, and wild strawberry showed no increase in cadmium levels due to wastewater irrigation. Cadmium concentrations in all species were lower in the wastewater-irrigated area than in the control area because of the greater biomass production resulting from the irrigation. This provides evidence that under low cadmium applications in a wastewater disposal system the actual cadmium concentration of some herbaceous vegetation may be decreased. None of the species sampled in the old gamelands area showed a significant increase in cadmium levels as a result of the wastewater irrigation.

Differences in cadmium concentrations between the vegetation species were found in both areas. Goldenrod had higher cadmium levels than white spruce or wild strawberry in the old field area. In the old gamelands area, red maple had higher cadmium levels than white oak or wild sarsaparilla. Thus it is evident that a single species, rather than a composite of several species, must be sampled when examining the impact of wastewater irrigation on the cadmium content of forest vegetation.

Soil cadmium status was not significantly affected by wastewater irrigation in either area, except for the increase in cadmium concentration at the 0 to 5 cm depth in the irrigated old gamelands area. Cadmium levels decreased with depth in the soil profile in both treated and control areas.

5. CONCLUSIONS

On the basis of information currently available it appears that the introduction of cadmium into forest ecosystems as an indirect result of industrial pollution or through the planned application of municipal wastewater and sludge will have a minimal effect. Forest vegetation species appear to vary considerably in their susceptibility to cadmium contamination. There is also some evidence that the introduction of high levels of cadmium into a forest ecosystem may influence some of the natural levels of biological processes that are critical to the maintenance of the ecosystem.

ACKNOWLEDGMENTS

This research was supported in part by funds provided by the Northeastern Forest Experiment Station through the Pinchot Institute of Environmental Forestry Research, Consortium for Environmental Forestry Studies. Partial support was also provided by the Office of Water Research and Technology, U.S. Department of the Interior, as authorized under the Water Resources Research Act of 1964, Public Law 88-379.

REFERENCES

Bond, H., Lighthart, B., Shimabuku, R., and Russell, L. (1976). "Some Effects of Cadmium on Coniferous Soil and Litter Microcosms," *Soil Sci.* **121,** 278–287.

Buchauer, M. J. (1973). "Contamination of Soil and Vegetation near a Zinc Smelter by Zinc, Cadmium, Copper, and Lead," *Environ. Sci. Technol.* **7**(2), 131–137.

Chaney, W. R., Kelly, J. M., and Strickland, R. C. (1978). "Influence of Cadmium and Zinc on Carbon Dioxide Evolution from Litter and Soil from a Black Oak Forest," *J. Environ. Qual.* **7**(1), 115–119.

Jackson, D. R. and Watson, A. P. (1977). "Disruption of Nutrient Pools and Transport of Heavy Metals in a Forested Watershed near a Lead Smelter," *J. Environ. Qual.* **6**(4), 331–335.

Jordan, M. J. and Lechevalier, M. P. (1975). "Effects of Zinc Smelter Emissions on Forest Soil Microflora," *Can. J. Microbiol.* **21,** 1855–1865.

Lamoreaux, R. J. and Chaney, W. R. (1977). "Growth and Water Movement in Silver Maple Seedlings Affected by Cadmium," *J. Environ. Qual.* **6**(2), 201–205.

McKee, J. E. and Wolf, H. W. (1963). *Water Quality Criteria,* 2nd Ed., California State Water Resources Control Board, Publi. 3-A, Sacramento, Calif., 548 pp.

Melsted, S. W. (1973). "Soil-Plant Relationships (Some Practical Considerations in Waste Management)." In *Recycling Municipal Sludges and Effluents on Land.* National Association of State Universities and Land-Grant Colleges, Washington, D.C., pp. 121–128.

Menzies, J. D. and Chaney, R. L. (1974). "Waste Characteristics." In *Factors Involved in Land Application of Agricultural and Municipal Waters* U.S. Department of Agriculture, ARS, Beltsville, Md., pp. 18–36.

Parker, G. R., McFee, W. W., and Kelly, J. M. (1978). "Metal Distribution in Forested Ecosystems in Urban and Rural Northwestern Indiana," *J. Environ. Qual.* **7**(3), 337–342.

Ruhling, A. and Tyler, G. (1973). "Heavy Metal Pollution and Decomposition of Spruce Needle Litter," *Oikos* **24,** 402–416.

Shacklette, H. T. (1972). *Cadmium in Plants.* U.S. Geol. Surv. Bull. 1314-G. Washington, D.C., 28 pp.

Sidle, R. C. and Kardos, L. T. (1977). "Transport of Heavy Metals in a Sludge-Treated Forested Area," *J. Environ. Qual.* **6**(4), 431–433.

Sidle, R. C. and Sopper, W. E. (1976). "Cadmium Distribution in Forest Ecosystems Irrigated with Treated Municipal Waste Water and Sludge," *J. Environ. Qual.* **5**(4), 419–422.

Smith, W. H. (1973). "Metal Contamination of Urban Woody Plants," *Environ. Sci. Technol.* **7,** 631–636.

Sopper, W. E. and Kardos, L. T. (1973). "Vegetation Responses to Irrigation with Treated Municipal Wastewater." In W. E. Sopper and L. T. Kardos, Eds., *Recycling Treated Municipal Wastewater and Sludge through Forest and Cropland.* The Pennsylvania State University Press, University Park, Pa., pp. 271–294.

Tyler, G. (1972). "Heavy Metals Pollute Nature, May Reduce Productivity," *Ambio* **1**(2), 52–59.

Tyler, G. (1978). "Leaching Rates of Heavy Metal Ions in Forest Soil," *Water, Air, Soil Pollut.* **9**(2), 137–148.

INDEX

ENERGY UTILIZATION AND ENVIRONMENTAL HEALTH
Richard A. Wadden, Editor

METHODOLOGICAL APPROACHES TO DERIVING ENVIRONMENTAL AND OCCUPATIONAL HEALTH STANDARDS
Edward J. Calabrese

FOOD, CLIMATE AND MAN
Margaret R. Biswas and Asit K. Biswas, Editors

CHEMICAL CONCEPTS IN POLLUTANT BEHAVIOR
Ian J. Tinsley

RESOURCE RECOVERY AND RECYCLING
A. F. M. Barton

QUANTITATIVE TOXICOLOGY
V. A. Filov, A. A. Golubev, E. I. Liublina, and N. A. Tolokontsev

ATMOSPHERIC MOTION AND AIR POLLUTION
Richard A. Dobbins

INDUSTRIAL POLLUTION CONTROL—Volume I: Agro-Industries
E. Joe Middlebrooks

BREEDING PLANTS RESISTANT TO INSECTS
Fowden G. Maxwell and Peter Jennings, Editors

NEW TECHNOLOGY OF PEST CONTROL
Carl B. Huffaker, Editor

COPPER IN THE ENVIRONMENT, Parts I and II
Jerome O. Nriagu, Editor

ZINC IN THE ENVIRONMENT, Parts I and II
Jerome O. Nriagu, Editor

THE SCIENCE OF 2,4,5-T AND ASSOCIATED PHENOXY HERBICIDES
Rodney W. Bovey and Alvin L. Young

NUTRITION AND ENVIRONMENTAL HEALTH—Volume I: The Vitamins
Edward J. Calabrese

CADMIUM IN THE ENVIRONMENT, Parts I and II
Jerome O. Nriagu, Editor